W0234389

Lecture Notes in Physics

Lecture Notes in Physics

Edited by H. Araki, Kyoto, J. Ehlers, München, K. Hepp, Zürich
R. Kippenhahn, München, H. A. Weidenmüller, Heidelberg
and J. Zittartz, Köln

168

Heavy-Ion Collisions

Proceedings of the International Summer School
Held in La Rábida (Huelva), Spain, June 7–18, 1982

Edited by G. Madurga and M. Lozano

Springer-Verlag
Berlin Heidelberg GmbH 1982

Editors

G. Madurga
M. Lozano
Departamento de Física Atómica y Nuclear
Facultad de Física, Universidad de Sevilla
Sevilla, Spain

ISBN 978-3-540-11945-6   ISBN 978-3-540-39525-6 (eBook)
DOI 10.1007/978-3-540-39525-6

2153/3140-543210

PREFACE

This volume contains the lectures and seminars presented by the invited speakers at the International Summer School on Heavy-Ion Collisions held in La Rábida (Huelva), Spain from June 7 to 18, 1982.

The organizers chose not to select the topics in such a way as to cover all the branches of heavy-ion physics in balanced proportion. Rather we invited researchers from a variety of fields and left the choice of subjects largely to their initiative. Since most of the invited speakers kindly consented to come, the result was a fairly even representation of the most active aspects of the field.

It is an experimental fact that international schools, planned primarily for beginners or novice researchers, tend to become "mixed states" with a component of the symposium or conference which cannot be overlooked: experienced researchers see such schools as an opportunity of meeting with their colleagues and hearing lectures directly from top-rank teachers, and the lecturers, in turn, feel compelled not just to give a clear account of the well-established theories, but also to treat the latest findings and their own contributions to the field with special care. This is by no means a drawback for the younger participants: for many it is the first time that they are surrounded by such a large group of active researchers in one and the same field from different countries and experience the close contact of living together for one or two weeks. This was the case at the La Rábida school: the complete cooperation and friendly attitude of all the participants during both academic and nonacademic activities created from the beginning a wonderfully pleasant atmosphere which cannot come through in these proceedings but will be remembered by all who experienced it.

The organizers are therefore indebted to all of the participants for making the school so enjoyable. Special thanks are due to the speakers for the high quality of their contributions and for their success in reaching this far corner of Europe on time (even when public transportation did not cooperate), so that the scheduled program could be followed unchanged.

The school was made possible by the generous support of the Rector of the University of Seville and the Vice-Rector in charge of the "Universidad Hispano-Americana Nuestra Señora de La Rábida," where the school took place. The entire staff of these buildings did their best to make our stay there fruitful and enjoyable.

Financial assistance in the organization of the school was received from the Instituto de Estudios Nucleares, Madrid, through the Grupo Interuniversitario de Física Teórica, G.I.F.T., and from the I.C.E. of the University of Seville. Our gratitude for financial support also goes to the Diputación Provincial de Huelva and the Compañía Sevillana de Electricidad.

Finally, the graduate students of our Department of Atomic and Nuclear Physics, to whom we specially dedicated the school, have confirmed beyond any doubt through their enthusiastic and tireless cooperation during the school and always that the *1982 LA RABIDA SCHOOL* was worth the organizing effort.

Sevilla, August 1982   G. Madurga
                                                       M. Lozano

TABLE OF CONTENTS

OPENING TALK

Gonzalo MADURGA

Facultad de Física, Universidad de Sevilla.

Ladies and gentlemen! Friends and colleagues!

Welcome to this LA RABIDA INTERNATIONAL SUMMER SCHOOL ON HEAVY ION COLLISIONS!

Within ten years all over Spain, most particularly in this place, and all through America we are going to commemorate the 500th anniversary of the discovery of the New World. This means that 490 years ago Christopher Columbus with a handful of brave sailors from this town, Palos de la Frontera, and from nearby places started on board three ships (the Santa María, the Pinta, and the Niña) the exciting adventure of sailing towards the West with the hope of coming back from the East. At the Monastery of La Rábida, a few meters from us, Columbus spent some months of reflection immediately before initiating his voyage. Here he met the Franciscan fathers who were instrumental for securing him the access to Queen Isabella so as to obtain from her femenine intuition the possibility of recruiting in this area experienced seemen. And here he left his son with the Franciscans while he went to discover the New World.

Many of you are familiar, no doubt, with a cartoon which appeared in The New Yorker and was reproduced by A. Bromley at the Nashville International Conference on Reactions between Complex Nuclei (1974): Columbus offers to Queen Isabella his plans written down on a parchment. The royal answer "Three ships is a lot of ships. Why can't you prove the world is round with one ship?" is meant to describe the approach of the U.S. administration concerning heavy ion facilities (in 1974) as compared with the Soviet attack on the production of the supertransuranic species. A highly schematic view of that Soviet strategy was represented in a Map of Isotopes shown by Flerov at the 1973 Munich Conference on Nuclear Physics: while sailing from the mainland towards the Island of Stability through the See of Instability, the Santa María, the Pinta, and the Niña display in their banners the symbols of the heaviest projectiles available at the time (Kr, Xe, Ge); a new boat with the U symbol is being built meanwhile on the coast.

Thus the Columbian remembrances of this place, where we are, have been related to the search of superheavies long ago. They also lead us to recall something which is at the basis of heavy ion physics. Chistopher Columbus and his seemen were looking for something else when they found the New World. Similarly the production of superheavies, -the hope of finding an island of stability- was much of a stymulus for physicists 10 years ago to have heavy ion accelerators built and to develope new and more accurate detection techniques. However their efforts in this respect have been rather unsuccessful. The superheavies seem to be very elusive (though you will probably hear these days Professor Greiner tell us more optimistic news on the subject). Heavy ion physicists did find, however, an unexpected variety of new landscapes, a New World in Nuclear Physics.

The mainland has been expanded at both sides with proton-rich and neutron-rich nuclei far from stability. The knowledge of particular nuclei has been deepened upon by raising them to very high spin states unaccesible with light projectiles, thus openning the exciting fields of yrast line, band crossing, back bending, superfluidity of nuclear mater...

A certain surface transparency of systems like $^{12}C + ^{12}C$ is deduced from the observation of a nuclear rainbow in elastic and inelastic scattering.

The friction between nuclei while sliding or sticking around each other in deep inelastic collisions is reasonably well described with almost classical methods.

As heavier and faster projectiles become available and more exhaustive detection techniques are developed, an extremely rich typology of reactions between the deep inelastic collisions and complete fusion emerges: prompt fission, pre-equilibrium particle emission, incomplete fusion, one-step multifragmentation, fusion window ... are new concepts frequently not fully characterized or agreed upon by different authors.

The formation of a neck is suggested as a hipothesis to explain the enhancement of subbarrier fusion with heavy systems. On the contrary an extra push (or an extra-extra push) is required by some theoreticians as an explanation for the low fusion cross-section of very heavy systems at high energy.

Today physicists measure not only the overall temperature of the composite nucleus from the spectrum of light particles emitted, but they think they detect a hot spot formed in the target's impact zone. The still recent idea of spectator parts in the proyectile and target, as clearly distinguished from the participant fire ball, seems today to be inadequate, as well as the intranuclear cascade model.

This is only a very rapid, incomplete hint at some of the new landscapes opened us by the adventure of heavy ion collisions, you will hear about much more in detail during the next days.

All this has been done, and is being done, by the joint effort of experimentalists and theoreticians; and, what is very important, with the working together of scientist from different countries, their collaboration in national or international institutions, their discussions and exchanges of ideas and information in international conferences, their teaching and learning at international courses like the one we open today.

Sadly enough the achievements of science have been used some times, and are still being used, for destruction of lives, of property and peace. But I hope we can truly say that the scientific endeavour has more frequently contributed to foster the peaceful and friendly coexistence between different peoples, especially when it has been undertaken with an academic mood, with a sheer eagerness to learn and to help others to learn. I hope and wish, I am sure, this is the spirit which will dominate all of us here during these two weeks, so this summer school will have its modest positive contribution not only towards learning physics (this is our explicit and direct purpose) but also towards producing a more friendly, human world.

Thanks for coming here. Welcome to La Rábida!

CALCULATION OF EFFECTIVE INTERACTIONS

F. Brieva.

Departamento de Física, Universidad de Chile,

Casilla 5487, Santiago, Chile.

1.   Introduction

Effective nucleon-nucleon interactions have played an important role in the
past years to obtain a microscopic description of both nuclear structure and the
scattering of nucleons and nuclei from nuclei [1-6]. The idea has been to calculate, via some workable model which includes the basic physical effects, the
effective force between two nucleons in the nucleus from the interaction between
free nucleons. This free internucleon force is assumed to be known. It corresponds to any of the realistic nucleon-nucleon potentials [7-9] determined empirically by fitting nucleon-nucleon scattering below 350 MeV and the deuteron
properties. Although these realistic forces are in principle equivalent for a
two nucleon system they give different results in many-body calculations due to
their different off-shell behaviour. Within these ambiguities the calculated
effective interactions have produced reasonable results [1-6].

One of the most important applications where the effective interactions
have been successful   is the generation of the average single-partice potential. This single-particle potential, which is identified with the optical model
potential [10,11] for unbound nucleons , is obtained by folding the effective
internucleon force with the nucleon matter density describing the nucleus [4].
An extension of this idea   has led to calculate the average nucleus-nucleus
optical potential [5-6]. This is obtained by double-folding the effective force
with the densities of the interacting nuclei.

The effective internucleon interaction is identified with the coordinate-
space representation of an operator t which satisfies a Lippmann-Schwinger type
integral equation [12],

$$t(z) = V + V \Lambda(z)t(z) \quad . \tag{1}$$

In eq. (1) z is a complex energy parameter, V is the realistic internucleon force
and $\Lambda(z)$ is the propagator for the nucleon pair. This operator t gets different
names: transition matrix (t-matrix) or Bethe-Goldstone reaction matrix [13] (g-
matrix) according to the explicit propagator $\Lambda(z)$ considered. We shall refer to
the t operator as the t-matrix. Some well known examples for the $\Lambda(z)$ propagator

are:

$$\Lambda(z) \rightarrow \Lambda_0(z) = \frac{1}{z - K} \quad , \tag{2}$$

for free nucleon-nucleon scattering. In eq. (2) K is the relative kinetic energy operator. For two interacting nucleons in nuclear matter, the $\Lambda(z)$ propagator becomes [13)

$$\Lambda(z) \rightarrow \Lambda_{NM}(z) = \sum_{q_1,q_2} \frac{Q(q_1,q_2)}{z - e(q_1) - e(q_2)} \quad , \tag{3}$$

with $Q(q_1,q_2)$ the Pauli operator projecting into unoccupied states and $e(q)$ is the single-particle energy. A generalization of eq. (3) to describe the propagation of nucleons belonging to two different nuclei   has been discussed in ref. [14).

The t-matrix operator (eq. (1)) has general properties. It is non-local in the coordinate-space representation and it depends on the energy. If it is calculated in nuclear matter, the t-matrix will show a density dependence as well. Moreover the t-matrix will be either real or complex depending on the value of the energy parameter z. Although the properties of the t-matrix are very general, the non-local characteristics of the effective interaction have been systematically overlooked [2,4,6). The main reason is the technical numerical difficulties one faces when solving eq. (1) with a realistic internucleon force. The difficulties increase when calculating the t-matrix from a hard core potential. The standard approach (see for example refs. [4,6)) has been to define a local effective interac̲ tion which reproduces some definite t-matrix elements in the momentum-space representation (usually the on-shell matrix elements). Then it is expected that this local interaction gives a fair account of the neglected off-shell effects.

The role of the true non-local effective force is not well understood. Since the problem is technical in a great deal we would like to discuss a general method [15) to calculate the t-matrix elements either in the coordinate or momentum-space representation. The method is independent on whether the internucleon force has a hard core or not. This is a considerable advantage over existent approaches since we can compare directly the results obtained from different nucleon-nucleon forces.

2.   A method to evaluate the t-matrix

To solve eq. (1) we take t-matrix elements in the mixed coordinate-momentum representation,

$$\langle \vec{r}|t(z)|\vec{k}\rangle = V(r) \left[\langle \vec{r}|\vec{k}\rangle + \langle \vec{r}|\Lambda(z)t(z)|\vec{k}\rangle \right] \quad . \tag{4}$$

The internucleon potential has been assumed local and it may be energy dependent. The $\vec{r}$ and $\vec{k}$ coordinates refer to the relative distance and momentum for the nucleon pair respectively. If a hard core is absent from the interaction V the momentum-space representation is generally used to solve eq. (1) (see ref.[16] for example). In order to proceed with the present method, let us make the standard partial wave expansion for the t-matrix,

$$<\vec{r}|t(z)|\vec{k}> = \left(\frac{2}{\pi}\right)^{1/2} \sum_{\ell,m} i^{\ell} t_{\ell}(r,k;z) Y_{\ell m}(\hat{r}) Y^{*}_{\ell m}(\hat{k}) \quad , \tag{5}$$

with $Y_{\ell m}$ the spherical harmonics. Then the $t_{\ell}(r,k;z)$ component is solution of the following integral equation,

$$t_{\ell}(r,k;z) = V(r) \left[j_{\ell}(kr) + \int_{0}^{\infty} \Lambda_{\ell}(r,r';z) t_{\ell}(r',k;z) r'^{2} dr' \right] \quad , \tag{6}$$

with $j_{\ell}(x)$ a spherical Bessel function. For simplicity we assume that the propagator is diagonal in momentum-space to derive eq. (6) from eq. (1). Now eq. (6) is transformed into a set of N linear equations by discretizing the right hand side integral, namely

$$\sum_{j=1}^{N} B_{\ell}(r_{i},r_{j};z) t_{\ell}(r_{j},k;z) = j_{\ell}(kr_{i}) \quad . \tag{7}$$

The B_{ℓ} is a $N \times N$ matrix with matrix elements

$$B_{\ell}(r_{i},r_{j};z) = \frac{1}{V(r_{i})} \delta_{ij} - c_{j} \Lambda_{\ell}(r_{i},r_{j};z) \quad , \tag{8}$$

where c_{j} is a factor which depends on the method used to discretize the integral in eq. (6). The $B_{\ell}(r_{i},r_{j};z)$ matrix elements are finite, even for a hard core interaction, provided that $V(r_{i})$ is different from zero and $r_{i},r_{j} \neq 0$. The potential is zero when it changes sign and for large separation distances between the interacting nucleons. However both problems are easily overcome in practice. We choose the r_{i} coordinates such as we never pick the point where the potential changes sign. Also for the last coordinate the potential is very small indeed but different from zero. The solution to eq. (7) can be written as

$$t_{\ell}(r_{i},k;z) = \sum_{j=1}^{N} b_{\ell}(r_{i},r_{j};z) j_{\ell}(kr_{j}) \quad , \tag{9}$$

with $b_{\ell}(r_{i},r_{j};z)$ the matrix elements of the inverse matrix B_{ℓ}^{-1}. It is clear that det $(B_{\ell}(z)) \neq 0$ unless the energy parameter z has the value corresponding to a bound state.

The momentum-space representation of the t-matrix is now easily calculated.

If the t-matrix elements are expressed as

$$\langle \vec{k}'|t(z)|\vec{k}\rangle = 4\pi \sum_{\ell,m} t_\ell(k',k;z)Y_{\ell m}(\hat{k}')Y^*_{\ell m}(\hat{k}) \quad , \tag{10}$$

then the $t_\ell(k',k;z)$ are given by

$$t_\ell(k',k;z) = \frac{1}{2\pi^2}\int_0^\infty j_\ell(k'r)t_\ell(r,k;z)r^2 dr \quad . \tag{11}$$

The integral in eq. (11) does not present major difficulties since the $t_\ell(r,k;z)$ reflects the range of the V interaction.

The coordinate-space representation of the t-matrix, or effective force, is expressed as

$$\langle \vec{r}|t(z)|\vec{r}'\rangle = \sum_{\ell,m} t_\ell(r,r';z)Y_{\ell m}(\hat{r})Y^*_{\ell m}(\hat{r}') \quad , \tag{12}$$

where the $t_\ell(r,r';z)$ are given by (from eq. (9))

$$t_\ell(r,r';z) = \sum_{j=1}^N b_\ell(r,r_j;z)\frac{\delta(r_j - r')}{r_j r'} \quad . \tag{13}$$

In fact the $t_\ell(r,r';z)$ are automatically calculated when solving eq. (7) without further complications. Eq. (13) is a convenient way to express the non-local properties of the effective force. If, for example, the calculation of the b_ℓ matrix elements shows that

$$b_\ell(r_i,r_j) \simeq 0 \quad \text{for} \quad r_j \neq r_i \quad , \tag{14}$$

then

$$t_\ell(r,r';z) \simeq b_\ell(r,r;z)\frac{\delta(r - r')}{rr'} \quad , \tag{15}$$

namely a local effective interaction.

3. **An example: the free t-matrix**

To calculate the t-matrix elements for free nucleon-nucleon scattering we take the $\Lambda(z)$ propagator as the free propagator given by eq.(2). Therefore the $\Lambda_\ell(r,r';z)$ component in eq. (6) is given by

$$\Lambda_\ell(r,r';z) = -\frac{M}{\hbar^2}q\, j_\ell(qr_<)h_\ell^{(+)}(qr_>) \quad , \tag{16}$$

where M is the nucleon mass, $h_\ell^{(+)}(x)$ is a Hankel function of the first kind [17], $r_<(r_>)$ are the smaller (larger) lengths of r and r' and q is given by

$$q = \left(\frac{M}{\hbar^2} z\right)^{1/2}. \tag{17}$$

The integral equation for $t_\ell(r,k;z)$ has to be modified to take into account the tensor coupling present in the nucleon-nucleon potential. The resulting integral equation is

$$t_{\ell\ell'}^{JST}(r,k;z) = i^{\ell'-\ell} V_{\ell\ell'}^{JST}(r) j_{\ell'}(kr) +$$

$$+ \sum_{L=|J-S|}^{J+S} i^{L-\ell} V_{\ell L}^{JST}(r) \int_0^\infty \Lambda_L(r,r';z) t_{L\ell'}^{JST}(r',k;z) r'^2 dr'. \tag{18}$$

In eq. (18) J is the total angular momentum, S and T are the total spin and isospin for the nucleon pair, $|J - S| \le \ell, \ell' \le (J + S)$ and $V_{\ell\ell'}^{JST}(r)$ is the potential in the corresponding antisymmetrized state. Eq. (18) can be solved following exactly the scheme discussed in section 2. In the general case, eq. (18), we get a set of 2N linear equations for N coordinate points.

We have chosen to illustrate the method by presenting some results for the momentum and coordinate-space representation of the t-matrix in the uncoupled $^1S_0$ quantum state. The z energy parameter was set at 100 MeV. For the internucleon force we have taken the $^1S_0$ component from the Hamada-Johnston [7] and Reid [8] potentials. They provide good examples of a hard and a soft core interaction respectively. The Reid hard core potential gives very similar results to the Hamada-Johnston force.

In fig. 1 we show the real and imaginary parts of $t_\ell(k,k';z)$ as a function of k'. The k coordinate has been left as a parameter for the different curves. For clarity we have plotted for the real part of the t-matrix the curves corresponding to $k = 0,1,2,3$ fm^{-1} and $k = k'$. For $k = 1,2,3$ fm^{-1} the curves are plotted only up to $k' = k$ since the remaining part can be obtained from the symmetry property $t_\ell(k,k';z) = t_\ell(k',k;z)$. We observe that both the hard and the soft core forces give qualitatively similar off-shell matrix elements for $k,k' < 2$fm^{-1}. Above this value the curves differ significantly. The off-shell matrix elements from the soft core interaction are very small above $k,k' \sim 8$fm^{-1}. The results from the hard core force are considerably larger for high values of the momentum. This is a consequence of the presence of the hard core which introduces high momentum components in the t-matrix. Similar results may be observed for the imaginary parts of the t-matrix. As a general remark we can say that fig. 1 shows clearly how two phenomenologically equivalent nucleon-nucleon forces (they give essential-ly the same on-shell value) produce different off-shell t-matrix elements. The

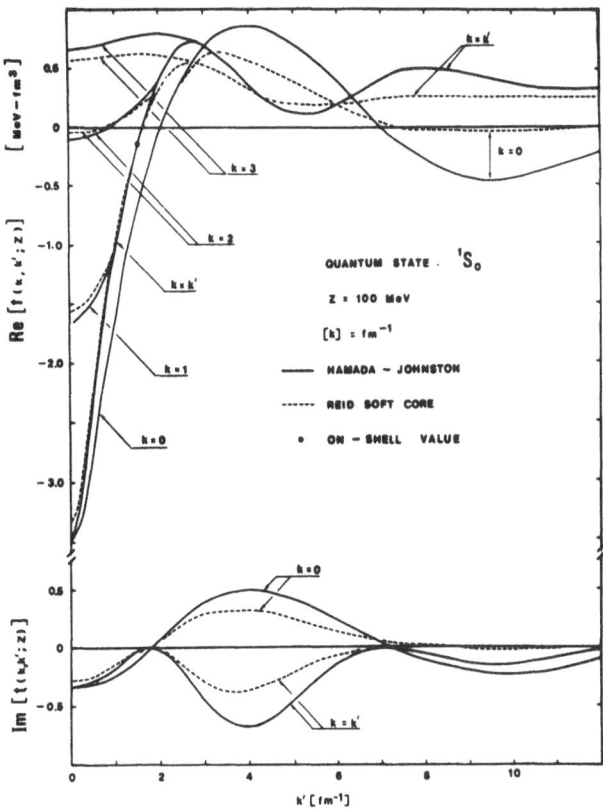

Fig. 1 : Off-shell behaviour of the free t-matrix for
nucleon-nucleon scattering.

careful calculation of these off-shell effects will help to distinguish the role of
the different realistic internucleon forces in many-body systems.

With regard to the effective interaction for free nucleon-nucleon scattering
we have obtained two very definite results concerning its real and imaginary parts.
For the real part of the t-matrix we have calculated that it is essentially local
and energy independent for r,r' > 0.5fm from a soft core force and for r,r' > 0.9fm
from a hard core interaction. All the non-local effects are relevant in the core
region. In particular, when the hard core interaction is used to calculate the
effective force we obtained δ-like singularities in the core region. This fact
makes these effective forces unreliable to work with from a numerical view point.

In fig. 2 we have plotted the imaginary part of $t_\ell(r,r';z)$ for the $^1S_0$ state.
It has been calculated from the Reid soft core potential. The main result here is
that the imaginary part of the effective force is very non-local in the coordinate
representation. Also we have checked that it has a considerable energy dependence.
This result casts some doubts on some imaginary effective forces [4,6] which are
calculated under a locality assumption. Further it shows that a local parameteri

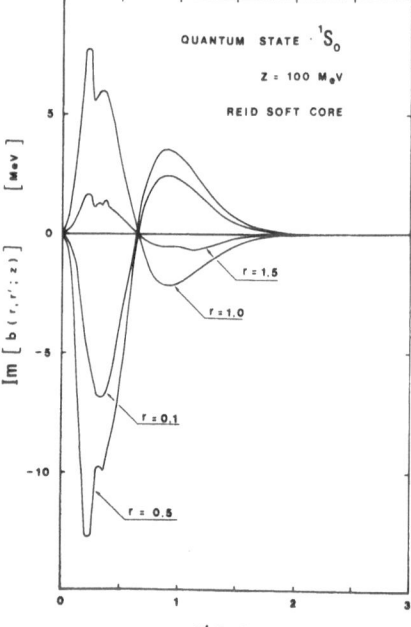

Fig. 2 : Non-local behaviour of the
imaginary part of the effecti
ve force for free nucleon-
nucleon scattering.

zation of the free t-matrix [18] may not be very realistic. The calculated imaginary part of the effective force from a hard core interaction is null in the core region, it presents a singularity at the core radious and then it behaves fairly normal outside the core region.

4. Summary

We have discussed a powerful method to evaluate effective interactions for nucleon-nucleon scattering. The application of this method shows very precisely the different off-shell behaviour of different internucleon potentials. The scope of the method is very wide. For example, the evaluation of the free t-matrix and its application to intermediate energy nucleon-nucleus scattering. Also the evaluation of non-local effective forces in nuclear matter. Some of these calculations are under way.

References

1. K. A. Brueckner and H. Weitzner, Phys. Rev. 110 (1958) 431.
2. P. J. Siemens, Nucl. Phys. A141 (1970) 225.
3. J. W. Negele, Phys. Rev. C10 (1970) 1260.
4. F. A. Brieva and J. R. Rook, Nucl. Phys. A291 (1977) 317.

5.   G. R. Satchler and W. G. Love, Phys. Rep. $\underline{55c}$ (1979) 183.

6.   A. Faessler, T. Izumoto, S. Krewald and R. Sartor, Nucl. Phys. $\underline{A359}$(1981)509

7.   T. Hamada and I. D. Johnston, Nucl. Phys. $\underline{34}$ (1962) 382.

8.   R. V. Reid, Ann. of Phys. $\underline{50}$ (1968) 411.

9.   M. Lacombe et al, Phys. Rev. $\underline{D12}$ (1975) 1495;

                        Phys. Rev. $\underline{C21}$ (1980) 861.

10. H. Feshbach, Ann. Phys. (N. Y.) $\underline{5}$ (1958) 357; $\underline{19}$ (1962) 287.

11. C. Mahaux and H. A. Weidenmüller, Shell-Model Approach to Nuclear Reactions
     (North-Holland, Amsterdam, 1969).

12. M. L. Goldberger and K. M. Watson, Collision Theory, (Wiley, N. Y., 1964).

13. H. A. Bethe and J. Goldstone, Proc. Roy. Soc. $\underline{A238}$ (1957) 554.

14. T. Izumoto, S. Krewald and A. Faessler, Nucl. Phys. $\underline{A341}$ (1980) 319; Nucl.
     Phys. $\underline{A357}$ (1981) 471.

15. F. Brieva, submitted to Phys. Lett.

16. M. I. Haftel and F. Tabakin, Nucl. Phys. $\underline{A158}$ (1970) 1.

17. A. Messiah, Quantum Mechanics (North-Holland, Amsterdam, 1968).

18. W. G. Love and M. A. Franey, Phys. Rev. $\underline{C24}$ (1981) 1073.

POTENTIAL MODEL DESCRIPTION OF HEAVY ION SCATTERING
USING SPLINE TECHNIQUES

A M Kobos[+] and R S Mackintosh

The Open University, Milton Keynes, UK

[+] On leave from the Institute of Nuclear Physics, Cracow, Poland.
Also at Nuclear Physics Laboratory, Oxford, UK.

1.   RATIONALE FOR POTENTIAL MODELS OF ELASTIC SCATTERING

It is important that elastic scattering be well understood. In principle, it is the simplest direct reaction both for experiment and for
theory. If elastic scattering is not understood, then spectroscopically
interesting reactions, with the inevitable complication of the direct
involvement of two nuclear states instead of one, certainly cannot be
regarded as being reliably understood. In fact, it has recently emerged
that there are profound differences between alternative potentials and
their associated phase shifts each corresponding to fits of elastic
scattering data that would be commonly adjudged, by current standards,
to be good fits. We do not believe that there is, in general, any reliable description of elastic scattering even at a phenomenological level.

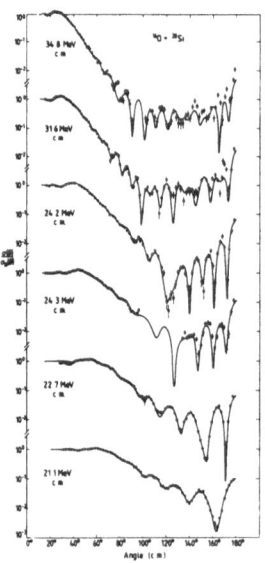

Fig.1 Fits to $^{16}O + {}^{28}Si$ elastic scattering with real M3Y folded
potential + surface correction and the Woods-Saxon volume +
derivative imaginary potential.

One reason for this is that there has been no incentive to produce precise, wide angular range, elastic scattering data. With customary models or potentials, even imprecise wide range data has been unfittable, e.g. $^{16}O + ^{28}Si$; figure 1 shows what <u>can</u> be done when uncustomary models are used - this will be described later when we discuss model independent methods that render such data all-too-easily fittable. We believe that better elastic scattering data should be part of a program for putting direct reactions with heavier ions on some sort of sound footing. We shall discuss the contribution made so far to the phenomenology of elastic scattering by "model independent" potentials, in doing this we make some tentative explorations of the nature and applicability of potential models of heavy ion scattering.

We briefly refer now to one of the recurring themes of this work. Model independent fits (not necessarily with splines!) always lead to potentials with some degree of oscillatory structure, varying from very little for some recent ^{16}O fits to be described, to quite large oscillations for ^6Li and protons. To what extent these are artifacts of the method remains a serious and unsettled question. It should be said, however, that in at least two cases we have specific and plausible arguments for the appearance of oscillatory structure. In the case of proton scattering (which will not be discussed in detail in this paper) we have shown how oscillations might rise inevitably in an ℓ-independent representation of fundamentally ℓ-dependent pickup coupling processes [1, 2]. For the case of ^6Li, we shall mention later in this talk how we have related oscillatory structure to breakup of the ^6Li projectile. This is relevant here because it shows one reason why model independent optical model phenomenology is worth doing. It is the only calculation that can be normally put in a search code to the point of getting "perfect fits" - breakup calculations could feasibly not be put in a search code. Although breakup contributions [3] certainly improve the fit, phenomenology tells us that there is an enormous range of models (or potentials; even phase shifts) which would fit the data "as well" (necessarily a subjective judgement). <u>Nevertheless</u>, a connection can be made between features characteristic of precise fits, and the general properties of the contributions induced in the potential by specific reaction processes.

Before explaining more specifically how we come to these conclusions, we must first specify the spline interpolation method at the heart of our procedure.

2.   THE SPLINE INTERPOLATION PROCEDURE

The basic use of "spline functions" is a means of finding a function
$y(x)$ which has certain values y_i at points x_i (where x_i are members of
some set, not necessarily uniformly spaced $x_1 \ldots\ldots x_N$) and which is,
in some sense, optimally smooth for other points x between x_1 and x_N.
For cubic splines which we are concerned with, the function $y(x)$ is
piecewise continuous; it is a cubic within each interval x_i to x_{i+1}
and it has continuous second derivatives at each "knot" as the points
defined by x_i and called. The spline interpolation procedure is linear
and unique, so that if $y^{(1)}(x)$ is the interpolate for $y_i^{(1)}$ and $y^{(2)}(x)$
is the interpolate for $y_i^{(2)}$, then $y^{(1)}(x) + y^{(2)}(x)$ is the interpolate
for values $y_i^{(1)} + y_i^{(2)}$ at the knots. This leads to the definition of
a linearly independent "spline basis" which simplifies the error analysis
and facilitates a simple inversion procedure to be described below. We
define $U^{(j)}(r)$ as the spline interpolate through knot values y_i defined
by $y_i = \delta_{ij}$. Clearly, the interpolate through values $\bar{y}_i$ at knots x_i
is then given by

$$U(r) \quad = \quad \sum_i \bar{y}_i U^{(i)}(r) \qquad\qquad 1.$$

By specifying values $\bar{y}_i$ at x_i, we specify $U(r)$, so a rather form-indep-
endent means of parameterizing a nuclear potential $V(r)$ is through its
values $V(r_i)$ at knots r_i. By means of an appropriate choice of knots,
commensurate with the underlying physical degrees of freedom, we have
a means of defining a parameterized potential containing little theoretical prejudice and which is suitable for incorporating in optical
model search codes.

3.   APPLICATIONS TO PROTON ^{6}Li AND ^{16}O SCATTERING

Our first application of the spline model was to proton scattering as
we have described in ref[1]. Very good fits were obtained and the
potentials obtained had very strong oscillatory structure. This was
interpreted as being necessary in an ℓ-independent potential which
was equivalent to a potential which had a strong ℓ-dependent component.
We could then attribute this ℓ-dependence to the effect of channel
coupling. We mention this work since it has features which occur in
all subsequent applications of spline models. These are
(1)   The presence or oscillatory structure in the potentials. Is this
      structure real or an artefact? Although we did have a specific
      rationale for oscillations in the case of protons the fact remains

that for one case, we have managed to get a rather smooth potential.
Very likely, the amplitude of oscillatory components varies with
the projectile, precision of the data, and so on.

(2) The determination of a local potential cannot be an end in itself,
in the way that determining a nuclear density might be. Historic-
ally, some applications of splines to nuclear scattering have been
inspired by similar electron scattering studies. This has sometimes obscured the fact that the nuclear potential is much more a
model entity than the nuclear charge density. The hope is that by
studying local equivalents to the underlying complex, non-local,
ℓ-dependent nuclear interaction, it will be possible to get further
understanding of the processes occuring in nuclear scattering.

These features occur in the spline fits to ^{6}Li and ^{16}O scattering [4]
with which we were able to confirm in a model independent manner the
fact that the M3Y folding model [5] gives much too deep a potential for
^{6}Li ions. We were not able to interpret the oscillations that were
found in the potentials. Subsequently, a connection between oscillations
in the ^{6}Li potential and breakup processes has been proposed, and we
shall mention it below. Some idea of the quality of fits which we can
get for ^{6}Li scattering can be seen in figure 2. Unfortunately these
fits correspond to an enormous range of possible potentials, and there
may be little point in further work on this case until the angular range
of the experimental data is greatly increased. We hope it will soon
be apparent why this may be worth doing. In figure 3, we show the
potentials for the ^{40}Ca target corresponding to the fit in figure 2.
The oscillatory features are still evident on the logarithmic scale.
We shall now discuss two cases where we have been able to reduce the
amplitudes of the oscillations in the potential with only marginal loss
of quality of fit.

4. $\underline{^{16}O + ^{40}Ca \text{ at } E_{c.m.} = 35.7 \text{ MeV}}$

This reaction provided a test case for two important questions, i.e.
To what extent do error bars given by the error analysis exclude potentials outside these error bars? and, How inevitable are strong oscillations
in the potential?

The first question is most easily answered - we easily found various
potentials that fitted the data, that were outside each other's error

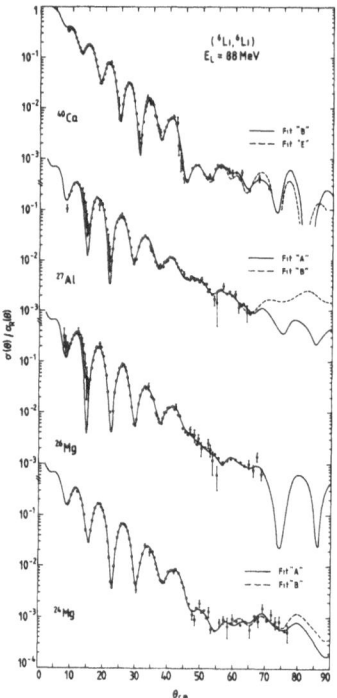

Figure 2   Fits to the scattering of 88 MeV ^{6}Li from various nuclei.

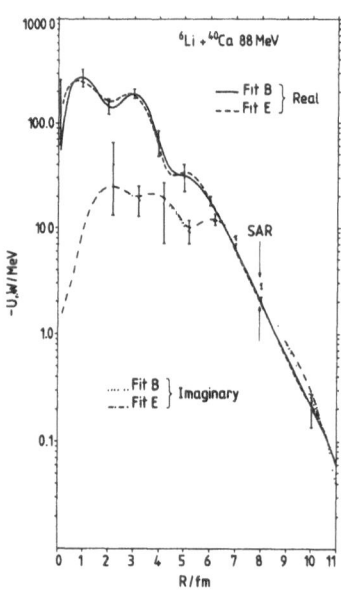

Figure 3   Real and Imaginary ^{6}Li + ^{40}Ca potential at 88 MeV.

bands, and were outside the error bands of a previously published potential [6]. Thus, the error bands for spline model optical potentials
cannot be interpreted in the same way as those for charge densities
where scattering model potential ambiguities (which for potentials often
involve the imaginary component) do not enter.

The second question, however, required some degree of reformulation of
the model. Instead of taking an overall spline potential such as that
of equation 1, we took the potential to be one of two forms:

(smooth potential) + (spline model correction)

or

(smooth potential) x (spline model modulating factor).

By this means, we hoped to get potentials with much lower amplitude
oscillations.

In this case, we took for the smooth potential an optimised Woods-Saxon
potential and the imaginary potential was a superposition of volume and
surface Woods-Saxon terms with independent geometrical parameters. A
spline model corrective term was required for the real part only. A
typical fit with an additive spline term is shown in figure 4.

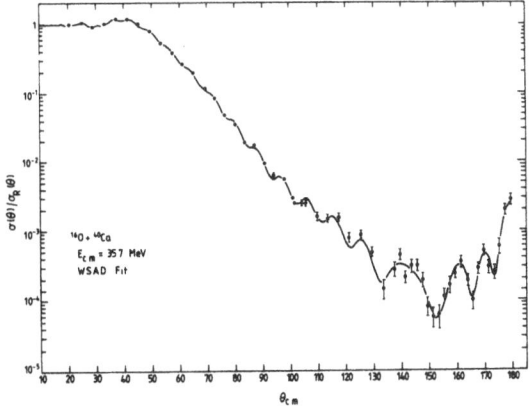

Figure 4   A fit to data of Kubono [7] for elastic scattering of ^{16}O
from ^{40}Ca.

In figure 5 we show the correction required for the Woods-Saxon potential.
The solid line is just the additive spline term and the dashed line is
obtained by subtracting the Woods-Saxon term from the Woods-Saxon term
which was modulated with the multiplicative spline function.

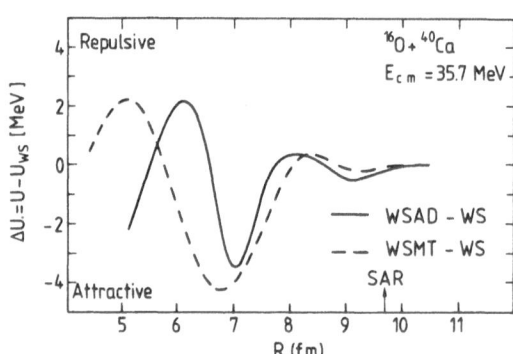

Figure 5   Radial dependence of correction to Woods-Saxon potential obtained by fitting additive or multiplicative correction terms.

The two corrections are quantitatively similar and bear some resemblence
to various calculations of coupled channel polarization effects.   The
oscillations in the overall potential are now very small, much smaller
than those in ref. 6 as can be seen in figure 6, where the total potentials corresponding to the correction terms of figure 5 are shown, together
with the imaginary term.   It appears that the potential is marginally
surface transparent.

5. $^{16}O + ^{28}Si$ Between 21 MeV and 35 MeV c.m.

There is a huge literature on this reaction and we shall not attempt to
survey it here. We have already shown, figure 1, fits obtained with the
real M3Y folded potential supplemented by spline parameterized surface
correction and the Woods-Saxon volume + derivative imaginary potential.
None of the previous attempts have satisfactorily done the following
four things simultaneously
(i)   explain the unexpected rise in differential cross section near 180°.
(ii) get the details of the angular distribution correct in the import-
      ant mid-angle range where interference between amplitudes is

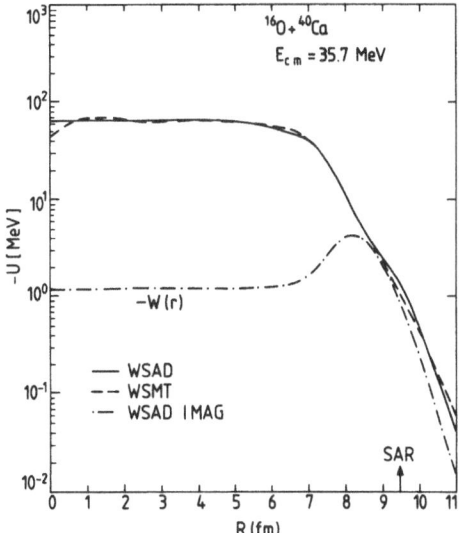

Figure 6 The total real potentials for the same cases as figure 5, together with the imaginary part corresponding to the additive case

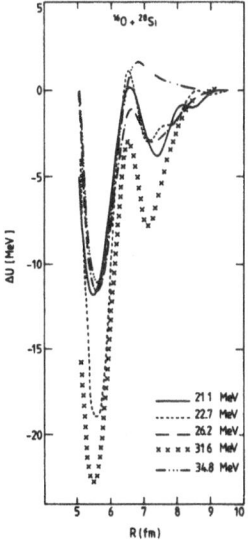

Figure 7 The surface correction to the M3Y folded potential determined for each energy by optimizing the spline function.

evident. We regard this as <u>at least</u> as important as (i).

(iii) give some account of the energy dependence of the back angle cross
     section. At the best this would mean explaining the details of
     the 180° excitation function, but our explanation of the energy
     dependence actually shows why this might be so difficult.

(iv) the potential should be as closely related as possible to what is
     expected from folding models,

The fits shown in figure 1, therefore, are our attempt to get potentials
which simultaneously satisfy the above criteria as well as possible. We
parameterized the potential as follows. The real part consisted of the
renormalized by 0.96 M3Y double folding potential [5] with a spline
additive component (set to zero for nuclear c.m. separation less than
5fm) the parameters of which are searched upon. The imaginary part was
the sum of volume and suface Woods-Saxon terms with independent geometrical parameters (this proved essential). We found potentials that
came quite close to satisfy the above four conditions.   In figure
7, we present the real correction term for all energies. It is remarkably similar in form at all energies and amounts to a relatively small
modification to the M3Y folding model potential.

From figure 8, which shows the overall real potential at one energy and
the imaginary part at all energies, it is apparent that the potential
is markedly surface transparent. Even more striking is the marked but
steady change in surface absorption with energy. It is this which is
vital for understanding the 180° excitation function, but the difficulty
of parameterizing the change in geometry with energy makes it clear
that a precise fit to the 180° excitation function is difficult. Perhaps
it would be simpler to parameterize the imaginary part in the physically
based ℓ-dependent form [8] to which we may have found the local equivalent.

Again, we have found a ^{16}O spline model potential in which the oscillations are present but moderate in amplitude. Our fitting procedure
has fitted previously unfittable data, and could well exploit data of
higher precision. We next describe a method by which the physical
meaning of potentials so-determined might be explored.

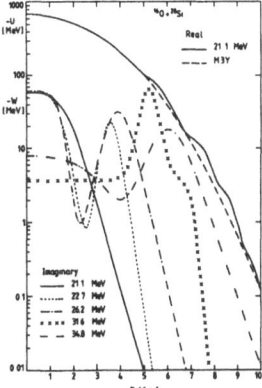

Figure 8   The real potential at 21.1 MeV compared to the M3Y form, and
the total imaginary potential at each energy.

6.   SPLINE MODEL INVERSION PROCEDURE

It has always been apparent that the justification for spline model fits
to elastic scattering lies in the information about physical processes
which might be extracted from the local potentials which we find.   We
now describe a technique which should be useful in providing a connection
between theory and local potential phenomenology.   It involves a simple
phase shift-to-potential inversion procedure based on the use of splines.
Thompson and Nagarajan [3] have shown how the fits to ^{6}Li scattering
can be greatly improved by including the effect of the breakup of the
^{6}Li ion.   Their calculation involved an adiabatic model which was of
such complexity to exclude the possibility of putting it in any search
code.   Although the fit was greatly improved, in a subjective sense,
by the inclusion of breakup, it is nevertheless true that χ^2 must have
been orders of magnitude higher than that of a spline fit which, as we
know, would still have been subject to considerable uncertainty.   Nevertheless, there is a way in which the effect of the coupling to breakup
channels can be linked to phenomenological potentials.   What we have done
is determine the potential which, added to the "bare" folding model
potential, would give the same phase shifts as the complete calculation
including breakup.   In other words, we find a potential representation
of breakup effects.   Having such a potential representation, we are then
in a position to compare it with features such as various bumps or departures from folding model predictions which occur in the precise fit

phenomenological potentials.

A brief outline of the inversion method we use is as follows. Having
determined the s-matrix corresponding to the bare folding model potential,
we then perturb this potential in turn by some suitable multiple of each
of the spline basis functions $U^{(j)}(r)$ which we introduced previously.
In this way we determine the corresponding perturbations in the s-matrix.
Recall that $U^{(j)}(r)$ is a smooth function which is unity at the jth knot
and zero at other knots, so it is large only near the jth knot, and
identically zero only at the other knots. We then carry out a matrix
inversion to determine what superposition of spline basis functions
would give the same change in the s-matrix as the breakup effect. Because of the non-linearity in the response of the s-matrix to perturbations, an iterative procedure is necessary. We have usually included
more partial waves than basis states in the calculation so overdetermined
matrix methods are required.

The potential components which we determine by our iterative inversion
procedure as being equivalent to breakup for 156 MeV ^{6}Li ions on ^{12}C
are shown in figure 9, the real part, and figure 10, the imaginary part.

While we cannot make exact comparisons with our earlier phenomenology,
if only because the target and energy were incompatible, we can at once
identify two features of our model independent fits. Firstly, there
is a repulsive effect of 20 - 30% in the nuclear surface region, which
we identify with the phenomenological finding that for ^{6}Li (and not
for most other projectiles) the M3Y folding potential has to be multiplied by 0.6 or 0.7 [5]. Secondly, the new potential component has
quite strong oscillations. As was the case for the phenomenological
imaginary potential shown in figure 3, there is a peak about 2fm within
the SAR and a dip at about 3fm within the SAR. To what extent this
might be coincidence, only time will tell - it is however very suggestive
and in any case illustrates very well the manner in which local potentials derived from precise model independent fits can be used as a source
of understanding of the processes involved in nuclear collisions. We
envisage many other applications of our simple iterative inversion procedure. While we have by no means explored the limits of applicability
of our procedure, the present indications are that it works better for
lighter ions and higher energies.

In summary, then, we have shown that two distinct uses of spline functions
can contribute to our understanding of nuclear scattering. Firstly,

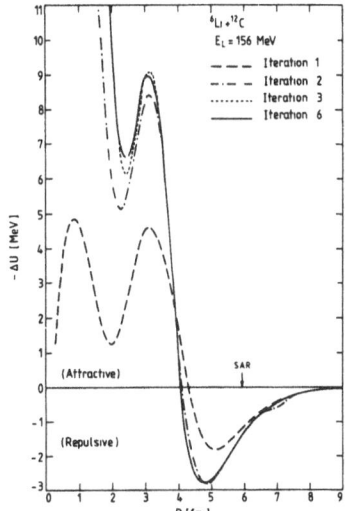

Figure 9   The real potential contribution to the ^{6}Li + ^{12}C potential
at 156 MeV resulting from coupling to the L = 0, 1, 2, breakup states.
The procedure had converged by the sixth iteration.

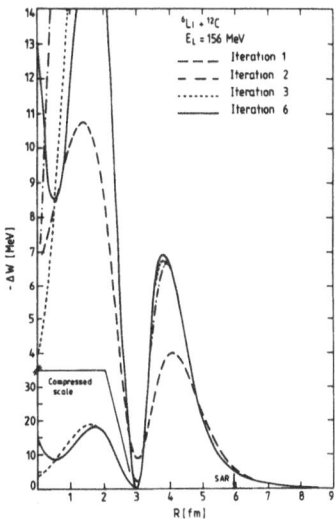

Figure 10   The imaginary potential for the same case as figure 9.

they provide a rather model independent (theoretically unprejudiced) way of determining the nuclear optical potential. We remark that it is often useful to put back some theoretical prejudice by finding, model independently, the correction to some smooth potential.

However, we have also shown the limited usefulness of fully model-independent potentials for scattering of heavy ions, as they may prove to be very different and extremely difficult to interpret.

The second application is as part of a very simple iterative inversion procedure which should be very useful in providing a link between calculations of processes which are not naturally represented as potentials, and potentials derived by the exact fit phenomenology.

We are grateful to G R Satchler for discussions, P E Hodgson for facilities and the SERC for a grant.

References

1.   A M Kobos and R S Mackintosh, Ann.Phys. __123__ (1979) 296.

2.   A M Kobos and R S Mackintosh, Acta Physica Polonica __B12__ (1981) 1029

3.   I J Thompson and M A Nagarajan, Phys.Lett. __106B__ (1981) 163

4.   R S Mackintosh and A M Kobos, Phys.Lett. __92B__ (1980) 59

5.   G R Satchler and W G Love, Phys.Rep. __55__ (1979) 183

6.   S Krewald, A Djaloeis and S Gopal, Phys.Rev. __C24__ (1981) 966

7.   S Kubono, P D Bond and C E Thorn, Phys.Lett. __81B__ (1979) 140

8.   R A Chatwin, J S Eck, D Robson, A Richter, Phys.Rev. __C1__ (1970) 795

FOLDING MODELS FOR ELASTIC AND INELASTIC SCATTERING*

G. R. Satchler
Oak Ridge National Laboratory
Oak Ridge, TN 37830, U.S.A.

1. Motivation

We are accustomed to the use of many simple models in nuclear physics. In the present context, the most widely used models are the optical model potential (OMP) for elastic scattering, and its generalization to non-spherical shapes, the deformed optical model potential (DOMP) for inelastic scattering. These models are simple and phenomenological; their parameters are adjusted so as to reproduce empirical data. Nonetheless, there are certain, not always well-defined, constraints to be imposed. The potential shapes and their parameter values must be "reasonable" and should vary in a smooth and systematic way with the masses of the colliding nuclei and their energy. Without these constraints, the potentials tell us very little, and they will be essentially useless for other purposes, such as in DWBA calculations.

One way of satisfying these constraints, without going back to a much more fundamental theory, is through the use of folding models. Remember that the basic justification for using potentials of the Woods-Saxon shape for nucleon-nucleus scattering, for example, is our knowledge that a nuclear density distribution is more-or-less constant in the nuclear interior with a diffuse surface. When this is folded with a short-range nucleon-nucleon interaction, the result is a similar shape with a more diffuse surface. Folding procedures allow us to incorporate many aspects of nuclear structure (although the nuclear 'size' is one of the most important), as well as theoretical ideas about the effective interaction of two nucleons within nuclear matter. It also provides us with a means of linking information obtained from nuclear (hadronic) interactions with that from other sources, as well as correlating that from the use of different hadronic probes. For example, inelastic electron scattering measurements may provide a transition density $\rho_{tr}(r)$ for an excitation. Frequently this is represented by a simple collective (deformed density) model when the transition is strong, whose shape may be of the form

$$\rho_{tr}^{C}(\underline{r}) = \delta_{L}^{C} \; \frac{d\rho_{C}(r)}{dr} \; Y_{L}^{M}(\hat{\underline{r}}), \tag{1}$$

where $\rho_{C}(r)$ is the ground state charge density distribution and δ_{L}^{C} is an amplitude or 'charge deformation length' for the 2^{L}-pole transition. The interaction for hadronic inelastic scattering is frequently represented in an analogous way (the DOMP) where the transition <u>potential</u> $U_{tr}(r)$ is taken to be

$$U_{tr}(\underset{\sim}{r}) = \delta_L^N \frac{dU(r)}{dr} Y_L^M(\hat{\underset{\sim}{r}}) \tag{2}$$

where $U(r)$ is the optical potential for elastic scattering of that particular hadron, and δ_L^N is the corresponding 'nuclear' 2^L-pole deformation length.

The use of these models (1) and (2) (and their extensions) has led to much discussion as to how the δ^N and δ^C are related and, indeed, how the δ^N obtained with different probes are to be related. (One popular prejudice[1]) is that the deformation lengths should be the same, $\delta_L^C = \delta_L^N$.) Folding provides some insight into this problem[2-4]) and indeed may be used to construct the transition potential $U_{tr}(\underset{\sim}{r})$ directly from the transition density by folding with a suitable interaction. As usual, such extensions of a model also open up new uncertainties. For example, the charge density (1) arises predominantly from proton excitations, while neutron and proton excitations contribute comparable amounts to the hadronic transition potential (2). However, we may turn this situation into an advantage and use it to give information about the neutrons.

Folding is still a model, and, at best, an approximation. It is a reasonable model for nucleon scattering. At low energies, it is closely related to the Brueckner-Hartree-Fock model for bound states. At high energies, it can be justified in terms of the impulse approximation. Extensions of the Brueckner-type theories provide support at intermediate energies. The situation is not so clear for composite projectiles, especially the scattering of two heavy ions. However, here the situation is helped because such systems usually exhibit strong absorption. Under such circumstances, the details of the interaction when the two nuclei overlap are largely irrelevant; all that matters is the interaction for peripheral collisions where again folding may be reasonable. Ultimately, the justification of the folding procedure must be two-fold: (i) it must be seen to work in practice (i.e. to explain experimental data), and (ii) it should be justified from a more fundamental theory. This is the common fate of physical models.

2. Further Theoretical Background

2.1. Coupled Equations and the Feshbach Projection Theory for the Optical Potential

We have spoken so far as though the inter-nucleus potential was defined uniquely. This is not so; there are various ways in which we may reduce the complicated $(A_1 + A_2)$-body scattering problem to an effective two-body problem. Failure to remember this has sometimes led to confusion. A conventional optical model potential $U(R)$ for two nuclei A_1, A_2 is one which appears in a one-body Schrödinger equation

$$\left[-\frac{\hbar^2}{2\mu_\alpha} \nabla^2 + U(R) \right] \chi(\underset{\sim}{R}) = E \chi(\underset{\sim}{R}), \tag{3}$$

where μ_α is the reduced mass of the pair, R is the separation of their centers of mass, and E is the CMS energy of relative motion. The solution $\chi(\underset{\sim}{R})$ with the appropriate boundary conditions describes the elastic scattering of $A_1 + A_2$. Usually U(R) is assumed to be local (although its parameter values may vary with bombarding energy) and to have a smooth, simple functional form like that of the Woods-Saxon shape. Occasionally, some L-dependence is introduced, but only if one is forced to consider such a generalization by some intractable data. The primary consideration is simplicity.

How do we justify the use of an equation like Eq. (3) which makes no reference to the many internal degrees of freedom and the corresponding non-elastic channels except insofar as they lead to absorption from the elastic channel and require U to be complex? One standard way to make it plausible that such an equation may adequately describe elastic scattering is to use the projection operator formalism of Feshbach.[5]

Suppose that the complete Hamiltonian for the colliding pair is given by

$$H = H_1 + H_2 + T + V \tag{4}$$

where H_1 and H_2 denote the internal Hamiltonians for the isolated nuclei, T is the kinetic energy of <u>relative</u> motion in the center-of-mass system, and V is the coupling interaction between the ions. At this point, we ignore any explicit reference to antisymmetrization between nucleons in different ions, although we assume that the wavefunctions for the internal states of the individual nuclei are antisymmetric. We denote a complete set of states for the internal Hamiltonians of nuclei A_1 and A_2 by ψ_i and ϕ_j respectively, where

$$(H_1 - \varepsilon_{1i}) \psi_i = 0, \qquad (H_2 - \varepsilon_{2j}) \phi_j = 0. \tag{5}$$

If the complete wavefunction is denoted by $\psi^{(+)}$, we note that measurements of elastic or inelastic scattering provide us with information about the projection of the complete wavefunction onto the channel subspace defined by the particular measurement. We denote these R-dependent projections (amplitudes) by

$$\chi_{ij}(\underset{\sim}{R}) \equiv \langle \psi_i \phi_j | \psi^{(+)} \rangle, \tag{6}$$

which means that we can express the total wavefunction as

$$\psi^{(+)} = \sum_{i,j} \chi_{ij}(\underset{\sim}{R}) \psi_i \phi_j. \tag{7}$$

A set of coupled equations governing the $\chi_{ij}(\underset{\sim}{R})$ can be found by inserting the above expansion into the complete Schrödinger equation

$$(H-E) \psi^{(+)} = 0. \tag{8}$$

This yields

$$[T + (\psi_i \phi_j | V | \psi_i \phi_j) - E_{ij}] \chi_{ij}(\underset{\sim}{R}) = \sum_{[k, \ell \neq i,j]} (\psi_i \phi_j | V | \psi_k \phi_\ell) \chi_{k\ell}(\underset{\sim}{R}). \tag{9}$$

where

$$E_{ij} = E - \varepsilon_{1i} - \varepsilon_{2j} \text{ and } (i,j=0...\infty).$$

The rounded brackets denote integration over the __internal__ coordinates of the two nuclei and serve to remind us that the matrix elements remain a function of $\underset{\sim}{R}$.

Now Eqs. (9) represent an infinite set of coupled equations which, although complete, are not very useful. Practical calculations require a severe truncation of these to a __small__ set of equations coupling a few states of particular interest; this is what is involved in the so-called 'coupled-channels' method. The use of an equation like Eq. (3) in fact involves truncation to just that __one__ channel associated with the ground states, $i = j = 0$. The price we pay for any such truncation is that the interaction V should be replaced by an __effective__ interaction. This becomes clear in Feshbach's formalism[5] in which all __explicit__ couplings to channels other than the one of interest are transformed away and their effects incorporated into an effective interaction. For that channel in which both nuclei remain in their ground states this procedure results in an equivalent one-body Schrödinger equation for the $\chi_{00}(\underset{\sim}{R})$,

$$[T + U_{op} - E] \chi_{00} = 0 \tag{10}$$

with the effective interaction or optical potential operator U_{op} given by

$$U_{op} = (\psi_0\phi_0|V|\psi_0\phi_0) + (\psi_0\phi_0|VQ \frac{1}{E-H_{QQ}+i\varepsilon} QV|\psi_0\phi_0)$$

$$= U_F + \Delta U. \tag{11}$$

U_F is defined as the first term in Eq. (11)

$$U_F = (\psi_0\phi_0|V|\psi_0\phi_0), \tag{12}$$

and is called the folded potential for elastic scattering. (Note that this same folded potential and its generalization to excited states and off-diagonal couplings also appears in Eqs. (9).) Q is the operator which projects off the ground states of the two nuclei, so that only excited states appear as intermediate states in ΔU. By its construction, the exact elastic scattering amplitude may be obtained from χ_{00} as $R \to \infty$.

A similar result is obtained when the truncation is to a model space which includes one or a few excited states as well as the ground states. U_{op} becomes a matrix in the model space and provides the effective interactions to be used in the corresponding truncated coupled-channels problem. In particular, it yields (complex) corrections ΔU to the interchannel couplings in addition to the folded terms that already appear in the Eqs. (9).

The folded potential U_F is real (provided that V is real). The remaining term ΔU, which we may refer to as a 'dynamic polarization potential', arises from coupling to all the other states, is much more difficult to calculate even approximately, and, in general, is complex, non-local, energy- and angular-momentum dependent. (In practice, because of the strong short-range repulsion in the bare nucleon-nucleon interaction, the V itself will be an effective interaction or G-matrix which itself includes some 'polarization' corrections, primarily those associated with the short-range correlation between the interacting nucleons and hence high excitation energies. In principle, this leaves open the possibility of some double counting.)

In phenomenological approaches it is this object U_{op} which is approximated by a local, complex model potential U(R) in Eq. (3). However, although this formalism provides us with the _form_ of the optical model, it does not guarantee that any given simple representation of U_{op} will be adequate. This question may be explored both empirically (is a given model successful in fitting data?) and by studying the structure of ΔU.

2.2. Other Approaches to the Nucleus-Nucleus Interaction Energy

A particular point to be stressed is that with the potential U_{op} (and, by implication, an equivalent model potential U) defined in this way, the solution $\chi(R)$ of Eq. (3) represents $\chi_{oo}(R)$ of Eqs. (9) and (10) and describes the relative motion of the two nuclei _while_ _they_ _both_ _remain_ _in_ _their_ _respective_ _ground_ _states_. This may be a _very_ small component of the total wavefunction in that region of space where the two nuclei overlap appreciably; in that case, the strong absorption into other channels manifests itself through $\chi_{oo}(R)$ becoming very small for $R \lesssim R_1 + R_2$.

This is to be contrasted with most of the potentials calculated microscopically for heavy-ion collisions (see Refs. 6-10, for example). These calculations may use the energy-density approach, the Thomas-Fermi approximation, the liquid-drop model, the proximity theorem, etc., but they all attempt to follow to a greater or lesser degree the readjustments that the two nuclei must make as they begin to interact and overlap; distortion of the nuclear shapes, reaction to the Pauli principle, effects of the saturating nature of the nuclear forces, etc. Such an interaction energy function does not determine just the χ_{oo} component of the wavefunction (7) but is to be used in a description of the motion of a wave packet which includes a wide range of excited states of the separated systems. It is not to be identified with U_{op} of Eq. (11), and, in principle, it should not be used in an equation like Eq. (3).

It is possible that the equations for elastic scattering from these other approaches may be formally recast into the same _form_ as the optical model Eq. (3) (see Pal, _et al._[11]) for a recent discussion of this). The corresponding effective optical potential $\tilde{U}$ appearing in such an equation will be different from U_{op}. It will contain not only the interaction energy function just referred to but also correction terms. For example, in one model situation[12] there are corrections corresponding to changes in the kinetic energy due to changes in the inertial properties as the system begins to interact. The corresponding scattering solutions $\tilde{\chi}(\underset{\sim}{R})$ from such equations will differ from $\chi_{oo}(\underset{\sim}{R})$. Before approximations are made, we must have $\tilde{\chi}(\underset{\sim}{R}) \rightarrow \chi_{oo}(\underset{\sim}{R})$ asymptotically because the various theories describe the same elastic scattering, but these wavefunctions will differ in the interaction region at small R.

The actual physical processes that are to be described in the two types of approaches are, of course, the same. In the Feshbach theory they manifest themselves in the dynamic polarization potential ΔU; in some ways this formalism is deceptively simple, being most convenient to use when the effects of a few non-elastic channels are to be studied explicitly. The other approaches provide vehicles for including, perhaps in intuitively more obvious ways, various bulk physical effects. Nonetheless, the 'potential' functions they generate should not simply be identified as optical potentials without bearing in mind the considerations already mentioned.

In practice, of course, approximations are made in both kinds of approach. Frequently an adiabatic aproximation is used when calculating an ion-ion interaction energy. This implies that the collision is slow enough for the two nuclei to readjust, both as they begin to interact and overlap and as they separate, so that they return to the elastic channel; there is no 'absorption' into non-elastic channels and the calculated interaction energy is real. The absorption into other channels that actually occurs must then arise from the correction terms mentioned above. In practice, these are usually replaced by a phenomenological absorptive potential. On the other hand, the folding model evaluates only the U_F term of Eq. (11), identifying it with the real part of the potential and implicitly assuming that the polarization potential ΔU is predominantly imaginary and can be represented by a phenomenological imaginary potential. This corresponds to a sudden approximation in which excitations ('polarization') of the system tend not to de-excite back to the elastic channel, but result in absorption.

2.3. Folding Model Potentials

The idea of a folded potential has a long history. The electrostatic potential $U_C(r)$, due to a charge distribution $\rho_C(r)$, is given by a single folding,

$$U_C(\underset{\sim}{r}) = \int \frac{\rho_C(r')}{|\underset{\sim}{r}-\underset{\sim}{r}'|} d\underset{\sim}{r}', \tag{13}$$

while the electrostatic interaction between two charge distributions is given by a double folding,

$$U_C(\underset{\sim}{R}) = \int\int \frac{\rho_{C1}(r_1)\rho_{C2}(r_2)}{r_{12}} d\underset{\sim}{r}_1 d\underset{\sim}{r}_2 \tag{14}$$

where $r_{12} = \underset{\sim}{R} + \underset{\sim}{r}_2 - \underset{\sim}{r}_1$ (see Fig. 1). In the nuclear case, the Coulomb interaction $1/r_{12}$ is replaced by a nucleon-nucleon (effective) interaction v_{12}. The

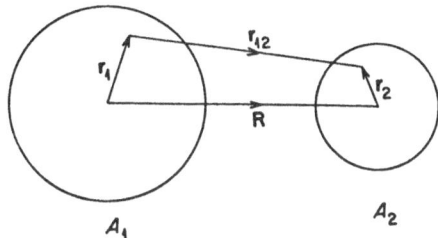

Fig. 1. Coordinates used in folding calculations

first term $U_F(R)$ in Eq. (11) was identified as a folded potential,

$$U_F(R) = \iint \psi_0^*(\xi_1)\phi_0^*(\xi_2) \sum_{ij} v_{ij} \, \psi_0(\xi_1)\phi_0(\xi_2) d\xi_1 d\xi_2 \tag{15}$$

where nucleon i is in nucleus 1, nucleon j is in nucleus 2, and ξ_1 and ξ_2 refer to the <u>internal</u> coordinates of these two nuclei, respectively. If v_{ij} is a local interaction, $v_{ij} = v(r_{ij})$, then the potential (15) may be reduced to the analogue of Eq. (14):

$$U_F(R) = \iint \rho_1(r_1)\rho_2(r_2)v(r_{12})dr_1 dr_2, \tag{16}$$

where $\rho_i(r_i)$ is the one-body density for the ith nucleus, usually assumed to be spherically symmetric. This we call a <u>double-folded potential</u>. The properties and applications of this expression have been discussed in detail elsewhere (see Refs. 14-16, where many other references are also given). We just mention here that many results are more easily understood in momentum space by using Fourier transforms. If we denote the transform of a function $f(r)$ by

$$\tilde{f}(q) = \int dr \, \exp(iq \cdot r)f(r), \tag{17}$$

then in momentum space Eq. (16) becomes

$$\tilde{U}_F(q) = \tilde{v}(q) \, \tilde{\rho}_1(q) \, \tilde{\rho}_2(-q). \tag{18}$$

Then, for example, if the densities ρ_1 and ρ_2 are spherically symmetric,

$$U_F(R) = \frac{1}{2\pi^2} \int q^2 \, dq \, j_0(qR) \, \tilde{v}(q)\tilde{\rho}_1(q)\tilde{\rho}_2(q). \tag{19}$$

When only small momentum transfers are involved in the scattering, it is helpful to expand the various factors in powers of q in order to see which are their important characteristics. For example, if $f(r)$ is scalar in Eq. (17),

$$\tilde{f}(q) = J_0[1 - \tfrac{1}{6} \langle r^2 \rangle q^2 + \tfrac{1}{120} \langle r^4 \rangle q^4 + \ldots], \tag{20}$$

where J_0 is the volume integral of $f(r)$ and $\langle r^n \rangle$ is its nth radial moment.

A simple generalization of the formula (16) is applicable to <u>inelastic</u>

scattering.[14-16] The transition potential for exciting one of the nuclei is obtained by replacing the corresponding ground-state density by the appropriate transition density (for example, like the simple collective model one of Eq. (1)). The resulting folded transition potential is non-spherical (if $L \neq 0$). There is no reason to believe that its radial shape will be exactly like the radial derivative of the diagonal potential $U_F(R)$ of Eq. (16), as is assumed in the simple DOMP model of Eq. (2). Indeed, it is known[15,16] that the folded shape depends upon the multipolarity, whereas the DOMP prescription (2) gives a shape that is independent of the multipole order.

We may break the double folding into two steps and rewrite Eq. (16):

$$U_F(R) = \int d\underline{r}_1 \; U_{2N}(\overline{\underline{r}}_{12})\rho_1(\underline{r}_1), \tag{21}$$

where

$$U_{2N}(\overline{\underline{r}}_{12}) = \int d\underline{r}_2 \; \rho_2(\underline{r}_2)v(\underline{r}_{12}), \tag{22}$$

and $\overline{\underline{r}}_{12} = \underline{r}_{12}-\underline{r}_2 = \underline{R}-\underline{r}_1$. Then U_{2N} is a folded potential for the interaction of nucleon 1 with nucleus 2 (analogous to Eq. (13)). Clearly Eq. (22) is the appropriate folding model for <u>nucleon</u> scattering from a nucleus.[13]

3. Single-Folded Potentials

3.1. Introduction

In practice, the single-folded form (21) has often been invoked for the scattering of composite nuclei, but with U_{2N} replaced by an empirical, phenomenological nucleon-nucleus A_2 optical potential. Usually this is what is meant by a single-folding model. Clearly the procedure is symmetrical in the two nuclei A_1 and A_2; one may equally well take an empirical nucleon-nucleus A_1 optical potential U_{1N} and fold it with the density distribution for nucleus A_2. Since the potentials U_{1N} and U_{2N} are empirical, there is no <u>á priori</u> guarantee that the two results will be identical.

The phenomenological nucleon-nucleus optical potential U_{2N} will be complex if it is obtained from analysis of nucleon-nucleus scattering. However, it should not be concluded that single-folding with the imaginary part will yield the correct imaginary potential for nucleus-nucleus scattering, even if use of the real part is successful. The imaginary part of U_{2N} accounts for absorption due to the excitation of one nucleus by a free nucleon; it does not account for excitation of the other nucleus in which the interacting nucleon resides, or effects due to the presence of the other nucleons in that nucleus. In the simplest case of the scattering of the loosely-bound deuteron, the absorption due to break-up of the deuteron is not included; the single-folding model gives an absorptive potential which is much too weak.[17] (Such break-up can also modify the effective real potential.[18 19])

Of course, if the potential U_{2N} (or its imaginary part) is <u>not</u> taken from empirical nucleon scattering data, but is adjusted to fit nucleus-nucleus scattering data, as in the alpha scattering case discussed below, then it is reasonable (though not necessarily correct!) to use the same adjusted interaction for application to inelastic scattering.

The other ingredient of the single folding model is the density distribution for the nucleus into which the interaction is being folded (or transition density, if an inelastic excitation is being considered). The tail of the folded potential, important for the scattering of strongly absorbed particles, is very sensitive to the extent of this density distribution, raising the possibility of learning something about the matter distribution provided some interaction potential can be shown to be valid. This is important because our other source of data on nuclear densities is electron scattering which primarily tells us about the proton distributions. Hadrons, on the other hand, interact as strongly with the neutrons. Only for nuclei with N = Z is it reasonable to assume that neutron and proton distributions are the same (and even then there can be small differences because of the Coulomb forces). It is sometimes assumed that the distributions are in the ratio N/Z when N ≠ Z, or $\rho_n(r) = (N/Z)\rho_p(r)$. Although reasonable, this assumption must be used with caution; it is believed, for example, that neutron distributions in the ground state extend to slightly larger radii than do the protons, and this can have quite large effects.

The mean square radius (MSR) is a useful characteristic of the ground-state densities. Although the folded potential does depend to some extent upon the detailed surface shape of the density, it is largely determined by the MSR.[15] In the case of 2^L-pole excitations, this role is taken by the r^{L+2} moment of the transition density. These results can be easily understood[14-16] in momentum space, using the expansion (20).

A variety of representations of the ground-state density have been used. There is the standard Fermi (= Woods-Saxon!) shape, or the convoluted form of Helm (itself a folding model) which has some analytic advantages.[20] Independent-particle models have also been used, either from Hartree-Fock calculations or more phenomenological shell-model calculations.[14,15] These provide a way of incorporating our prejudices about shell effects and about the behavior of the neutrons, for example.

Transition densities may be constructed in similar ways. One popular form is like that given by Eq. (1), except for the appearance of the matter, rather than the charge, distribution. Another is the Tassie or hydrodynamical model whose radial part differs from Eq. (1) by an additional factor of r^{L-1}; for L > 2 and isoscalar excitations,

$$\rho_{tr}(\underset{\sim}{r}) = \alpha_L \ r^{L-1} \ \frac{d\rho(r)}{dr} \ Y_L^M(\hat{\underset{\sim}{r}}). \tag{23}$$

This has some appeal if the transition is strong (such as for a giant resonance), for it has been shown[21] to be correct for a single transition that exhausts the classical energy-weighted sum rule (EWSR).

3.2. Alpha Scattering

Single folding has been applied with some success to alpha-nucleus elastic and inelastic scattering using either model N-α potentials or ones derived from data on nucleon-alpha scattering.[14] When we are concerned only with data for forward scattering angles (or small momentum transfers), the most relevant properties of the effective interaction are its strength (volume integral) and mean square radius (see Eq. (20)). In these circumstances, a simple form like a Gaussian may be used. A recent example[22] adjusted the strength -(V+iW) and range α of a Gaussian N-α interaction to fit the forward angle (θ ≲ 30°) elastic scattering of 140-MeV alphas. As had been found in earlier studies,[14] there is a remarkable consistency in the results (Fig. 2) with $V \approx 36$ MeV, $W \approx 23$ MeV, and $\alpha \approx 1.94$ fm.

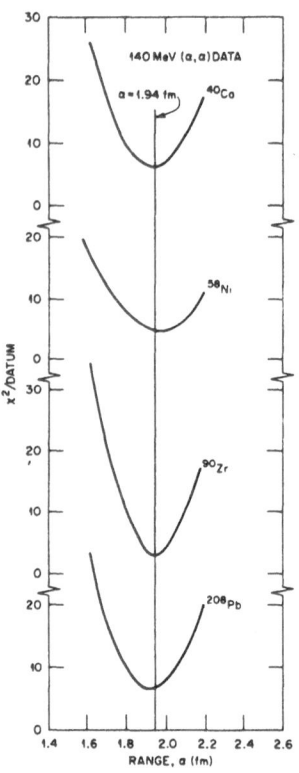

Fig. 2. Variation of χ^2 with the range of the Gaussian nucleon-alpha interaction used in a single-folding model when real and imaginary strengths are optimized to fit elastic data at forward angles.[22]

This empirical interaction was then applied to inelastic data for the excitation of both low-excited states and giant-resonance states. The Tassie form (23) was adopted for the transition densities and the amplitudes α_L deduced by comparison with the data. Figure 3 shows an example for the lowest 3⁻ state in ^{208}Pb; the amplitude α_3 required corresponds to 18% of the EWSR, which is exactly what one would deduce from the measured B(E3) value for this transition if it is assumed that the _proton_ transition density is just (Z/A) times the matter density. This illustrates how folding allows one to directly correlate information from different sources in a relatively unambiguous way. (The excellent agreement, of course, has to be somewhat fortuitous. Equation (23) is still a model; the proton and neutron contributions do not have exactly the same shape, and the proton part is not in exact agreement with the measured charge transition density. On the other hand, when a transition density of the form of Eq. (1) was used, the B(E3) value deduced from the (α,α´) data was 50%

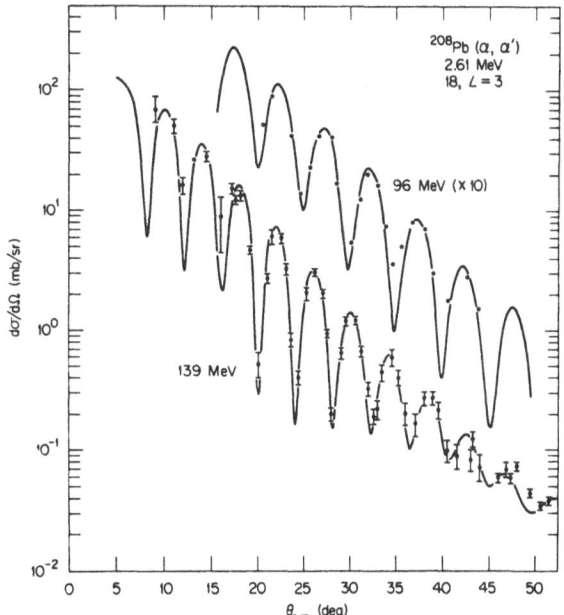

Fig. 3. Comparison of single-folding model predictions,[22] using 18% of the energy-weighted sum rule with measured cross sections for the 3⁻ state in ^{208}Pb.

larger than the measured one. We can conclude that the Tassie model (23) is more appropriate.)

3.3. Heavier Ions

The single-folding approach to the optical potentials for heavy ions starts with some standard phenomenological nucleon optical potential obtained from nucleon-nucleus scattering. It invariably overestimates the strength of the real potential (near the strong absorption radius, where it matters most) by a factor of about two.[23] Several reasons have been offered in explanation of this,[24] including the improper treatment of any density-dependence present in the underlying nucleon-nucleon effective interaction.

It has been suggested[25] that single folding of this kind may be adequate for estimating both the real and imaginary contributions to the optical potential from one or a few loosely-bound 'valence nucleons', even if it does fail for the bulk of the more tightly bound nucleons. In this approach, the nucleus is divided into 'core' and 'valence' parts. The core contribution is taken to be the same as the phenomenological potential which describes scattering from the corresponding core nucleus, while the valence nucleons are treated 'microscopically' within the single-folding model. Successful applications have been made[25] to the elastic and inelastic scattering of ^{17}O and ^{18}O where, of course, the core nucleus is taken to be ^{16}O. The relatively weak binding of the valence nucleons results in their contribution to the interaction being of longer range than that due to the core nucleus. Thus, they may play a disproportionately large role in generating the potential at the large distances that are important for strongly absorbed scattering. In particular, they may have important effects on the Coulomb-nuclear interference pattern when the nucleus is excited.

A recent example of the use of this model is shown in Fig. 4. The curves represent a preliminary analysis by Rhoades-Brown of Oak Ridge data[26] for the excitation of 120 MeV ^{18}O to its first 2^+ state by scattering from ^{208}Pb. The calculations are almost free of adjustable parameters. The complex core potential is

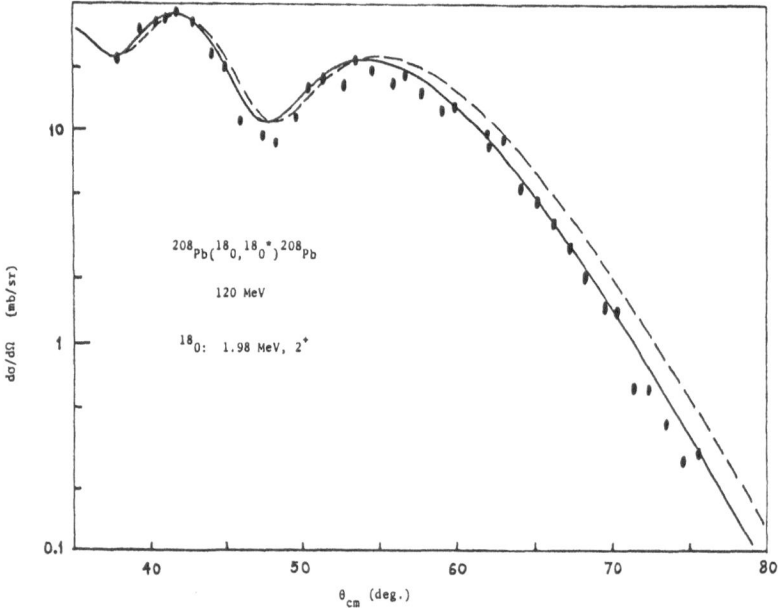

Fig. 4. Coupled-channels predictions using the hybrid folding model of Landowne, et al.[25] for exciting the first 2^+ state of ^{18}O when scattering at 120 MeV from ^{208}Pb, compared to measured cross sections.[26] The solid curve includes reorientation due to a hexadecapole moment of the 2^+ state, the dashed curve does not. (Both include reorientation due to the quadrupole moment.)

taken from analysis of 130-MeV ^{16}O + ^{208}Pb elastic data. The interaction of the valence neutrons of ^{18}O with the target is an empirical (complex) optical potential taken from analysis of n + ^{208}Pb scattering. The valence neutron wavefunctions result from shell-model calculations.[25] These alone are not sufficient to account for the 2^+ excitation; for example, being neutrons they would result in a vanishing B(E2) value unless they are assigned an effective charge. This implies polarization of the core protons. This core polarization is included[25] by using the standard prescription (e.g., Eq. (2)) of deforming the core potential by an amount required to explain the observed B(E2) for this transition. The coupling is sufficiently strong that a first-order DWBA is not sufficient for a description of this excitation. Particularly important is the re-orientation of the 2^+ state after its initial excitation (i.e. transitions between the various magnetic substates). The quadrupole component of this reorientation interaction is constructed in the same way,[25] normalized to the observed electric quadrupole moment of the 2^+ state. In addition, it was found advantageous to include a hexadecapole interaction as well. Although there is no independent information on the electric hexadecapole moment of this 2^+ state, the values chosen are consistent with our expectations for nuclei in this region. The main effect of the additional reorientation due to this term is to decrease the cross section at the larger angles (see Fig. 4). (It is a common feature of this and other analyses of these data that such a term seems to be required. Consequently, we may obtain a measure of this hexadeca-pole interaction and, through folding, of the underlying mass moment.)

The agreement with the data is remarkable considering that no parameters were adjusted. The standard type of DOMP analysis has great difficulty in reproducing accurately the Coulomb-nuclear interference minimum near 50° without destroying the fit at other angles. Even the agreement shown in Fig. 4 probably could be improved by small parameter adjustments; this would be valid because all of the empirical quantities used to determine the input have appreciable experimental errors associated with them.

4. The Double-Folding Model

4.1. The Ingredients

The double-folding model has been reviewed in detail several times in the last few years (for example, see Refs. 15, 27, and 28). We only mention a few points here.

The discussion in Sec. 3.1 on the nuclear densities or transition densities applied equally here. The other ingredient in the formula (16) is the effective nucleon-nucleon interaction v. At one time (see Ref. 13 for example), it was popular to use for v a simple potential, such as a Gaussian, which fits low- energy nucleon-nucleon scattering. However, this has been shown[23] to overestimate the heavy-ion optical potential by roughly a factor of two. (It has also been shown[29] to predict (p,p′) cross sections which are too large.)

It is desirable to use an effective interaction v which is based upon a realis-tic nucleon-nucleon force. Thus, it is some kind of G-matrix[27] that will depend upon the environment in which the two interacting nucleons are embedded. Conse-quently, in general, we can expect different effective interactions to be appropri-ate for nucleon-nucleus and the various nucleus-nucleus scattering problems. None-theless, it seems worthwhile seeing how far we can go with the simple assumption that at least the real part of the effective interaction is essentially the same in all these situations, and to find what kind of corrections we may have to make in particular cases.

A 'high-energy' approach to this problem was used by Dover and Vary.[16,28] In the high-energy limit (or impulse approximation), v would become the (complex) t-matrix for free space nucleon-nucleon scattering. However, for the low-energy ($\lesssim$ 20 MeV per nucleon) collisions that are encountered particularly in heavy-ion scattering, there are very large corrections to be made for the effects of the nuclear medium in which the two interacting nucleons are embedded (Pauli principle, off-shell propagation and Fermi motion of the nucleons). One 'low-energy' ap-proach[30] assumes that the effective interaction v is similar to the G-matrix for two nucleons bound near the Fermi surface. One consequence of this assumption is that v is real, so that the imaginary interaction has to be treated phenomenologically. An example of this real interaction which has been widely (and successfully) used[15,30] has become known as the M3Y interaction. The part that is independent of spin and isospin has the simple form

$$v_{M3Y}(r) = \left[7999 \frac{e^{-4r}}{4r} - 2134 \frac{e^{-2.5r}}{2.5r} \right] - 262 \, \delta(\underset{\sim}{r}),$$ (24)

where the ranges are in fm and the strengths in MeV. The other spin, isospin-dependent terms have similar forms. The last, zero-range term in Eq. (24) is a pseudo-potential that represents the effects of knock-on exchange between the incident and the struck nucleon. (This is believed to be the leading correction when the formula (16) is properly antisymmetrized with respect to the interchange of nucleons between the two nuclei A_1 and A_2. The individual nuclear wavefunctions ψ and ϕ are themselves assumed to be antisymmetric.)

Other work on nucleon-nucleus scattering[27,31] has taken into account that the projectile nucleon is in the continuum <u>above</u> the Fermi sea. This yields a complex G-matrix; however, it seems probable that at least the absorptive processes in the scattering of composite nuclei, especially heavy ions, are very different so that it may still be preferable to treat the imaginary potential phenomenologically. More recent work[32] has recognized that, in the collision of two composite nuclei, both of the two interacting nucleons are embedded within two pieces of nuclear matter that are moving with respect to each other. This work provides both real and imaginary components of the interaction. The imaginary part provides the absorptive potential due to the bulk properties of the overlapping nuclei; it is supplemented by additional surface terms, calculated according to the ΔU term of Eq. (11), which arise from the finite extent of the two pieces of nuclear matter and correspond to the excitation of collective surface vibrations.[33]

It is worth remarking that replacing the bare nucleon-nucleon potential by an effective or G-matrix interaction is a process analogous to the derivation of the effective interaction U_{op} in Sec. 2.1. The G-matrix transformation itself incorporates some polarization effects, namely the short-range correlations between pairs of nucleons and, in principle, allows us to use simpler nuclear wavefunctions such as those of the shell model. Then a full derivation of the operator U_{op} of Sec. 2.1 would go explicitly through this two-stage projection from the original many-body problem (see, for example, Ref. 34).

From its origin as an <u>effective</u> interaction or G-matrix, it is clear that, in general, this v should be both energy-dependent and density-dependent.[†] For

---

[†]The M3Y interaction (24) does not depend upon density or energy (except for a weak energy-dependence of the strength of the zero-range pseudo-potential). From its construction[30] within a truncated space of oscillator functions, one can see that it represents a certain average over a range of kinetic energies and densities.

example, the volume integral of the effective interaction (including, implicitly, the knock-on exchange[35]) for a nucleon scattering from symmetric nuclear matter was found by Jeukenne, et al.[31] to be represented approximately by

$$J_o(\rho,E) \approx F(E)(1 - d\rho^{2/3}),$$ (25a)

with

$$d = 2.03 \text{ fm}^2, \qquad F(E) = (903 - 7.67E + 0.22E^2)\text{MeV fm}^3$$ (25b)

and the density ρ in fm^{-3}. ('Normal' nuclear matter has $\rho \approx 0.17$ fm^3; then $1 - d\rho^{2/3} \approx 0.38$, a large reduction from the value in free space.)

The usual way of handling the density dependence is to use the local density approximation. In this we assume that the effective interaction between two nucleons embedded in a nucleus at some position where the density has a certain value is given by the G-matrix for infinite nuclear matter with the same density value. This ignores the effects on G of density gradients. Also, because the force has a finite range, two positions, r_1 and r_2, are involved; frequently one uses the density at the center of gravity, $\frac{1}{2}(r_1 + r_2)$. This kind of uncertainty is further aggravated in heavy ion double folding because there are two densities, $\rho_1(r_1)$ and $\rho_2(r_2)$. The simplest assumption is a sudden (or 'frozen density') approximation in which the two undisturbed densities are simply superposed. Then one uses $\rho = \rho_1 + \rho_2$ in Eq. (25). This may be adequate at high energies (relative energy per nucleon high compared to the Fermi energy) where the Pauli principle has little effect and there is little time for the two nuclei to readjust as they collide. It seems less appropriate at lower energies; unfortunately, attempts to do better are quite complicated.[32,33] Fortunately, the region of strong overlap is not experienced in most heavy-ion collisions because of the strong absorption, and the frozen density approximation is more plausible for the low-density surface region. For a recent application to heavy-ion scattering, see Ref. 36. In the example of alpha scattering at 140 and 172 MeV discussed below, one might argue that the energy per nucleon is moderately high and that the tight binding of the alpha discourages its polarization.

4.2. Applications to Heavy-Ion Scattering

The M3Y interaction (24) has been used extensively to construct the real parts of optical potentials for the elastic scattering of heavy ions (see Refs. 15 and other references quoted there). For example, analyses of some 156 sets of data for some 67 systems at energies mostly between 5 and 15 MeV per nucleon were reported in Refs. 15. Except for ^{6,7}Li and ^{9}Be, these data were fitted by renormalizing the real folded potential by a factor N which was found to be close to unity, $\overline{N} = 1.06 \pm 0.11$. Although the imaginary potentials had to be chosen phenomenologically, good data do allow one to determine the strength required for the real part with little ambiguity.

Figure 5 shows another example of the use of the M3Y interaction. The experimental deformed potential contours were deduced[37] from measurements on excitations of ^{160}Gd up to its 12^+ state by the inelastic scattering of ^{40}Ar. The folded (non-spherical) potential was generated by using a non-spherical, quadrupole-deformed density for ^{160}Gd.

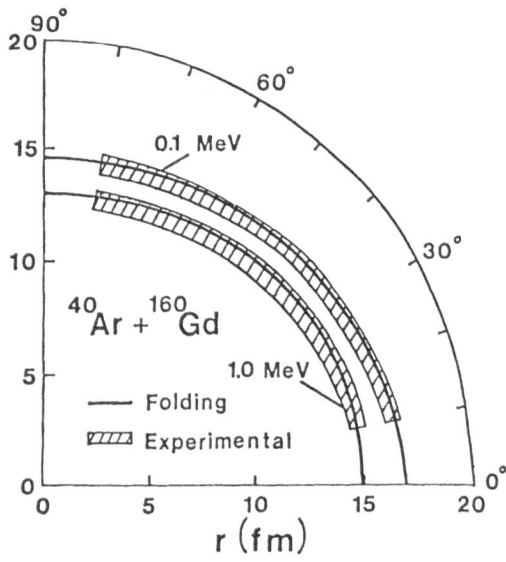

Fig. 5. Non-spherical optical potential contours[37] for ^{40}Ar + ^{160}Gd, referred to the body-fixed axes for ^{160}Gd, compared to folding-model calculations using the M3Y interaction.

The exceptional cases of ^{6,7}Li and ^{9}Be scattering require an optical potential whose real part is only about half the strength predicted by the M3Y interaction. One feature that all three ions share is a dissociation energy much smaller than for other stable nuclei, so one is tempted to attribute the failure to the effects of break-up. This seems to have been confirmed, at least for ^{6}Li, by some recent calculations.[19] A similar overestimate of the real potential when the M3Y interaction is used has been observed[38] for the scattering of the light ions d,t, and ^{3}He, which are also weakly bound.

4.3. Density Dependence and Alpha Scattering

It was noted in Sec. 3.2 that a simple nucleon-alpha potential of Gaussian form was adequate, within the single-folding model, to reproduce the observed alpha scattering at forward angles. Use of the M3Y interaction (24) in the double-folding model (plus a phenomenological imaginary potential) also provides agreement with data at 140 and 172 MeV in the diffraction region at small angles,[39] with only a few percent renormalization required for an optimum fit. This indicates that the model is providing the correct real potential for peripheral collisions.

It has long been known[40] that observation of alpha scattering at energies high high enough ($\gtrsim$ 100 MeV) and angles large enough to exhibit the 'nuclear rainbow' refractive effects can resolve the discrete ambiguities associated with the interior depth of the optical potential. Such data indicate that a real depth of ~ 140 MeV is required. However, the folded potential is ~ 230 MeV deep in the interior; consequently it will generate incorrect scattering at large angles. This is indicated in Fig. 6. Reducing the strength of the interaction (by N ~ 0.5) to give the

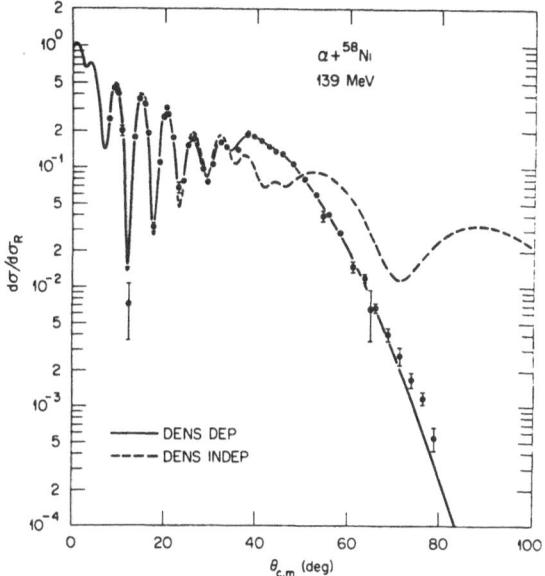

$\alpha + {}^{58}\text{Ni}$
139 MeV

—— DENS DEP
----- DENS INDEP

$\theta_{c.m}$ (deg)

Fig. 6. Comparison of measured elastic scattering of $\alpha + {}^{58}\text{Ni}$ with double-folding calculations using either the M3Y interaction (dashed curve) or the same interaction modified by a density-dependent strength (full curve).

correct large-angle scattering completely destroys the agreement at smaller angles because the potential in the surface is then too weak.

A natural way to rectify this situation is to introduce the expected density dependence, like that of Eq. (25), which reduces the strength in the interior relative to that in the low-density surface region. Explicit calculations[39,41] confirmed that this was so, as shown by the example in Fig. 6. For the illustrative calculations of Ref. 39, the M3Y interaction (24) was simply multiplied by a density-dependent factor. This factorization is not strictly correct but makes the calculations easier. A factorized dependence on the densities, different from (25a), was adopted for computational convenience[42]

$$v(r,\rho) = f(\rho)\, v_{M3Y}(r), \tag{26a}$$

with

$$f(\rho) = C(1 + \alpha\, e^{-\beta\rho}), \qquad \rho = \rho_1(r_1) + \rho_2(r_2). \tag{26b}$$

The (energy-dependent) constants in $f(\rho)$ were chosen to reproduce the magnitude and density dependence of the volume integral of the G-matrix of Jeukenne, et al.[31] at the appropriate energy per nucleon. An overall renormalization of the real folded potential by a factor of N was allowed in fitting the 11 sets of alpha-scattering data. Quite consistently the optimum value was $N \approx 1.3$. With this, the factor $N\, f(\rho)$ is unity for a density of $\rho \approx 0.15$ fm^{-3}, nearly equal to the 'normal nuclear matter' density of $\rho_0 \approx 0.17$ fm^{-3}. This is somewhat larger than had been expected,[15,42] but it is difficult to interpret this result precisely, in the light of the hybrid nature of the simple model interaction (26). We also note that the G-matrix of Ref. 31 is only given for $\rho \lesssim \rho_0$, whereas larger values of $\rho = \rho_1 + \rho_2$ are encountered when the alpha is within the target nucleus. Further, the use of $\rho = \rho_1 + \rho_2$ may introduce too strong a dependence on density. The two densities are actually moving with respect to each other, thus weakening the repulsive effect of the Pauli exclusion principle compared to that for the static superposition $\rho_1 + \rho_2$. Further studies with a more sophisticated interaction would be of interest.

References

*Research sponsored by the Division of Basic Energy Sciences, U.S. Department of Energy, under contract W-7405-eng-26 with the Union Carbide Corporation.

1. N. Austern and J. S. Blair, Ann. Phys. (New York) 33, 15 (1965).

2. A. M. Bernstein, Adv. Nucl. Phys. 3, 325 (1969).

3. R. S. Mackintosh and L. J. Tassie, Nucl. Phys. A222, 187 (1974); R. S. Mackintosh, Nucl. Phys. A266, 379 (1976).

4. A. M. Bernstein, V. R. Brown, and V. A. Madsen, Phys. Lett. 103B, 255 (1981); 106B, 259 (1981).

5. H. Feshbach, Ann. Phys. (New York) 19, 287 (1967).

6. W. U. Schroder and J. R. Huizenga, Ann. Revs. Nucl. Sci. 27, 465 (1977).

7. D. M. Brink, J. de Phys. C-5, 47 (1976).

8. U. Mosel, in Heavy-Ion Collisions (ed. R. Bock: North-Holland).

9. J. Blocki, J. Randrup, W. J. Swiatecki, and C. F. Tsang, Ann. Phys. (New York) 105, 427 (1977).

10. H. J. Krappe, J. R. Nix, and A. Sierk, Phys. Rev. C 20, 992 (1979).

11. D. Pal, J. R. Rook, and A. M. Kobos, Nucl. Phys. A348, 45 (1980).

12. U. Mosel, Particles & Nuclei 3, 297 (1972).

13. G. W. Greenlees, et al., Phys. Rev. 171, 1115 (1968); C 1, 1145 (1970); C 2, 1063 (1970).

14. R. C. Barrett and D. F. Jackson, Nuclear Sizes and Structure (Clarendon Press, Oxford, 1977).

15. G. R. Satchler and W. G. Love, Phys. Reports 55C, 183 (1979); G. R. Satchler, Nucl. Phys. A329, 323 (1979).

16. F. Petrovich, Nucl. Phys. A251, 143 (1975); P. J. Moffa, C. B. Dover, and J. P. Vary, Phys. Rev. C 16, 1857 (1977).

17. F. G. Perey and G. R. Satchler, Nucl. Phys. A97, 515 (1967).

18. P. W. Keaton and D. D. Armstrong, Phys. Rev. C 8, 1692 (1973).

19. I. J. Thompson and M. A. Nagarajan, Phys. Lett. 106B, 163 (1981); R. S. Mackintosh and A. M. Kobos, to be published, and paper 3 in this volume.

20. H. J. Krappe, Ann. Phys. (New York) 99, 142 (1976).

21. E. I. Kao and S. Fallieros, Phys. Rev. Lett. 25, 827 (1970); H. Ui and T. Tsukamoto, Prog. Theor. Phys. 51, 1377 (1974).

22. F. E. Bertrand, G. R. Satchler, D. J. Horen, J. R. Wu, A. D. Bacher, G. T. Emery, W. P. Jones, and D. W. Miller, Phys. Rev. C 22, 1832 (1980).

23. G. R. Satchler, Phys. Lett. 59B, 121 (1975).

24. L. D. Rickertsen and G. R. Satchler, Phys. Lett. 66B, 9 (1977); F. Petrovich, D. Stanley, and J. J. Bevelacqua, Phys. Lett. 71B, 259 (1977); W. G. Love, Phys. Lett. 72B, 4 (1977).

25. S. Landowne, R. Schlicher, and H. H. Wolter, Nucl. Phys. A373, 141 (1982).

26. E. E. Gross, J. R. Beene, K. A. Erb, M. P. Fewell, D. Shapira, M. J. Rhoades-Brown, G. R. Satchler, and C. E. Thorn, Bull. Am. Phys. Soc., Washington meeting, April, 1982, and to be published.

27. H. V. von Geramb, Microscopic Optical Potentials (Springer-Verlag, Berlin, 1979).

28. C. B. Dover and J. P. Vary, in Symposium on Classical and Quantum Mechanical Aspects of Heavy-Ion Collisions, Heidelberg, 1974 (Springer-Verlag, Berlin, 1975).

29. E. C. Halbert and G. R. Satchler, Nucl. Phys. A233, 265 (1974); G. R. Satchler, Zeit. f. Phys. 260, 209 (1973).

30. G. Bertsch, J. Borysowicz, H. McManus, and W. G. Love, Nucl. Phys. A284, 399 (1977); G. R. Satchler and W. G. Love, Phys. Lett. 65B, 415 (1976).

31. J. P. Jeukenne, A. Lejeune, and C. Mahaux, Phys. Rev. C 16, 80 (1977); F. A. Brieva and J. R. Rook, Nucl. Phys. A291, 317 (1977); F. A. Brieva, H. V. von Geramb, and J. R. Rook, Phys. Lett. 79B, 177 (1978).

32. A. Faessler, T. Izumoto, S. Krewald, and R. Sartor, Nucl. Phys. A359, 509 (1981).

33. S. B. Khadkikar, L. Rikus, A. Faessler, and R. Sartor, Nucl. Phys. A369, 495 (1981).

34. V. A. Madsen, in Nuclear Spectroscopy and Reactions, Part D (ed. J. Cerny, Academic Press, New York, 1975).

35. J. P. Jeukenne and C. Mahaux, Zeit. f. Phys. A302, 233 (1981).

36. F. J. Vinas, M. Lozano, and G. Madurga, Phys. Rev. C 23, 780 (1981).

37. R. E. Neese, M. W. Guidry, R. J. Donangelo, and J. O. Rasmussen, Phys. Lett. 85B, 201 (1979).

38. J. Cook and R. J. Griffiths, Nucl. Phys. A366, 27 (1981); J. Cook, to be published.

39. A. M. Kobos, B. A. Brown, P. E. Hodgson, G. R. Satchler, and A. Budzanowski, Nucl. Phys. A (in press).

40. D. A. Goldberg, S. M. Smith, and G. F. Burdzik, Phys. Rev. C 10, 1362 (1974).

41. H. J. Gils, E. Friedman, Z. Majka, and H. Rebel, Phys. Rev. C 21, 1245 (1980).

42. W. G. Love, Phys. Lett. 72B, 4 (1977); F. Petrovich, D. Stanley, and J. J. Bevelacqua, Phys. Lett. 71B, 259 (1977).

RELATIONS BETWEEN THE SIMULTANEOUS AND SEQUENTIAL
TRANSFER OF TWO NUCLEONS*

G. R. Satchler
Oak Ridge National Laboratory
Oak Ridge, TN 37830, U.S.A.

1. Introduction

The transfer of two nucleons between projectile and target in a direct or peripheral reaction such as (p,t) or (^{16}O,^{14}C) may occur in one-step or two-steps. These we refer to as 'simultaneous' and 'sequential' transfers, respectively. In the former, the interaction acts once and both nucleons are transferred. In the latter, the interaction acts once to transfer one nucleon, the system then propagates in one or more intermediate states and is followed by a second action of the interaction to transfer the second nucleon. This process may be symbolized for the above examples as (p,d;d,t) and (^{16}O,^{15}N;^{15}N,^{14}C), implying the intermediate formation of a deuteron or the nucleus ^{15}N. (Of course, the intermediate system may exist in more than one state of excitation.)

In terms of a perturbation theory expansion, such as the distorted-wave Born series, simultaneous transfer is possible in first order while sequential transfer requires second order. This is illustrated in Fig. 1. We are accustomed, perhaps, to thinking that a first-order process is more likely than a second-order one. However, a closer examination can prepare us for the possibility that this may not be so. The nuclear forces are predominantly two-body in character; hence, in first-order (Fig. 1a) only one of the two nucleons experiences an interaction. The possibility of finding that the other nucleon has also transferred arises only because its state within the projectile is not orthogonal to the state in the target into which it transfers. This process corresponds to a kind of quantum-mechanical tunneling from the projectile to the target.[1] However, in the two-step process (Fig. 1b) each nucleon is transferred under the direct influence of an interaction with the target; intuitively, this might seem more plausible. It requires an explicit calculation to determine which process is most likely in a given case, and such calculations are often beset with uncertainties. Nonetheless, it seems clear that the one-step and two-step amplitudes are frequently comparable in magnitude for light-ion reactions,[2] while the two-step may dominate in reactions with heavy ions.[3] (The existence of strong Q-window effects, especially with heavy ions,[4] may enhance the sequential process when there is a large mismatch between the entrance and exit channels. The gap may be bridged more easily in two steps, with the interaction acting twice.) Consequently, it is not safe to ignore the existence of sequential transfer. However, one of the main reasons for studying two-nucleon transfers is to learn about the two-nucleon overlaps (existence and extent of pairing correlations, etc.). So we wish to know if this information is still

A(a,b)B

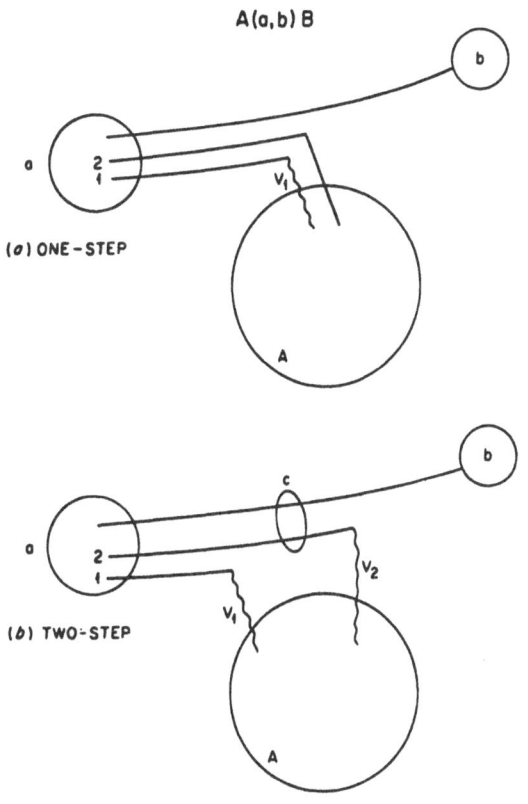

(a) ONE-STEP

(b) TWO-STEP

Fig. 1. (a) Simultaneous transfer, (b) sequential transfer. The wavy line corresponds to one action of the projectile-target interaction.

available when the reaction proceeds by two, sequential, one-nucleon transfers. Our purpose here is to gain some insight into the relationship between the two amplitudes by using a simple approximate form of the theory. For simplicity, we shall discuss a light-ion reaction and, to be specific, we choose the (t,p) reaction (or the inverse (p,t) reaction). Similar considerations apply to other reactions, and we include some remarks concerning (α,d) reactions. This work (done in collaboration with W. T. Pinkston) is described in a paper shortly to be published in Nuclear Physics A. Hence, here we shall only give an abbreviated and schematic outline of the theory, leaving the reader to consult the forthcoming paper for the details.

2. The (t,p) or (p,t) Reaction

Apparently contradictory evidence exists for the relative importance of the two-step or sequential-transfer mechanism in (t,p) or (p,t) reactions. Indeed, the results of a confusing variety of sophisticated calculations are available, each of which treats properly some, but not all, aspects of these transitions. Among these aspects are: (i) the use of a realistic triton wavefunction and the associated realistic interaction;[5-8] (ii) an exact treatment of the finite-range of the interactions (as opposed to use of a zero-range approximation), in both the one-step[5-9] and the two-step[8-11] amplitudes; (iii) accounting for the non-orthogonality correction to the two-step term,[2,9,11,12] which tends to cancel the one-step amplitude;[12-14] (iv) allowing the intermediate n-p system of the two-step process to exist in continuum states, both spin-singlet and triplet, as well as in the bound, triplet deuteron ground state[12,15] (sometimes this is called the 'deuteron break-up effect'). In addition, there is a sensitivity of the results to the optical potentials employed, and a particular sensitivity of the cross section

magnitudes to the nuclear wavefunctions used to construct the two-nucleon overlaps.[6,7]

In view of the complexity of the theoretical description of this transfer process, the agreement with measured cross sections obtained by calculations which
only include some of these aspects must be viewed with some caution. In this situation it is valuable to have an understanding of any general, albeit approximate,
features that the amplitudes may possess, independent of these details, which may
illuminate the results of more detailed calculations.

The central point of the present study is to use a closure approximation for
the intermediate nuclear states that appear in the second-order distorted-wave-Born
expression for the two-step amplitude in order to show that the simultaneous and
sequential processes tend to depend in the same way upon the nuclear structure of
the initial and final states[3,16] and to exhibit similar selection rules. Various
additional simplifying assumptions, such as zero-range, can be used to demonstrate
more dramatically[17] the basic similarity of the two amplitudes in a way that is
not obscured by partial-wave expansions, angular momentum algebra, etc.[3] While
these may have large effects upon the magnitude of the amplitudes, they do not af-
fect the underlying structure. They do help to explain why the sequential and
simultaneous processes tend to yield similar angular distributions.

A closer examination of the spin angular momenta involved modifies the initial
conclusion that the selection rules are the same in both cases. For example, in
the two-neutron transfer (t,p) reaction, if the n-p system or 'deuteron' associated
with the intermediate states of the two-step amplitude appears in spin-singlet or
spin-triplet states with equal weight, the selection rules for one- and two-step
are identical. However, constraining it to be in a triplet state only (such as the
common assumption that it is the physical deuteron ground state) acts as a spin-
filter and determines that the two neutrons are transferred with a particular mixture of total spin $S = 0$ or 1. This mixture will differ from that for the one-step
term. In this way, inclusion of singlet deuterons might change the results for
vector analyzing powers and might modify conclusions that have been drawn from
measurements of these quantities.[18]

2.1. First- and Second-Distorted-Waves Born Approximations

We do not refer explicitly to spins for the schematic presentation given here;
see Ref. 17 for details. Consider the A(t,p)B reaction, so that B = A+2n. The
first-order (or simultaneous transfer) amplitude has the form[19]

$$T^{(1)} = \int \chi_p^{(-)*} (\underline{R}_p)(B|A)(V_1 + V_2)\psi_t \, \chi_t^{(+)} (\underline{R}_t),$$ (1)

where the χ_p and χ_t are the usual distorted waves describing the centre-of-mass
motion of the proton and triton, while V_1 and V_2 are the residual interactions

acting on neutrons 1 and 2, and ψ_t describes the internal state of the triton. Also, $(B|A)$ is the overlap between the initial and final nuclear states,

$$(B|A) \equiv \int \psi_B^*(\underline{r}_1,\underline{r}_2,\xi)\psi_A(\xi)d\xi = \phi^{BA}(\underline{r}_1,\underline{r}_2), \tag{2}$$

say. All the nuclear structure information, including pairing correlations, is manifest through this ϕ^{BA}, which is often called the two-nucleon form factor for the transition.

The second-order (or sequential transfer) amplitude for the processes A+t → C+d → B+p has the form[19]

$$T^{(2)} = \sum_C \int \chi_p^{(-)*}(\underline{R}_p)(B|C)V_2\psi_d\, G_C^{(+)}(\underline{R}_d,\underline{R}_d')\psi_d^*V_1(C|A)\psi_t\, \chi_t^{(+)}(\underline{R}_t), \tag{3}$$

where ψ_d is the internal state of the intermediate deuteron and G_C is the Green function describing the relative motion (in the Coulomb + optical potential) of the intermediate d+C system. For example, if we ignored the optical potential and the Coulomb field so that the system propagated in plane waves with momentum k_d, we would have

$$G_C(\underline{R},\underline{R}') \rightarrow \exp(ik_d s)/s, \qquad s = |\underline{R}-\underline{R}'|. \tag{4}$$

The wavenumber k_d could be re-interpreted as the local value $k_d(\overline{R})$ at the position $\overline{R} = \frac{1}{2}|\underline{R}+\underline{R}'|$ in the potential, thus introducing a 'local energy approximation'[3] for the effects of the optical and Coulomb potentials.

2.2. Closure for the Two-Step Amplitude

Now we do not see a two-nucleon overlap appearing in the two-step amplitude (3), but a sum of products of one-nucleon overlaps weighted by G_C, where

$$(C|A) = \int \psi_C^*(\underline{r}_1,\xi)\psi_A(\xi)d\xi = \phi^{CA}(\underline{r}_1), \tag{5}$$

say, with a similar expression for $(B|C)$. The question is, "Is the information on two-neutron correlations that is carried by $\phi^B(r_1,r_2)$ in the first-order $T^{(1)}$ lost if it should happen that the second-order $T^{(2)}$ is important?". The answer is "No" if the important intermediate states C are sufficiently close in energy that the propagators G_C do not vary much. For then we may replace each G_C by an average value $\overline{G}_C$ and then use the closure property of a complete set of states,

$$\sum_C |C)(C| = 1, \tag{6}$$

to collapse the sum over C in Eq. (3):

$$\sum_C (B|C)G_C(C|A) \approx \overline{G}_C \sum_C (B|C)(C|A) = \overline{G}_C(B|A). \tag{7}$$

The result is the appearance of the same two-neutron overlap as that which occurs in the one-step amplitude $T^{(1)}$. Consequently, when the approximation (7), is sufficiently good, nuclear structure effects such as pairing correlations will manifest themselves in much the same way in one-step or two-step processes. In this way, one can understand how analyses of measured cross sections, using the theory for a one-step transfer alone, may yield useful information on the relative behavior of spectroscopic factors for different states and different nuclei,[20] even when sequential transfer may be important.

Although it is not necessary for the present argument, it is instructive to introduce further the usual zero-range approximation[19] for the one-step process,

$$V\psi_t \approx D(r)\delta(\rho)$$

where $r = |r_1 - r_2|$ is the separation of the two neutrons, while ρ is the position of the proton relative to the centre of mass of the two neutrons. We also take the usual zero-range approximation[19] for each of the one-neutron transfers in the two-step process. Then the amplitudes (1) and (3), with the closure (7), become

$$T^{(1)} \approx \int \chi_p^{(-)*}(R')\phi^{BA}(r_1,r_2)D(r)\chi_t^{(+)}(R)dr\ dR, \tag{8}$$

$$T^{(2)} \approx D_0(d,p)D_0(t,d) \int \chi_p^{(-)*}(R')\phi^{BA}(r_1,r_2)\overline{G}(r_1,r_2')\chi_t^{(+)}(R)dr\ dR, \tag{9}$$

where $r_2' = r_2A/(A+1)$, $R' = RA/(A+2)$ and $R = \frac{1}{2}(r_1 + r_2)$. The D_0 are the usual zero-range normalization constants. There is a remarkable similarity of structure in the two amplitudes, which differ only by the appearance of the short-ranged overlap function $D(r)$ in $T^{(1)}$ and the average Green function $\overline{G}$ in $T^{(2)}$. In the plane-wave approximation (4), $\overline{G}$ is also a function only of r (if A >> 1 so we can neglect the small recoil correction). (Of course, $\overline{G}$ is complex, whereas D is usually taken to be real.) Consequently, we should not be surprised to find that one-step and two-step transfers have rather similar angular distributions.

2.3. Further Comments on the Closure Approximation

One can easily construct model situations in which the procedure (7) becomes exact. For example, consider two-neutron transfer to a closed-shell target in which the intermediate states ψ_C are pure single-particle j-states and the residual state ψ_B has a pure j^2 configuration. Only one intermediate state can contribute, namely the corresponding single-particle state in nucleus A+1 with the same j, and Eq. (7) becomes exact.

The expression (3) assumes a single intermediate state for the intermediate n-p system, in this case the ground state of the deuteron. More generally, both intermediate nuclei may be in more than one state and expressions like (3) need to be supplemented by an additional sum over the states of the "deuteron". In a heavy-ion reaction, the intermediate nuclei may be comparable and hence should be treated on the same footing. This was done by Feng, et al.[3] when they applied the closure approximation. The tendency has been to treat light-ion reactions differently, by

assuming that the intermediate light particle remains in its ground state. This
has been a matter of expediency; excited states of light ions are unbound and thus
introduce a continuum of intermediate $\geq$ 3-body states which greatly increases the
complexity of the calculation. This has been investigated[12,15] for the intermediate 'deuteron' in (p,t) reactions.

The closure approximation (7) may be applied to the states of c, as well as
those of C, in the process A+a → C+c → B+a, if the important contributions still
come from a sufficiently narrow band of energies (i.e. narrow enough that the Green
function G_{Cc} does not vary much). This results in the appearance of a second two-
nucleon overlap or form factor (b | a), which again is the same as that which occurs
in the one-step amplitude $T^{(1)}$. Then the nature of the intermediate nucleus c
plays no rôle and, for example, angular momentum selection rules are the same for
one-step and two-step transfer. It is plausible that this is approximately valid
for heavy-ion reactions[3] where the important intermediate states are likely to be
bound states of moderate excitation energy. It is not so obvious for light-ion
reactions. The explicit calculations[12,15] for ^{48}Ca(p,t) at 20 and 40 MeV included
the n-p scattering states, both singlet and triplet. Important contributions came
from the continuum states at 40 MeV, but the effect (on the cross section) was
small at 20 MeV. The intermediate spin-singlet n-p state is always associated with
continuum states of relative motion. Hence, the transition would be dominated by
triplet intermediate states if the contribution of the continuum were negligible.
This constitutes a spin-filter through which the two-step process must pass and introduces differences in the spin selection rules compared to the one-step process.
In the simplest triton wavefunction,[21] the two neutron spins are coupled to S = 0
and the triton spin - $\frac{1}{2}$ is due to the odd proton. Then we have S=0 transfer in the
one-step amplitude. After recoupling the spins within the triton, we find a linear
combination of s_d = 0 and 1 for the proton plus one of the neutrons. The assumption of an intermediate triplet deuteron in the two-step process selects s_d = 1;
i.e. it filters out the s_d=0 component. The resulting n+n+p function is no longer
pure S = 0; it also contains S=1 components for the two neutrons. However, if the
singlet and triplet deuteron states were included with equal weights, the filter is
removed and only S=0 transfer is possible for the sequential transfer also.

3. The (α,d) Reaction

A similar analysis has been made of the (α,d) reaction.[17] Two sequential
processes are available here, (α,t;t,d) and (α,h;h,d) (where h ≡ ^{3}He). Only s_d=1
transfer is possible for the simultaneous transfer because of the zero spin of the
alpha. If we assume an S-state of maximum symmetry for the alpha, only spin - $\frac{1}{2}$
states are allowed for the triton and helion. Thus, although they act as spin - $\frac{1}{2}$
'spin filters', they impose no further constraint on the spin transfer. The spin-
transfer selection rules are the same for simultaneous and sequential transfer.

There could be, however, an <u>isospin</u> filtering effect which would result from any difference in propagation of the t and h particles in unsymmetric, charged nuclear matter (i.e. if they had different Green functions).

By introducing similar zero-range and closure approximations, the one-step and two-step amplitudes are reduced to the same forms as Eqs. (8) and (9) for the (t,p) reaction. Again, the two amplitudes are seen to depend upon nuclear structure in the same way. The same two-nucleon overlap, analogous to Eq. (2), appears in both. In particular, there does not appear any phase difference between them that depends upon the spin of the residual nuclear state.

The latter point is especially relevant because of some observations of the Pittsburgh group[22] on the $^{208}Pb(\alpha,d)^{210}Bi$ reaction. Consider excitation of the $(g_{9/2}h_{9/2})_J$ multiplet of states with J = 0,1,...9. Simple angular momentum considerations, together with the tendency of (α,d) reactions to favor large L-transfers, lead to the prediction that the one-step cross sections should show a regular stepped pattern when plotted against J, with the rise at each step increasing as J increases. DWBA calculations[22] confirm this (see Fig. 2a). The discussion of the present paper suggests that a similar pattern should be seen for the two-step process by itself and again explicit calculations[22] confirm this (Fig. 2a).

The measured cross sections[22] show a rather different, saw-tooth pattern (Fig. 2). The predicted steady rise with L transfer is observed; however, the cross sections for the even-J members of the J=(L-1,L) pairs are larger than those for the odd-J members. Coupled-channels calculations[22] including direct and sequential transfers reproduced these trends and seemed to indicate that the saw-tooth pattern resulted from successively constructive and destructive interferences of direct and sequential amplitudes due to a J-dependence in their relative phase. No such J-dependent phase factor appears in the present treatment[†], although this treatment is able to account for the trends shown by the individual one-step and two-step amplitudes in terms of the reaction dynamics and the LS-jj transformation coefficients involved in the two-nucleon overlap factor.[17] Indeed, the integrand of the two-step amplitude differs from that for the one-step only by the replacement of the light-ion overlap function $D(r)$ by $\bar{\bar{G}}_0(r,R)$, the average of the triton and helion Green functions.[17] The symmetric part of the two-nucleon overlap $\Phi(\underline{r}_2,\underline{r}_1)$, from which any J-dependence must arise, is common to both amplitudes.

Our analysis makes it difficult for us to understand either the calculations presented in Refs. 22 or the experimental data, which seem to be nicely explained by the coupled-channels calculations. Cross sections for exciting the same states by $^{208}Pb(^3He,p)^{210}Bi$ show a similar behavior.[23] The $(^3He,p)$ reaction differs from (α,d) in that S=1 and S=0 transfer are both possible. However, nuclear structure

---

[†]Although a coupled reaction channels code was used, the calculations of Refs. 22 were made in such a way as to correspond exactly to the second-order distorted-wave Born approximation discussed here, including the use of the zero-range approximation.

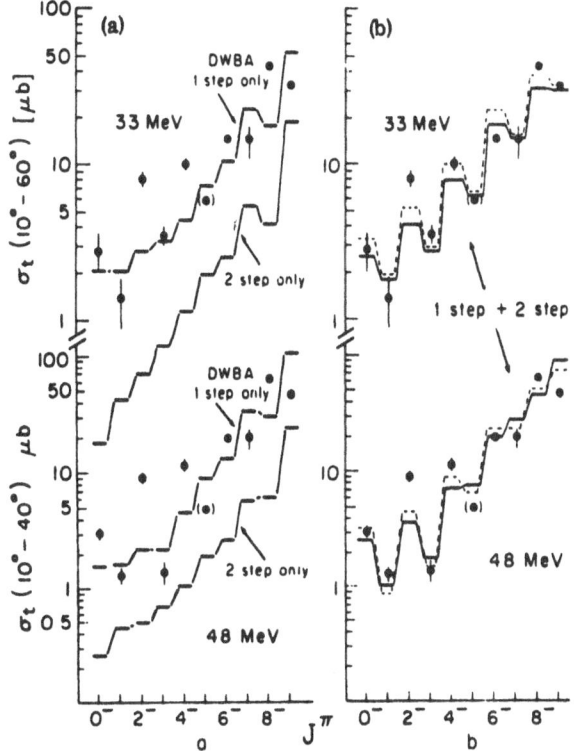

Fig. 2. Integrated cross sections, versus J, for the ^{208}Pb$(\alpha,d)^{210}$Bi reaction exciting the $(h_{9/2}g_{9/2})_J$ multiplet. (a) Individual one-step and two-step cross section; (b) coherent sum of one-step and two-step; as reported in Ref. 22.

factors make the S=0 contribution much smaller than that for S=1, so that, except for kinematic effects, one should expect the two reactions to behave in a very similar way.

Our approach differs from that using coupled channels in that we do not consider intermediate states explicitly but make the closure approximation. However, with the assumption of a single $(g_{9/2}h_{9/2})$ configuration for the final state in the ^{208}Pb$(\alpha,d)^{210}$Bi case, the result of making the closure approximation yields the same result as restricting the intermediate states to be the ^{209}Pb and ^{209}Bi ground states, as was done in the coupled-channels calculations.[22]

4. Some Remarks on Non-Orthogonality and Second-Order Processes

In a rearrangement collision A(a,b)B, in which (b,B) are a pair of nuclei (or partition) different from (a,A), the internal states in the initial and final channels are not orthogonal;[19] for example,

$$(b,B \mid a,A) \equiv \int \psi_b^*(\xi_b)\psi_B^*(\xi_B)\psi_a(\xi_a)\psi_A(\xi_A)d\xi_a d\xi_A \neq 0. \tag{10}$$

Since $\xi_B = \xi_B(\xi_a,\xi_A,r_{aA})$, etc., this overlap remains a function of the a+A separation r_{aA}. Such overlaps vanish asymptotically $(r_{aA} \to \infty)$, but remain finite in the region where the two nuclei overlap. This is just the region that contributes to transition amplitudes like (1) or (3). As a consequence, in general Eq. (3) is not the correct second-order Born approximation to the transition amplitude; there is a "non-orthogonality (NO) correction" term. The expression for the NO term[14,19] depends upon the particular choice, post or prior, made for the interactions in each of the two steps in the amplitude T$^{(2)}$ of Eq. (3). If prior is chosen for the a+A → c+C step and post for the c+C → b+B step, the NO term vanishes. For other choices it does not. For example, frequently post is chosen for both steps for sequential stripping (like the (t,p) reaction) induced by light ions because then a zero-range approximation can be invoked for both of the steps (as was done in deriving Eq. (9)). Then the NO term has the form

$$T^{(NO)} = - \sum_{c,C} \int x_b^{(-)*} \psi_b^*(B \mid C) \psi_c (V_1 + V_2) \psi_c^*(C \mid A) \psi_a x_a^{(+)}. \tag{11}$$

This has a structure intermediate between that for $T^{(1)}$ and $T^{(2)}$ of Eqs. (1) and (3); but note the overall minus sign. Indeed, if we sum over a complete set of states of both nuclei c+C using Eq. (6) and its analogue for nucleus c, then $T^{(NO)}$ becomes[13] equal and opposite to the one-step amplitude $T^{(1)}$. The two-step term $T^{(2)}$ (which should be summed over the same complete set of intermediate states) is then left as the lowest-order contribution to the transition. In practice, however, it is not feasible to use a complete set. Formally, it implies non-convergence of the Born series because of the mathematical difficulties associated with the 3- and more-body states that appear when the intermediate states are unbound. One standard approach[19] is to regard our calculations as being done within a truncated model space, limited to 2-body channels, and with effective interactions. Thus, a complete set is ruled out from the start. Nonetheless, the tendency remains for $T^{(NO)}$ and $T^{(1)}$ to interfere destructively, sometimes quite strongly (see Refs. 9 and 12-15 for examples), and this enhances the importance of the two-step process described by $T^{(2)}$.

One final note: the separation of the second-order amplitude into $T^{(2)}$ and $T^{(NO)}$ is artificial. Despite the intuitive appeal of interpreting $T^{(2)}$ as "the two-step process", one cannot physically distinguish these two terms from each other (nor they from the first-order term $T^{(1)}$). This is emphasized by the lack of uniqueness in the separation into two parts; for example, if we had chosen the prior-post interaction form for $T^{(2)}$, there would be no $T^{(NO)}$ term and hence no possibility of the cancellation of $T^{(1)}$. This lack of uniqueness also renders ambiguous even the conceptual division into one-step and two-step. Only the total amplitude has direct physical meaning; its breakdown into pieces is merely a consequence of the way we do calculations. Nonetheless, the discussion given earlier can be valuable as an aid to understanding the results of those calculations. Of course, the division within any particular calculation is (or should be!) unambiguous. Many calculations have ignored the possibility of NO corrections (for example, the (α,d) calculations referred to earlier [22]), usually as a matter of computational convenience. However, the closure argument given above, albeit invalid, remains as a warning that NO effects may not be negligible, and the results of explicit calculations[9,12-15] support this conclusion.

References

*Research sponsored by the Division of Basic Energy Sciences, U.S. Department of Energy, under contract W-7405-eng-26 with the Union Carbide Corporation.

1. I am indebted to R. M. Drisko for this remark.

2. For example, see N. Hashimoto and M. Kawai, Prog. Theor. Phys. 59, 1245 (1978).

3. For example, see D. H. Feng, T. Udagawa and T. Tamura, Nucl. Phys. A274, 262 (1976).

4. For example, see R. Bass, Nuclear Reactions with Heavy Ions, Springer-Verlag, Berlin, 1980.

5. M. A. Nagarajan, M. R. Strayer and M. F. Werby, Phys. Lett. 68B, 421 (1977).

6. T. Takemasa, T. Tamura and T. Udagawa, Nucl. Phys. A321, 269 (1979).

7. D. H. Feng, M. A. Nagarajan, M. R. Strayer, M. Vallieres and W. T. Pinkston, Phys. Rev. Lett. 44, 1037 (1980).

8. P. P. Tung, K. A. Erb, M. W. Sachs, G. B. Sherwood, R. J. Ascuitto and D. A. Bromley, Phys. Rev. C 18, 1663 (1978).

9. M. Igarashi and K. I. Kubo, Phys. Rev. C 25, 2144 (1982).

10. L. A. Charlton, Phys. Rev. C 14, 506 (1976).

11. N. Hashimoto, Progr. Theor. Phys. 59, 1562 (1978).

12. N. Hashimoto, Progr. Theor. Phys. 63, 858 (1980).

13. P. D. Kunz and E. Rost, Phys. Lett. 47B, 136 (1973).

14. T. Udagawa, H. H. Wolter and W. R. Coker, Phys. Rev. Lett. 31, 1507 (1973); U. Gotz, M. Ichimura, R. A. Broglia and A. Winther, Phys. Repts. 16C, 115 (1975).

15. N. Hashimoto, Progr. Theor. Phys. 59, 804 (1978).

16. R. Schaeffer and G. F. Bertsch, Phys. Lett. 38B, 159 (1972).

17. W. T. Pinkston and G. R. Satchler, Nucl. Phys. A383, 61 (1982).

18. Y. Aoki, H. Iida, S. Kunori, K. Nagano, Y. Toba and K. Yagi, Phys. Rev. C 25, 1050 (1982).

19. G. R. Satchler, Direct Nuclear Reactions, Oxford University Press, Oxford, 1982.

20. For example, see R. A. Broglia, O. Hansen and C. Riedel, Adv. Nucl. Phys. 6, 287 (1973).

21. M. A. Nagarajan, M. R. Strayer and M. F. Werby, Phys. Lett. 67B, 141 (1977).

22. W. W. Daehnick, M. J. Spisak, J. R. Comfort, H. Hufner and H. H. Duhm, Phys. Rev. Lett. 41, 639 (1978); W. W. Daehnick, M. J. Spisak and J. R. Comfort, Phys. Rev. C 23, 1906 (1981).

23. R. J. Peterson, R. E. Anderson and M. J. Fritts, Zeit. Phys. A302, 63 (1981).

POLARIZATION PHENOMENA IN HEAVY ION
TRANSFER REACTIONS

F.D. SANTOS

Centro de Física Nuclear da Universidade de Lisboa
Av. Gama Pinto 2, 1699 Lisboa Codex, Portugal

1 — INTRODUCTION

The study of polarization phenomena in heavy ion reactions is
a fairly recent research subject that started about ten years ago.
Generally speaking the polarization phenomena in nuclear reactions are
determined by the interactions that involve the nuclear spin.   The
saturation of spin in nuclei with increasing mass number decreases the
relative importance of the spin dependent interactions in heavy ion
reactions as compared with light ion reactions. Thus the contribution
of spin to nuclear dynamics tends to decrease with increasing mass
number.

On the other hand we note that the polarization observables
have a very specific dependence on the angular momentum structure of
the nuclear states which can make them very useful for spectroscopic
studies. This aspect of polarization is very well documented in light
ion reactions. It is therefore appropriate to mention,   although
briefly, some features of polarization phenomena in light ion reactions
which will be useful for our discussion of heavy ion reactions.

Finally we note that polarization in heavy ion reactions has
become a topic of increasing research interest because   of   the
availability of polarized heavy ion beams such as ^{6}Li, ^{7}Li and ^{23}Na.
Here we shall be mainly concerned with reactions induced by polarized
Li beams.

2 — POLARIZATION IN LIGHT ION REACTIONS

2.1 — J-dependence

One of the most useful and well known applications of
polarization in nuclear spectroscopy is based on the so called sign

rule j dependence[1,2] that is observed, for instance, in $(\vec{d},p)$ reactions. Consider a reaction $A(\vec{d},p)B$ on a spinless target. The angular distributions of $i\,T_{11}$ for two final states with spin $J_B = \ell_2 + \frac{1}{2}$ and $J_B = \ell_2 - \frac{1}{2}$ (ℓ_2 is the orbital angular momentum of the captured neutron in the residual nucleus B) are related by the approximate formula

$$i\,T_{11}(J_B = \ell_2 + \frac{1}{2}) = -\frac{\ell_2}{\ell_2 + 1}\,i\,T_{11}(J_B = \ell_2 - \frac{1}{2}). \qquad (1)$$

Spin assignements can be made by measuring $i\,T_{11}$ angular distributions and comparing with results of calculations using the distorted wave Born approximation (DWBA).

The j-dependence described by eq.(1) is a simple consequence of the angular momentum coupling involved in a (d,p) reaction and is common to other transfer reactions[3] where the transferred particle is in an s state in the projectile.

Also in a (d,p) reaction the T_{22} angular distributions at small angles show a very pronounced ℓ and j-dependence[4] which is useful to determine both the ℓ and j-values of the transferred neutron and to study configuration mixing in the residual nucleus. A fair amount of detailed information on nuclear structure effects in polarization phenomena involving light ions and in particular on j dependent effects can be found in the proceedings[5,6] of the last two conferences on "Polarization Phenomena in Nuclear Reactions".

2.2 — Structure of the projectile

Instead of studying the structure of the target and residual nucleus we can also explore the possibility of extracting information about the projectile. Following this idea Knutson and Haeberli[7] have shown that the tensor analyzing powers of sub-Coulomb (d,p) reactions can be used to determine the asymptotic D to S state ratio η in the deuteron. The deuteron wave function is not spherically symmetric and because of the correlation between spin and deformation the probability that a neutron capture takes place in a sub-Coulomb (d,p) reaction depends on the orientation of the deuteron spin. The presence of the deuteron D-state enhances the $(\vec{d},p)$ cross section when the deuteron spins are parallel or antiparallel to the vector $\vec{d}$ in Fig. 1 and decreases the cross section for deuterons whose spins are perpendicular to $\vec{d}$.

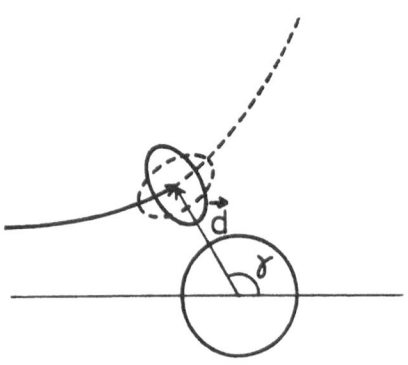

Fig.1 - Semiclassical model for a
sub-Coulomb (d,p) reaction where
the incident deuteron and outgoing
proton follow a Coulomb trajectory
as shown by the solid and dashed
lines. When the deuteron has spin
alignment axis parallel to the
vector $\vec{d}$ (full line ellipse) there
is more overlap with the target and
therefore the transfer is more
probable.

The effect of the deuteron D state at low energy is determined by the
value of a single parameter[8] D_2, related to the asymptotic D/S state
ratio by

$$D_2 = \eta/\alpha^2 \tag{2}$$

where $\alpha^2 = MB_d/\hbar^2$ (M is the nucleon mass and B_d is the deuteron binding
energy).

Recently a considerable effort has been devoted to the
determination of η from sub-Coulomb (d,p) reactions.[9] Another method
to determine η is by analytical extrapolation in the $\cos\theta$ plane of
the tensor analysing powers[10] in $\vec{d}$-p elastic scattering. It is
important to obtain an accurate experimental determination because η is
a significant theoretical quantity as regards the nuclear force. In
fact η is to a large extent determined by the one pion exchange (OPE)
force. As pointed out recently by Ericson et al.[11] a precise
determination of η has implications on the value of the πN coupling
constant and for the size of quark bags.

In the case of $(\vec{d},t)$ and $(\vec{d},{}^3He)$ reactions the analysis of
the tensor analyzing powers leads to a determination of η in the 3
nucleon bound system[12]. Recently α-particle D-state effects were also
identified in $(\vec{d},\alpha)$ reactions[13]. The asymptotic D/S ratio η in all
these reactions is a manifestation of the tensor part of the nuclear
force. The relation between the OPE tensor force and η is approximately
given by[13]

$$\eta = \frac{\sqrt{8}\,S}{B} \frac{\int_o^\infty V_T(r)u_o^N(r)r^3\,dr}{\int_o^\infty u_2^N(r)r^3\,dr} \tag{3}$$

which can be applied to the A = 2,3 and 4 nucleon bound systems. In eq.(3), B is the binding energy and S is a spin-isospin factor (S(d) = 1, S(t) = S(^{3}He) = -1, S(α) = -2) that determines the relative sign of the tensor analyzing powers in ($\vec{d}$,p), ($\vec{d}$,t), ($\vec{d}$,^{3}He) and ($\vec{d}$,α) reactions. u_L^N are the bound state radial wave functions normalized so that asymptotically they behave as $u_L^N(r) \xrightarrow{r \to \infty} e^{-\alpha r}$ and $V_T(r)$ is the radial part of the OPE tensor potential.

3 – POLARIZATION IN HEAVY ION ELASTIC SCATTERING

3.1 – Shape effects

The effects of a deformed nuclear surface on the interaction between two heavy ions is a question that can be answered through the study of reactions initiated by polarized beams.[14] The basic ideas are the same that were first applied in (d,p) reactions. Because of the correlation between spin and a non-vanishing spectroscopic quadrupole moment the influence of deformation on the nucleus-nucleus interaction can be investigated by comparing the results of experiments with aligned and non-aligned beams. The surface of a rotational ellipsoid is given by

$$R(\theta) = R_0 \left[1 + \beta \, Y_2^0(\theta) \right] \tag{4}$$

where θ is the angle between the symmetry axis and the radius vector and β describes the spectroscopic deformation. For instance in the case of ^{6}Li, ^{7}Li and ^{23}Na the values of β estimated from the electric quadrupole moment are $\beta(^6$Li$) = 0.00$, $\beta(^7$Li$) = - 0.16$ and $\beta(^{23}$Na$) = 0.12$. The fact that ^{6}Li is approximately spherical and ^{7}Li is strongly deformed (oblate) should manifest itself in the elastic scattering of aligned ^{6}Li and ^{7}Li beams. This effect is clearly illustrated in the measurements of Dreves et al.[15] for elastic scattering on ^{58}Ni, shown in Fig. 2. σ_0 is the cross section for an unpolarized beam and $\sigma_{a\ell}$ is the cross section for a polarized beam aligned perpendicular to the reaction plane. The relation between σ_{al} and the tensor analyzing powers T_{20} and T_{22} referred to the helicity coordinate system defined by the Madison Convention[16] is

$$A_{yy} = -(\sqrt{3} \, T_{22} + \frac{1}{\sqrt{2}} \, T_{20}) = \sqrt{2} \, ^T T_{20} = \sqrt{2} \, \frac{\sigma_{al} - \sigma_0}{\sigma_0} . \tag{5}$$

Here $^T T_{20}$ is referred to a coordinate system where the Z axis is along $\vec{k}_{in} \times \vec{k}_{out}$. The data points for A_{yy} in the case of ^{7}Li deviate from zero for angles greater than the strong absorption angle θ_{sa} where

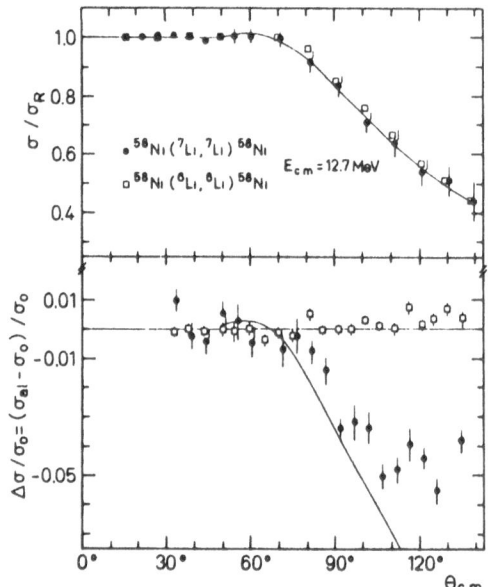

strong absorption sets in $(\sigma/\sigma_R < 1)$. The values of A_{yy} are negative for $\theta > \theta_{sa}$ because the oblate shape of ^{7}Li implies that $\sigma_{al} < \sigma_0$. The tensor analyzing powers in the elastic scattering of ^{7}Li can be interpreted using a phenomenological tensor potential of the form $\beta \dfrac{dV}{dr} T_r$ where V is the usual Saxon-Wood central optical potential and $T_r = (\vec{s} \cdot \vec{r})^2 - s(s+1)/3$ is the spin tensor potential.

Fig.2 - Angular distributions[15] of σ/σ_R and $(\sigma_{al} - \sigma_0)/\sigma_0$ for the elastic scattering of ^{6}Li and ^{7}Li on ^{58}Ni at 12.7 MeV. The curves are optical model calculations[15].

3.2 — Folding models

The theoretical interpretation of the spin dependence of the interaction between two heavy ions has been done mainly within the framework of the folding model.[17] In particular the spin-orbit potentials of ^{6}Li and ^{7}Li have been extensively studied using that model.[18] The calculations are based on the d + α and t + α cluster structures which are well developed in the ground states. The vector analyzing power is to a large extent determined by the spin-orbit potential and the tensor analyzing powers by tensor potentials. In the case of ^{7}Li the folding model based on a t + α cluster structure predicts tensor potentials of rank k = 2 and 3.

Recently Nishioka et al.[19] have shown that effects due to the projectile excitation[20] in heavy ion elastic scattering are strongly spin-dependent and generate an effective spin-orbit potential which may account for discrepancies that were found between phenomenological and folded spin-orbit potentials.

4 — POLARIZATION IN HEAVY ION TRANSFER REACTIONS

Polarization measurements in heavy ion reactions have been

used to investigate the reaction mechanism and in particular   to
distinguish between quasi-elastic and deep inelastic processes.[21]   In
the case of quasi-elastic processes we may expect that the DWBA
constitutes a useful framework for the analysis   of   the   reaction.
Because of the short de Broglie wavelength of heavy ion projectiles we
can use a semiclassical approach to the DWBA description and in this
way obtain approximate expressions for the sign and magnitude of the
polarization as a function of few parameters[22].

Here we shall consider the study of the analyzing powers of
quasi-elastic transfer reactions within the theoretical framework of
the DWBA. Other polarization observables are also of interest in   the
study of heavy ion transfer reactions, but will not be considered here.
As examples we mention the polarization of the outgoing particle[23] and
polarization transfer coefficients.[24]

Polarization phenomena in deep inelastic processes have been
investigated by measuring the vector polarization of ^{12}B nuclei in
$(^{14}N, ^{12}B)$ reactions as a function of scattering angle and reaction
Q-value.[25] In these experiments the polarization of ^{12}B nuclei   is
determined by observing the β-ray asymmetry along the normal to the
scattering plane. Another method is to observe the circular polarization
of γ rays emitted in reactions induced by projectiles and targets of
varying masses.[26]

4.1 — DWBA and semiclassical models

The DWBA transition matrix for a reaction A(a,b)B   where
$a = b + x$ and x is the transferred particle is a sum of   reduced
transition amplitudes

$$\beta^{LM}_{j_1 j_2} = \int d^3R \int d^3r \, \chi_b^{(-)*}(\vec{r}_b) \, F^{LM}_{j_1 j_2}(\vec{R} - \vec{r}, \vec{r}) \, \chi_a^{(+)}(\vec{r}_a) \qquad (6)$$

where L is the orbital angular momentum transfer in the reaction.   In
eq.(6) χ_a and χ_b are the distorted waves and

$$F^{LM}_{j_1 j_2}(\vec{r}', \vec{r}) = i^{\ell_1 - \ell_2} \sum_{m_1 m_2} (\ell_2 m_2 \, \ell_1 m_1 | LM) R_{\ell_2 j_2}(r') \, Y^{m_2 *}_{\ell_2}(\hat{r}') V_{bx}(\vec{r}) \, R_{\ell_1 j_1}(r) \, Y^{m_1}_{\ell_1}(\hat{r}) \qquad (7)$$

is the transfer form factor. ℓ_1, j_1 are the orbital and total angular
momentum of x in the projectile, ℓ_2, j_2 are the orbital and total
angular momentum of x in the residual nucleus and $R_{\ell_1, j_1}, R_{\ell_2, j_2}$ are
the corresponding normalized radial bound state wave functions. We
follow the notation of ref.27 where the present formalism is developed

in greater detail. Since the integrand in eq.(6) is strongly localized in the displacement $\vec{R}$ between the heavy ion cores b and A we use, for the product of distroted waves, the ansatz[27]

$$\chi_b^{(-)^*}(\vec{r}_b) \chi_a^{(+)}(\vec{r}_a) = e^{\frac{i}{\hbar} \vec{r} \cdot \vec{P}} \chi_b^{(-)^*}(\frac{m_A}{m_B} \vec{R}) \chi_a^{(+)}(\vec{R}). \tag{8}$$

In this expression m_i is the mass of particle i and $\vec{P}$ is a "recoil momentum" that plays the role of the momentum conjugate with the variable $\vec{r}$. Hence we identify $\vec{P}$ with the most probable transfer momentum of x relative to b in the projectile. It can be determined using a semiclassical argument proposed by Brink.[29] We write[27]

$$\vec{P} = p_1 \, \vec{n} \tag{9}$$

where $\vec{n}$ is a unit vector in the reaction plane tangent to the projectile trajectory at the point of transfer,

$$p_1 = -\frac{Q}{v} - \frac{1}{2} m_x v, \tag{10}$$

and v is the relative velocity between the heavy ions at that point.

By substituting eqs.(8), (9) and (10) into eq.(6) we obtain

$$\beta_{j_1 j_2}^{LM} = \int d^3R \, \chi^{(-)^*}(\frac{m_A}{m_B} \vec{R}) \chi_a^{(+)}(\vec{R}) \int d^3r \, F_{j_1 j_2}^{LM}(\vec{R} - \vec{r}, \vec{r}) \, e^{\frac{i}{\hbar} p_1 \vec{r} \cdot \vec{n}} \tag{11}$$

4.2 — The analyzing powers i T_{11}, T_{2q} and i T_{3q} of the ^{58}Ni($^7\vec{\text{Li}}$, ^{6}Li)^{59}Ni reaction

The cross section and analysing power data of Tungate al.[29] for the ^{58}Ni($^7\vec{\text{Li}}$, ^{6}Li)^{59}Ni at 20.3 MeV was analyzed using eq.(11). Neglecting the effect of two step processes the DWBA transfer form factor involves the $<^6\text{Li}|^7\text{Li}>$ overlap which is a linear combination of p 3/2, p 1/2 and f 5/2 one-particle states. Since the p 3/2 state has the largest spectroscopic amplitude[30] ($S_{p\ 3/2} = 0.3792$) we have initially assumed that the neutron is transferred from a pure p 3/2 state in ^{7}Li. Calculations using this model and shown in Fig. 3 reproduce satisfactorily the Q-value dependence observed in i T_{11}.

Generally we find that the analyzing powers T_{kq} with k odd are very sensitive to the momentum p_1 and therefore have a strong Q dependence. Those with k even have a weaker dependence on p_1 and Q but are specially sensitive to the deformed shape of the projectile. Although i T_{11} is independent of θ over a wide range of angles near the

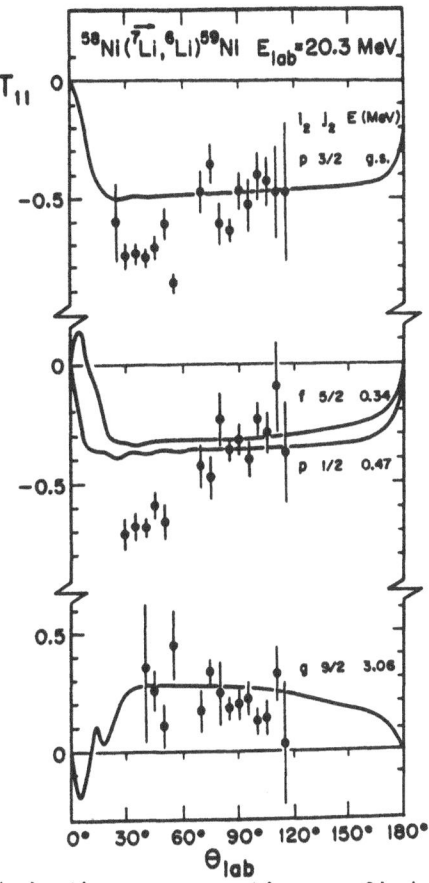

Fig.3 - Comparison between DWBA calculations[27] and i T[11] data[29] for the ^{58}Ni ($^7\vec{\text{Li}}$, ^{6}Li) ^{59}Ni reaction at 20.3 MeV. The neutron final state was assumed to be p 3/2, p 1/2 and g 9/2. For the unresolved first excited state[29] the calculated i T[11] is also shown for a f 5/2 final state. E are the excitation energies.

peak in the cross section we find that at smaller angles there is a pronounced ℓ_2 and j_2 dependence that can be useful in spectroscopic studies.

For the T_{2q} it is particularly significant to consider the observable A_{yy} defined in eq. (5). Since the T_{2q} are almost entirely determined by the quadrupole deformation we conclude that because of the oblate shape of ^{7}Li, A_{yy} tends to be positive and independent of θ. This result is exact ($A_{yy} = (5\sqrt{2})^{-1}$) for a transfer to an s state in the absence of spin-dependent distortion. For $\ell_2 > 0$, A_{yy} is still approximately constant near the cross section peak and its value is strongly dependent on ℓ_2 and j_2 because of selection rules that define the range of allowed values of L. This ℓ_2 and j_2 dependence is much stronger than the one observed at small angles in the cross sections of ^{50}Cr(^{7}Li, ^{6}He)^{51}Mn reactions[31].

The sign rule j-dependence described by eq.(1) is not present in i T[11] for (^{7}Li, ^{6}Li) reactions since the transferred particle is not in an s state. However the same type of j-dependence is found in the tensor analysing powers i T_{3q} as shown in Fig. 4. In fact in a

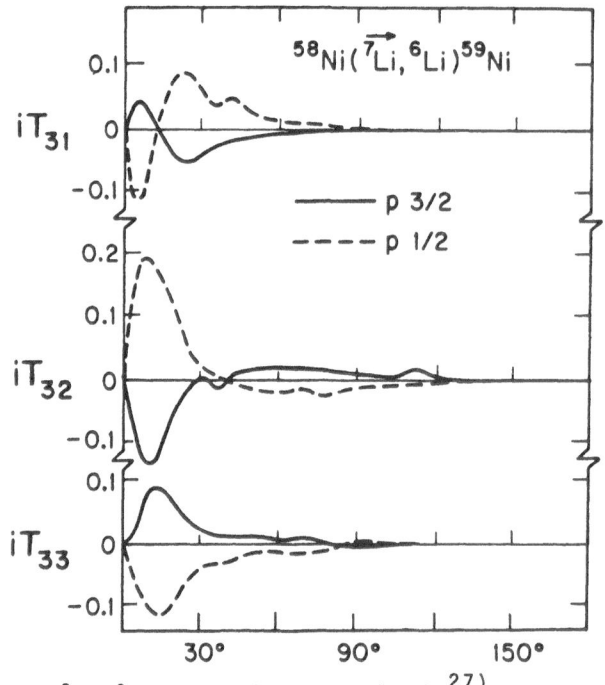

$^{58}Ni(^{7}\overrightarrow{Li},^{6}Li)^{59}Ni$

—— p 3/2
- - - - p 1/2

Fig.4 - J-dpendence of $i T_{3q}$ predicted by DWBA calculations for the same reaction of Fig.3 to the ground state of ^{59}Ni, assuming it to be a p 3/2 state (full line) and a p 1/2 sate (dashed line).

transfer from a p state we obtain[27)]

$$i T_{3q}(J_B = \ell_2 + \tfrac{1}{2}) = - \frac{\ell_2}{\ell_2+1} \, i T_{3q}(J_B = \ell_2 - \tfrac{1}{2}). \qquad (12)$$

4.3 — Peripheral model

The essencial features of the analyzing power data in the $^{58}Ni(^{7}Li, ^{6}Li)^{59}Ni$ reaction can also be interpreted using a considerably simplified peripheral model. Since the distorted waves tend to peak at the distance of closest approach $\vec{d}$, represented in Fig.5, we assume that

$$\chi_b^{(-)*}(\frac{m_A}{m_B} \vec{R}) \, \chi_a^{(+)}(\vec{R}) = \delta(\vec{R} - \vec{d}) . \qquad (13)$$

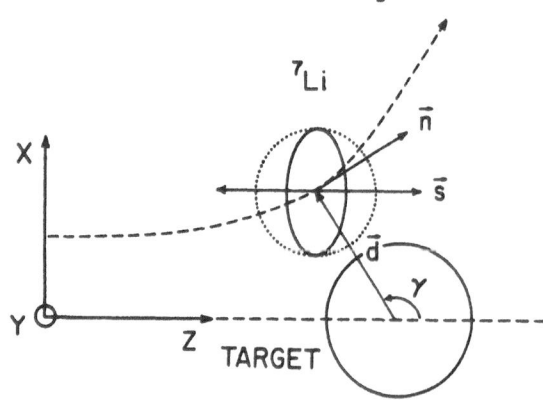

Fig.5 - The full and dotted lines represent ^{7}Li with spin alignment axis parallel to the z axis and perpendicular to the reaction plane. $\vec{n}$ is tangent to the Coulomb orbit.

Substituting this approximation into eq.(11) we obtain

$$\beta^{LM}_{j_1 j_2} = \int d^3r \; F^{LM}_{j_1 j_2}(\vec{d} - \vec{r},\vec{r}) \; e^{\frac{i}{\hbar} p_1 \vec{r}.\vec{n}} . \qquad (14)$$

In this simplified model the dynamics of the transfer process is entirely described by the vectors $\vec{d}$ and $p_1 \vec{n}$.

In view of the peripheral nature of the transfer reaction at low energy we now represent the radial bound state wave functions by the asymptotic Hankel functions.[32] With this representation it is possible to obtain closed analytical expressions for the reduced transition amplitudes and therefore for the analyzing powers T_{kq}. The formulas for $i \; T_{11}$ and T_{2q} in the particular case of a transfer from a p 3/2 to a s1/2 state are given in ref.27 and are in good agreement with the DWBA calculations of section 3.2.

Up to now it was assumed that the transferred neutron is in a pure p 3/2. The effect of a p 1/2 admixture in the $<^6Li \mid {}^7Li>$ overlap is determined by the difference between the reduced transition amplitudes corresponding to these two states. Since the transfer is preferentially localized at large R we assume that

$$\beta^{LM}_{1/2 \; j_2} = \frac{N_{1,1/2}}{N_{1,3/2}} \; \beta^{LM}_{3/2 \; j_2} \qquad (15)$$

where $N_{\ell_1 j_1}$ are the asymptotic normalization constants of the radial bound state wave functions. In a transfer to an s state this model gives[27]

$$i \; T_{11} = C_1 \; i \; T_{11}(3/2) \qquad (16a)$$

$$T_{2q} = C_2 \; T_{2q}(3/2) \qquad (16b)$$

where $T_{kq}(3/2)$ are the analyzing powers for a pure p 3/2 configuration. The constants C_1 and C_2 in eq.(16) are given by

$$C_1 = (1 + 4\sqrt{5} \; R/11 + 10R^2/11)/(1 + R^2), \qquad (17a)$$

$$C_2 = (1 + 4\sqrt{5} \; R)/(1 + R^2), \qquad (17b)$$

where $R = S_{1,1/2} \; N_{1,1/2}(S_{1,3/2} \; N_{1,3/2})^{-1}$ and S_{ℓ_1,j_1} are the spectroscopic amplitudes. Given that[27] $R \cong 0.93$ we obtain $C_1 = 1.36$ and $C_2 = 5.00$. This result shows that the effect of the p 1/2 state is considerably larger in T_{2q} than in $i \; T_{11}$.

3.4 — Full finite range DWBA calculations

In order to test the DWBA predictions based on the semiclassical model of section 3.2 and also the results of section 3.3 concerning the effect of p 1/2 mixing we performed full finite range DWBA calculations. We have used the code Ptolemy[33] which allows for configuration missing both in the projectile and in the residual nucleus.

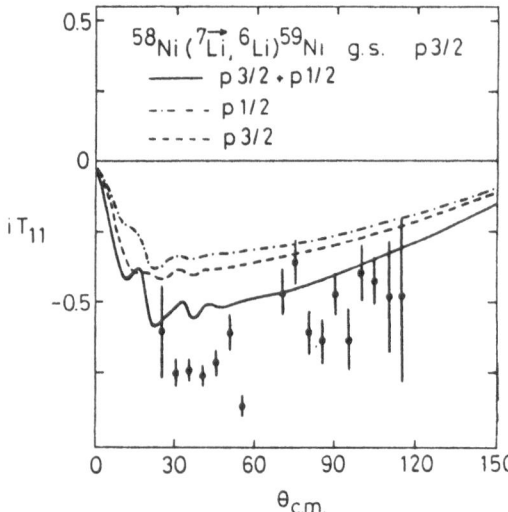

Fig.6 - Full finite range DWBA calculations[33] for $i\,T_{11}$ in the ^{58}Ni (^{7}Li, ^{6}Li)^{59}Ni ground state p 3/2 transition using Li optical model potentials from ref.34. The dashed and point--dashed curves assume that the transferred neutron is in a pure p 3/2 and p 1/2 state in ^{7}Li. The full curve is obtained with p 3/2 + p 1/2 configuration mixing with spectroscopic amplitudes from ref. 30.

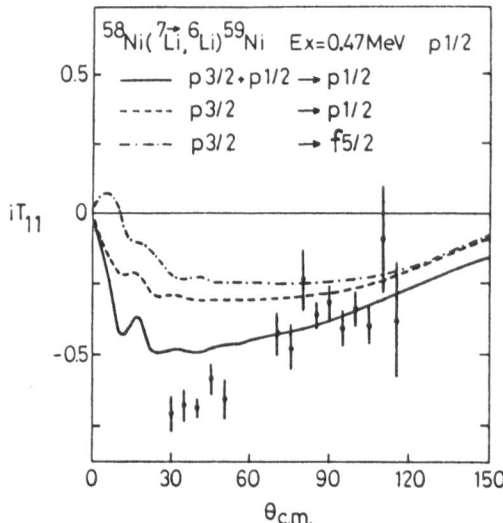

Fig.7 - DWBA calculations as in Fig.6 for the Ex = 0.47 MeV, p 1/2 state. The curves have the same meaning as in Fig.6 except that here the point--dashed curve corresponds to a transition from a pure p 3/2 state to a pure f 5/2 state as in Fig.3.

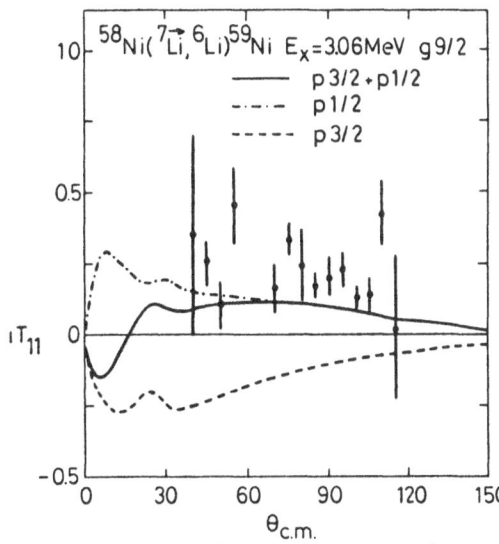

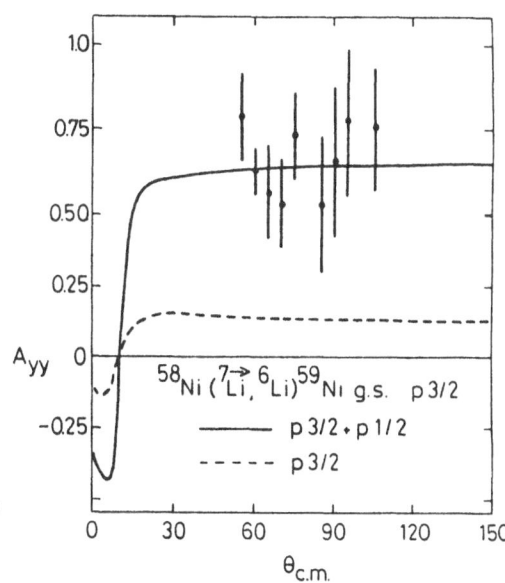

Fig.8 - DWBA calculations as in Fig.6 for the E_x = 3.06 MeV, g 9/2 state. The curves have the same meaning as in Fig. 6.

Fig.9 - DWBA calculations of A_{yy} for the ground state p 3/2 transition. The curves have the same meaning as in Fig.6.

As shown in Figs. 6 to 8 the observed Q value dependence could not be reproduced when it is assumed that the transferred neutron is initially in a pure p 3/2 state. In the particular case of the g 9/2 transition we obtained values for i T_{11} that were systematically negative using various Li optical model potentials. A change of sign in i T_{11} could only be obtained by the inclusion of the p 1/2 component of the ^{7}Li ground state. Far away from the optimum Q-value for transfer the reaction is not strongly localized in space. Small differences in the binding potentials can induce large effects in i T_{11} as shown in Fig. 8. Although the agreement with experiment is improved we note that the remaining discrepancies are probably due to neglected effects in the DWBA calculations such as projectile excitation and spin-dependent distortion.

Results of DWBA calculations for the A_{yy} angular distribution in the ground state p 3/2 transition, shown in Fig. 9, are in good agreement with the data[36]. The large effect on A_{yy} induced by the p 1/2 component predicted by eqs. (16) and (17), is indeed confirmed. This in an example in heavy ion transfer reactions where an accurate measurement of A_{yy} may give precise information on the asymptotic ratio between two projectile states.

5 — CONCLUSIONS

In general the differential cross section of heavy ion transfer reactions are structureless and do not show pronounced ℓ and j-dependent effects. The situation is quite different in the angular distributions of the analyzing powers. The interpretation of polarization data in the ^{58}Ni($^7\vec{\text{Li}}$, ^{6}Li)^{59}Ni reaction, using the DWBA theory, indicates that the T_{kq} angular distributions can have a strong ℓ and j-dependence which may be useful to obtain spectroscopic information on the projectile and residual nucleus. We find that the T_{kq} with k odd have a much stronger Q-value dependence than those with k even. On the other hand the T_{2q} are to a large extent determined by the deformed shape of the overlap batween the projectile and ejectile. Hence, at low energy, they can be used to determine the asymptotic ratio batween projectile states in the same way as it is done, for instance, in ($\vec{\text{d}}$,p) reactions. The analyzing power A_{yy} is particularly relevant for this type of investigations. In the particular case of (^{7}Li, ^{6}Li) reactions there is no sign rule j-dependence in iT_{11} since the neutron is transferred from a p state. However the same type of j-dependence is predicted in iT_{3q} (eq.(12)).

Full finite range calculations for the ^{58}Ni($^7\vec{\text{Li}}$, ^{6}Li)^{59}Ni reaction show that the inclusion of the p 1/2 component in the $<^6$Li | ^{7}Li$>$ overlap is essential to explain the observed Q-dependence of iT_{11}. This result indicates that the semiclassical model of ref. 29 represents an oversimplification of the transfer process. There is at present a very limited amount of analysing power data to well defined final states in heavy ion transfer reactions. Further measurements would be very useful to test the reaction models presented here.

Finally we note that the present DWBA analysis presupposes a considerable simplification of the reaction mechanism. In fact we have not included the possibility of projectile excitation which is known to play an important role in Li induced reactions. Furthermore we neglected the spin dependent terms in the optical potentials. These refinements of the theory have to be taken in account if we are to obtain precise spectroscopic information from the analysis of polarization data in transfer reactions.

I am very thankful to A.M. Gonçalves, D. Fick, G. Tungate and D.M. Brink for many helpful discussions. The hospitality of the University of Wisconsin where part of the calculations were performed is gratefully acknowledged.

REFERENCES

1. H.C. NEWNS, Proc. Phys. Soc. London, Sect. A 66 (1953) 477.

2. T.J. YULE and W. HAEBERLI, Nucl. Phys. A117 (1968) 1.

3. F.D. SANTOS, in Proceedings of the IV International Symposium on Polarization Phenomena in Nuclear Reactions, edited by W. Gruebler and V. König (Birkhauser Verlag, Basel, 1976) p. 205.

4. F.D. SANTOS, Phys. Rev. 3C (1976) 1145.

5. Proceedings of the IV International Symposium on Polarization Phenomena in Nuclear Reactions, edited by W. Gruebler and V. König (Birkhauser Verlag, Basel, 1976).

6. Proceedings of the V International Symposium on Polarization Phenomena in Nuclear Reactions, edited by G.G. Ohlsen et al. (American Institute of Physics Conference Proceedings, New York, 1981).

7. L.D. KNUTSON and W. HAEBERLI, Phys. Rev. Lett. 35 (1975) 558.

8. R.C. JOHNSON and F.D. SANTOS, Part. Nucl. 2 (1971) 285.

9. K. STEPHENSON and W. HAEBERLI, Phys. Rev. Lett. 45 (1980) 520; R.P. GODDARD, L.D. KNUTSON and J.A. TOSTEVIN, to be published.

10. R.D. AMADO, M.P. LOCHER and M. SIMONIUS, Phys. Rev. C17, (1978) 403; M.E. CONZETT, F. HINTERBERGER, P. VON ROSSEN, F. SEIBER and E.J. STEPHENSON, Phys. Rev. Lett. 43 (1979) 572; F.D. SANTOS and P.C. COLBY, Nucl. Phys. A367 (1981) 197.

11. T.E.O. ERICSON and M. ROSA-CLOT, Phys. Lett, 110B (1982) 193.

12. L.D. KNUTSON, B.P. HICHWA, A. BARROSO, A.M. EIRÖ, F.D. SANTOS and R.C. JOHNSON, Phys. Rev. Lett. 35 (1975) 1570; J.L. FRIAR, B.F. GIBSON, D.R. LEHMAN and G.L. PAYNE, to be published in Phys. Rev. C.

13. F.D. SANTOS, S.A. TONSFELDT, T.B. CLEGG, E.J. LUDWIG, Y. TAGISHI and J.F. WILKERSON, to be published in Phys. Rev. C.

14. E. STEFFENS, in Proceedings of the V International Symposium on Polarization Phenomena in Nuclear Reactions, edited by G.G. Ohlsen et al. (American Institute of Physics Conference Proceedings, New York, 1981) p. 1001.

15. W. DREVES, P. ZUPRANSKI, P. EGELHOF, D. KASSEN, E. STEFFENS, W. WEISS and D. FICK, Phys. Lett. 78B (1978) 36.

16. Proceedings of the III International Symposium on Polarization Phenomena in Nuclear Reactions, edited by H.H. Barschall and W. Haeberli (University of Wisconsin Press, Madison, 1971) p. 25.

17. W.J. THOMPSON, in Proceedings of the Rochester Symposium on Heavy-Ion Elastic Scattering, edited by R.M. Devries (Rochester, 1977) p. 321.

18. F. PETROVICH, D. STANLEY and L.A. PARKS, Phys. Rev. 17C (1978) 1642.

19. H. NISHIOKA, R.C. JOHNSON, J.A. TOSTEVIN and K.I. KUBO, to be published.

20. V. HNIZDO, K.W. KEMPER and J. SZYMAKOWSKI, Phys. Rev. Lett. 46 (1981) 590.

21. G. GRAW, in Proceedings of the V International Symposium on Polarization Phenomena in Nuclear Reactions, edited by G.G.Ohlsen et al. (American Institute of Physics Conference Proceedings, New York, 1981) p. 1027.

22. P.D. BOND, Phys. Rev. 22C (1980) 1539.

23. F. POUGHEON, P.ROUSSEL, M. BERNAS, F. DIAF, B. FABBRO, F. NAULIN, E. PLAGNOL, G. ROTBARD, Nucl. Phys. A325 (1979) 481.

24. I. KOENIG, D. FICK, R. BOTTGER, P. EGELHOF, H. INGWERSEN, S. KOSSIONIDES, K.H. MÖBIUS, D.PRESINGER, E. STEFFENS, in Proceedings of the V International Symposium on Polarization Phenomena in Nuclear Reactions, edited by G.G. Ohlsen et al. (American Institute of Physics Conference Proceedings, New York, 1981) p. 919.

25. N. TAKAHASHI, in idem, p. 1016.

26. W. TRAUTMANN, in Proceedings of the International Symposium on Continuous Spectra of Heavy Ion Reactions, edited by T. Tamura et al. (Harwood Academic, 1979).

27. F.D. SANTOS and A.M. GONÇALVES, Phys. Rev. 24C (1981) 156.

28. *D.M. BRINK*, in Les Houches, Session XXX (North-Holland, Amsterdam, 1978) p.1; *H. HASAN and D.M. BRINK*, J. Phys. <u>65</u>, (1979) 771:

29. *G. TUNGATE, R. BÖTTGER, P. EGELHOF, K.H. MÖBIUS, Z. MOROZ, E. STEFFENS, W. DREVES, D. FICK, D.M. BRINK and T.H. HILL*, Phys Lett. <u>95B</u> (1980) 35.

30. *S. COHEN and D. KURATH*, Nucl. Phys. <u>A101</u> (1967) 1.

31. *J.E. KIM and W.W. DAEHNICK*, Phys. Rev. <u>23C</u> (1981) 742.

32. *F.D. SANTOS*, Nucl. Phys. <u>A212</u> (1973) 341.

33. *M.H. MACFARLANE and S.C. PIEPER*, Argonne National Laboratory Report ANL-76-11 Rev. 1, 1978, unpublished.

34. *R.I. CUTLER, M.J. NADWORNY and K.W. KEMPER*, Phys. Rev. <u>15C</u> (1977) 1318.

35. *F.D. SANTOS*, to be published.

36. *G. TUNGATE*, private communication.

NUCLEAR CHARGE AND MATTER DISTRIBUTIONS

P. E. HODGSON

Nuclear Physics Laboratory, Oxford, U.K.

1. Introduction

The radial distributions of the nuclear charge and nuclear matter, or equivalently of the neutrons and protons, are among the most basic of nuclear properties. Substantial advances in our knowledge of these distributions have been made in recent years, both experimental and theoretical. Experimentally, the accurate measurements of the differential cross-sections for electron elastic scattering and of the energies of muonic atom transitions, analysed by model-independent techniques, have given very precise charge distributions for many nuclei. At the same time, the elastic scattering of energetic protons by nuclei have been analysed by the Glauber theory and by the optical potential method of Kerman, McManus and Thaler to yield nuclear matter distributions of somewhat lower accuracy.

Parallel with this experimental work, several theories have been used to calculate the nuclear charge and matter distributions from our knowledge of nuclear structure. Among these, the mean field theories and Brueckner-Hartree-Fock theory enable nuclear densities to be calculated from various effective nucleon-nucleon potentials. These methods are of great generality and give an excellent understanding of overall trends but frequently encounter difficulties in accounting for the densities of particular nuclei. Another method, based on the shell model, uses empirical data for each nucleus to obtain precise fits to the distributions obtained from the analyses of experimental data. This method has been developed over the last few years in Oxford, and is the subject of the present review.

The earlier stages of this work have been reviewed at previous meetings (Hodgson, 1976, 1979ab, 1981) and so this review will concentrate on the more recent developments. Originally, the work grew out of a series of studies of the systematic behaviour of nuclear single-particle states (Millener and Hodgson 1973; Malaguti and Hodgson, 1973, 1976; Malaguti 1978). These showed that the centroid energies of single-particle states, determined from analyses of nucleon transfer and knock-out reactions, can be expressed as the eigenvalues of one-body potentials with depths depending rather simply on the atomic weight and on the nuclear asymmetry parameter. As a check of these potentials, the corresponding charge distributions were calculated by summing the squares of the single-particle wavefunctions, weighted by the appropriate occupation probabilities. The RMS radii of the resulting charge distributions agreed well with the experimental values, and the radial distributions themselves proved to be unexpectedly accurate, so that they were successfully used to calculate proton and heavy-ion potentials by a folding procedure (Kujawski and Vary, 1975; Vary and Dover, 1973).

The success of this work encouraged us to develop the method as a precise technique for calculating nuclear density distributions. As a severe test of the method, we first analysed the data for ^{208}Pb, because two model-independent analyses of electron elastic scattering were available, and also much information from nucleon transfer reactions to and from the single particle occupied and unoccupied states near the Fermi surface. The model-independent charge distribution showed considerable structure, and it was found possible to fit this structure with the single particle potential (SPP) calculations (Hodgson, 1979a). This established the usefulness of the method, but just as we were preparing an account of this work for publication some new experimental results became available that gave a significantly different model-independent charge distribution. The work on ^{208}Pb was therefore set aside until the experimental situation clarified.

One difficulty with the ^{208}Pb analysis was the lack of information on the more deeply-bound states. In the calculations it was assumed that the potential corresponding to these states is the same as that for the surface states for which data is available, but it is not certain that this is correct and indeed it is not consistent with the energy dependence of the potential that is found for the deeper states in lighter nuclei for which experimental binding energies have been determined (Bear and Hodgson, 1978).

With this in mind, the next nucleus was selected to satisfy the requirement that experimental binding energies are available for all the single-particle states. This, together with the need for the nucleus to be as heavy as possible to avoid complications due to centre-of-mass motion and deformation, indicated ^{58}Ni as a suitable choice. A detailed analysis was made of this nucleus, and the quantum-mechanical basis of the SPP method was described in the paper giving the results (Malaguti et al 1978). Subsequently the method was applied to ^{39}K, ^{40}Ca and ^{48}Ca, and also to the even zirconium isotopes (Malaguti et al, 1979ab). These analyses are briefly summarised in §2.

All this work concerns nuclear charge distributions, but the same technique is also applicable to nuclear matter distributions, and these are particularly useful for calculating optical potentials by a folding procedure. The matter distributions may be calculated using the potentials found to give the best fit to the charge distributions, with depths adjusted to the neutron as well as to the proton separation energies. The calculations are somewhat less certain because there are no model-independent matter distributions for comparison.

It was expected that the matter distributions obtained in this way are particularly reliable in the far surface region, because the potentials are adjusted to fit the experimental separation energies, and this is the most critical region for optical potentials, especially those of composite particles and heavy ions. To study this possibility, some nuclear matter distributions were calculated for several isotopic sequences and compared with the critical radii obtained from analyses of alpha-particle elastic scattering near the Coulomb barrier. It was found possible to reproduce the departures from the $A^{1/3}$ variation of the interaction radii through the various isotopic sequences, thus confirming the usefulness of the method (Lozano, Madurga and Hodgson, 1979). Subsequently the effective radii obtained in this way were used to improve the global heavy-ion optical potential of Christensen and Winter, giving improved fits to a wide range of heavy-ion elastic scattering data (Lozano, Madurga and Hodgson 1980).

The matter distributions calculated by the SPP method may be compared with proton elastic scattering just as the charge distributions are compared with electron elastic scattering, although the accuracy of the comparison is less due to our less complete knowledge of the nuclear as compared with the electromagnetic interaction. The comparison with

proton scattering is best made at high energies to minimise the effect
of this lack of knowledge, for then the Glauber and
Kerman-McManus-Thaler theories are applicable. Such comparisons were
made with data on the elastic scattering of 800 MeV protons by several
nuclei and gave additional information on the nuclear matter
distributions (Ray and Hodgson, 1979). All this work on nuclear matter
distributions is summarised in §3.

These analyses treat the neutrons and protons as independent
particles moving in rather similar potentials. In reality the neutrons
and protons interact strongly with each other and this interaction is
responsible not only for the similarity of the neutron and proton
potentials but also for the isovector term in the nucleon potential.
This term provides a way of treating the neutrons and protons together,
by defining an isovector potential in terms of the neutron and proton
densities. It is then possible to iterate the potentials and the
densities until self-consistency is attained (Brown, Massen and Hodgson,
1979). This method has been used to analyse the charge and matter
distributions of some of the oxygen and calcium isotopes, and is
described in more detail in §4.

Some recent work on nucleon orbit radii is described in §5.

2.   Analyses of Nuclear Charge Distributions

As mentioned briefly in the introduction, the SPP method assumes
that each nucleon is moving independently in a one body potential, and
the proton and neutron density distributions are then obtained by
summing their probability distributions, weighted by the occupation
numbers of each orbit.

The essential feature of this method is that it is constrained at
every point to fit selected experimental data for the nucleus whose
charge and matter distributions are being calculated. Thus the
potentials used to generate the single particle wavefunctions are
adjusted to give the centroid separation energies and occupation numbers
as determined by one-nucleon transfer and knock-out reactions, and also
to fit the charge distributions as determined by electron elastic
scattering and muonic atom analyses. The resulting distributions thus
have a secure experimental basis, and because of the known systematic
behaviour of single particle states they may be extended to a wide range
of nuclei with some confidence.

The important feature of these calculations is that they unify the
nuclear data obtained from nucleon transfer and knock-out reactions on
the one hand, and from electron scattering and muonic atom data on the
other. This increases the overall accuracy of our knowledge of nuclear
charge distributions paticularly in the region of the nuclear surface.
Examination of the uncertainties in the model-independent analyses of
electron scattering and muonic atom data shows that the resulting charge
distributions are well-determined in the inner surface or knee region
where the density begins to fall from its interior value but are
relatively poorly determined in the interior and far surface regions.
The distributions calculated by the SPP method, however, are
particularly accurate in the far surface region because in this region
they depend on the tails of the wave functions which are fitted to the

experimental separation energies for each orbit.

The charge and matter distributions are given by expressions of the form

$$\rho(r) = \sum_i a_i \mid \psi_i(r) \mid^2 \qquad (2.1)$$

Where $\psi_i(r)$ is the wavefunction of the particle in the ith state, a_i the occupation numbers and the summation runs over all occupied states. The wavefunctions are bound solutions of the radial Schrodinger equation for a potential

$$V(r) = V_c(r) + Uf(r) + \left(\frac{\hbar}{m_\pi c}\right)^2 U_s \frac{df(r)}{dr} L.\sigma \qquad (2.2)$$

where $V_c(r)$, the electrostatic potential due to a uniformly-charged sphere, is included only for protons, and the form factor $f(r) = [1+\exp\{(r-R)/a\}]^{-1}$.

The proton charge distribution is obtained by folding the point distribution (2.1) with the charge distribution $\rho_p^c(r)$ of the proton itself (Chandra and Sauer, 1976)

$$\rho_{ch}(r) = \int \rho_p^c(\underline{r}') \ \rho_p(\mid\underline{r}-\underline{r}'\mid)d\underline{r}' \qquad (2.3)$$

The contribution to the nuclear charge distribution from the neutrons is evaluated similarly; this is a small but not negligible correction.

The occupation numbers a_i are taken to be (2J+1) for the deeper states and for the surface states the experimental values determined from nucleon transfer reactions are used wherever possible.

Local potentials and wavefunctions are used in these calculations for convenience, although it is known that they are partly non-local in character. The effect of using effective local potentials is to enhance the wavefunction in the nuclear interior, subject of course to the overall normalisation. An approximate relation between the local and non-local wavefunctions has been obtained by Perey (1962) in the form

$$\psi_{NL}(r) = \psi_L(r) \ \exp \ \frac{\{m\beta^2 \ V(r)\}}{2\hbar^2} \qquad (2.3)$$

where β is the non-locality parameter and m the reduced mass, and this is used to correct the calculated local wavefunctions.

Since the potential is allowed to vary from state to state to fit the measured binding energies, the resulting wavefunctions are not orthogonal. This inconsistency in the calculations is removed using the Gram-Schmidt orthogonalisation procedure. In practice this requires the extra assumption that the occupation numbers of all the states of a particular NLJ are unity except possibly that of the least bound one.

Shell model considerations indicate that this approximation is a very reasonable one. Calculations with and without orthogonalisation show very little difference, as might be expected from the small differences between the potentials corresponding to the different states.

The early calculations of the charge distribution of ^{208}Pb gave a good fit to the model-independent charge distributions provided (1) the occupation number of the $3s_{1/2}$ orbit is reduced from 2 to 1.8; this is physically very reasonable and can be verified experimentally in principle, though it is difficult in practice, (2) the radius parameter is allowed to depend weakly on the principal quantum number N. The resulting fit is shown in fig.1. As mentioned previously this work had to be abandoned because new data gave a significantly different model-independent charge distribution. Further work on ^{208}Pb has been carried out recently.

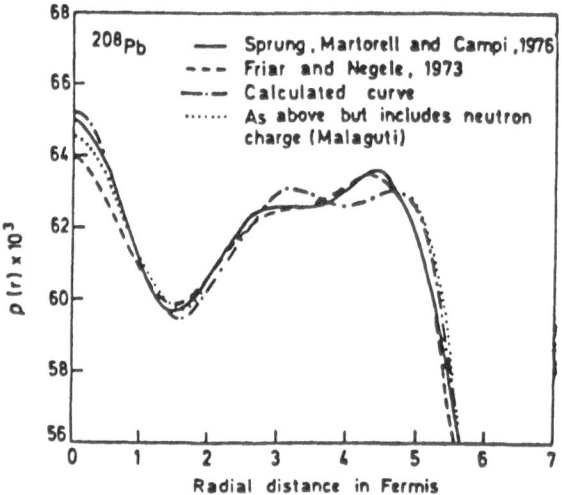

Fig.1. Model-independent charge distributions of ^{208}Pb compared with SPP model calculations with r_n=1.21, 1.23, and 1.244 for states with N=1, 2 and 3.

The first complete analysis was made for ^{58}Ni. A model-independent charge distribution had been determined by Sick et al (1975), and the mean separation energies of most of the single-particle states have been determined experimentally. The centroid energies of the single-particle states are defined as the weighted mean of the energies of the hole and particle ($T_<$ and $T_>$) fragments, according to the prescription of Baranger (1970) and Clement (1969). The experimental occupation numbers are used for the $2p_{3/2}$ and $1f_{7/2}$ states, and (2J+1) for the remainder. The fine adjustment of the potential to the charge distribution was made by adjusting the radius R of the potential, the non-locality parameter, the centroid of the 1d states and the $2p_{3/2}$ occupation number (within the range permitted by the experimental data), subject to the RMS charge radius being constrained to 3.775fm. Subsequent work on the calcium isotopes showed that there is a correlation between the diffuseness parameter a_1 of the central potential and the occupation number of the p-state. Since a full parameter search varying both these parameters as well as those already listed proved ambiguous, a simplified calculation was made with the non-locality parameter fixed to 0.85 and the central potential of the 1d states to 60 MeV. The parameter search in the subspace $\{R, a_1\}$ was then made for many fixed values of n_p. The best fit is obtained for $n_p \approx 1.5$, which is not acceptable experimentally. It must therefore be fixed to the optimum experimental value, and the resulting fit with the remaining parameters allowed to vary is shown in Fig.2.

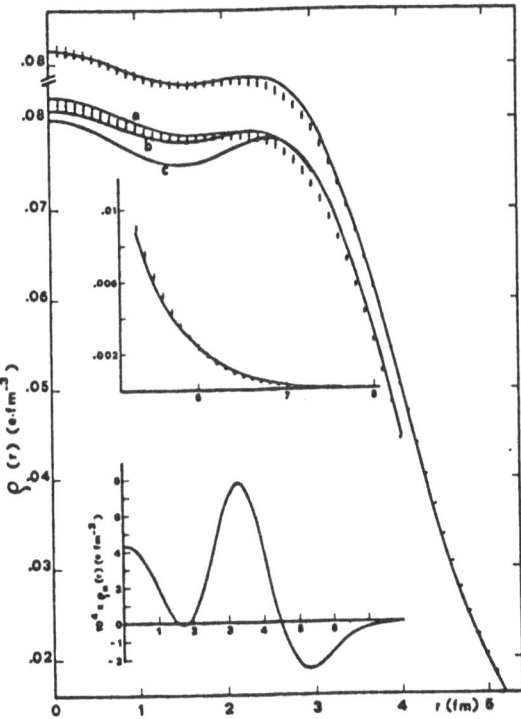

Fig.2. Comparison between SPP and model-independent charge distributions for ^{58}Ni. The tail is shown on an expanded scale in the upper insert, and the contribution of the neutron density in the lower. The curves a, b and c are sensitivity studies showing the effects of parameter variation (Malaguti et al, 1978).

The next series of analyses were made for ^{39}K, ^{40}Ca and ^{48}Ca, partly because adequate data are available on the single-particle states and partly because it is interesting to see if the method can account for charge distribution differences between isotopes, which are often better known than the charge distributions themselves. The analysis was made in essentially the same way as before, varying R, a_1, n_p ($2p_{3/2}$ occupation number) and n_s ($2s_{1/2}$ occupation number). The resulting charge distributions are compared with the experimental ones in Figure 3.

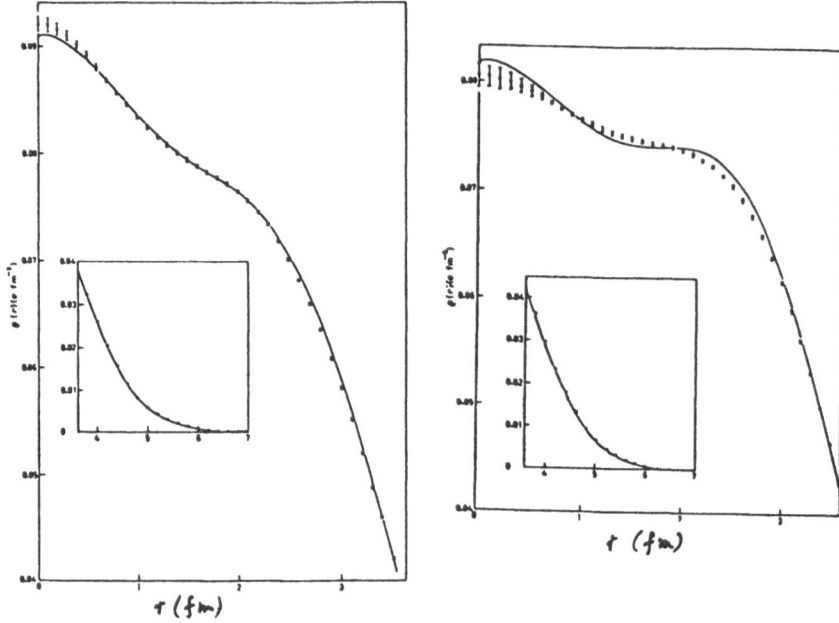

Fig.3. Comparison between SPP and model-independent charge distributions for ^{39}K and ^{48}Ca (Malaguti et al, 1979a).

A further opportunity to compare the SPP method with the experimental charge distribution differences between isotopes was provided by the Mainz data on the zirconium isotopes (Rothhaas, 1978). For these nuclei, the potentials for the deep states were calculated from the non-local potential fitted to the surface states, with the non-locality parameter $\beta=0.85$fm. The variation of these potentials along the isotopic sequence was calculated using an isospin potential $U_1=36$ MeV. Several calculations were made to explore the sensitivity of the charge densities to variations of the binding energies, the occupation numbers of the surface states, the potential form factor and the non-locality parameter. These showed that the results are rather insensitive to changes in the binding energies within the statistical uncertainties for the surface states and within 3 to 4 MeV for the deeper states. Further calculations were therefore made varying only the form factor parameters R_1 and a_1, and the occupation number of the $2p_{1/2}$ proton state, the others occupation numbers being fixed to their experimental values. The resulting charge distributions and charge distribution differences are compared with the experimental data in Fig.4.

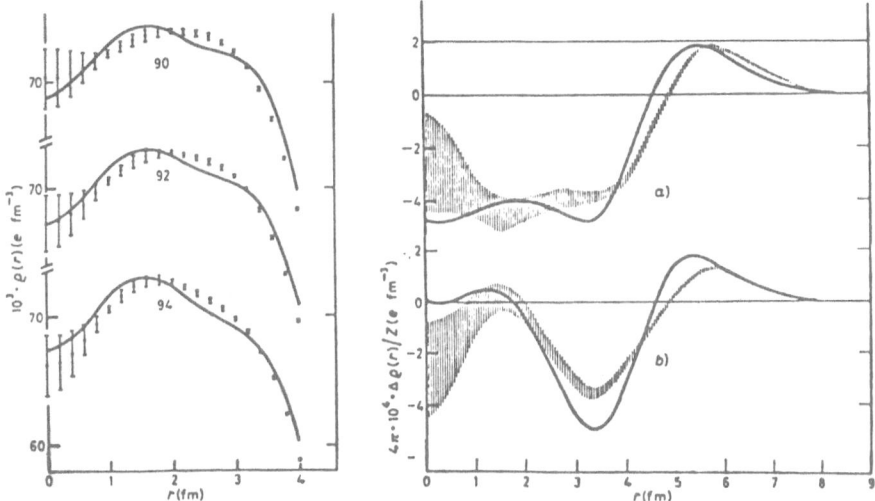

Fig.4.   Comparison between SPP calculations and experimental data.
(A) Charge distributions of zirconium isotopes (B) Charge distribution differences (a) ^{92}Zr-^{90}Zr, (b) ^{94}Zr-^{92}Zr (Malaguti et al, 1979b).

It is notable that the two curves are rather different, but both are well fitted by the SPP calculations. The first difference is what would be expected if the two extra neutrons produced just a uniform expansion of the core protons, and this is what might be expected from the strong neutron-proton attractive force. The second pair of extra neutrons act in a somewhat different way, however, and leave the central density practically unchanged while pulling out the rest of the protons.

3.   Nuclear Matter Distributions

The methods described in the previous section for calculating nuclear charge distributions can also be used to calculate nuclear matter distributions. These matter distributions cannot be compared with experimental data so accurately as the charge distributions because the nuclear interaction is not so well understood as the electromagnetic interaction.   The theories become more reliable at higher energies, so the most accurate comparisons are made using the Glauber and Kerman, McManus and Thaler theories to calculate the elastic scattering cross-sections of energetic protons. Earlier work (Ray et al, 1978; Ray, 1979) used functional forms for the matter distribution, and adjusted their parameters to optimise the fit to the proton elastic scattering data.   Such functional forms are not however very satisfactory, especially in the interior and for surface regions, so the analysis has been repeated using matter distributions calculated by the SPP method (Ray and Hodgson, 1979).

These calculations were made for ^{40}Ca, ^{48}Ca, ^{58}Ni and ^{208}Pb for which detailed studies of the charge distributions are available, and also for ^{64}Ni, ^{116}Sn and ^{124}Sn.   The resulting differential cross-sections for the elastic scattering of 800 MeV protons by these nuclei are compared with some of the experimental data in Fig.5. The only parameter allowed to vary in the calculation of the matter distribution was the radius of the neutron potential, and the calculated cross-sections in the Figures are labelled by their value of $\Delta r_{np} \equiv \langle r_n^2 \rangle^{\frac{1}{2}} - \langle r_p^2 \rangle^{\frac{1}{2}}$. On the whole the data is well fitted, and comparison between the experimental and theoretical cross-sections enables Δr_{np} to be estimated to about ±0.05fm.

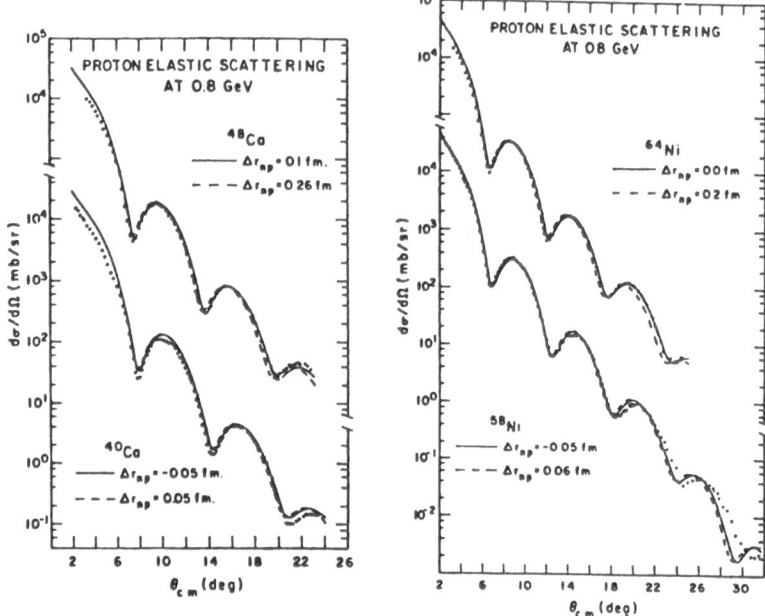

Fig.5. Comparison between experimental differential cross-sections for proton elastic scattering and those calculated from SPP matter distributions (Ray and Hodgson, 1979).

These calculations provide neutron density distributions, and some of these are compared in Fig.6 with the corresponding neutron densities obtained by the density matrix expansion calculations of Negele (1970) and Negele and Vautherin (1972). Particularly precise comparisons can be made for the differences between the neutron density distributions of isotopes, and one of these are compared in Fig.7 with the empirical neutron densities of Ray (1979).

More recently, the nuclear matter distributions calculated from a standard potential have been compared with all the available experimental matter distributions (Streets et al, 1982). In this work, the parameters of the potential were chosen to give the best fits to the binding energies and charge radii of ^{40}Ca and ^{208}Pb, and a linear interpolation formula with an asymmetry term used to obtain the values for other nuclei. The matter distributions obtained in this way are compared with the experimental data for the nickel isotopes in Fig.8 and for ^{90}Zr and ^{208}Pb in Fig.9.

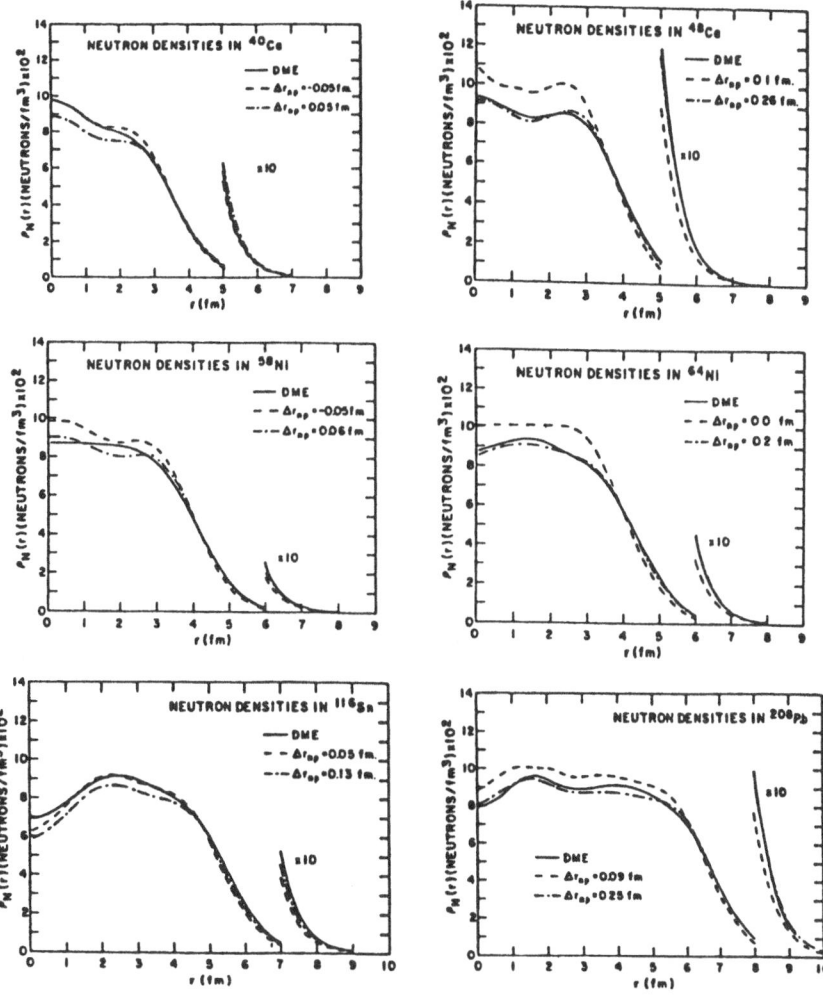

Fig.6. Comparison between neutron density distributions calculated by the density matrix expansion (DME) and SPP methods with different values of Δr_{np} (Ray and Hodgson, 1979).

The nuclear matter distributions calculated by the SPP method are particularly reliable in the far surface region because the potentials of the surface states are adjusted to fit the measured centroid separation energies. They are therefore very suitable for calculating optical potentials by the folding model, particularly for light and heavy ions whose elastic scattering is dominated by the far surface potential. This application of the SPP method was first made by Vary and Dover (1973) and their work showed the usefulness of the method.

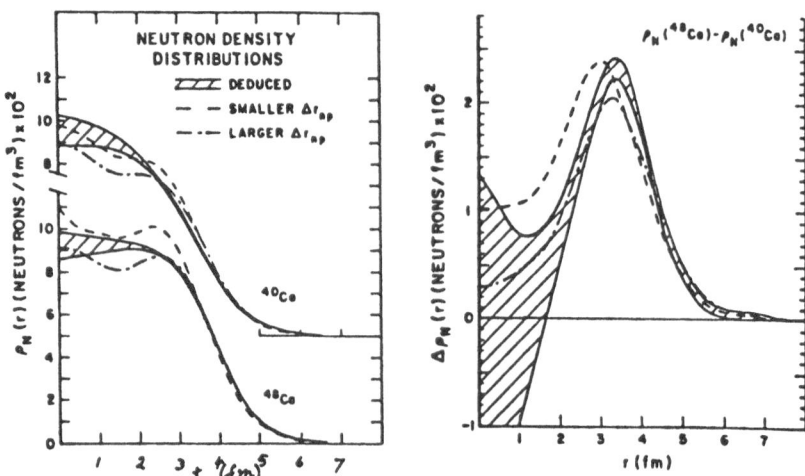

Fig.7. Comparison between SPP and experimental neutron density distributions for ^{40}Ca and ^{48}Ca (a) Density distributions (b) Density distribution differences (Ray and Hodgson, 1979).

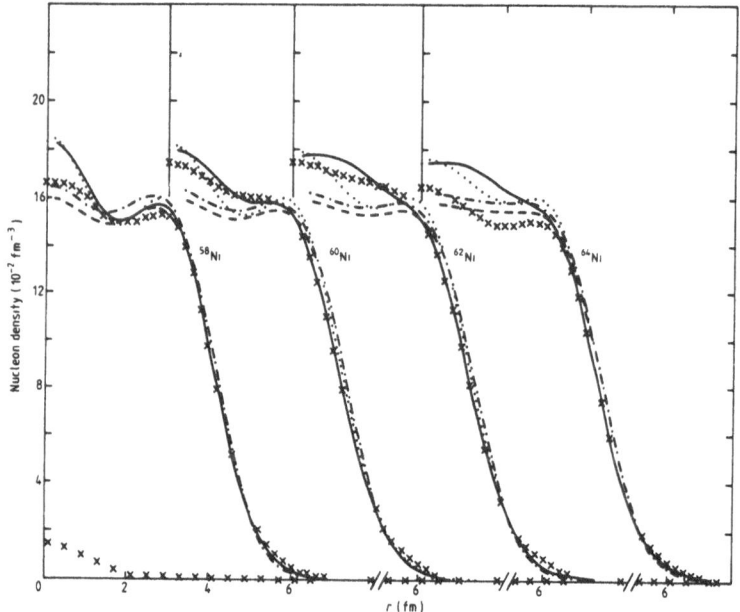

Fig.8. Matter distributions for the nickel isotopes compared with SPP and Hartree-Fock calculations with the standard potential. The experimental points (xxx) are the results of a model independent analysis of 1 GeV proton elastic scattering by Shaginyan and Starodubsky (1979). The SPP calculations were made with the occupation numbers of the simple shell model (———) and with those of Chung and Wildenthal (1978) for ^{28}Si, ^{32}S and ^{34}S, and Brown (1979) for ^{48}Ti (····). The Hartree-Fock calculations used the same occupation numbers and the Skyrme 3 interaction (- - -). Where an appreciable difference occurs, the results of scaling the latter calculation to fit the experimental charge radius are also shown (-·-·). The estimated uncertainty in the model independent analysis is shown at the foot of the figures.

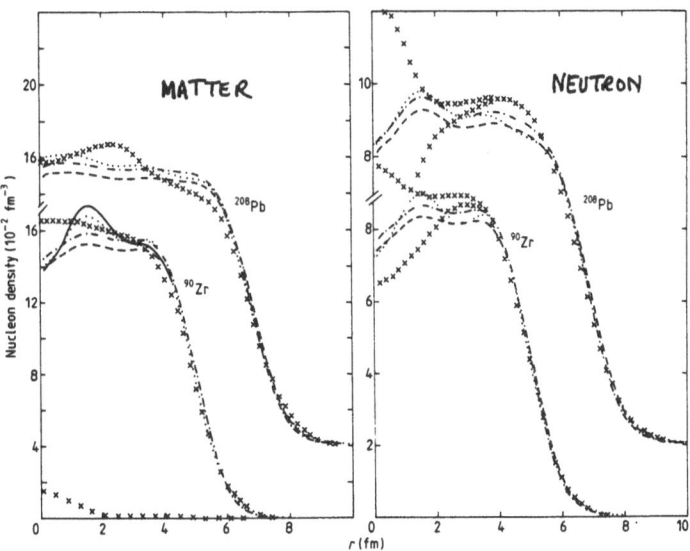

Fig.9. As Fig.8 for ^{90}Zr and ^{208}Pb.

Since that time, SPP density distributions have been used in several calculations of alpha-particle and heavy-ion potentials. In the case of nucleons, it is possible to find global optical potentials that describe to good accuracy the scattering from many nuclei over a range of energies. This not so easy for light and heavy ions because their interactions are much more sensitive to the nuclear surface, so that the scattering is sensitive to details of nuclear structure. It is therefore interesting to see whether some information from SPP matter densities can be used to improve the optical potentials.

The first studies were made of the elastic scattering of alpha particles by several isotopic sequences at energies near the Coulomb barrier (Badawy et al, 1978). The cross-sections determine the radius R at which the nucleon density is 2×10^{-3} nucleons fm^{-3}. These radii are approximately proportional to A$^{1/3}$, but there are small systematic differences due to the structure of the nuclei concerned. In particular, the quantity RA$^{1/3}$ is found to increase slowly through some isotopic sequences and to decrease through others. This may be understood qualitatively in terms of the shell structure of the nuclei, since we expect the nuclear radius to increase more rapidly than A$^{1/3}$ when a major shell is just beginning to be filled than when it is nearly full.

To see if this effect could be reproduced by SPP densities, Lozano et al (1979) fitted them in the tail region by the expression

$$\rho(r) = \rho_0\{1+\exp[(r-R_\rho)/a_\rho]\}^{-1} \approx \rho_0 \exp\{R_\rho-r)/a_\rho\} \qquad (3.1)$$

Setting ρ_0=0.14 gives

$$R = R_\rho + 4.25\ a_\rho \qquad (3.2)$$

Badawy et al showed that the real part of the alpha-particle potential can be obtained from the nuclear density distribution using the relation

$$R(0.2) = R_\rho + 5.54\ a_\rho + 2.37 \qquad (3.3)$$

where R(0.2) is the radius at which the real potential is -0.2 MeV. Thus

$$R(0.2) = R + 1.29\ a_\rho + 2.37 \qquad (3.4)$$

This enables the values of R obtained by Badawy et al to be compared with those calculated by the SPP method. As shown in Fig.10, this comparison shows the same overall trends for each isotropic sequence. It is therefore possible to use the SPP densities to calculate improved optical potentials.

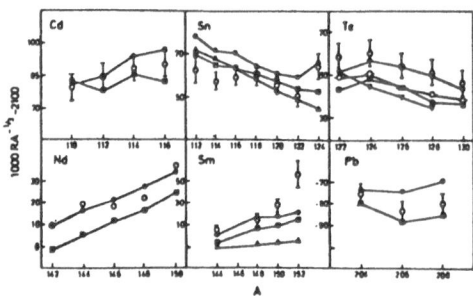

Fig.10. Nuclear radius parameter as a function of A for several isotopic
sequences. The experimental data are that of Tabor et al (1976)
for Nd and of Badawy et al (1978) for the other nuclei, and are
indicated by open circles. The solid symbols show the results
of SPP calculations and are connected by lines to show the
trends of the variations with A (Lozano et al, 1979).

This method was extended to heavy ions by Lozano et al (1980). They started from a global optical potential developed by Christensen and Winther (1976) with parameters adjust to fit a range of heavy-ion elastic scattering data and altered it in two ways; firstly it was allowed to become complex, and secondly the radius parameter was related to the radius at which the nuclear density has a certain fixed value. The resulting potential has the form

$$V(r) = \frac{R_2+R_2}{R_1+R_2}\ \{50\ \exp\ (\frac{R_1+R_2-r}{0.63}) + i\ W\ \exp\ (\frac{R_1+R_2-r}{a_i})\} \qquad (3.5)$$

The parameters W and a_i were determined by optimising the fit to twenty reactions, using Christensen and Winther's expression for the radius

$$R_i = 1.233 \, A_2^{1/3} - 0.978 \, A_i^{-1/3} \tag{3.6}$$

This gave W_o=21.38 MeV and a_i=0.686fm.

The nuclear structure effects in the nuclear radii were then introduced by defining

$$R_i = R_i \, (\bar{\rho}) - \eta \, A_i^{-1/3} \tag{3.7}$$

where $R_i(\bar{\rho})$ is the radius at which the matter density has the value $\bar{\rho}$. The optimum values of the parameters $\bar{\rho}$ and η were determined by fitting several sets of data, giving $\bar{\rho}$=0.04, η=0.903. With these values, the overall fit to a large range of data is significantly improved. This is further evidence that the global heavy-ion optical potential can be improved by the incorporation of nuclear structure information.

A more fundamental calculation using the SPP densities has been made by Viñas et al (1980). They used a double-folding model and calculated several heavy-ion potentials with the density-dependent interaction.

$$V(r,r') = -U_o(1-S_{ij}\alpha) \, [1-c\rho(r)^{1/3}\rho(r')^{1/3}] \exp[-(\tfrac{r-r'}{a})^2] \tag{3.8}$$

where r and r' are the positions of the two nucleons and s_{ij}=+(-1) for the interaction between like (unlike) nucleons. The parameters of this interaction were chosen to give the best agreement with nuclear densities, binding energies and radii as described by Viñas and Madurga (1977). The resulting potentials were found to give excellent fits to several heavy-ion elastic scattering cross-sections.

4. Self-Consistent Calculations

One of the main difficulties of the calculations described so far is that while the charge distributions can be compared with the experimental data to high accuracy, the neutron distributions are much less reliable, as they are calculated on the assumption that the neutron and proton potentials are the same, apart from an adjustment of the neutron potential radius to fit the proton elastic scattering data. In this respect the Hartree-Fock method is more satisfactory, since the neutron and proton wavefunctions are together iterated to self-consistency, and it is then very difficult to believe that the neutron distribution is not as reliable as the proton distribution, when the latter fits the experimental data.

In order to introduce a similar self-consistency into the SPP method, the nucleon potentials were written in a form that depends on the densities. The isospin term in the nucleon-nucleus potential was replaced by $V_1\rho_1(r)/\rho_o(r)$, where

$$\rho_0(r) = \rho_n(r) + \rho_p(r)$$

and $\qquad\qquad\qquad\qquad\qquad\qquad\qquad$ (4.1)

$$\rho_1(r) = \rho_n(r) - \rho_p(r).$$

Thus the proton and neutron potentials become

$$V_p(r) = \left(V_o + V_1 \frac{\rho_1(r)}{\rho_o(r)}\right) f(r) + V_{so}(r) + V_c(r)$$

and $\qquad\qquad\qquad\qquad\qquad\qquad\qquad$ (4.2)

$$V_n(r) = \left(V_o - V_1 \frac{\rho_1(r)}{\rho_o(r)}\right) f(r) + V_{so}(r)$$

In the first stage of the calculations these coupled equations are solved iteratively for a closed-shell nucleus using standard values of V_1, R_{s1}, a_s and a, and adjusting V and R to fit the single-particle centroid energies of orbits near the Fermi surface and the RMS charge radius.

To calculate the densities of other nuclei formed by adding one or more nucleons the potentials are written

$$V_p(r) = \{V_o \, \rho_o(r) + V_1 \, \rho_1(r)\} \, F(r) + V_{so}(r) + V_c(r)$$
$$\qquad\qquad\qquad\qquad\qquad\qquad\qquad (4.3)$$
$$V_n(r) = \{V_o \, \rho_o(r) - V_1 \, \rho_1(r)\} \, F(r) + V_{so}(r)$$

where $F(r)=f(r)/\rho_o^C(r)$ and $\rho_o^C(r)$ is the density of the closed shell or core particles. These equations are solved in the same way as before by iteration (Brown et al, 1979).

This formalism is particularly appropriate for studying changes in the proton and neutron density distributions relative to core nuclei such as ^{16}O and ^{40}Ca, and indeed in many experiments it is these distributions that are determined most accurately. There are no free parameters for the density differences as they are all fixed by the fit to the core nuclei. It is then possible to see if configuration mixing is able to account for the detailed structure of the charge density and also the RMS radii. The method is sensitive mainly to the changes in the configuration relative to the core nuclei. The model also includes the effects of polarisation of the core by the valence particles.

This method has been applied to the oxygen and calcium isotopes, and some of the results for the latter will be presented here. The first calculations were made using the occupation probabilities (90% $1f_{7/2}$ particles plus 10% $2p_{3/2}$ particles) of McGrory et al (1970), and the results for RMS radii are shown by the open circles in Fig.11, normalised at ^{40}Ca by choosing R=4.614fm. The isotopic dependence is small and smooth, in disagreement with experiments. The corresponding density changes shown by the curves IIA in Fig.12 are however in fair agreement with experiment.

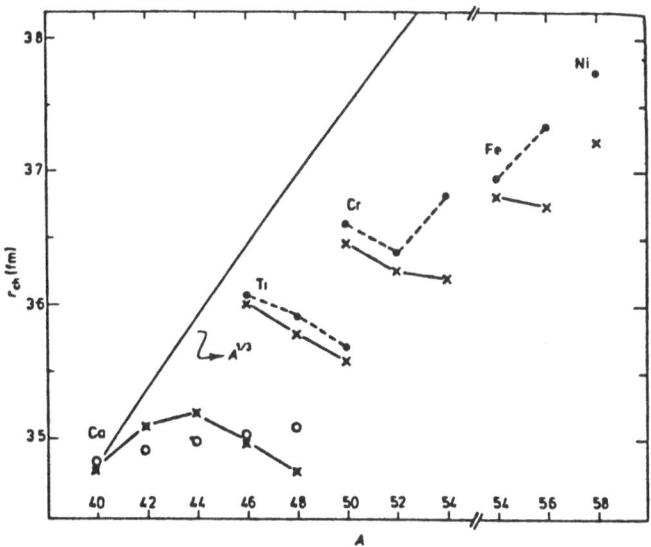

Fig.11. RMS charge radii for A=40-58 nuclei. The experimental (●) and theoretical (oIIA; xIIB) values are shown for each isotopic sequence. Calculation IIA refers to a closed shell configuration and IIB allows for excitation out of the sd shell (Brown et al, 1979).

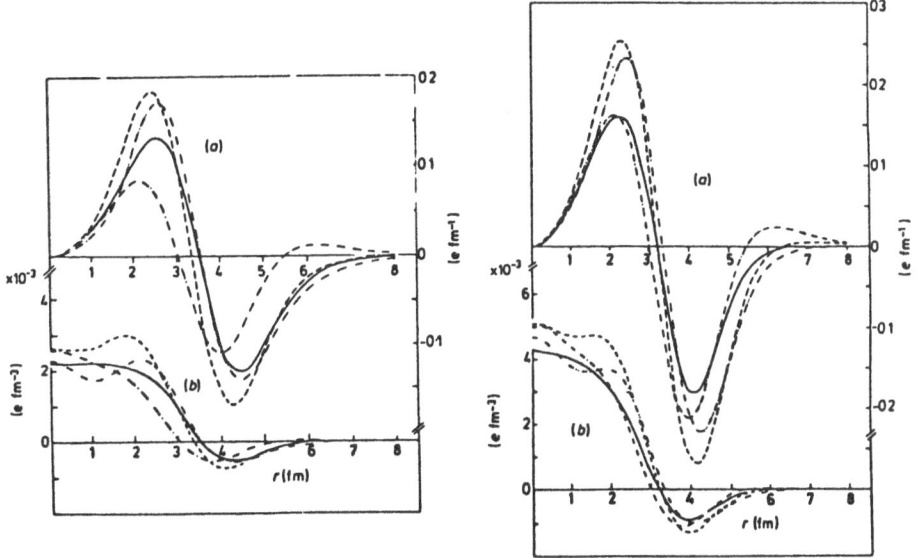

Fig.12. Comparison of experimental (full curves) and theoretical (broken curves) charge density differences between ^{42}Ca and ^{40}Ca and between ^{44}Ca and ^{40}Ca. The experimental curves are from the model-dependent fit to the electron scattering data by Frosch et al (1968). The three theoretical curves correspond to a closed sd shell configuration (-·-·,IIA) to a non-closed shell configuration with 10% 2p$_{3/2}$ (——IIB) and to a non-closed shell configuration with 20% 2p$_{3/2}$ (-··-,IIC). (a) $4\pi r^2 \Delta\rho(r)_{ch}$; (b) $\Delta\rho(r)_{ch}$ (Brown et al, 1979).

The calculations were then repeated with the following one-parameter wavefunctions for $n > 2$

$$|^{40+n}\text{Ca}> = \alpha|(\nu f_{7/2}p_{3/2})^n) + \beta|(\pi d_{3/2})^{-2} (\pi f_{7/2}p_{3/2})^2 (\nu f_{7/2}p_{3/2})^n> \quad (4.4)$$

and for ^{40}Ca,

$$|^{40}\text{Ca}> = \alpha|\text{o}> + \beta|(\pi d_{3/2})^{-1} (\pi f_{7/2}p_{3/2})^1 (\nu d_{3/2})^{-1} (\nu f_{7/2}p_{3/2})^1>(4.5)$$

where $|\text{o}>$ is the closed-shell configuration. The same fp configurations were used as in the previous calculations. The numbers of proton holes in the $d_{3/2}$ orbit relative to ^{40}Ca is then

$$\Delta(n) = 2\beta^2(^{40+n}\text{Ca}) - \beta^2(^{40}\text{Ca}) \quad (4.6)$$

The $\Delta(r)$ were then chosen to reproduce the experimental RMS radii for the nuclei from ^{42}Ca to ^{48}Ca and the corresponding charge distribution differences are compared with the experimental data and with the previous calculations in Fig.12. These calculations including core excitation (labelled IIB) are in much better agreement with the data than the closed-shell calculations (IIA), especially in the important surface region. Also included in Fig.12 are the results of a calculation with 20% admixture of $2p_{3/2}$ particles to show the sensitivity to this orbit.

The values of $\Delta(n)$ are compared with determinations of the number of proton holes in the sd shell obtained from analyses of various stripping and pickup reactions. Since $\Delta(n)$ is the number of proton holes relative to ^{40}Ca, the number of proton holes in $\Delta(n)$, $\delta \approx .7$, is added to $\Delta(n)$ to give an estimate of the absolute number. It is apparent from the table that there is not a satisfactory agreement among these quantities. This is being investigated further.

5. Nucleon Orbit Radii

Several recent analyses of the cross-sections of nucleon transfer reactions (Chapman et al 1979) have given values of the RMS radii of the probability distributions of the nucleons in particular quantum states. These RMS radii can also be calculated from a single-particle model, so it is interesting to see how well such values agree with those obtained experimentally and what conclusions can be drawn from the comparison.

To do this, we use the global single-particle potential BH of Bear and Hodgson (1978), whose parameters were chosen to give a good overall fit to the centroid binding energies of many single-particle states in nuclei from ^{40}Ca to ^{208}Pb. This potential has the standard form

$$V(r) = V_c(r) + V \, f(r) + \left(\frac{\hbar}{m_\pi c}\right)^2 V_s \frac{1}{r} \cdot \frac{df}{dr} \underline{L \cdot \sigma} \quad (1)$$

where $V_c(r)$ is the Coulomb potential (for protons only) and the form factor $f(r) = (1+\exp[(r-R)/a])^{-1}$ with $R = r_0 A^{1/3}$.

The potential depth for surface states is

$$V = V_o^{p,n} \text{ for } -15 < E < 0 \tag{2}$$

and for deep states

$$V = V_o^{p,n} - \beta(E+15) \quad \text{for E} <-15 \text{ MeV} \tag{3}$$

where the superscripts indicate neutron or proton potentials. These depend on the asymmetry potential V_1 and can be expressed in the forms

$$V_o^p = V_o + [(N-Z)/A]V_1$$
$$V_o^n = V_o - [(N-Z)/A]V_1 \tag{4}$$

The form factor parameters were fixed to the values $r_0=1.236$fm., $a=0.62$fm. for the central potential and $V_s=7$ MeV., $r_s=1.1$fm., $a_s=0.65$ for the spin-orbit potential. The fit to a range of nuclei gave $V_0=55.7$ MeV., $V_1=39.3$ MeV, and $\beta=0.51$. The Perey non-locality correction was applied with range parameter 0.87fm. and the wavefunctions were orthogonalised as described by Malaguti et al (1978).

This potential was used to calculate the RMS radii of some single-particle states in the tin isotopes, and the results are compared in Table 2 with those found experimentally. The corresponding binding energies of the single-particle states are given in Table 3. Also included in the tables are the results of Hartree-Fock calculations made by Warwick et al (1979) using the single-particle mean field potential of Campi et al (1974). The calculated RMS radii in column A agree fortuitously well with the experimental values: the average difference between theory and experiment is only 0.03fm., about a quarter of the quoted experimental uncertainties, suggesting that these are systematic rather than statistical. The Hartree-Fock results are appreciably larger, around the limits of the quoted experimental uncertainties.

This comparison between theory and experiment can be extended to include the accurately-known RMS total point charge radii, and Table 4 shows that the calculated values are significantly lower than the experimental ones. Further calculations of the radii of the individual orbits were therefore made, adjusting the radius parameter in each case to give the measured value, and the results are given in columns B of Tables 2 and 3. These RMS radii are about 0.08fm. higher than before, on the average, but stilll within the quoted uncertainties and much closer to the Hartree-Fock values. This comparison suggests that the experimental orbital radii are somewhat low compared with the very precise total RMS radii.

A possible reason for this discrepancy is the potential assumed for the deep states, which is not known experimentally for tin. To test this, calculations were made with the same potential for all states, and this was found to increase the total radii by about 0.03fm. This is in the right direction, but only by about a fifth of the amount needed to give the experimental values. The calculated radii are thus insensitive to the potentials chosen for the deep states.

The discrepancy could also be resolved by altering the assumed value of the spectroscopic factors of the transitions used to determine the orbit radii. Warwick et al (1979) found that a 10% change in the spectroscopic factor corresponds to a 1% change in the orbit radius. A radius increase of 0.08fm. is about 1.6%, requiring a 16% decrease of the spectroscopic factor. Since the spectroscopic factors used by Warwick et al are normalised to the sum rule limit this would imply that about 14% of the spectroscopic strength is spread among higher states not included in their analysis.

The comparison in Table 3 shows that the calculated binding energies are considerably larger than the corresponding separation energies, and this is still the case when allowance is made for the known fragmentation of strength among some of the lower states. A further set of calculations was therefore made in which the binding energy for each state was adjusted to the experimental separation energy, always re-adjusting the radius parameter to give the experimental total RMS charge radius. The results of these calculations are given in column C of Table 2, and are on average about 0.22fm. greater than the experimental values, corresponding to a spreading of about 30% of the spectroscopic strength. This is however certainly an overestimate, since the spreading increases the centroid separation energy, and this reduces the need to increase the radius.

We therefore conclude that the available experimental data for proton states in the even tin isotopes, when analysed by the single-particle model, imply that the spectroscopic factors used by Warwick et al are about 20% too high, and that their orbital radii should be increased by an average of about 0.15fm. This is very similar to the 2-6% discrepancy between calculated and experimental orbital radii noted for lighter nuclei by Brown et al (1979).

Similar calculations were made for the neutron orbits and the results are compared with the experimental data of Chapman et al (1979) in Tables 5 and 6. As before, the calculations in column A correspond to the BH potential, and those in column B correspond to potentials with neutron radius parameters set to the values that give the measured total charge radii. Column C gives the results of calculations with the potential depths adjusted to give the measured separation energies. In this case the calculated RMS radii are on the average only 0.02fm. greater than the experimental values, well within the quoted uncertainty.

Acknowledgement

This work has been carried out in collaboration with Franco Malaguti, Arnaldo Ugguzoni, Ettore Verondini, Manolo Lozano, Gonzalo Madurga, Alex Brown, Stelios Massen, Lanny Ray, Sally Owen and Jonathan Streets, and it is a pleasure to thank them for their stimulating discussions and essential contributions.

Table 1

Comparison between the numbers of proton holes in the sd shell obtained from electron scattering and from pickup and stripping reactions (Brown et al, 1979)

Nucleus	$\Delta+\delta$	(h,d)	(h,d)	(d,h)	(t,α)
^{40}Ca	0.7	0.4	0.27	0.73	0.3
^{42}Ca	1.7	0.9	1.12	1.03	0.4
^{44}Ca	1.8	1.9	1.98	1.01	0.6
^{46}Ca	1.05	0.4	–	–	0.2
^{48}Ca	0.3	0.15	0.44	0	0

Table 2

Comparison of calculated RMS point charge radii of proton states with experimental and Hartree-Fock values (Warwick et al 1979). The calculated values are obtained (A) from the BH potential, (B) after adjustment of proton radius parameter to give experimental total RMS radius and (C) after adjustment of potential depth to give measured separation energies.

Nucleus	State	Experimental	Calculated			Hartree-Fock
			A	B	C	
^{112}Sn	$1g_{9/2}$	$4.95^{+0.17}_{-0.12}$	4.98	5.06	5.10	5.16 ± 0.04
	$2p_{1/2}$	$4.51^{+0.17}_{-0.12}$	4.55	4.62	4.74	4.59 ± 0.04
	$2p_{3/2}$	$4.54^{+0.15}_{-0.11}$	4.53	4.58	4.75	4.57 ± 0.04
^{116}Sn	$1g_{9/2}$	$5.01^{+0.16}_{-0.11}$	4.99	5.10	5.15	5.18 ± 0.04
	$2p_{1/2}$	$4.46^{+0.15}_{-0.11}$	4.55	4.64	4.77	4.60 ± 0.04
	$2p_{3/2}$	$4.55^{+0.14}_{-0.10}$	4.53	4.58	4.78	4.57 ± 0.04
^{178}Sn	$1g_{9/2}$	$5.00^{+0.15}_{-0.14}$	5.00	5.12	5.17	5.21 ± 0.04
	$2p_{1/2}$	$4.52^{+0.14}_{-0.11}$	4.54	4.63	4.78	4.60 ± 0.04
	$2p_{3/2}$	$4.46^{+0.14}_{-0.12}$	4.53	4.57	4.79	4.58 ± 0.04
^{120}Sn	$1g_{9/2}$	$4.99^{+0.16}_{-0.13}$	5.01	5.13	5.18	5.22 ± 0.04
	$2p_{1/2}$	$4.49^{+0.14}_{-0.11}$	4.54	4.62	4.79	4.63 ± 0.04
	$2p_{3/2}$	$4.57^{+0.14}_{-0.11}$	4.51	4.56	4.80	4.59 ± 0.04
^{124}Sn	$1g_{9/2}$	$5.02^{+0.14}_{-0.11}$	5.03	5.15	5.22	5.22 ± 0.04
	$2p_{1/2}$	$4.58^{+0.14}_{-0.10}$	4.53	4.61	4.80	4.61 ± 0.04
	$2p_{3/2}$	$4.50^{+0.13}_{-0.11}$	4.49	4.55	4.81	4.60 ± 0.04

Table 3

Comparison of experimental and calculated proton binding energies (MeV).
The columns A and B are as in Table 1.

Nucleus	State	Experimental[*]	Calculated		Hartree-Fock
			A	B	
^{112}Sn	$1g_{9/2}$	7.51	9.69	11.84	8.87
	$2p_{1/2}$	8.05	10.76	12.45	12.29
	$2p_{3/2}$	8.32	12.22	13.88	14.11
^{116}Sn	$1g_{9/2}$	9.27	11.49	14.23	10.51
	$2p_{1/2}$	9.61	12.52	14.67	13.86
	$2p_{3/2}$	9.87	13.96	16.90	15.74
^{118}Sn	$1g_{9/2}$	10.01	12.36	15.69	11.13
	$2p_{1/2}$	10.33	13.37	16.33	14.58
	$2p_{3/2}$	10.60	14.79	18.78	16.46
^{120}Sn	$1g_{9/2}$	10.67	13.21	17.46	11.84
	$2p_{1/2}$	10.98	14.20	18.03	15.39
	$2p_{3/2}$	11.27	16.08	20.46	17.17
^{124}Sn	$1g_{9/2}$	12.10	14.85	21.27	13.27
	$2p_{1/2}$	12.42	16.39	21.66	16.63
	$2p_{3/2}$	12.76	18.83	24.09	18.53

[*]   These figures are the separation energies of Gove and Wapstra
(1972). The centroid energies of known fragments are approximately
0.5 MeV higher, a difference that is not significant for the
present purpose.

Table 4

Comparison between the calculated (A) charge RMS radii and the experimental results of Ficenec et al (1972).

Nucleus	Calculated	Experiment
^{112}Sn	4.499	4.586±0.005
^{116}Sn	4.506	4.619±0.005
^{118}Sn	4.511	4.634±0.005
^{120}Sn	4.514	4.646±0.005
^{124}Sn	4.520	4.670±0.005

The calculated values are found from the point proton rms radii by folding in the proton radius.

Table 5

Comparison of Neutron orbit radii. The columns A, B and C are as in Table 1, keeping neutron and proton radius parameters equal.

Nucleus	State	Experimental[*]	Calculated		
			A	B	C
^{113}Sn	$3s_{1/2}$	5.31 ± 0.12	5.30	5.28	5.28
	$1g_{7/2}$	5.00 ± 0.12	4.99	5.09	5.07
^{115}Sn	$3s_{1/2}$	5.34 ± 0.12	5.33	5.30	5.36
	$2d_{3/2}$	5.29 ± 0.12	5.25	5.27	5.31
	$1h_{11/2}$	5.27 ± 0.12	5.34	5.40	5.39
^{117}Sn	$3s_{1/2}$	5.37 ± 0.12	5.35	5.32	5.43
	$2d_{3/2}$	5.38 ± 0.12	5.27	5.30	5.36
	$1h_{11/2}$	5.29 ± 0.12	5.36	5.44	5.43
^{119}Sn	$3s_{1/2}$	5.52 ± 0.12	5.37	5.34	5.47
	$2d_{3/2}$	5.42 ± 0.12	5.29	5.32	5.39
	$1h_{11/2}$	5.44 ± 0.12	5.39	5.48	5.47

[*] Chapman et al 1979.

Table 6

Comparison of experimental and calculated neutron binding energies.
The columns A and B are as in Table 1.

Nucleus	State	Experimental[*]	Calculated	
			A	B
^{113}Sn	$3s_{1/2}$	10.80	9.37	11.06
	$1g_{7/2}$	10.88	10.45	12.32
^{115}Sn	$3s_{1/2}$	10.32	9.26	11.16
	$2d_{3/2}$	9.82	8.97	11.02
	$1h_{11/2}$	9.60	8.19	10.75
^{117}Sn	$3s_{1/2}$	9.57	9.16	11.25
	$2d_{3/2}$	9.41	8.87	11.15
	$1h_{11/2}$	9.43	8.14	10.95
^{119}Sn	$3s_{1/2}$	9.33	9.05	11.32
	$2d_{3/2}$	9.31	8.78	11.23
	$1h_{11/2}$	9.24	8.10	11.12

[*]   From separation energies of Gove and Wapstra (1972).

References

I.Badawy, B.Berthier, P.Charles, M.Dost, B.Fernandez, J.Gastebois and S.M.Lee, Phys.Rev. C17 978, 1978.

M.Baranger, Nucl.Phys. A149 225, 1970.

K.Bear and P.E.Hodgson, J.Phys. G4 L287, 1978.

B.A.Brown, S.E.Massen and P.E.Hodgson, J.Phys. G5 1655, 1979.

B.A.Brown, 1979 (Private Communication).

X.Campi, D.W.L.Sprung and J.Martorell, Nucl.Phys. A223 541, 1974.

H.Chandra and G.Sauer, Phys.Rev. C13 245, 1976.

R.Chapman, M.Hyland, J.L.Durell, J.N.Mo, M.Macphail, H.Sharma and N.H.Merrill, Nucl.Phys. A316 40, 1979.

A.Chaumeaux, V.Layly and R.Schaeffer, Ann.Phys. (New York), 116 247, 1975.

P.R.Christensen and A.Winther, Phys.Lett. 65B 19, 1976.

W.Chung and B.H.Wildenthal, (Private Communication) 1978.

C.F.Clement, Phys.Lett. 28B 398, 1969.

J.R.Ficenec, L.A.Fajardo, W.P.Trower and I.Sick, Phys.Lett. 42B 213, 1972.

J.L.Friar and J.W.Negele, Nucl.Phys. A212 93, 1973.

R.F.Frosch, R.Hofstadter, J.S.McCarthy, G.K.Nöldeke, K.J.Van Dostrum, M.R.Yearian, B.C.Clark, R.Herman and D.G.Ravenball, Phys.Rev. 174 1380, 1968.

N.B.Gove and A.H.Wapstra, Nucl.Data Tables A11 127, 1972.

P.E.Hodgson, Lecture Notes in Physics Vol.55 Ed. S.Boffi and G.Passatore (Springer-Verlag 1976) p88; Lecture at G.I.F.T. Seminar on Heavy-Ion Reactions, Seville 1979a; Atomki Közlemenyck 21 165, 1979b; Notas de Fisica 4 169, 1981.

E.Kujawski and J.P.Vary, Phys.Rev. C12 1271, 1975.

M.Lozano, G.Madurga and P.E.Hodgson, Phys.Lett. 82B 170, 1979; (Preprint) 1980.

J.B.McGrory, B.H.Wildenthal and E.C.Halbert, Phys.Rev. C2 186, 1970.

F.Malaguti, Nucl.Phys. A308 125, 1978.

F.Malaguti, A.Uguzzoni, E.Verondini and P.E.Hodgson, Nucl.Phys. A297 206, 1978.

F.Malaguti, A.Uguzzoni, E.Verondini and P.E.Hodgson, Nucl.Phys. A297 287, 1978; Nuovo Cim. 49A 412, 1979a; Ibid 53A 1, 1979b.

F.Malaguti and P.E.Hodgson, Nucl.Phys. A215 243, 1973; A257 37, 1976.

D.J.Millener and P.E.Hodgson, Nucl.Phys. A209 59, 1973.

J.W.Negele, Phys.Rev. C1 1260, 1970.

J.W.Negele and D.Vautherin, Phys.Rev. C5 1472, 1972.

A.S.Owen, B.A.Brown and P.E.Hodgson, J.Phys.G. 7 1057, 1981.

F.Perey, Proc.Conf.on Direct Interactions and Nuclear Reaction Mechanisms, Padua, 125, 1962.

L.Ray and P.E.Hodgson, Phys.Rev. C20 2403, 1979.

L.Ray, W.R.Coker and G.W.Hoffmann, Phys.Rev. C18 2641, 1978.

L.Ray, Phys.Rev. C19 1855, 1979.

H.Rothhaas, Dissertation Mainz, 1978.

V.R.Shaginyan and V.E.Starodubsky, 1979. Leningrad Nuclear Physics Institute Preprint No.457.

I.Sick, J.B.Bellicard, M.Bernheim, A.Bussière de Nercy, B.Frois, M.Huet, Ph.Leconte, J.Mongey, Pham Xuan Ho, D.Royer and S.Turck, Phys.Rev.Lett. 35 910, 1975.

D.W.L.Sprung, I.Martorell and X.Campi, Nucl.Phys. A268 301, 1976.

J.Streets, B.A.Brown and P.E.Hodgson, J.Phys.G.

S.L.Tabor, B.A.Watson and S.S.Hanna, Phys.Rev. C14 514, 1976.

J.P.Vary and C.B.Dover, Phys.Rev.Lett. 31 1511, 1973.

F.J.Viñas and G.Madurga, Anales de Fisica 73 92, 1977.

F.J.Viñas, M.Lozano and G.Madurga, (In Press) 1980.

A.Warwick, R.Chapman, J.L.Durell and J.N.Mo, Phys.Lett. 87B 335, 1979a.

A.Warwick, R.Chapman, J.L.Durell, J.N.Mo and S.Sen, Phys.Lett. 88B 55, 1979b.

WHAT DO WE LEARN FROM SELF-CONSISTENT MODELS ABOUT

NUCLEAR DENSITY DISTRIBUTIONS ?

F. TONDEUR

Université Libre de Bruxelles

and

Institut Supérieur Industriel de Bruxelles

Belgium

1.   INTRODUCTION

The detailed knowledge of nuclear density distributions is of great interest in several models of heavy ion collisions. Experiment has reached a very good precision in determining the shape of the charge densities in nuclei, even for far-unstable nuclei. However, only few accurate data have been obtained till now about neutron and matter distributions. People who need those distributions as an input for their calculations must then rely on a theoretical model of nuclear densities. The aim of the following discussion is to give a few elements which can help them in their choice, by comparing different models, with a special emphasis on self-consistent results.

Nuclear density distributions can be described by several characteristic quantities, like their RMS radii. Higher moments can also be used to give a detailed description of the density $\rho(\bar{r})$, but it is often more convenient to use the quantities ρ_c , $R_{1/2}$ and t, which are the central density, the half-density radius and the 90% - 10% surface thickness, although they can often only be defined approximately. In the present contribution, we shall also examine the surface asymmetry, defined as the difference between the inner part of the surface thickness (inside $R_{1/2}$) and the outer part (outside $R_{1/2}$).

2.   A FEW STRIKING EXPERIMENTAL RESULTS

A systematic analysis of experimental RMS charge radii has been done by Angeli and Csatlos (1). These authors observe what they call a fine structure, i.e. deviations from the average trend described by a simple formula :

$$\left< r^2 \right>^{1/2}_{average} = \sqrt{0.6} \ A^{1/3} \ (1.15 + 1.8 \ A^{-2/3} - 1.2 \ A^{-4/3}) \ \text{fm.} \quad (1)$$

The deviations from this formula are strongly correlated with the neutron magic numbers. Similar results have been obtained by Gupta (2) for the half-density radii and surface thicknesses of the charge densities. The fine structure has also been found in the study of long isotopic sequences including far unstable nuclei (3). The smallness of the shell effects obtained in nuclear radii by self-consistent

98

models (4 and sect. 5) suggest that the fine structure is not a pure shell effect ,
and it can indeed be explained by collective properties : the zero-point vibrations
or a static deformation of the nucleus (5, 6).

For a few nuclei, a detailed description of the charge distribution has been
obtained by model independent analyses of electron scattering data. We shall here
point out the results obtained (7) for ^{208}Pb, which are given in fig. 1. In this
nucleus, the density does not vary too much in the core, and an average core density
can be calculated with a good accuracy. It is then possible to determine $R_{1/2}$, t and
the surface asymmetry. The result we want to emphasise is that the surface of the
charge density in ^{208}Pb is slightly asymmetric, the inner part of the surface being
stiffer than the outer part, leading to a negative asymmetry. For light nuclei,
the surface asymmetry is more difficult to determine, because the definition of an
average core density is more ambiguous.

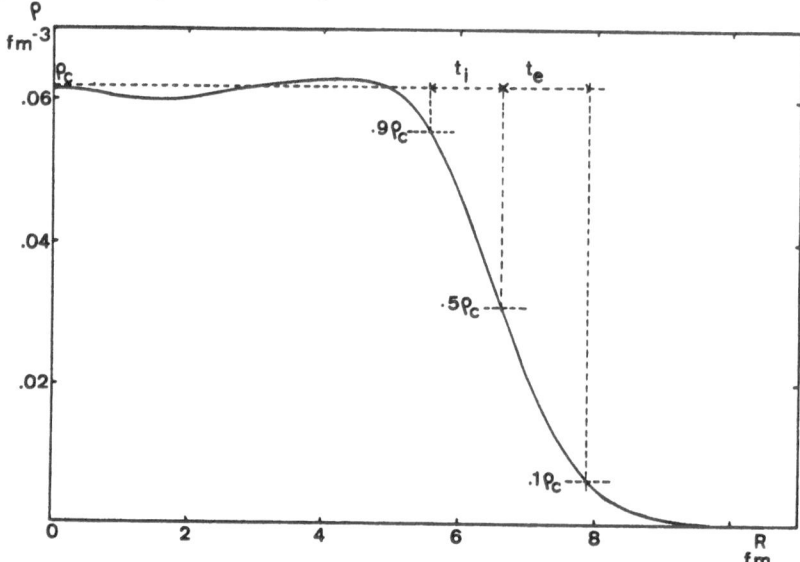

Fig. 1 : inner (t_i) and outer (t_e) parts of the surface thickness of the experimental charge distribution of ^{208}Pb. The estimation is : t_i = 1.06 ± 0.02 fm
and t_e = 1.27 ± 0.02 fm.

Such a detailed information does not exist for the neutron and matter distributions. For most nuclei, no experimental information at all is available. Moreover,
in many cases for which data have been obtained, the results are still controversial.
For the following discussion,we shall only point out the   recent determination of
the neutron skin thickness for ^{208}Pb (8) : $\langle R^2 \rangle_n^{1/2} - \langle R^2 \rangle_p^{1/2}$ = 0.14 ± 0.04 fm.
This value has been obtained after a careful analysis of 1 GeV proton scattering
data.

3.   BRIEF INTRODUCTION TO A FEW POPULAR MODELS

The oldest model for nuclear densities is the liquid drop model, which has been

generalised by Myers and Swiatecki (9) with the droplet model (DM). The DM, like
the old drop model, considers density distributions with a sharp surface, the surface
diffuseness being not explicitly taken into account, because its contribution to
the binding energy of the nucleus occurs at an order which is neglected by the DM.
On the other hand, the compression of the core and the possibility of different
radii for neutrons and protons are introduced in the DM. The predictions of the DM
about the densities thus only concern the core densities of protons and neutrons.

A more detailed description of the densities (though still macroscopic) is
obtained by the Thomas-Fermi model and other related approaches (energy density
method, extended Thomas-Fermi models, semiclassical models). This method is particularly useful when a Skyrme-type effective interaction is used: the binding energy
density of the nucleus can then be written as the integral of an energy density functional depending only on the density distributions and their derivatives. The equilibrium distributions are usually obtained by introducing parametrised trial densities and by minimising the total energy with respect to the parameters. The weak
point of the method is that it is not valid beyond the classical turning point,
leaving a doubt about the accuracy of the description of the surface.

The density distributions can also be obtained within the shell model, by summing the squares of the individual wavefunctions of the nucleons. This approach can
in principle include possible shell effects in the densities, and is expected to
give an accurate description of their tails. However, all problems are now shifted
to the description of the nucleon-nucleus shell model potential. The fit of a parametrised potential (e.g. Woods-Saxon potential) to the experimental "single particle" spectra can lead to an unsatisfactory result, partly because the "single particle" states in fact involve significant collective contributions (10). Shell model
potentials used to calculate density distributions should thus be mainly fitted in
order to reproduce the experimental distributions. Unfortunately, the extremely
scarce data concerning the neutron distributions do not allow at the present time
a systematic determination of the neutron potential on this basis. Even for the
protons a systematic fit of the shell model potential on the charge densities is
not available. Brown et al. (11) have proposed a generalisation of the shell model
method, which includes a partial self-consistency and emphasises the role of configuration mixing. The model is however dependent on experimental and theoretical
data which are not available for all nuclei.

Another natural extension of the shell model is the Hartree-Fock (HF) method.
For any given effective interaction between the nucleons, it can predict self-consistently the density distributions as well as the single particle potentials for any
nucleus. Although the limitation of the wavefunction to a single Slater determinant
(or to a BCS wavefunction in the HF+BCS approach) is still not completely satisfactory, we can obtain from this model more detailed and more reliable results than
with the DM, TF and ordinary shell models. Our aim is to describe several interes-

ting results obtained in HF+BCS calculations, and to compare them with the predictions of other models.

4. THE SELF-CONSISTENT MODEL

The model we have used is very similar to the HF calculations with a Skyrme interaction. This interaction allows to write the binding energy of the nucleus (except the Coulomb energy) as the integral of an energy density functional $\mathcal{H}$ which depends only on the densities ρ_q, kinetic densities τ_q and spin-orbit densities $\vec{J}_q$, defined as in (12), q standing for n or p. In a first step (13), we have parametrised directly $\mathcal{H}$ instead of the interaction :

$$\mathcal{H} = \frac{\hbar^2}{2m}\tau + a\rho^2 + b\rho^{7/3} + c\rho^{-1/3}(\rho_n - \rho_p)^2 + d\vec{J}\cdot\vec{\nabla}\rho + \eta|\vec{\nabla}\rho|^2 \tag{2}$$

Recently, a Skyrme force giving a nearly equivalent energy density has also been fitted. Its parameters are given in table 1 (sect. 5). In the following discussion, the results obtained with eq. (2) and with the equivalent force Tl will not be distinguished.

The main differences between this model and the commonly used Skyrme forces concern the incompressibility of nuclear matter (235 MeV with the present model) and the effective mass m^* which is set equal to the nucleon mass m. The choice of m^*/m leads to a better fit to the nuclear binding energies than $m^*/m = 0.7$ to 0.8 which is used in many Skyrme forces. Moreover, it allows to compare directly the HF mean field with the shell model potentials of common use.

Pairing correlations are introduced in the model with the BCS method, using a delta pairing force (14) of constant strength, which is expected to give more reliable extrapolations than the usual constant G approximation.

This model has been initially conceived as a HF mass formula, and its parameters have been mainly fitted on nuclear masses. However, the comparison of experimental and calculated radii (13) shows good results, better on the average than with the usual Skyrme forces.

The comparison of HF+BCS results with macroscopic models can be obscured by the shell effects which can be present in HF results. In order to allow a more conclusive comparison, we have also determined the average trends of HF results by introducing Strutinsky-type occupation numbers for the individual states of the nucleons at each step of the HF iterative calculation. We shall refer to this method as to the smooth self-consistent (SSC) model.

5. SELF-CONSISTENT RESULTS FOR SPHERICAL NUCLEI

As in previous works (15, 16), we shall first approximate the self-consistent density distributions by least square fitted Fermi distributions, in order to define the average core density which is set equal to the central density of the best Fermi distribution. The definition of ρ_{cn} and ρ_{cp} then enables us to determine $R_{1/2}$, t and the surface asymmetry for the proton and neutron distributions. Fig. 2 a, b, c, d, e show a simultaneous plot of HF+BCS results (dots) and SSC results (full lines) for the RMS radius, the neutron skin thickness , $R_{1/2}$, t and the surface asymmetry of nuclei near the stability line (more precisely along a line starting as N = Z up to Ca, going then straight to ^{140}Nd and finally parallel to N = 2Z through ^{200}Hg.

The shell effects observed in nuclear radii are quite weak, especially in RMS radii. Obviously, they do not explain the full fine structure found in experimental data. Moreover, in light nuclei, they contribute to it with the wrong sign :magic numbers correspond to a positive shell effect in the radius in ^{16}O, ^{40}Ca and ^{90}Zr, whereas a negative shell effect is found in ^{208}Pb and at N = 184. The shell effects in the surface diffuseness and in the radii are usually of opposite signs (and thus partly cancel each other in the RMS radii). In the surface diffuseness, the shell effects are more significant. In light and medium nuclei, they lead to a minimum of t at the shell closures. In heavy nuclei, this minimum seems to be related to the filling of highly degenerate subshells rather than to the shell closures.

The fact that the fine structure of nuclear radii is not due to shell effects but to collective effects is probably the principal limitation of HF calculations of nuclear densities. In view of the interest of a detailed knowledge of the density distributions, it would be very useful to establish how those collective effects modify the surface of the nucleus and if the significant shell effects found in the HF surface thickness actually subsist in nuclei.

The surface asymmetry and the neutron skin thickness obtained for ^{208}Pb in HF+BCS results are in rather good agreement with experiment. The model indeed predicts a negative surface asymmetry for the proton and charge distributions. A positive asymmetry is predicted for the neutrons. The neutron skin thiclness on the other hand is at the upper limit of the error bar given in section 2. We notice however that a better agreement can be obtained with the experimental value by re-fitting the parameters of the force. Table 1 gives a force (T6) which leads to a neutron skin thickness of 0.15 fm, with the same quality of fit to masses and radii as T1.

TABLE 1

Values of the parameters of two Skyrme forces used in the present work

(g is the power the the density dependence)

force	t_o	t_1	t_2	t_3	x_o	x_1	x_2	x_3	W_o	g
T1	−1794	298	−298	12812	.154	−.5	−.5	.089	110	1/3
T6	−1794.2	294	−294	12817	.392	−.5	−.5	.5	107	1/3

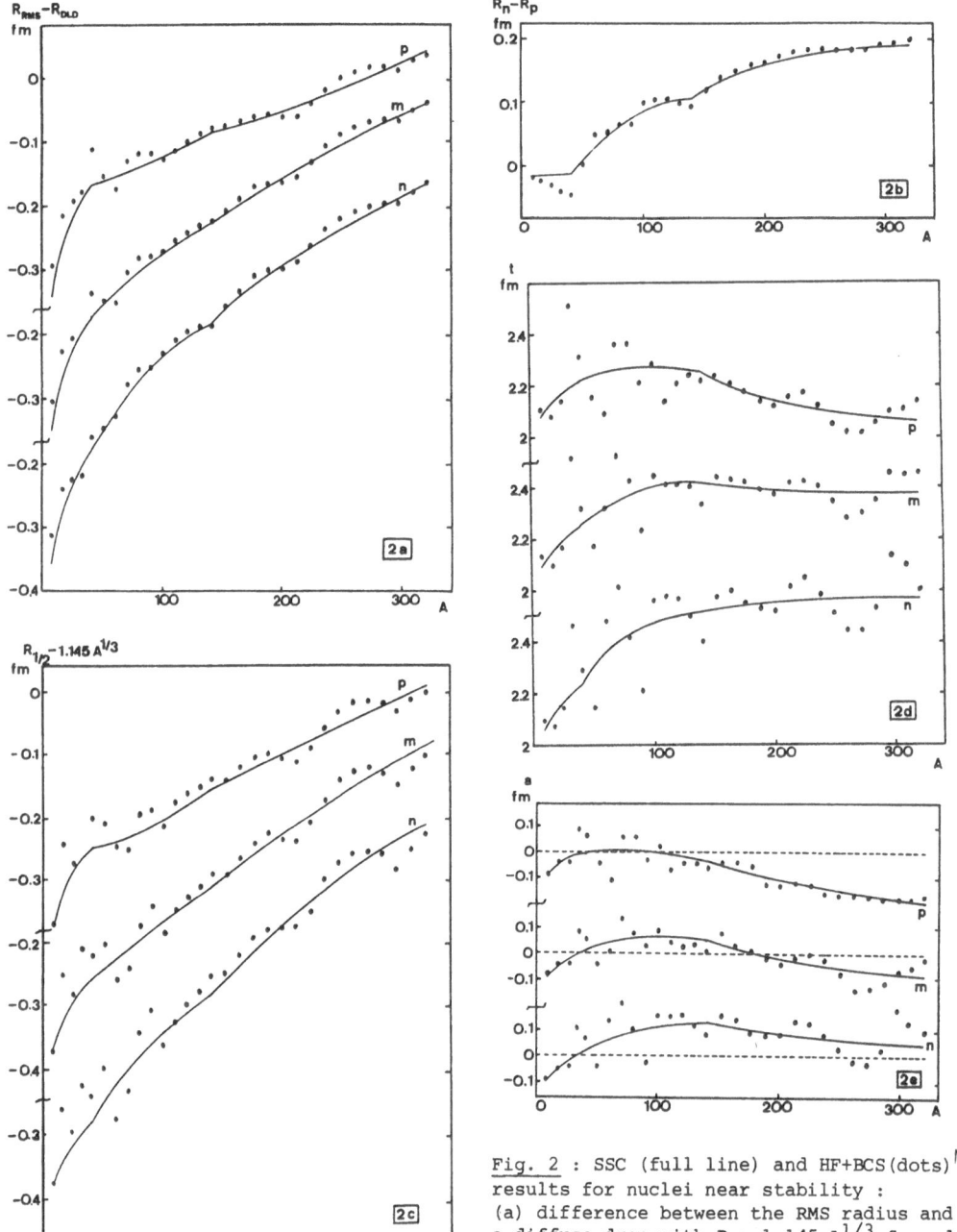

Fig. 2 : SSC (full line) and HF+BCS(dots)
results for nuclei near stability :
(a) difference between the RMS radius and
a diffuse drop with R = 1.145 A$^{1/3}$ fm and
t = 2.42 fm for neutrons, mass and proton
distributions ; (b) neutron skin thickness
(c) difference between the half-density radius and 1.145 A$^{1/3}$fm ; (d) surface thickness ; (e) surface asymmetry.

The average trends of the extrapolation of nuclear densities far from stabilily along isobars have also been examined with the SSC model (17). Defining I = (N-Z)/A, we can summarise these results by two main conclusions :

a) The average mass density in the core at a given A does not vary very much with I (but it varies with A !), ρ_{cn} and ρ_{cp} being nearly linear functions of I with opposite slopes ($\pm$ 0.05) which do not vary significantly with A. Moreover, the value of I for which $\rho_{cn} = \rho_{cp}$ is nearly constant (-0.04).

b) The constant value of ρ_c along isobars does not mean that $R_{1/2}$ and the RMS radius of the mass distribution are constant, because a significant increase of the surface thickness occurs at large I. Approximate formulae for the surface thickness have been given in (17).

6.   COMPARISON WITH OTHER MODELS

Several anomalies have been found (18) when comparing the SSC results with the DM. To give more consistency to this comparison, the DM parameters corresponding to the force we use have been determined self-consistently in HF calculations for semi-infinite nuclear matter. It appears that the DM overestimates the compression of nuclear matter by the surface tension and also overestimates the value of the neutron skin thickness. Both anomalies are of the same order of magnitude ($\sim$0.1 fm for the radii) and can lead to very significant errors. In the DM fits, the compression anomaly can be largely cancelled by an appropriate lowering of the density of nuclear matter, but the neutron skin anomaly cannot be removed satisfactorily. There are thus serious doubts concerning the accuracy and even the validity of the DM. We believe that the use of this model should be avoided.

It follows from the weakness of the shell effects in HF radii that the Thomas-Fermi and related models could in principle give nearly as good radii as HF calculations, except perhaps for a few light or medium nuclei. This is indeed the case, at least for the extended Thomas-Fermi model, as shown by the comparison of ETF and HF radii given by Bartel et al. (19). A systematic study of the surface diffuseness by the same group is under way; the preliminary results (20) confirm qualitatively the average trends obtained with the SSC model. However, like earlier TF models, the ETF model predicts a positive surface asymmetry for the protons in ^{208}Pb (21), when this degree of freedom is included in the model. For this point, the ETF model seems thus less satisfactory than HF. Of course, TF models are also unable to reproduce shell effects in the surface diffuseness.

We finally come to the comparison of HF results with the shell model densities. First, the analysis of the HF mean field gives us very useful indications concerning the nucleon-nucleus potential. Following the method given in (15), we define an average potential depth in the core, a half-value radius $R_{1/2}$, a 90% - 10% surface thickness and a surface asymmetry for the potential. The main results can be summarised as follows, the proton potential to which we refer including no Coulomb contribution :

a) The surface asymmetry of the potential well is strongly positive near the stability line, and has roughly the same value for protons as for neutrons (0.8 fm). It decreases far from stability.

b) The surface thickness of the potential includes significant shell effects (14).

c) On the average, the surface thickness of the neutron potential is approximately constant (3 fm), and is lower than the surface thickness of the proton potential, which increases significantly with $I = (N-Z)/A$, reaching 3.6 fm near the neutron drip line.

d) To the neutron skin in the density corresponds a "proton potential skin" in the potentials.

In order to see to what extent those properties can play a role in the calculation of shell model densities, we shall compare the HF results for the neutron surface diffuseness with the results obtained with a standard Woods-Saxon potential (Pot. 1 of ref. 22), i.e. a potential which has a symmetric surface, a constant surface thickness (the same for protons as for neutrons) and equal radii for the neutron and proton potentials. Pairing effects have been included in both models. The choice of the surface diffuseness of the neutron density for that comparison is due to the fact that this quantity shows the most important shell effects in HF results. The comparison is made in fig. 3 for nuclei near the stability line. It shows that

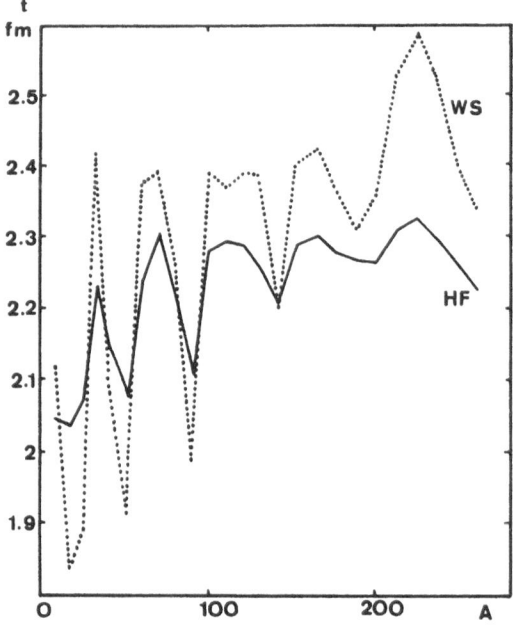

Fig. 3 : Surface thickness of the neutron density distribution near stability with the Hartree-Fock + BCS model (HF) and the Woods-Saxon + BCS model (WS).

the shell effects are amplified by a factor 2 to 3 in the Woods-Saxon model, compared with the HF model. The same conclusion arises from the analysis of radii : the
self-consistent treatment weakens the shell effects in the densities. As a consequence, shell model calculations of the densities should take into account the shell
effects in the potential , at least when a detailed description of the surface is
needed. This limits seriously the usefulness of such calculations, because these
shell effects can hardly be predicted without making a full self-consistent calculation .

Two other interesting quatities are the surface asymmetry of the charge density
in ^{208}Pb which is predicted to be positive with the Woods-Saxon model, and the neutron skin thickness of the same nucleus which is equal to 0.28 fm; both results are
not satisfactory, and show the importance of a correct description of the potential
well when a good description of the nuclear surface is wanted.

7. POSSIBLE IMPROVEMENTS OF THE HARTREE-FOCK MODEL FOR NUCLEAR DENSITIES

As shown in the discussion, the weakest point of HF models for the densities
is that they do not explain the fine structure of the RMS radii, whereas other
details of the surface like the surface asymmetry and the neutron skin thickness
can be rather well reproduced. There are strong indications that the fine structure
can be explained by zero-point vibrations and by static deformations. Both effects
can be estimated from experimental specrroscopic information, or from deformed HF
calculations, which are unfortunately very costly. Those collective effects can
also influence other characteristic properties of the densities, and it would be
very useful to know how the surface thickness, surface asymmetry and neutron skin
thickness can be modified. An estimate of the collective contributions to those
quantities would be of great interest before including systematically the experimental data about densities in the fits of the effective interactions to be used in
HF models.

8. CONCLUSIONS

The main conclusions of this discussion are the following :

a) There are major doubts concerning the accuracy and the validity of the droplet model; we believe that the use of this model should be avoided.

b) The Thomas-Fermi approach gives quite good results for the average trends
   of the densities, but it can be inaccurate in the description of the nuclear
   surface, and particularly of the surface asymmetry.

c) A great care must be taken when using the shell model to compute nuclear
   densities, particularly for the choice of the surface thickness of the potential, of its surface asymmetry and of its "proton skin" ; shell effects
   should be in principle included in the potential; unfortunately, those

effects are difficult to evaluate without making a full self-consistent calculation.

d) The Hartree-Fock model predicts large shell effects in the surface diffuseness, but the shell effects in radii are not sufficient to explain the fine
structure of RMS charge radii. On the other hand, it seems to give a good
description of the surface asymmetry and of the neutron skin thickness.

e) Collective effects (deformation, zero-point motion) must be added to the HF
model (and of course to the shell model too) if a detailed description of
the nuclear density distributions is required; a systematic study of the contribution of those effects to the surface properties is still missing.

Several points of the work reported here have been developed in collaboration with J.M. PEARSON (Université de Montréal), M. BRACK and H.B. HÅKANSSON (Universität Regensburg) to whom the author is indebted for valuable discussions and
information .

References

1) I. Angeli and M. Csatlos, Nucl. Phys. A288 (1977) 480
2) S.K. Gupta, private communication.
3) C. Thibault et al. , Phys. Rev. C23 (1981) 6
4) I. Angeli, M. Beiner, R. Lombard and D. Mas, J. Phys. G6 (1980) 303.
5) C. Thibault, Proc. 4th Int. Conf. on Nuclei far from Stability, Helsingør 1981,
   p. 47, CERN 81-09.
6) X. Campi, M. Epherre and G. Audi, see ref. 5, p. 62.
7) C. De Jager, H. De Vries and C. De Vries, Atomic Data and Nucl. Data Tables
   5-6 (1974) 479.
8) G. Hoffmann et al. , Phys. Rev. C21 (1980) 1488.
9) W. Myers and W. Swiatecki, Ann. Phys. 55 (1969) 395
10) V. Bernard and Nguyen van Giai, Nucl. Phys. A348 (1980) 75.
11) B. Brown, S. Massen and P. Hodgson, J. Phys. G5 (1979) 1655.
12) D. Vautherin and D. Brink, Phys. Rev. C5 (1972) 626.
13) F. Tondeur, Nucl. Phys. A303 (1978) 185.
14) F. Tondeur, Nucl. Phys. A315 (1979) 353.
15) F. Tondeur, J. Phys. G5 (1979) 1189.
16) F. Tondeur, J. Phys. G6 (1980) L 71.
17) F. Tondeur, Nucl. Phys. A383 (1982) 32.
18) F. Tondeur, J.M. Pearson and M. Farine, submitted for publication in Nucl.Phys.
    A (1982)
19) J. Bartel, P. Quentin, M. Brack, C. Guet and H.B. Hakansson, to be published
    in Nucl. Phys. A (1982).
20) M. Brack, C. Guet, H.B. Håkansson, A. Magner and V. Strutinski, see ref. 5,
    p. 65.
21) M. Brack, private communication.
22) F. Tondeur, Journal de Physique 34 (1973) 761.

PROBING THE NUCLEAR STRUCTURE WITH HEAVY-ION REACTIONS

Ricardo A. BROGLIA
Niels Bohr Institute
University of Copenhagen
Copenhagen, Denmark

and

*Oak Ridge National Laboratory**
Oak Ridge, TN 37830, U.S.A.

TABLE OF CONTENTS

* Research sponsored by the Division of Basic Energy Sciences, U.S. Department of
Energy, under contract W-7405-eng-26 with the Union Carbide Corporation.

Abstract

Nuclei display distortions in both ordinary space and in gauge space. It is suggested that it is possible to learn about the spatial distribution of the Nilsson orbitals and about the change of the pairing gap with the rotational frequency through the analysis of one- and two-nucleon transfer reactions induced in heavy-ion collisions.

1. Surface and Pairing Collective Modes

The I(I+1)-law displayed by the low-lying levels of the spectra of many nuclei has long been recognized to be the fingerprint of very coherent nuclear motion.[1] The particles align in the average field, filling first the magnetic substates with $|m| \sim 1/2$ or with $|m| \sim j$ (cf. in Fig. 1). These configurations define privileged directions z' in space. Rotational invariance is regained by allowing the system as a whole to rotate along an axis perpendicular to z'. A typical example of the resulting spectrum is shown in Fig. 2.

Rotational motion in ordinary space is an important example of coherent motion in finite many-body systems. Other degrees of freedom which are not amenable to a classical description can be understood by analogy to spatial rotations. The central tools to carry out this analogy are provided by the concept of average field and of elementary modes of excitation.

In Table 1 we use them to introduce the basic properties of pairing rotations. The violation of angular momentum and of particle number conservation (spontaneous symmetry breaking) is the mechanism responsible for the existence of quadrupole and pairing collective modes.

In the BCS state[2] the distinction between particles and holes is blurred, and a privileged orientation ϕ_0 in gauge space is defined. Because there is no restoring force in the ϕ variable, the system as a whole rotates restoring particle number conservation, as required by the original Hamiltonian. The members of these pairing rotational bands are the ground states of a series of superfluid isotopes (isotones).

It is not so much the regularity of the energy pattern which qualifies the collectivity of a mode, but the specificity with which the mode is excited by some external field and the strongly enhanced values of the associated cross sections (cf. Fig. 3). In particular, Coulomb excitation[3] and two-particle transfer reactions[4] are the specific probes for studying spatial and pairing rotations, respectively. In these processes, the nucleus increases its rotation frequency

$$\omega_{rot} = \dot{\omega} = \frac{1}{\hbar} \frac{\partial H_{rot}}{\partial I} = \frac{\hbar}{\mathcal{J}} I, \quad \phi_{rot} = \dot{\phi} = \frac{1}{\hbar} \frac{\partial H_{pair}}{\partial N} = \frac{\lambda}{\hbar}$$

as a whole, without changing its intrinsic structure.

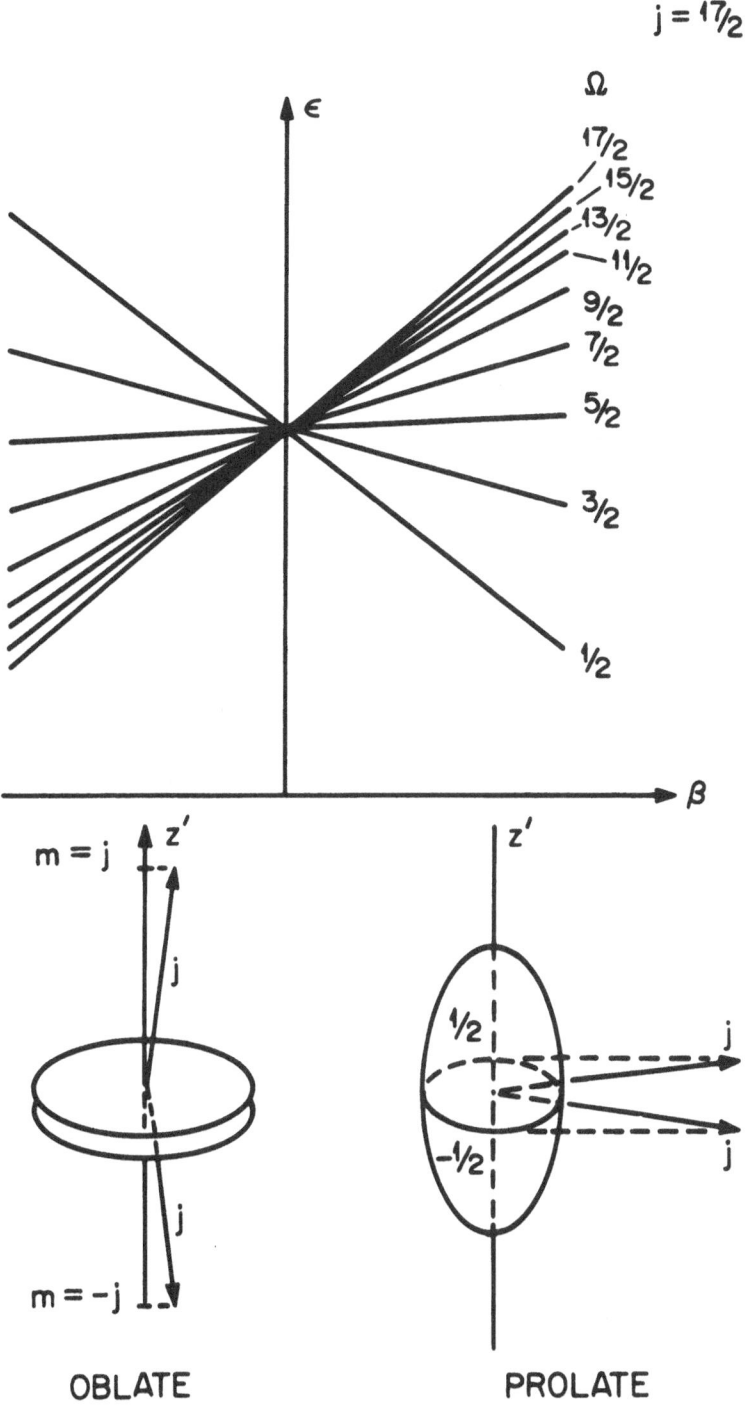

Fig. 1   TWO PARTICLE SYSTEM

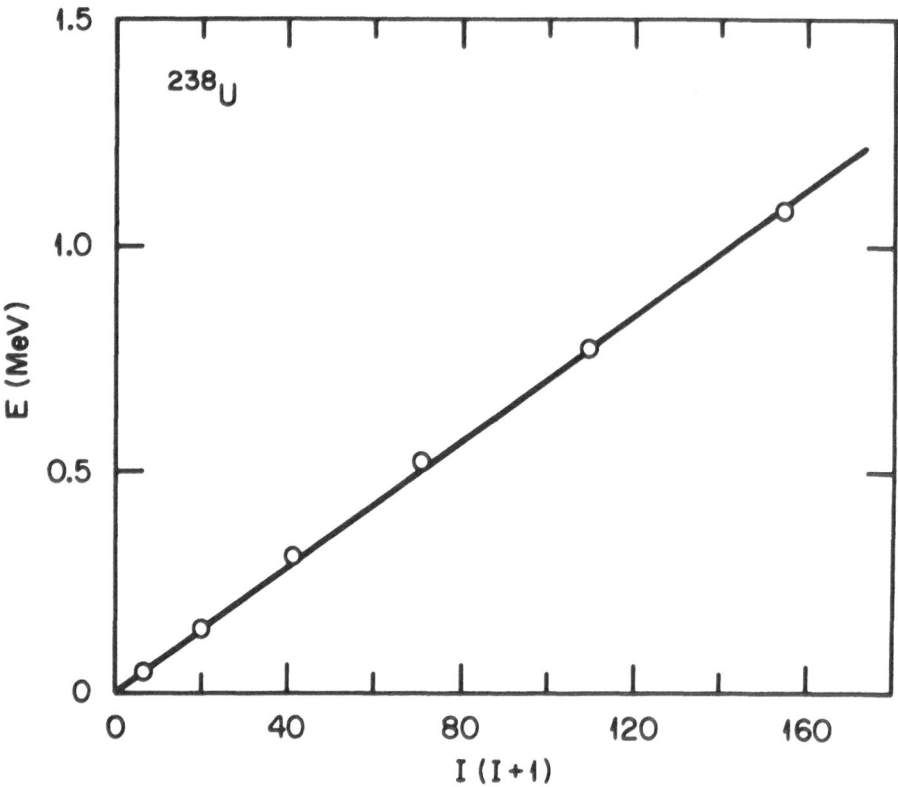

Fig. 2

The picture of a nucleus which displays distortion in gauge space is closely related to the model one has of a superconductor in the theory of condensed matter. In this case, it is not possible to observe the individual states of the rotational bands. This is because, due to the macroscopic size of the system, one can hardly study the transfer of single Cooper pairs at a time. However, the Josephson junction[5] displays in a clear way the collective rotational degrees of freedom in gauge space. This junction is built out of two superconductors separated by a thin insulator layer through which pairs of electrons can tunnel. Each of the superconductors can be viewed as a rotor where the number of particles plays the role of the angular momentum. The rotors are coupled together through the exchange of pairs of electrons which corresponds to the exchange of angular momentum. The pairs are created and annihilated by the two-particle transfer operators and the coupling Hamiltonian can be written as

$$H \sim e^{2i\phi_L} e^{-2i\phi_R} e^{2i\delta} + \text{h.c.}$$
$$\sim \cos 2(\phi_L - \phi_R + \delta),$$

(1)

where L and R label the superconductor to the left and to the right of the dioxide layer. The quantities ϕ_L and ϕ_R are the gauge angles of the two superconductors and δ a phase shift acquired by the electrons when tunneling through the barrier. The rate at which the Cooper pairs are exchanged is given by

$$\dot{N}_L(= \dot{N}_R) = \frac{i}{\hbar} [H, N_L] = -\frac{1}{\hbar} \frac{\partial H}{\partial \phi_L} \sim \sin 2 (\phi_L - \phi_R + \delta).$$

(2)

The rotational frequency of the rotors in normal space corresponds to the chemical potential of the superconductors

$$\dot{\phi} = \frac{1}{\hbar} \frac{\partial H}{\partial N} = \frac{\lambda}{\hbar}$$

(3)

Introducing $\phi = \dot{\phi}t = \frac{\lambda}{\hbar} t$ in the expression for N_L, we get

$$\dot{N}_L \sim \sin \left\{ \frac{2}{\hbar} (\lambda_1 - \lambda_2)t + 2\delta \right\}.$$

(4)

Thus, if there is a difference in the Fermi energy between the two superconductors which can be forced by applying an external voltage, there will be an a.c. current running between the superconductors. In terms of the voltage difference $V_1 - V_2$ one can write

$$\dot{N}_L \sim \sin 2 \frac{e}{\hbar} (V_1 - V_2)t + 2\delta .$$

(5)

The frequency of the a.c. current is proportional to the externally applied voltage, and the charge carried by the Cooper pair is 2e. The phenomenon described by

Table 1

Nuclei display Spontaneous
Symmetry Breaking In

Ordinary Space (Classical analogue exists)	Gauge Space (Quantal phenomenon)

The ground state wavefunction

$\|0\rangle = \sum_I d_I \|I\rangle$	$\|0\rangle = \sum_N d_N \|N\rangle$

describes a system which
has not a definite

Angular Momentum I	Number of Particles N

and the nucleus
as a whole rotates
(Goldstone boson)

in ordinary space	in gauge space

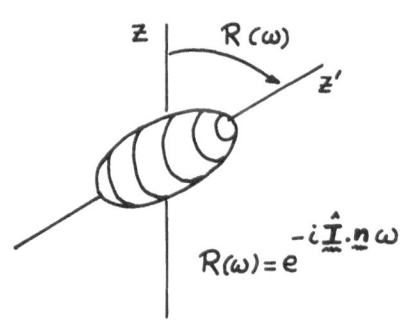

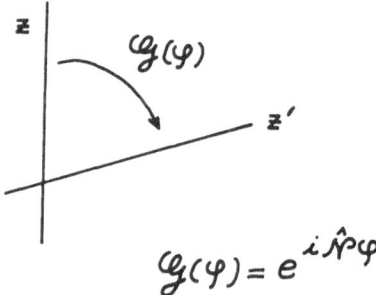

$\hat{I}$: angular mom.	$\hat{N}$: number of part.

restoring symmetry

The associated Regge
trajectory

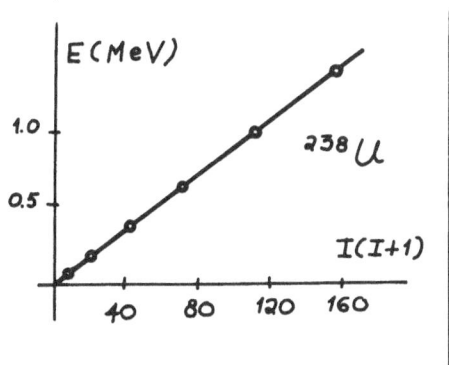

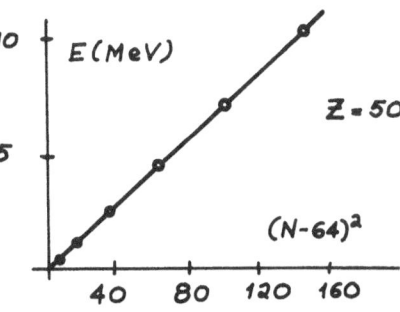

contains the levels
of a

quadrupole | pairing

rotational band

the members of this band
are specifically excited in

inelastic processes and
Coulomb excitation

$$\langle I+2|\hat{Q}|I\rangle^2 \sim Q_0$$

$$(\sim 200 \ B_{sp})$$

in two-nucleon
transfer processes

$$\langle N+2|\hat{T}|N\rangle^2 \sim \left(\frac{\Delta}{G}\right)^2$$

$$(\sim 10^2 \ \sigma_{2p})$$

The distortion of the
system is measured by

the quadrupole moment | the pairing gap

 Q_0

Δ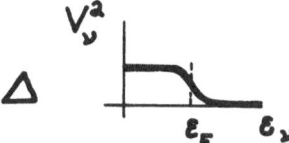

The Mayer-Jensen particles
which carry a definite

angular momentum	transfer quantum number

$$R a_{jm}^{\dagger} R^{-1} = \sum_{m'm} \mathcal{D}_{m'm}^{j}(\omega) a_{jm}^{\dagger}$$

$$R = e^{-i\,\vec{I}\cdot\hat{n}\omega}$$

$$\mathcal{G} a_{jm}^{\dagger} \mathcal{G}^{-1} = e^{-i\phi} a_{jm}^{\dagger}$$

$$\mathcal{G} = e^{-i\hat{N}\phi}$$

become

Nilsson particles	Bogoliubov-Valatin particle

$$a_{i\Omega}^{\dagger} = \sum_{j} W_{i\Omega}^{j} a_{j\Omega}^{\dagger}$$

$$\alpha_{jm}^{\dagger} = U_j \, a_{jm}^{\dagger} - V_j \, a_{\widetilde{jm}}$$

and the intrinsic state
$|0\rangle$ can be written as

Nilsson state	BCS state

$$|0\rangle = \prod_{i\Omega} a_{i\Omega}^{\dagger} |\rangle$$

$$|0\rangle = \prod_{jm} \alpha_{jm} \sim \prod_{jm} (U_j + V_j \, a_{jm}^{\dagger} a_{\widetilde{jm}}^{\dagger})$$

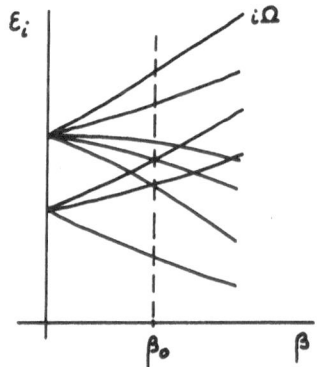

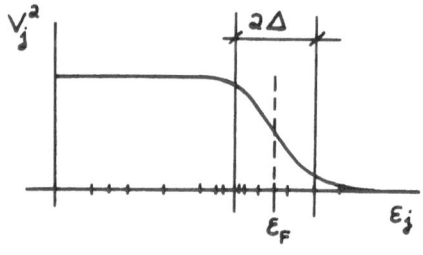

Eq. (5) is known as the a.c.-Josephson effect, and its experimental confirmation[6] has provided the most accurate values of $e/\hbar$ to date.

1a. Microscopic Structure of the Deformed State

The microscopic structure of the intrinsic state of a deformed nucleus[7] is determined by the interplay between the shell structure and the spontaneous symmetry-breaking phenomenon. The occupancy of the different orbitals gives a quantitative measure of this interplay, and can be determined in one-nucleon transfer reactions (cf. Fig. 4).

It is noted that distortions in space and in gauge space coexist. In fact, the quadrupole and the pairing field producing effects compete with each other. Pairing correlations are most prolific in a spherical situation because the degeneracy of the different orbitals is large. Even in a shape-deformed nucleus, the moment of inertia is considerably smaller than $\mathcal{J}_{rig}$ which is obtained by aligning the particles in the deformed field. Making use of a very simplified picture, one can say that when the nucleus rotates as a whole, only the mass at the tips will be dragged by the average field as the nucleons are in a superfluid state.

In a single-particle transfer process, one can obtain the spectroscopic factor S_j

$$C_j^2 \, S_{\ell j}^2 = \langle A+1| \, a_{\ell jm}^\dagger \, | A \rangle^2 \, ; \; C_j^2 = \begin{cases} 1 \text{ (stripping} \\ (2j+1) \text{ (pick-up)}, \end{cases} \tag{6}$$

The operator $a_{\ell jm}^\dagger$ creates a particle in the orbital ℓjm. The initial state $|A\rangle$ is assumed to be the ground state of an even-deformed nucleus. To calculate (6), one has to rotate $a_{\ell jm}$ into the intrinsic system in both gauge and normal space, i.e.,

$$
\begin{aligned}
a_{\ell jm}^\dagger &= \sum_\Omega D_{\Omega m}^j(\omega) \; e^{-i\phi} \; a_{\ell j\Omega}^{\dagger} \\
&= \sum_\Omega D_{\Omega m}^j(\omega) \; e^{-i\phi} \sum_i W_{\ell j}^{i\Omega} \; a_{i\Omega}^{\dagger} \\
&= \sum_{i\Omega} D_{\Omega m}^j(\omega) \; e^{-i\phi} \; W_{\ell j}^{i\Omega} \; (U_i \alpha_{i\Omega}^{\dagger} + V_i \alpha_{\widetilde{i\Omega}}).
\end{aligned}
\tag{7}
$$

where $W_{\ell j}^{i\Omega}$ is the amplitude of the spherical state ℓj in the Nilsson state $i\Omega$. The spectroscopic amplitude can now be calculated,

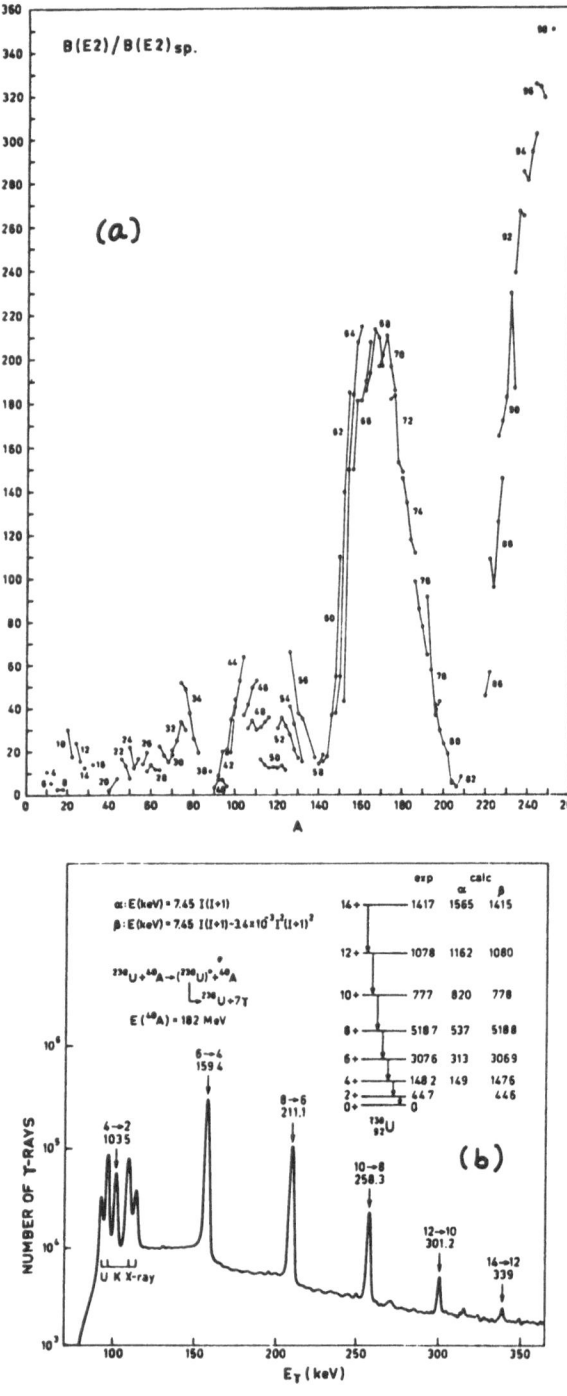

Fig. 3. (a),(b) (from ref. 1).

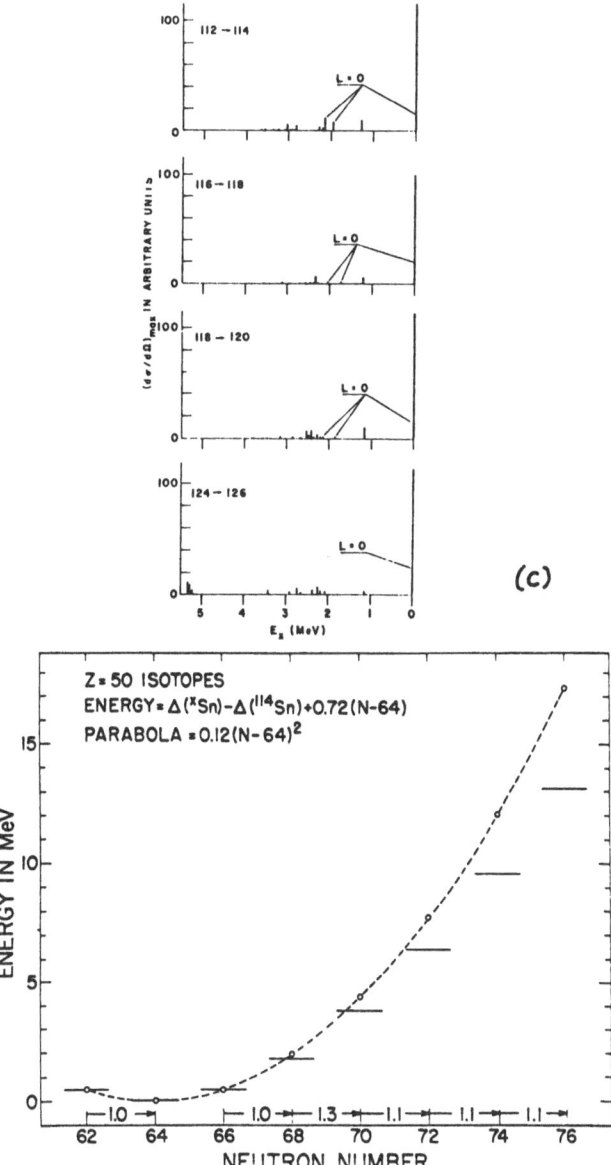

Fig. 3. (c) The Sn pairing rotational band (from ref. 4).

$$(S_{\ell j})^{1/2} \sim A' \langle I_f M_f K_f | a^{\dagger}_{\ell j m} | I_i M_i K_i \rangle A$$

$$= \int d\phi \, e^{i(A'-(A+1))\phi} \int d\omega \, D^{I_f \, *}_{M_f K_f}(\omega) D^j_{m\Omega}(\omega) D^{I_i}_{M_i K_i}(\omega)$$

$$\times \, W^{j\Omega}_{\ell j} \, U^2_j \tag{8}$$

$$\sim \delta(A'-(A+1)) \, \langle I_i M_i jm | I_f M_f \rangle \langle I_i K_i j\Omega | I_f K_f \rangle$$

$$\times \, W^{j\Omega}_{\ell j} \, U_i \, .$$

The quantities

$$S_{\ell j} \sim \left(W^{j\Omega}_{\ell j} \, U_i \right)^2 , \tag{9}$$

can be obtained from the one-particle transfer differential cross sections associated with the different members of the rotational band based on the Nilsson orbital $j\Omega$. They give the fingerprint of the Nilsson orbital (cf. Fig. 4(c)). Associated with the fact that all the angular momenta transferred in the reaction is carried by the particle, an integration over the Euler angles ω is performed in (8). Thus, the spectroscopic factor (9) does not provide direct information on the spatial distribution of the Nilsson orbitals (cf. Fig. 4(e)).

Because shape deformations are intimately related with a non-uniform distribution of mass in space, the topology of the Nilsson orbitals is an important prediction of the model. Possibilities to learn about the three-dimensional distributions of single-particle orbitals have been opened by the advent of heavy-ion machines capable of bringing two heavy nuclei into contact, as will be discussed below.

2. Mapping the Nuclear Surface

The most conspicuous features of the heavy-ion reactions which can be studied with the new generation of heavy-ion machines are collected in Table 2.

Let us make use of them to discuss what can be learned about the following "gedanken eksperiment": the collision at zero impact parameter between a deformed target and a spherical projectile at a bombarding energy close to that of the Coulomb barrier[8] (cf. Fig. 5(a)).

We assume that the interaction between the two ions is the sum of a monopole-monopole and monopole-quadrupole Coulomb term. It can be shown (cf., e.g., Ref. 9, Section II.7) that the angular momentum transferred to the rotor in a sudden collision is (for details cf. Appendix)

$$\Delta L \sim 2\hbar q \, \sin 2\theta_0 , \tag{10}$$

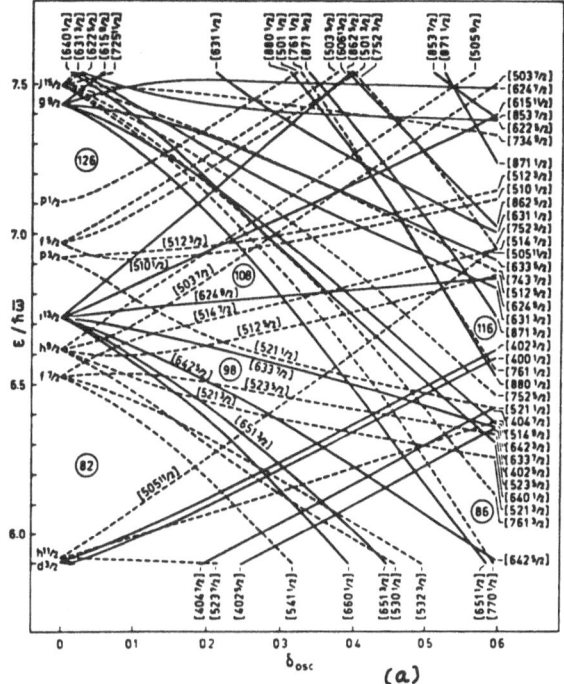

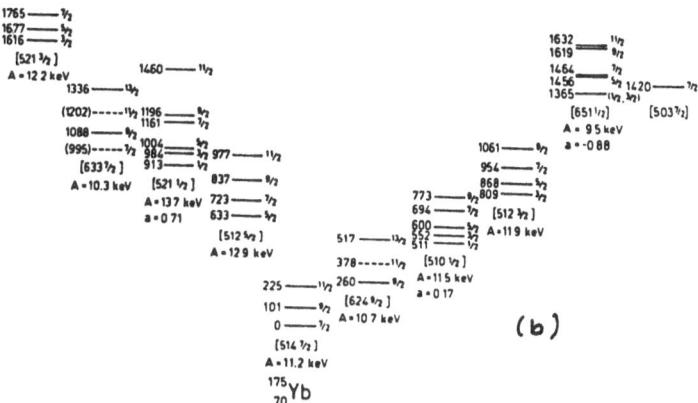

Fig. 4. (a) Neutron orbits in prolate potential (82 < N < 126) (from ref. 1).
(b) Spectrum of ^{175}Yb (from ref. 1).

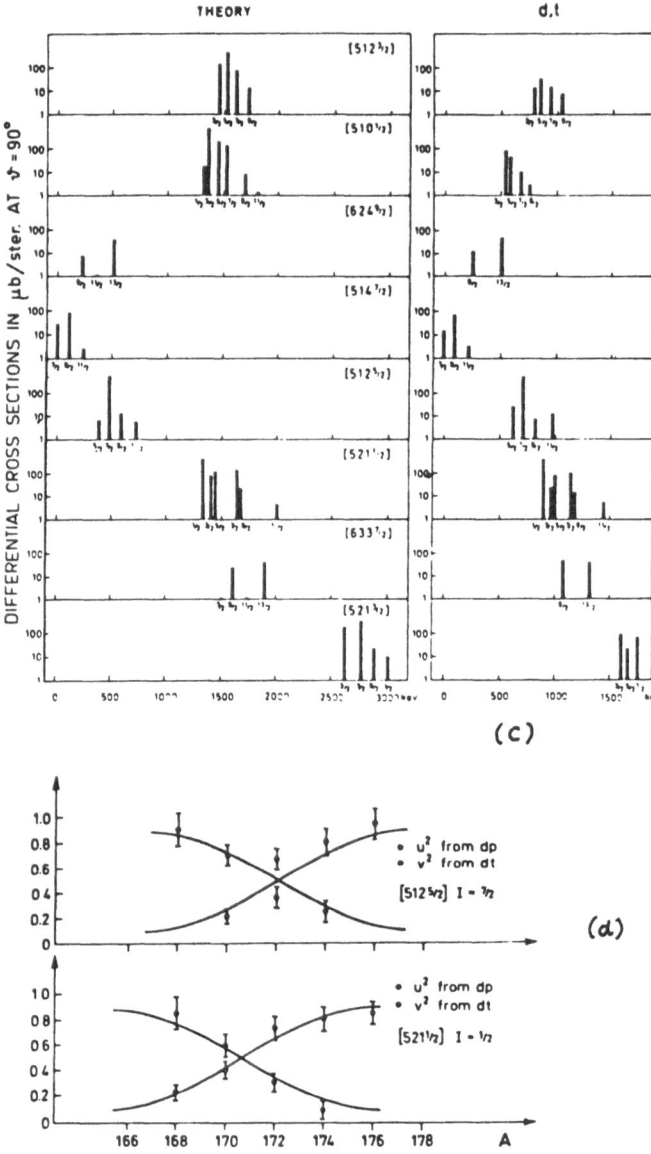

Fig. 4. (c) Fingerprints from intensities of (dp) reactions leading to $^{175}_{70}$Yb (from ref. 1). (d) Effect of pair correlations on transfer intensities (from ref. 1).

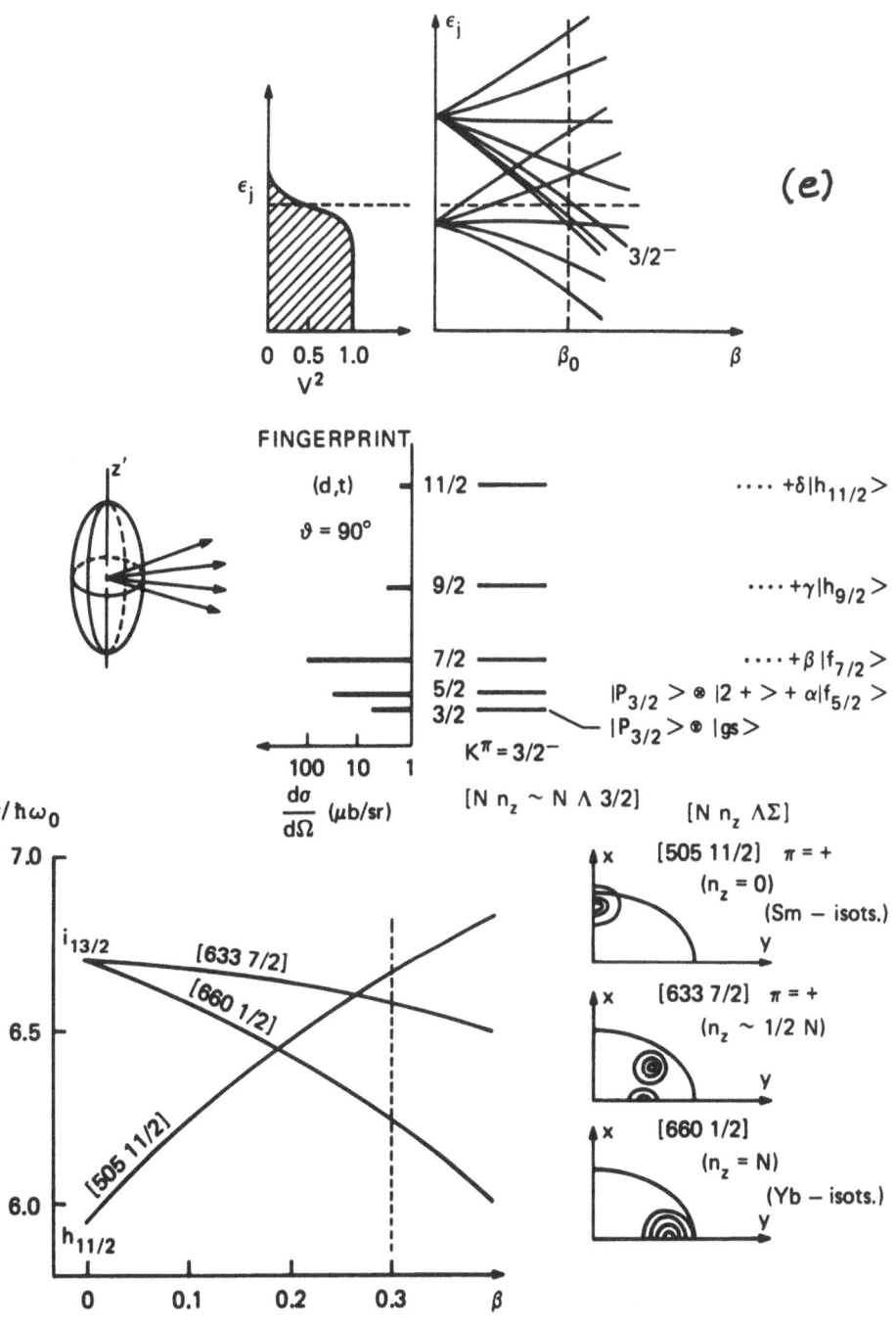

Fig. 4(e)

Table 2

a) Very short wavelength of relative motion

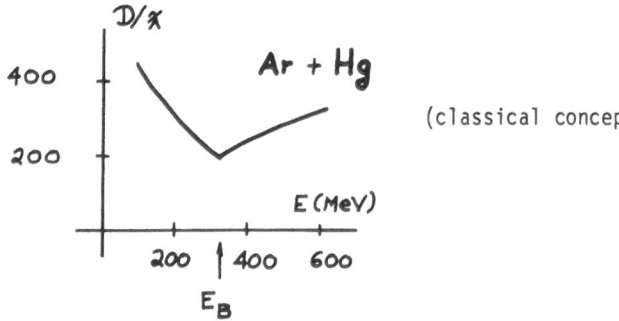

(classical concepts)

b) Large values of angular momenta in
relative motion

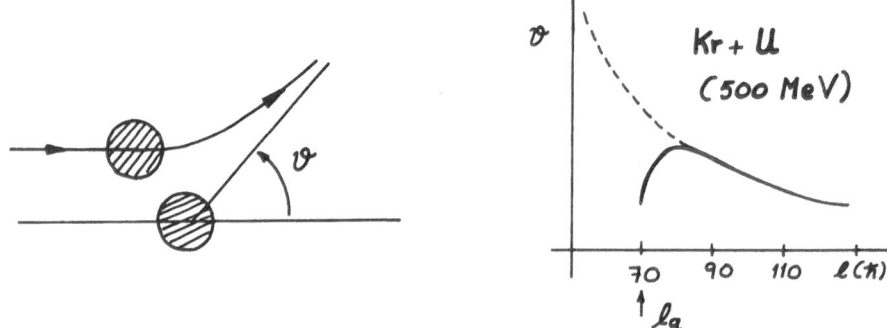

c) Strong localization of the ion-ion interaction

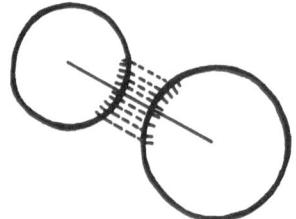

$$U_{aA}^N(s) = \iint dxdy\ e(s')$$

$e(s)$: int. en. per unit
area

s: surface-surface
distance

d) Strong time-dependent Coulomb and
nuclear fields

(reactions between nuclei in excited
states)

where Θ_o denotes the original orientation of the symmetry axis of the rotor with respect to the incoming beam, and where

$$q = \frac{Z_a \, e \, Q_o}{\hbar v \, b^2},$$

(10a)

is fixed by the charge eZ_a of the projectile, the static quadrupole moment Q_o of the target, the distance of closest approach b, and the relative velocity v at $t = -\infty$. Typical values of 2q for reactions involving very heavy ions are ~ 15-20.

As expected from classical arguments, the maximum angular momentum transfer is achieved for $\Theta_o = \pi/4$ or $\Theta_o = 3/4 \, \pi$. These are the orientations for which the difference between the Coulomb force acting on the two poles is maximum due to the $1/r^3$ coupling term, and consequently, when the torque excerted on the top is largest. For $\Theta_o = 0$ and $\Theta_o = 90°$ no angular momentum is given to the rotor. The relation between ΔL and Θ_o is schematically shown[*] in Fig. 5(b).

We now consider a transfer process of a nucleon from the spherical system a to the deformed system A. In general, the particle will carry only few units of angular momentum and most of the torque excerted on the rotor will be due to Coulomb interaction. Let us assume that in the transfer process the high-spin states of a given rotational band are preferentially excited. This result will indicate that the Nilsson orbital upon which this band is built is localized around $\Theta_o = 45°$ and 135° on the nuclear surface (cf. Fig. 5(c)). If in the transfer process only the low-spin states of a band are reached, one can expect that the associated Nilsson orbital is concentrated on the equator or on the poles.

Using the standard labeling for the Nilsson levels $[Nn_z\Lambda\Sigma]$, the orbitals with $n_z = N$ are located at the poles, have large intrinsic quadrupole moments, and are strong downsloping; i.e., their energy decreases as a function of the deformation. The orbitals with $n_z = 0$ have negative quadrupole moments and are spatially located around the equator. As a function of the deformation, the energy of these orbitals increases. Single-particle states with $n_z \sim \frac{1}{2} N$ have small quadrupole moments, their energy basically does not change as a function of the deformation, and the associated wavefunctions are localized midway between the equator and the poles. To obtain quantitative information on the spatial distribution of the Nilsson orbitals, the full three-dimensional scattering process has to be solved and the eventual spin-(scattering angle) correlations calculated. The main results of such a calculation are summarized in the next section.

---

[*]The existence of two different initial orientations leading to the same ΔL gives rise to quantal interference phenomena. Although these phenomena carry nuclear structure information which is interesting in its own right, we are going to deal only with average properties. It is noted that the interference effects are averaged out if a smoothing of $\Delta L \sim 2\hbar$ is introduced in the quantal results (cf. Appendix).

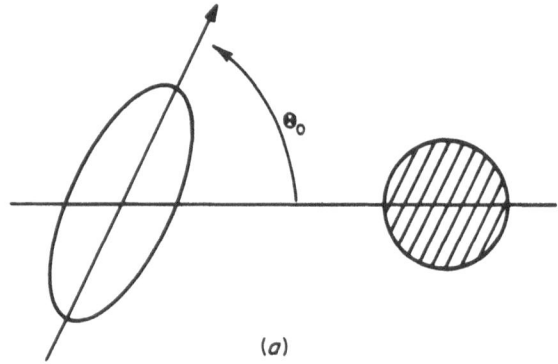

(a)

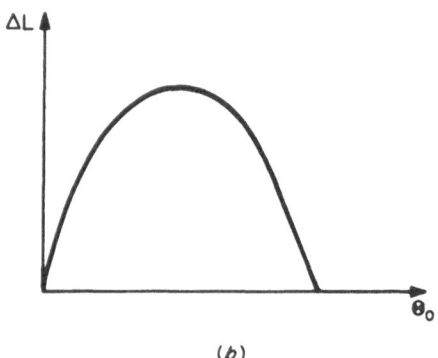

(b)

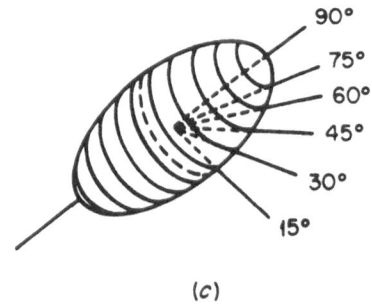

(c)

Fig. 5

2a. Scattering of a Deformed Nucleus in Three Dimensions

The Hamiltonian describing the process is[10]

$$H = T_{aA} + H_a + H_A + V_{aA} \tag{11}$$

where T_{aA} represents the relative kinetic energy, H_a and H_A are the intrinsic Hamiltonians for the projectile and target, and V_{aA} is the interaction between the two ions. For simplicity, we ignore the intrinsic degrees of freedom of the nucleus a. We consider the nucleus A to be an axially symmetric rotor with moment of inertia $\mathcal{J}$ with respect to a direction of rotation perpendicular to the symmetry axis. Thus,

$$H_A = \frac{1}{2\mathcal{J}} \left(P_\Theta^2 + \frac{P_\phi}{\sin^2\Theta} \right) . \tag{12}$$

The coordinate system used and the definition of the different variables is illustrated in Fig. 6. The ion-ion interaction is

$$V_{aA} = \frac{Z_a Z_A e^2}{r} + \frac{Z_a e Q_o}{2r^3} P_2(u) + U_{aA}^N(r,u). \tag{13}$$

The first two terms are the monopole-monopole and monopole-quadrupole contributions to the Coulomb interaction. The variable u denotes the cosine of the angle between the symmetry axis of the rotor and the coordinate of relative motion. The quantity Q_o is the intrinsic quadrupole moment of nucleus A.

The nuclear interaction U_{aA} is parametrized by a deformed Woods-Saxon function

$$U_{aA}^N(r,u) = \frac{V_o}{1 + \exp \dfrac{r - R_o - R_o(u)}{a}}, \tag{14}$$

with

$$R^A(u) = R_o^A \left(1 + \frac{5}{4\pi} \beta P_2(u) \right). \tag{15}$$

The nuclear deformation parameter β is related to Q_o by

$$\beta = \frac{4\pi}{3A_A e(R_o)^2} \frac{5}{4\pi} Q_o^A. \tag{16}$$

Making use of the Hamiltonian H and of the definitions (12)-(16), one can obtain the classical equations which govern the time dependence of the coordinates r, θ, ϕ, Θ, ϕ and their corresponding momenta. They form a set of coupled first-order differential equations which are integrated numerically. The initial conditions are specified by the center-of-mass energy E, the impact parameter ρ, which is fixed in the x-y plane, and a pair of angles Θ_o and ϕ_o defining the initial orientation of the rotor which we assume at rest at $t = -\infty$.

From the result of the numerical integration, one obtains the final scattering angle θ and the angular momentum I given to the rotor. A double-differential

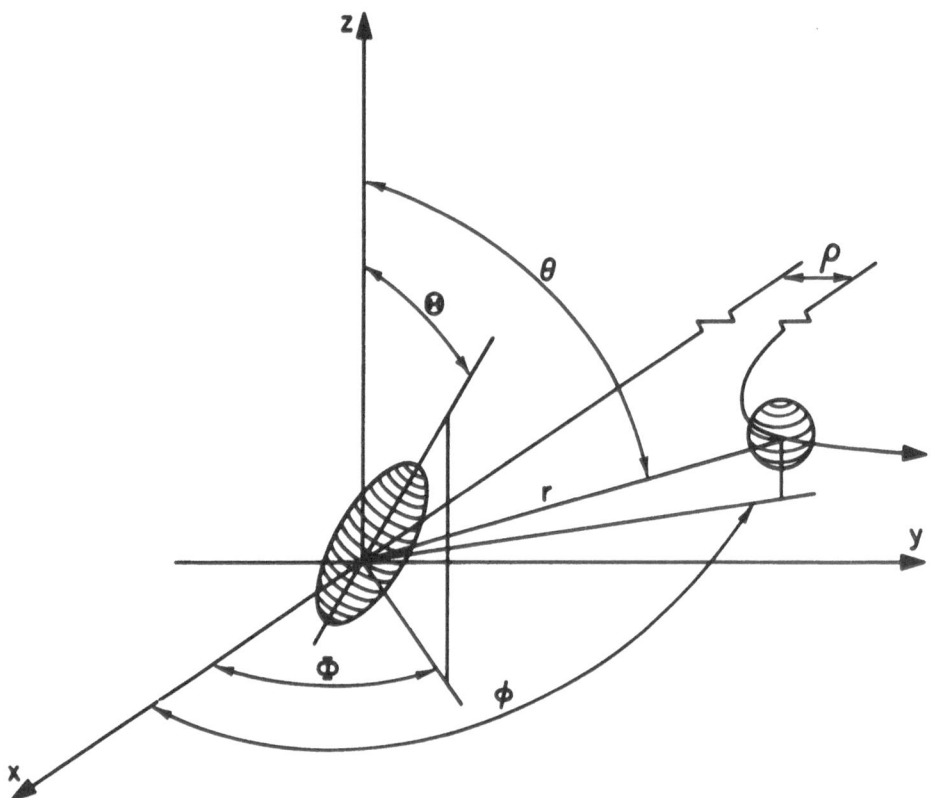

Fig. 6

inelastic cross section $d^2\sigma/d\theta dI$ can thus be constructed by sampling all impact parameters and initial orientations with the proper statistical weights

$$2\pi\rho\Delta\rho \; \frac{\sin\Theta_0 \; \Delta\Theta_0 \; \Delta\Phi_0}{4\pi}$$

associated with the steps $\Delta\rho$, $\Delta\Theta_0$, and $\Delta\Phi_0$ used in the calculations. To take into account the existence of many other channels not explicitly included in the calculations, we introduce an imaginary potential iW which follows the deformation of the system. For simplicity, we use

$$W(r,u) = \frac{W_0}{V_0} \; U^N_{aA}(r,u).$$

(17)

Typical values for (W_0/V_0) are ~ 0.2. The form chosen for W reflects the fact that in heavy-ion reactions the main contribution to the absorption comes from transfer processes.[11]

Because the nuclear potential is relatively weak compared to the Coulomb interaction for the very heavy systems considered here, one can treat the effects of W as a local mean-free path for the classical motion (cf., e.g., Ref. 9). Accordingly, we weigh each trajectory with a probability

$$P = e^{\frac{2}{\hbar} \int_{-\infty}^{\infty} W(r,u)dt},$$

(18)

where the integral is carried out along the trajectory. One thus obtains

$$\left(\frac{d\sigma}{d\theta dI}\right)_{inel+CE} = P \; \frac{d\sigma}{d\theta dI}$$

(19)

2b. Space Localization of the Single-Particle Orbitals

The double-differential cross section associated with transfer processes is built in a similar way as in the case of inelastic scattering, namely by integrating the equations of motion and accumulating the events at each scattering angle. In this case the number of events is multiplied by the transfer probability P_{transf}. Assuming one can treat the transfer process in first-order perturbation theory, an approximation which seems well justified in the present case, P_{transf} is given by the square of the integral of the transfer form factor along the trajectory

$$P_{transf} = \left(\frac{1}{\hbar} \int_{-\infty}^{\infty} f_T(r,u)dt\right)^2.$$

(20)

Note that the assumption has been made that the transfer process has optimum Q-value.

Calculations have been made where the transfer strength along the trajectory

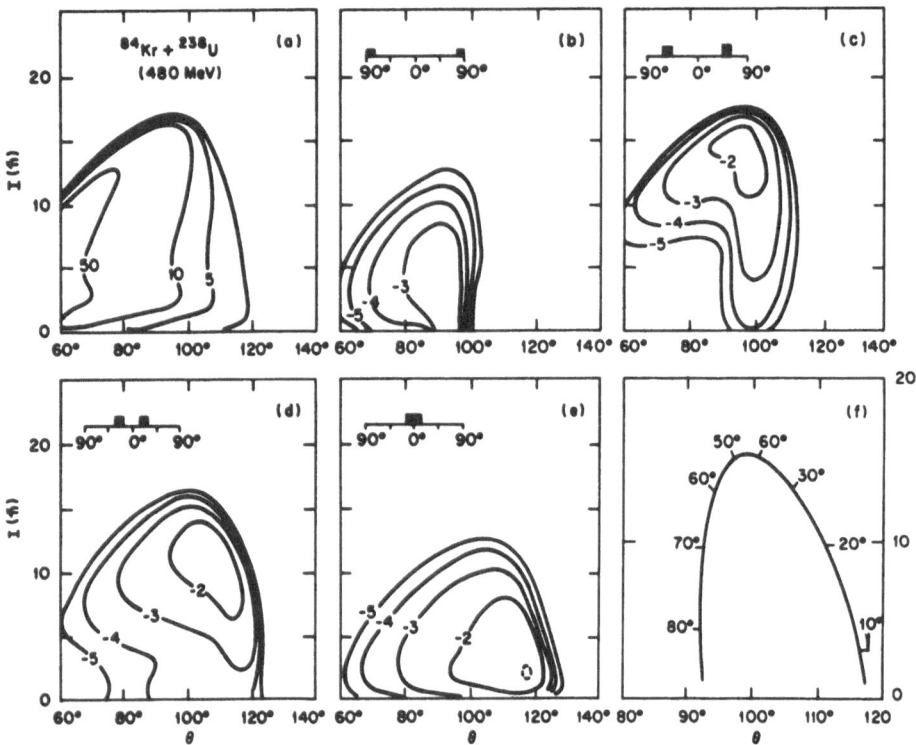

Fig. 7

differs from zero only when the angle between the relative distance vector and the rotor axis falls into pairs of rings 15° wide on the surface of the rotor at equal distances from the poles (cf. Fig. 5(c)). This defines six zones covering the nuclear surface. The results of the calculations are shown in Fig. 7. The associated inelastic cross section (19) is shown in Fig. 7(a). Very conspicuous correlations are found between the localization of the transfer form factor and the double-differential cross sections (cf. Figs. 7(b)-7(e)).

Actual transfer reactions will not occur in the particular zones used in the illustration. However, the correlation between scattering angle for any given form factor can be constructed in terms of the basic patterns of correlations associated with a set of narrow zones. Thus, measurements of $d^2\sigma/d\theta dI$ may be used to map out the amplitude of the single-particle wave functions along the surface of deformed nuclei.

2c. Study of the Pairing Phase Transitions in Rapidly Rotating Nuclei

When a deformed nucleus is subject to rapid rotations, changes in the I(I+1) rotational pattern can be observed. In particular, in the moment of inertia of the Yrast band as a function of the rotational frequency can take place (cf. Fig. 8).

In strongly rotating nuclei, the Corolis force acts as an external magnetic field in a superfluid system,[*] and it is expected that for some critical value of the rotational frequency, it becomes too expensive to rotate in the superfluid phase. In fact, $\mathcal{J}_{sup} \sim \frac{1}{2} \mathcal{J}_{rigid}$, where $\mathcal{J}_{sup}$ is the moment of inertia displayed by the ground-state rotational band at low I. The quantity $\mathcal{J}_{rigid}$ is the rigid moment of inertia and coincides with that of the independent particle motion. It is as in the superfluid phase, only the mass at the tips of the deformed nucleus is dragged by the walls, as will also be the case of a spheroidal bucket filled with superfluid HeII. Rotation breaks time-reversal invariance. The Coriolis force $H_c \sim \hbar \vec{\omega}_{rot} \cdot \vec{j}$ acts in opposite ways on each partner of a Cooper pair and for a value of H_c of the order of $2\Delta(\sim 1.7$ MeV), two quasi-particle states are excited.[12] This will happen first for Cooper pairs based on states with high-j and low-Ω[13] (cf., e.g., Fig. 9). For the case of ^{162}Er the candidate is the Cooper pair built on the [660 1/2] Nilsson orbital (cf. Fig. 4(e)). The observed critical frequency $(\omega_{rot})_{crit}$ is $\sim \frac{250}{\hbar}$ keV (cf. Fig. 8). Thus, $H_c \sim 250$ keV $\times$ 6 ~ 1.5 MeV ($\sim 2\Delta$) as expected. One can thus view the backbending phenomenon as resulting from the crossing of the ground-state band and of a band displaying a finite mass at $\omega_{rot} = 0$ and a rigid moment of inertia (cf. Fig. 10). It is noted that the fact that the moment of inertia of the rotational band based on the [660 1/2] two-quasiparticle state is close to $\mathcal{J}_{rig}$ does

---

[*]In this case, however, because of the shell structure, it is expected that the pairing phase transition will be connected with a rich variety of phenomena (band crossing, gapless superconductivity, pairing isomers, etc.)

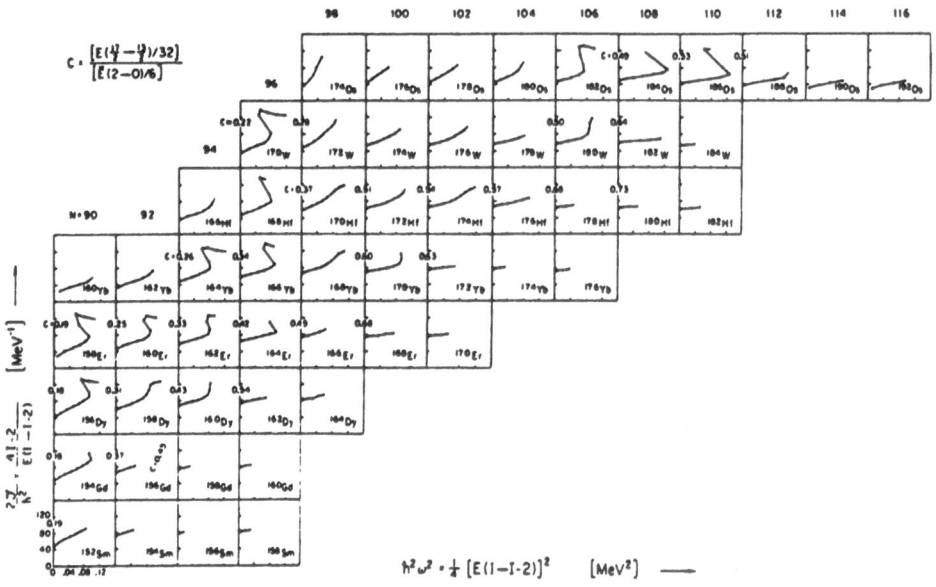

Fig. 8. Ground-band level energies in even-even rare-earth
nuclei (from Rev. of Mod. Phys. 47 (1975) 43)

not necessarily imply that $\Delta = 0$. The orbits which decouple from the rotational motion are strongly concentrated at the poles, leading to a large change in $\mathcal{J}$.

Although much has been learned during the past years, there are still important questions which need to be answered before achieving a consistent picture of back-bending. In particular, which is the dependence of Δ with ω along the Yrast line. In other words, which is the type of transition which takes place in going from the superfluid to the normal phases. Theoretically, one needs to have a good description of the shell model at high frequencies and to understand, among other things, the role of the multipole pairing components of the residual force and of the fluc-tuations of the pairing gap associated with pairing vibrations.[14]

Experimentally, there is only one possibility of measuring directly the pairing gap. It is through the transfer of a pair of particles (cf. Appendix Ref. 14 and Table 1). Furthermore, it has to be in a reaction induced in a heavy-ion col-lision. This is because reactions induced by light ions will not be able to excite the high-spin states we are interested in.

In the framework of the model discussed in Sections 2A and 2B, the two-nucleon double-differential cross section can be calculated by multiplying the events asso-ciated with each scattering angle by the two-nucleon transfer probability P_2. This quantity is obtained by integrating the two-particle transfer form factor along the different trajectories. It is noted that for a transition between a normal and a superfluid nucleus $P_2 \sim (\Delta(I)/G)^2$, where $\Delta(I)$ is the gap parameter and G the strength of the pairing force. From the ratio between $(d\sigma/d\theta dI)_{transf}$ and $(d\sigma/d\theta dI)_{inel+CE}$, one can thus extract $\Delta(I)$.

In the above discussion, we have used the standard pairing theory and charac-terized the pairing distortion by a single parameter. That is, we have assumed pairing to act only in s-waves. However, both in nuclei and in other Fermi liquids, pairing acting in channels with $\ell \neq 0$ can play an important role (cf. Table 3).

In situations where Nilsson orbitals with negative quadrupole moments are close to the Fermi surface, the pairing gap can display an important state dependence. This is because, due to angular momentum violation, the quadrupole component of the pairing force can contribute to the pairing gap.[15] This quantity is given by

$$\Delta_\nu = G_0 \sum_{\nu'} U_{\nu'} V_{\nu'} + G_2 q_\nu \sum_{\nu'} q_{\nu'} U_{\nu'} V_{\nu'} . \tag{21}$$

The quantity $q_\nu = 3n_z - N$ is the static quadrupole moment associated with the orbital $\nu = i\Omega$. The quantity $\Delta_2 = G_2 \sum_{\nu'} q_{\nu'} U_{\nu'} V_{\nu'}$ measures the quadrupole deformation around the Fermi surface, and G_2 is the quadrupole pairing coupling constant. Let us consider a situation like the one displayed in Fig. 11 where an orbital with negative quadrupole moment (i.e., the [505 11/2] Nilsson level) is close to the Fermi surface and is embedded in a set of downsloping states ($q_\nu > 0$). In this case, $\Delta_2 > 0$ and

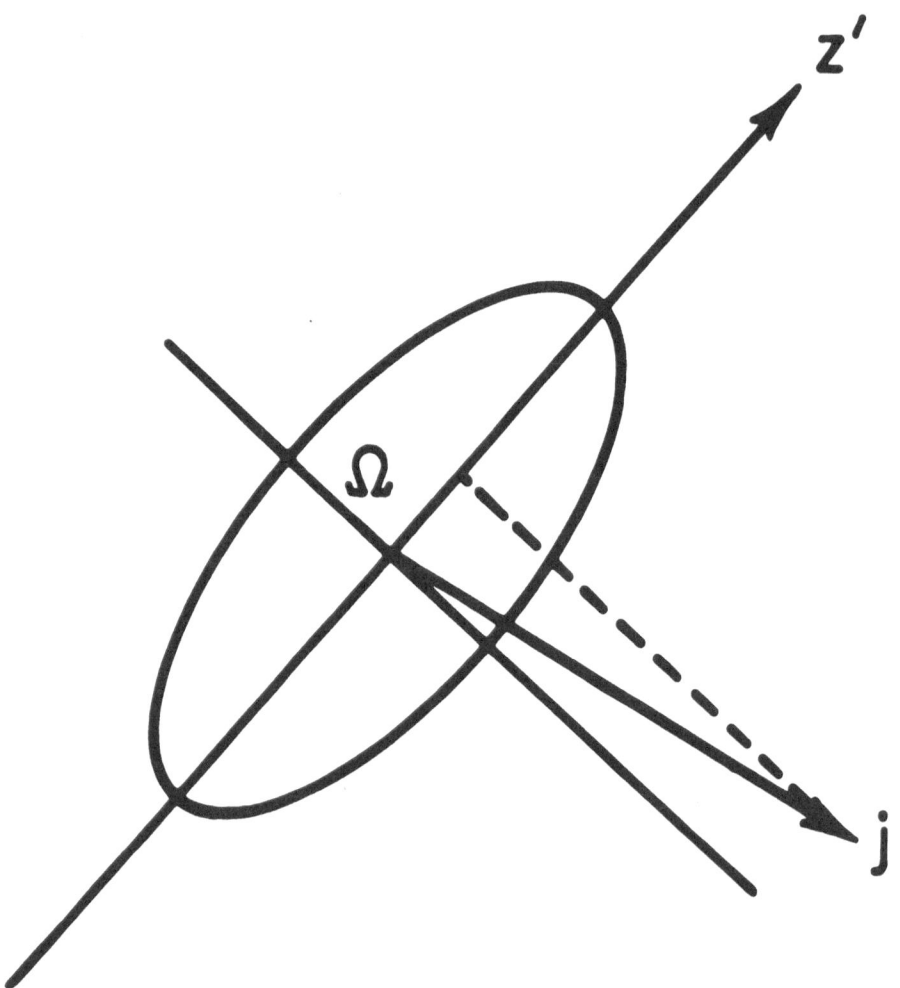

Fig. 9

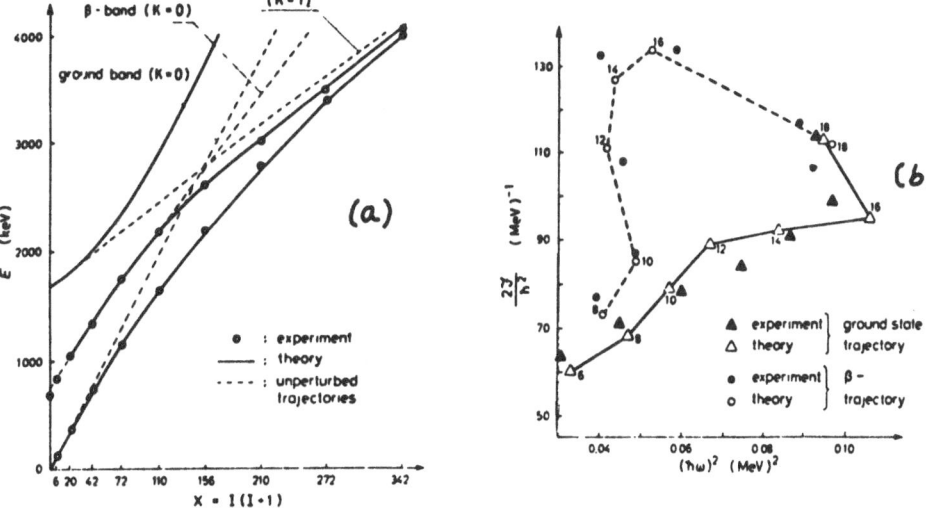

Fig. 10. (a) Energy trajectories for $^{90}_{64}Gd^{154}$ in comparison with the experimental data. (b) The same as (a) but in terms of different variables (from Phys. Lett. 50B (1974) 295).

Table 3

Pairing Correlations

ℓ-wave	System	Effect
s	electrons in metal nuclei	condensation
p	^3He	condensation
d	nuclei	fluctuations

$$\Delta_{i\Omega} \sim \begin{cases} 0 & [550\ 11/2] \\ \Delta & \text{other states.} \end{cases} \tag{22}$$

That is, the pairs of particles in the [550 11/2] feel a normal and not a super-fluid system (pairing isomer), a phenomenon that has its counterpart in condensed matter and is known as gapless superconductivity.[16]

This mechanism has recently been used[17] to explain anomalies observed in the critical rotational frequencies at which backbending takes place. The standard situation encountered is depicted in Fig. 12(a). The difference $\hbar\omega_{crit}(\text{even})-\hbar\omega_{crit}(\text{odd})$ is interpreted in terms of the decrease of the pairing gap in odd-nuclei due to the blocking of an orbital close to the Fermi surface. For the case of ^{162}Er and ^{161}Er, the situation is as shown in Fig. 12(b). In this case, the odd particle occupies the [505 11/2] orbital, and no change in Δ is expected in going from the even to the odd system.

Again, the specific way to test the value of the pairing gap at backbending is in a two-nucleon transfer reaction induced in a heavy-ion collision.

3. Variety of Nuclear Phenomena

In the previous section we have concentrated our attention on a particular aspect of nuclear structure which can specifically be probed in heavy-ion collisions. In this section we briefly list other nuclear structure problems for which heavy-ion reactions seem to be an ideal research tool (cf. Table 4).

3a. Excitation of Giant Resonances

The large Coulomb and nuclear fields available in heavy-ion collisions opens the possibility to study specific energy regions of the nuclear response by tuning the bombarding energy (cf. Fig. 13). In particular, one expects that heavy ions will become useful tools to study giant resonances with a good ratio of signal to background.[18]

3b. Excitation of Pairing Rotational Bands

Based on geometrical arguments one would expect, for medium heavy nuclei, two-nucleon transfer cross section $\sigma(2n)$ to be a factor $1/A \sim 10^{-4}$ smaller than single-nucleon transfer cross sections $\sigma(n)$. On the other hand, reactions involving two superfluid nuclei are estimated to be enhanced by a factor $(\Delta/G)^4 \sim 10^4$ as compared with the transfer of two uncorrelated nucleons, suggesting the possibility of multi-pair transfer.[19]

3c. Identification of High-ℓ Single-Particle Orbitals

The action integral

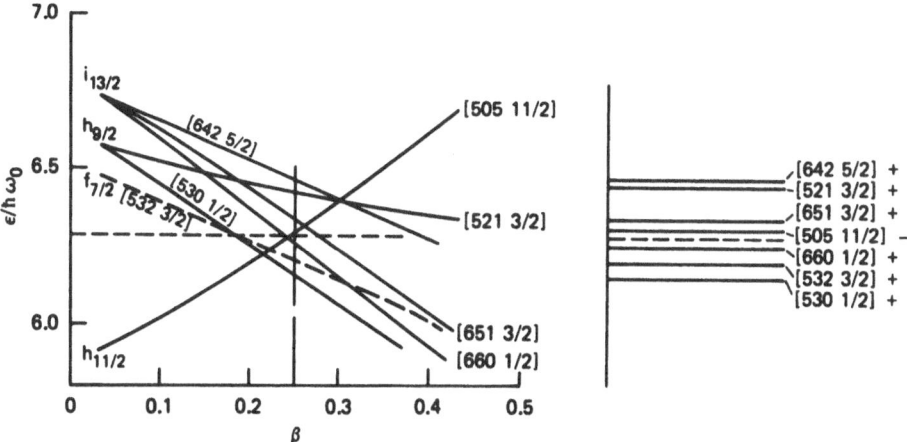

Fig. 11

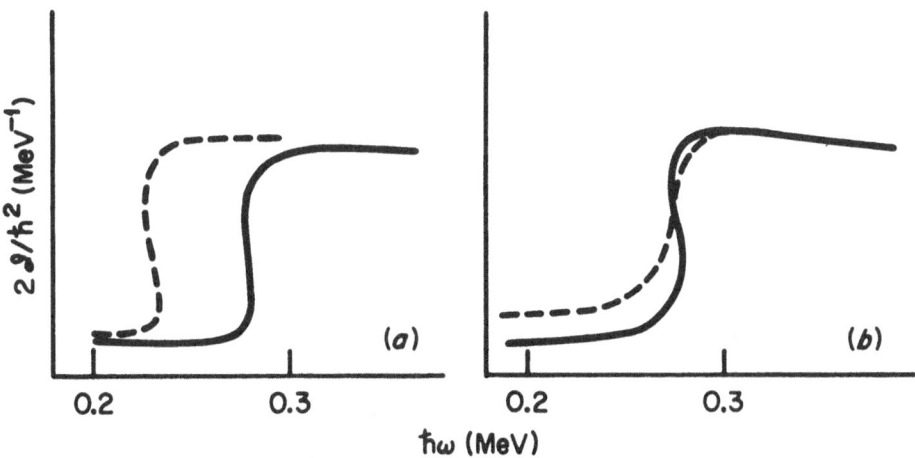

Fig. 12

Table 4

Some of the possibilities opened by new
generation of heavy-ion machines

local	QE	Mapping out the surface of deformed nuclei (Nilsson orbitals in 3D)
large U^c	QE	Study of the pairing phase transition in strongly rotating nuclei
large U^c short	QE	Excitation of giant resonances with good ratio of peak to background. (Excitation of magnetic modes)
massive	QE	Coherent excitation of pairing rotational bands
short to massive	hard coll.	Study of zero point motion induced by collective vibrations (isosc. surface, isov. N-Z imbalance)
large U^c	QE	Coulomb excitation of rotational bands to high-spin states
large	QE	Excitation of single-particle orbitals with high spin ($Q \ll 0$)
large	F	Study of nuclear shapes as a function of angular momentum and temperature (superdef., fission, γ-decay GDR)
large	F	Search for shape isomers (molecular states) in strongly rotating nuclei
massive	QE	Study of alpha correlations

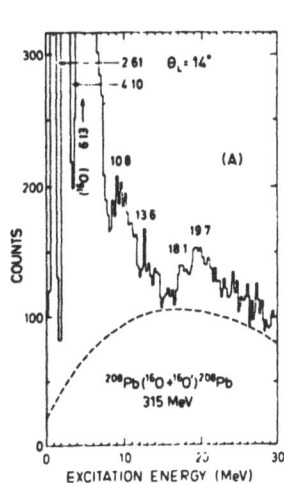
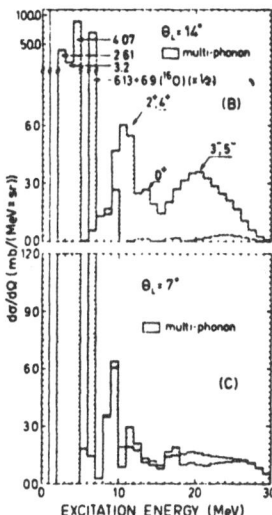

Fig. 13. (a) Double differential inelastic cross section
associated with the reaction $^{16}O + {}^{208}Pb$. In (a) the
experimental spectrum at the grazing angle $\theta_{lab} = 14^\circ$
is given. In (b) and (c) the predicted "direct reaction" double differential cross sections are shown
in mb/(MeV × sr). The darkly shaded area corresponds
to the events in which at least two harmonic oscillator
quanta were absorbed (from ref. 18).

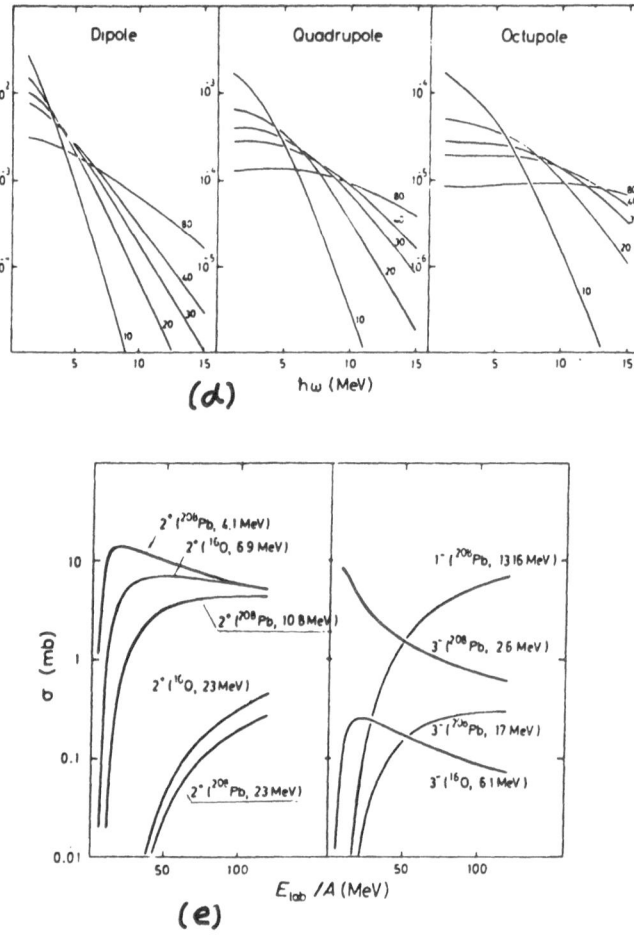

(d)

(e)

Fig. 13. (d) Probabilities of one-phonon Coulomb excitation as a
function of the energy of the mode for i = 1, 2 and 3. In
all cases a transition matrix element of one single particle unit was used for a trajectory leading to a distance of closest approach of 15 fm. The different curves
for each multipolarity are labelled by the bombarding
energy in MeV/nucleon (from ref. 18). (e) Coulomb excitation cross sections for some of the modes considered
in the calculations. Only contribution from trajectories
leading to distances of closest approach larger than
15 fm were included. The cross sections associated with
the giant dipole (13.16 MeV) and the giant isovector
quadrupole mode (23 MeV) of ^{208}Pb are also shown. They
exhaust 86% and 55% of their corresponding energy-
weighted sum rules, respectively (from ref. 18).

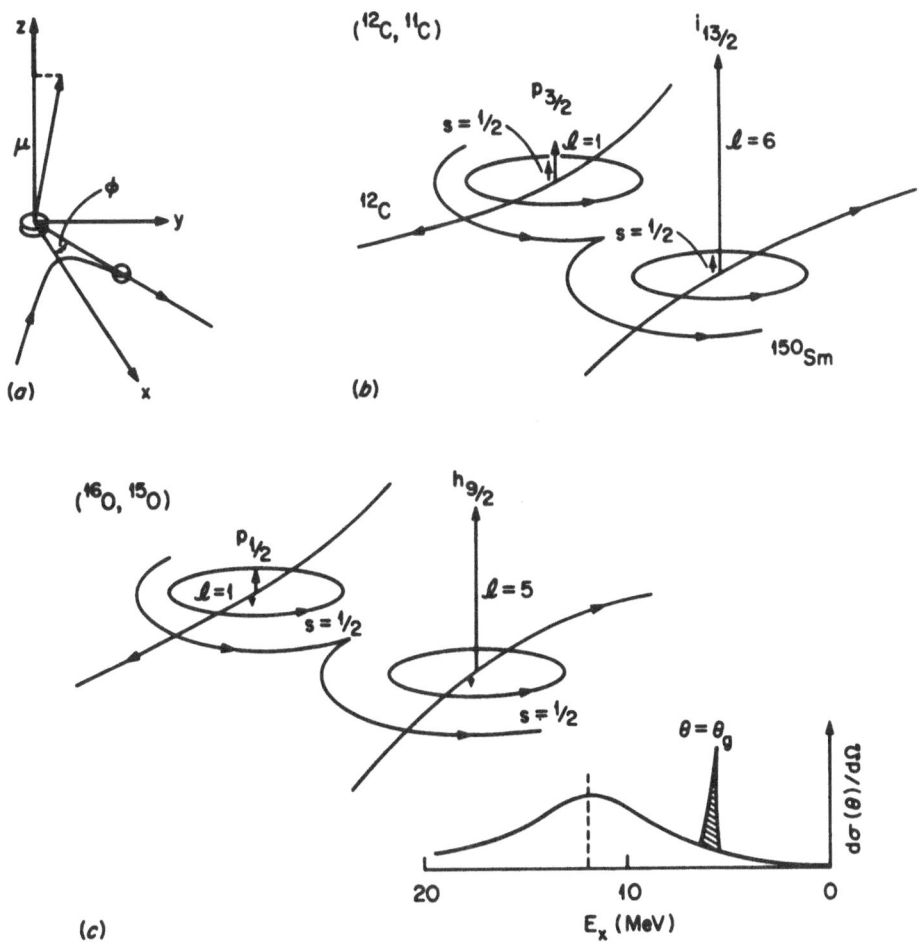

Fig. 14

$$\chi = \frac{1}{\hbar} \int_{-\infty}^{\infty} f(r(t)) \; e^{-\frac{i}{\hbar}(\Delta Et - \mu\phi(r(t)))} \; dt$$

measures the strength with which the reaction a+A → b+B associated with the form factor f(r(t)) takes place. Assuming $\chi^2 \ll \chi$, the process can be treated in perturbation theory and the cross section written as

$$\left(\frac{d\sigma}{d\Omega}\right)_{aA \to bB} \sim \chi^2 \left(\frac{d\sigma}{d\Omega}\right)_{elastic}$$

The integral in χ is carried over a classical trajectory (grazing trajectory) and ΔE and μ are the energy and angular momenta transferred in the reaction process (cf. Fig. 14(a)). The exponential factor measures the matching between the entrance and exit channels classical trajectories of relative motion. The optimal matching is achieved when the condition

$$\Delta E - \mu\dot{\phi} = 0$$

is fulfilled.

Let us consider the following reactions:

	Q (MeV)	$\mu(\Delta L)$ ($\hbar$)
$^{149}Sm(^{16}O,^{15}O)^{150}Sm$	- 9.8	-10
$^{149}Sm(^{12}C,^{11}C)^{150}Sm$	-12.8	-12
$^{149}Sm(^{13}C,^{12}C)^{150}Sm$	0.9	- 2

The last two columns give the optimum Q-value and the angular momenta needed to fulfill the matching condition $\Delta E - \mu\phi = 0$.

The two first reactions at bombarding energies of a few MeV per nucleon are optimal to excite single-particle orbitals with high angular momentum. One expects that the first reaction will predominantly excite orbitals where $j = \ell + 1/2$ and the second those were $j = \ell - 1/2$ (cf. Fig. 14(b) and 14(c)). Patterns of the type shown in Fig. 14(d) have already been observed.[20]

3d. Coulomb Fission and Beyond

With the availability of projectiles like ^{208}Pb, one can expect to multiply Coulomb excite the β-vibrations of, e.g., ^{234}U and eventually force the system into fission (cf. Fig. 15).

Evidence has recently been reported[21] on non-equilibrium effects in sequential fission induced by energetic collisions between heavy ions. This result suggests the evidence of a new phenomenon in which a direct descent towards fission occurs from the initial shape distribution of the primary fragments. The time between the

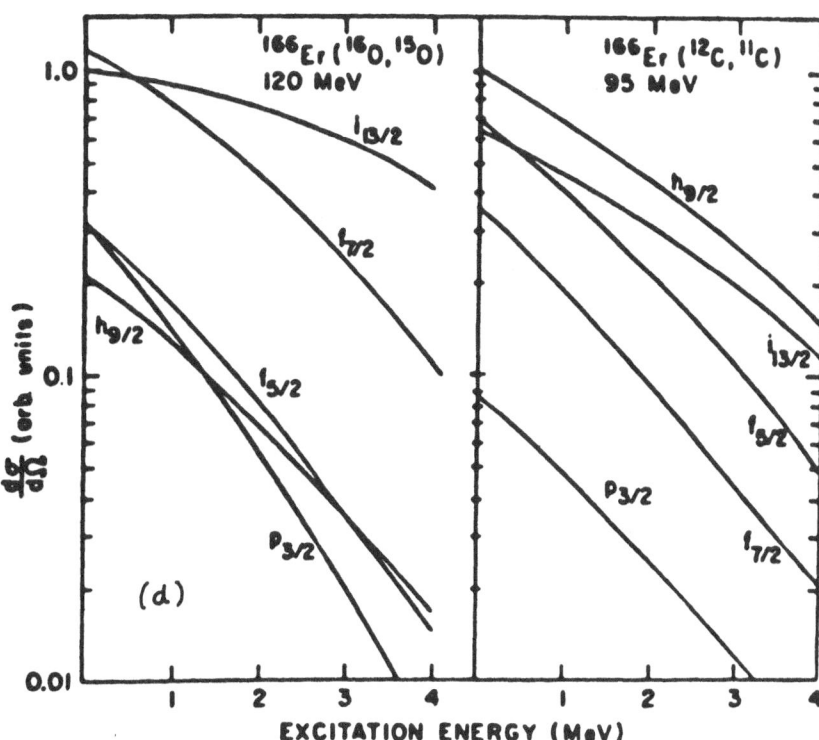

Fig. 14(d).Expected relative cross sections for active single-particle levels as a
function of excitation energy for the reactions $^{166}Er(^{16}O,^{15}O)$ and
$^{166}Er(^{12}C,^{11}C)$ calculated with the distorted-wave Born approximation code
PTOLEMY. Unit spectroscopic factors were assumed for all states (from
Phys. Rev. Lett. 46 (1981) 1565).

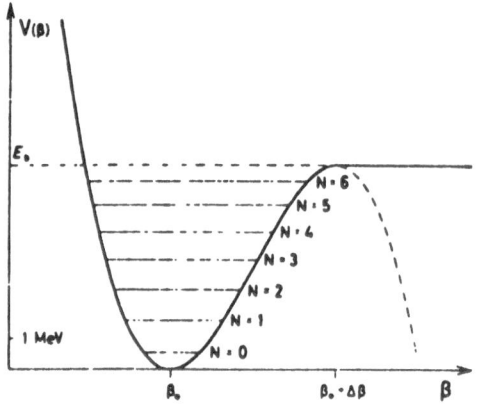

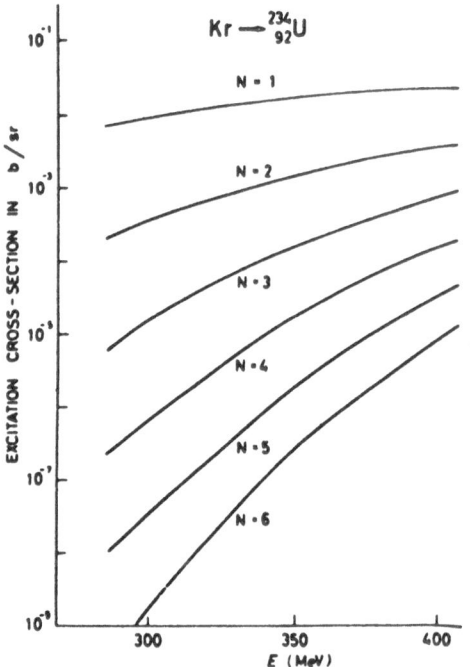

Fig. 15.   Cross section for the multiple Coulomb excitation of the beta vibration in
the deformed nucleus ^{234}U. In the lower part of the figure the differential cross section in barns/sr is given for backward scattering of various projectiles at the Coulomb barrier. The calculation was made in a
pure vibrational model including the finite energy loss. In the upper
part of the figure is given a schematic picture of the position of the
multiphonon beta-vibrational states used in the calculation (from ref. 9).

consecutive scission acts is estimated to be of the order of 10^{-21} sec.

3e. Coulomb Excitation of High-Spin States

Coulomb excitation below the Coulomb barrier is specifically suited to excite the rotational states (cf. Fig. 16). Because of the slow variation of the Coulomb field in space, it can only excite modes of low multipolarity. Similarly for bombarding energies $E < E_B$ the slow variation of the Coulomb field in time makes it impossible to excite states at high frequency, since the process in this limit becomes rapidly adiabatic.

In Fig. 17 we show results of a Coulomb excitation measurement of ^{238}U making use of ^{208}Pb as a projectile.

3f. Study of Shapes of Nuclei Just Before Fission

The isovector giant dipole resonance is influenced in a major way by the coupling to the surface degrees of freedom. In particular, in deformed nuclei the resonance splits in two peaks, as protons and neutrons can vibrate out of phase along the symmetry axis or in a plane perpendicular to it (cf. Fig. 18).

Dipole modes can be excited not only from the ground state of a nucleus, but also from the ensemble of states associated with a compound nucleus at finite temperature.[22] Measuring the γ-decay from the dipole resonance,[23] it is possible to construct the associated strength function and from its structure eventually infer the shape of nuclei subject to extreme conditions of rotation up to the fission threshold.

3g. Search for Shape Isomers in Strongly-Deformed, Rapidly-Rotating Nuclei

The liquid-drop model of nuclei has played an important role in the study of nuclear fission. To the interplay of surface tension and Coulomb repulsion the centrifugal force adds a new dimension and any nucleus will fission at sufficiently high angular momentum. The maximum amount of angular momentum a nucleus can accommodate before fissioning in two equal masses is $\sim 80\hbar$.

To obtain a detailed picture of the low-energy nuclear spectrum the shell structure has to be considered. The existence of low-lying collective modes of the fission products can lead to structure in the smooth potential energy surface predicted by the liquid drop,[24] as can be seen in Fig. 19. Due to the collective low-lying octupole vibration observed in ^{40}Ca and to the high angular moments with which the system rotates, two minima appear in the potential energy surface plotted as a function of the relative distance of the di-nuclear system $^{80}Zr = {}^{40}Ca + {}^{40}Ca$ and of the octupole deformation of the individual nuclei.

The Coulomb repulsion between the two pear-shaped ^{40}Ca systems can be balanced in two ways. Through a strong surface-surface attraction or by pulling away the centers of charge as much as possible. The first situation is achieved by having the two nuclei in contact at the base. This configuration is very compact and

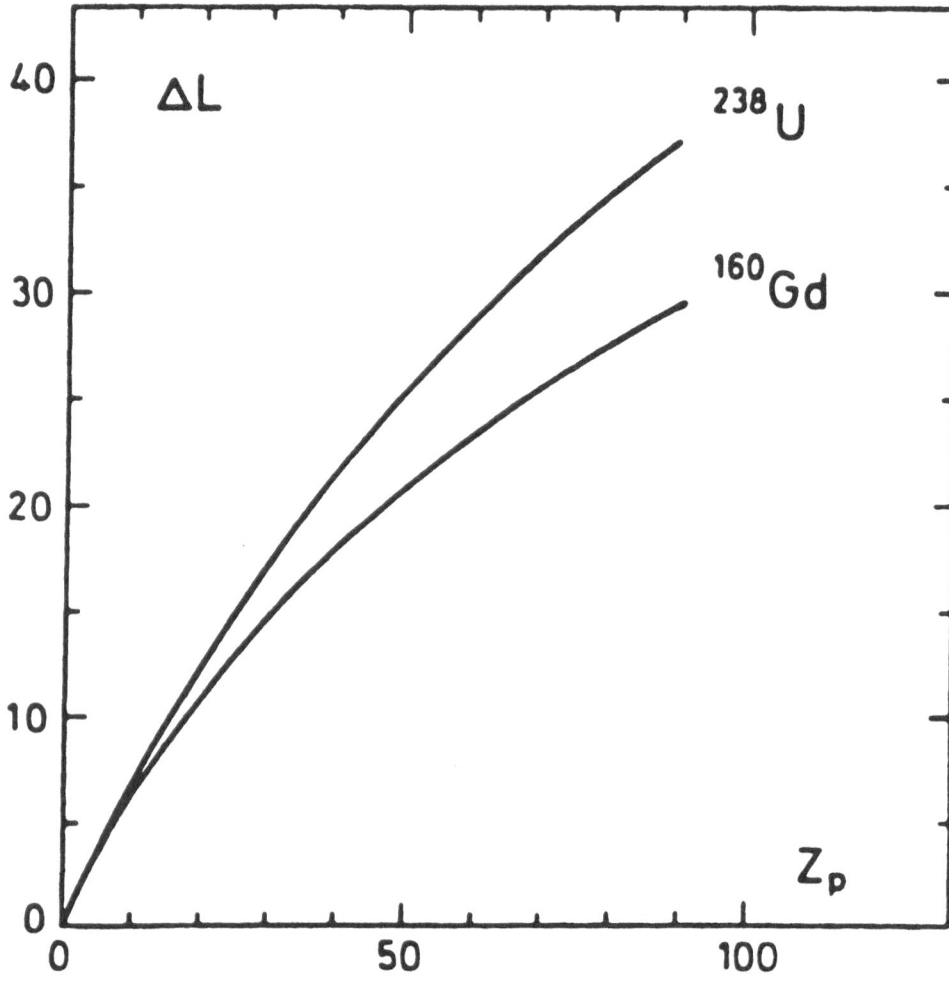

Fig. 16. Coulomb excitation of deformed nuclei. The expected
maximum angular-momentum transfer ΔL to rotational
motion in ^{238}U and ^{160}Gd is plotted as a function of
the projectile charge at a bombarding energy below
the Coulomb barrier so that no interference from
nuclear reactions is expected. With uranium ions on
a uranium target, one may thus expect to populate
states of spin 36 in the ground-state rotational
band. In estimating ΔL, the energy of excitation
has been neglected. This is expected to reduce the
maximum angular momentum by a few units (from ref. 9).

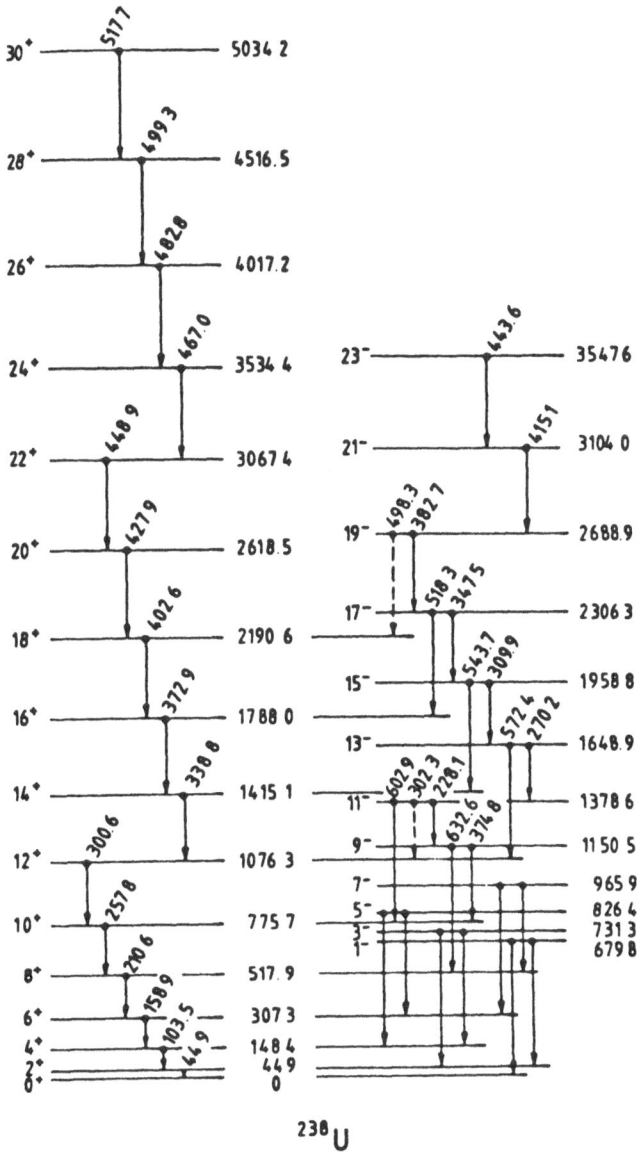

Fig. 17. Level scheme for ^{238}U showing the ground state band up to spin 30 and the octupole band up to spin 23. The transition energies have an error of 0.4 keV for J ≤ 24, with increasing spin this error increases up to a value of 1.0 keV for J = 30 (from Physica Scripta, Vol. 24 (1981) 337).

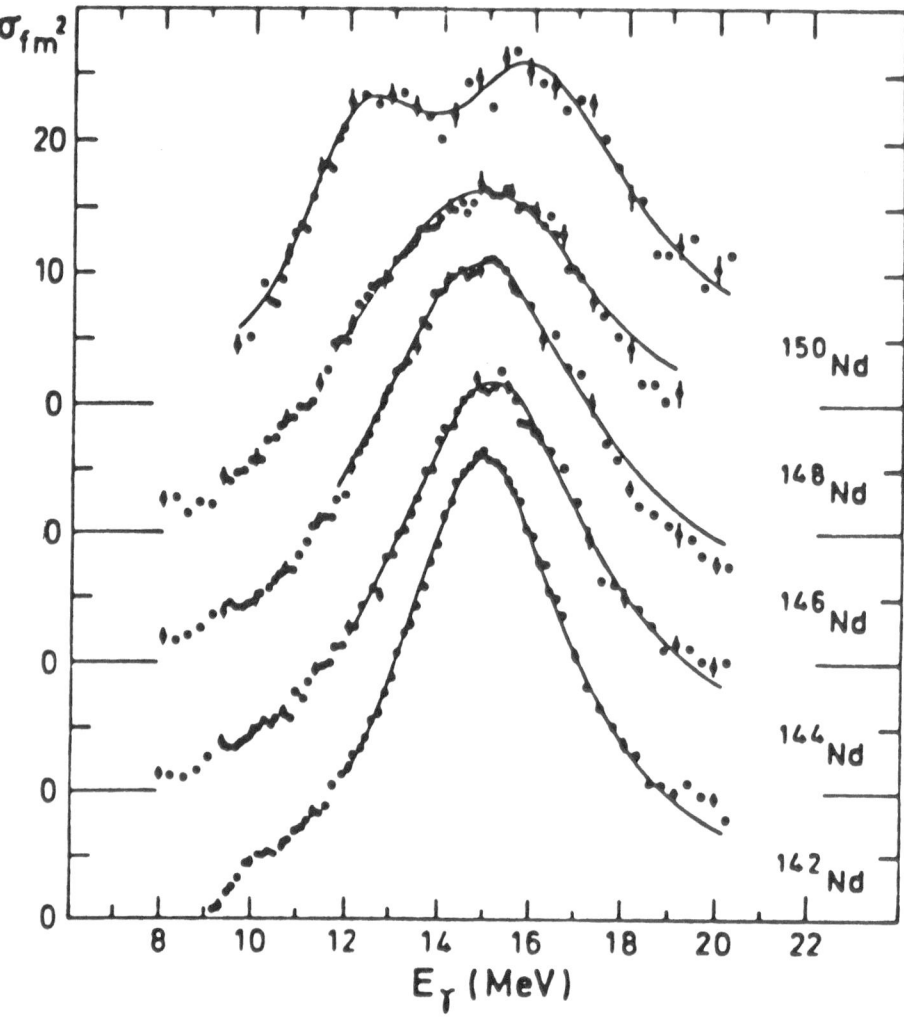

Fig. 18. Photoabsorption cross section for even isotopes of neodymium. The experimental data are from P. Carlos, H. Beil, R. Bergere, A. Oepretre, and A. Veyssiere, Nucl. Phys. A172 (1971) 437. The solid curves represent Lorentzian fits (from ref. 1).

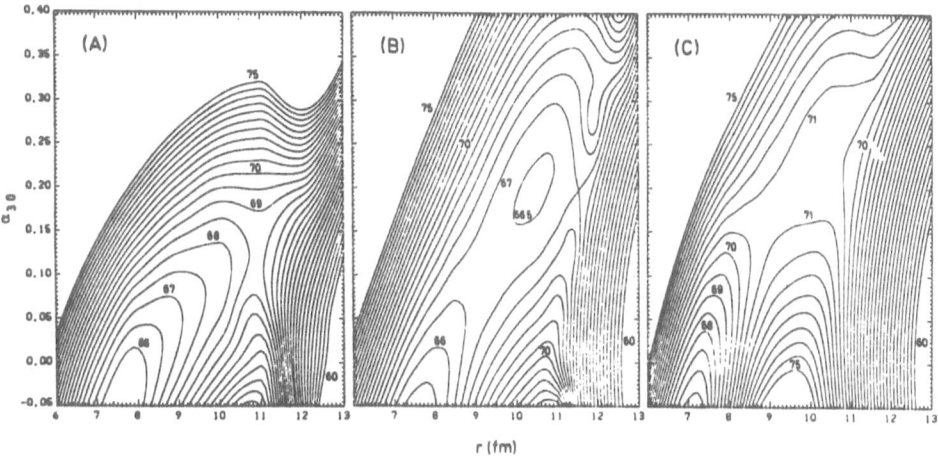

Fig. 19. Contour levels of the effective potential U_{eff} for
$\ell = 50$ minimized with respect to all deformation degrees of freedom except α_3 displayed in the $(\alpha_3 r)$
plane for the nucleus ^{80}Zr $= ^{40}$Ca $+ ^{40}$Ca. In (a) the
liquid-drop parameters were used for all multipoles
$\lambda(\leq 5)$. In (b) the octupole restoring force was extracted from a) Fig. 7, while C_2, C_4 and C_5 were
assumed to have the same liquid-drop values as in (a).
In (c) both C_2 and C_3 were extracted from a) Fig. 7,
while C_4 and C_5 have the liquid-drop values a) (from
Nucl. Phys. A349 (1981) 496).

coincides with the liquid drop minimum. The second configuration is achieved by having the two nuclei in contact at the tip. This configuration leads to a strongly-deformed system (shape isomer). It is an open question whether these strongly rotating shapes isomer exist and which are the optimal bombarding conditions to produce them. In the same context, it is a challenging question to what extent the shape isomer observed in ^{236}U reflects a low-lying collective octupole vibration in ^{118}Pd.

3h. Study of the Zero-Point Motion Induced by Collective Vibration

Collective vibrations play an important role in the damping of energy and angular momenta in deep-inelastic reactions.[25]

These collective modes induce zero-point fluctuations in the ground state. In classical terms, it means that for each impact parameter the surface of the two nuclei will be in a variety of positions.

Because the ion-ion potential depends on the position of the nuclear surfaces of the two interacting nuclei, one expects large variations of the energy and angular momentum loss. The analysis of data like the one displayed in Fig. 20 seems to offer a unique possibility of measuring the fluctuations in the position of the nuclear surface. Surface fluctuations seem also to play a central role in determining the fission cross section in collisions at energies below and above the Coulomb barrier.[26]

An isovector vibration induces fluctuations in the ground state ratio N/Z. In an energetic heavy-ion collision, it is possible that the projectile can cut off a part of the target. The mass distribution of this chunk of nuclear matter will carry information on the zero-point fluctuations associated with isovector modes.

3i. Four-Particle Correlations

The possible existence of alpha-correlations in nuclei is a subject that has attracted much attention. One of the intriguing questions is   the extent to which the alpha particle is transferred as a whole, rahter than a combination of a transfer of a pair of neutrons and a pair of protons.[27]

Transfer reactions induced by heavy ions allows for the transfer of four particles which can move both in target and projectile in the same potential. In this sense, heavy-ion reactions are the specific probes to study cluster correlation in nuclei. In Fig. 21 we show schematically the predictions of the pairing model for two- and four-particle transfer reactions around ^{208}Pb. Although the pairing spectrum is well established, preliminary attempts to observe the "alpha-vibrations" have not been successful.

Discussions with C. H. Dasso, S. Landowne and A. Winther are acknowledged.

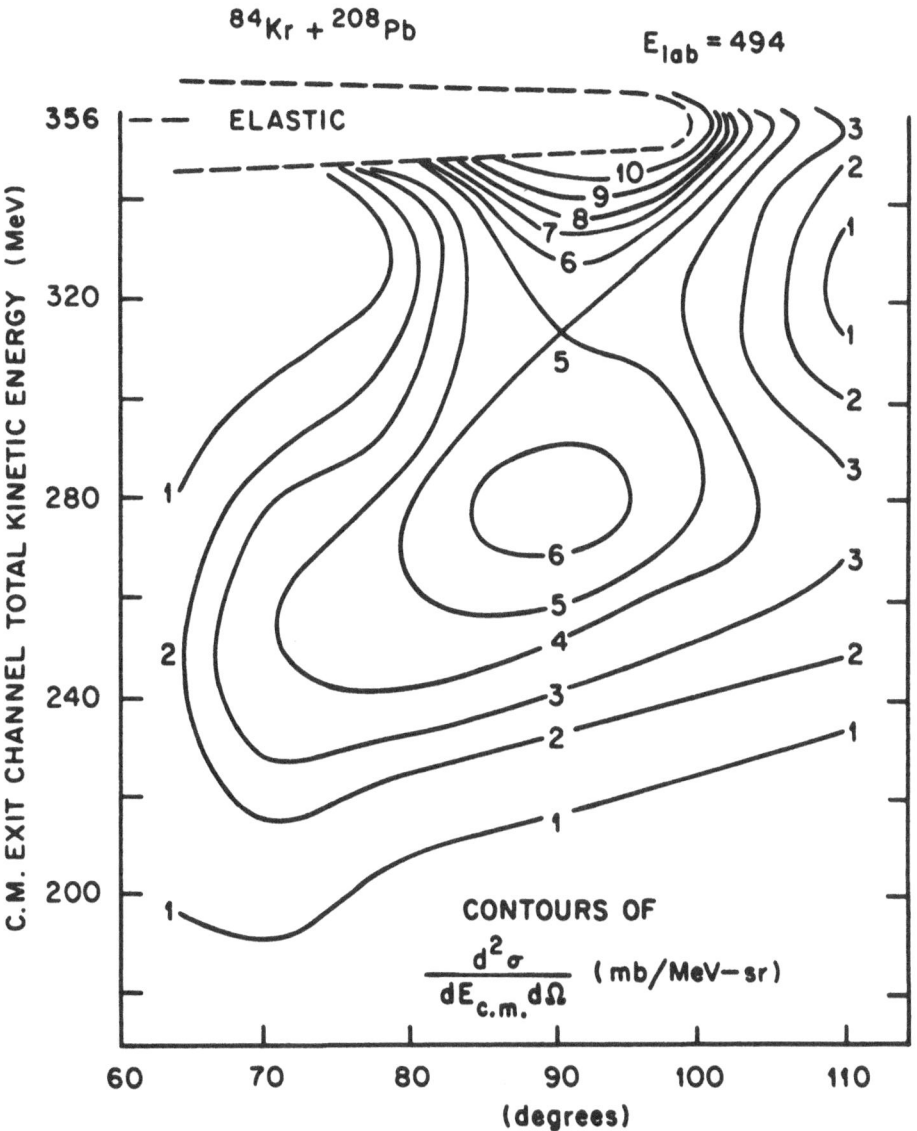

Fig. 20. Energy-angle distribution of the ^{84}Kr + ^{208}Pb reaction (E_{lab} = 494 MeV).

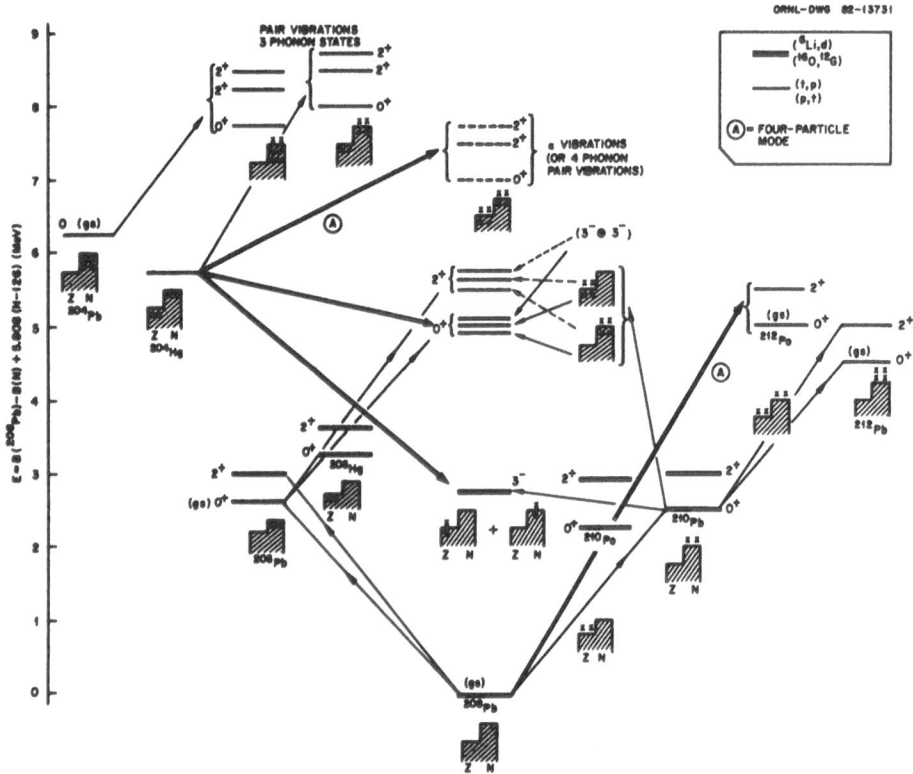

Fig. 21

Appendix

Scattering in two dimensions

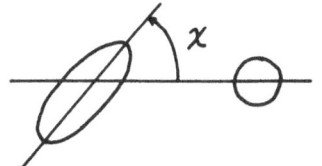

Interaction between
two nuclei: Coulomb

Solve problem: calculate probability
of exciting a given ang. mom.

$$H = \frac{P_r^2}{2m_0} + P_\chi^2 \left[\frac{1}{2m_0 r^2} + \frac{1}{2\mathcal{J}}\right] + \frac{Z_a Z_A e^2}{r} + \frac{Z_a e}{4r^3} Q_0 (3\cos^2 \chi - 1),$$

$$\begin{cases} \dot\chi = \frac{\partial H}{\partial P_\chi} = \left(\frac{1}{m_0 r^2} + \frac{1}{\mathcal{J}}\right) P_\chi \\[2em] -\dot P_\chi = \frac{\partial H}{\partial \chi} = -\frac{3Z_a e}{4r^3} Q_0 \sin 2\chi . \end{cases}$$

Order of magnitude estimates

$$P_\chi \sim \frac{3Z_a e}{4b^3} Q_0 \ \tau \sim \frac{3Z_a e}{4b^3} Q_0 \frac{b}{v} \sim \hbar \underbrace{\left(\frac{Z_a e Q_0}{\hbar v\, b^2}\right)}_{q} \ ,$$

$$P_\chi \sim \hbar q ,$$

$$\Delta\chi \sim \hbar q \left(\frac{1}{m_0 b^2} + \frac{1}{\mathcal{J}}\right)\frac{b}{v}$$

$$\sim q \left(\frac{1}{\eta} + \xi\right)$$

$$\tau \sim \frac{b}{v}$$

$$\boxed{v = \text{veloc. at } \infty}$$

$$b = \frac{Z_a Z_A e^2}{E}$$

$$\eta = \frac{Z_a Z_A e^2}{\hbar v} = \frac{b}{2\text{x}} \ (\text{Sommerfeld par.}) \qquad \text{x} = \frac{\hbar}{mv} = \frac{\hbar v}{2E}$$

$$E_R = \frac{\hbar^2}{2\mathcal{J}} I(I+1) \ ; \ (\Delta E_R)_{0\to2} = \frac{\hbar^2}{2\mathcal{J}} 6 = \frac{3\hbar^2}{\mathcal{J}}$$

$$\xi = \frac{\Delta E}{\hbar} \tau \sim \frac{3\hbar^2}{\hbar\mathcal{J}} \frac{b}{v} = \frac{3\hbar b}{\mathcal{J} v} \quad (\text{adiabaticity parameter})$$

$$\boxed{\frac{q}{\eta} \ll 1} \qquad \boxed{q\xi \ll 1} \qquad \boxed{\Delta\chi \sim 0}$$

$$\chi \sim \theta_0 \text{ throughout the collision.}$$

$$\frac{\partial H}{\partial r} = -\dot{P}r = -P_\chi^2 \frac{1}{m_o r^3} - \frac{Z_a Z_A e^2}{r^2} - \frac{3 Z_a e}{4\ r^4} Q_o(3\cos^2\chi - 1)$$

$$\frac{\partial H}{\partial p_r} = \dot{r} = \frac{Pr}{m_o}$$

$$m_o \ddot{r} = \overbrace{P_\chi^2 \frac{1}{m_o r^3}}^{A} + \overbrace{\frac{Z_a Z_A e^2}{r^2}}^{B} + \overbrace{\frac{3 Z_a e}{4\ r^4} Q_o(3\cos^2\chi - 1)}^{C}$$

$$\frac{A}{B} \sim \left(\frac{q}{\eta}\right)^2 \qquad\qquad \frac{C}{B} \sim \frac{q}{\eta}$$

Neglect   A

$$m_o \frac{d}{dt} \dot{r}^2 = \frac{2 Z_a Z_A e^2}{r^2} \dot{r} + \frac{3}{2} \frac{Z_a e}{r^4} Q_o(3\cos^2\chi - 1)\dot{r}$$

$$dt = \frac{dr}{\sqrt{\dfrac{2E}{m_o} - \dfrac{2 Z_a Z_A e^2}{m_o r} - \dfrac{1}{2m_o} \dfrac{Z_a e}{r^3} Q_o(3\cos^2\chi - 1)}}$$

$$\Delta L = p_\chi(\infty) = \frac{3 Z_a e Q_o}{4} \int_{-\infty}^{\infty} \frac{dt}{r^3} \sin 2\chi$$

$$\sim \frac{3 Z_a e Q_o}{4} \sin 2\theta_o \int_{-\infty}^{\infty} \frac{dt}{r^3}$$

$$\underbrace{\qquad\qquad\qquad}$$

$$\frac{8}{3} \frac{1}{b^2 v}$$

$$\boxed{\Delta L \sim 2\hbar q \sin 2\theta_o}$$

$$\boxed{(\Delta L)_{max} = \hbar 2q}$$

Solutions

$$f = \sin 2\theta_o$$

$$f = \frac{\Delta L}{(\Delta L)_{max}}$$

$\boxed{f < 1}$ (classically allowed region)

$$\theta_o^{(1)} = \frac{1}{2} \arcsin f$$

$$\theta_o^{(2)} = \frac{\pi}{2} - \frac{1}{2} \arcsin f$$

$\boxed{f > 1}$ (classically forbidden region)

Thus θ_o complex

$$\theta_o = \frac{\pi}{4} \pm \frac{i}{2} \operatorname{arcosh} f$$

$$\sin 2\theta_o = \sin\left(\frac{\pi}{2} \pm i \operatorname{arcosh} f\right)$$

$$= \cos(\pm i \operatorname{arcosh} f) = \cosh(\operatorname{arcosh} f) = f$$

Extreme Classical Model

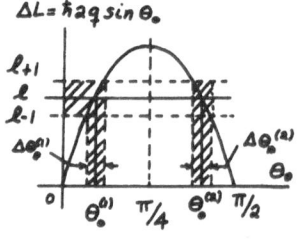

$\Delta L = \hbar a q \sin \theta_o$

ℓ_{+1}
ℓ
ℓ_{-1}
$\Delta \theta_o^{(1)}$
$\Delta \theta_o^{(2)}$
θ_o
$\theta_o^{(1)}$ $\pi/4$ $\theta_o^{(2)}$ $\pi/2$

$$\begin{cases} \theta_o^{(1)} = \frac{1}{2} \arcsin f \\[2ex] \theta_o^{(2)} = \frac{\pi}{2} - \frac{1}{2} \arcsin f \end{cases}$$

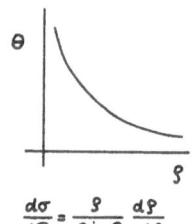

$$\frac{d\sigma}{d\Omega} = \frac{\mathcal{S}}{\sin\theta} \frac{d\mathcal{S}}{d\theta}$$

$$P_1(\Delta L) = \frac{1}{\pi/2} \Delta\theta_o^{(1)} = \frac{1}{\pi/2} 2\left(\frac{\partial\theta_o^{(1)}}{\partial L}\right) = \frac{4}{\pi} \frac{1}{2} \frac{1}{\sqrt{1-f^2}} \frac{1}{(\Delta L)_{max}}$$

$$|P_1| = \frac{2}{\pi} \frac{1}{\sqrt{1-f^2}} \frac{1}{(\Delta L)_{max}}$$

In the same way

$$|P_2| = \frac{2}{\pi} \frac{1}{\sqrt{1-f^2}} \frac{1}{(\Delta L)_{max}}$$

$$P(\theta_o) = \begin{cases} \left(\frac{\pi}{2}\right)^{-1} & (2D) \\[2ex] \sin\theta_o & (3D) \end{cases} \qquad (0 < \theta_o < \frac{\pi}{2})$$

(prob. for isotropic distr. in space)

$$\int_0^{\pi/2} \frac{1}{(\pi/2)} d\theta = 1 \quad ; \quad \int_0^{\frac{\pi}{2}} \sin\theta d\theta = 1.$$

$$|P_1| + |P_2| = \frac{\sin(\frac{1}{2}\arcsin f) + \cos(\frac{1}{2}\arcsin f)}{(\Delta L)_{max}\sqrt{1-f^2}} = \frac{\sqrt{1+f}}{(\Delta L)_{max}\sqrt{1-f^2}}$$

$$\boxed{\begin{aligned} P(f) &= P(\Delta L) = \frac{1}{(\Delta L)_{max}} \frac{1}{\sqrt{1-f}} \\[2ex] f &= \frac{\Delta L}{(\Delta L)_{max}} \end{aligned}}$$

$P(\Delta L)$

f

rainbow effect

ΔL

θ_o

large variation of θ_o, small var. of ΔL

$\dfrac{1}{\partial L/\partial\theta_o} \to \infty$

Semiclassical approximation

$$\boxed{f < 1}$$

$$P(\Delta L) = \left| \sqrt{|P_1|}\ e^{i\phi_1} + \sqrt{|P_2|}\ e^{i\phi_2} \right|^2$$

$$\phi_1 \atop \phi_2 = \pm \frac{\pi}{4} - \frac{1}{\hbar} \int_{-\infty}^{\infty} (x\dot{p}_x + r\dot{p}_r)\,dt$$

$$\frac{1}{\hbar} \int_{-\infty}^{\infty} x\dot{p}_x \,dt \sim 2q\theta_0 \sin 2\theta_0 = 2q\theta_0 f$$

$$\frac{1}{\hbar} \int_{-\infty}^{\infty} r\dot{p}_r \,dt \sim B + 2q\cos^2 2\theta_0$$

$$\boxed{\phi_1 \atop \phi_2 = \pm \frac{\pi}{4} - 2q\theta_0\ f \qquad -2q\cos^2 2\theta_0 + const}$$

$$P(\Delta L) = \underbrace{|P_1| + |P_2|}_{\text{class. result}} + \underbrace{2\sqrt{|P_1||P_2|}\ \cos(\phi_1 - \phi_2)}_{\text{interf.}}$$

$$\boxed{\begin{array}{c} P(\Delta L) = \dfrac{1}{(\Delta L)_{max}} \left[\dfrac{1}{\sqrt{1-f}} + \sqrt{\dfrac{2f}{1-f^2}}\ \cos(\phi_1 - \phi_2) \right] \\[2mm] \phi_1 - \phi_2 = \dfrac{\pi}{2} + 2q\left(f\ arc\ cosf - \sqrt{1-f^2} \right) \end{array}}$$

$\boxed{f > 1}$

$P(\Delta L) = \left| \sqrt{|P_1|} \; e^{i\phi} \right|^2 = |P_1| \; e^{i(\phi - \phi^*)}$

ϕ complex

$\boxed{P(\Delta L) = |P_1| e^{-2 \Im m\phi}}$ choose $\Im m\phi > 0$

$\phi = \pm \frac{\pi}{4} - 2qf\theta_0 - 2q \cos^2 \theta_0$

$\theta_0 = \frac{\pi}{4} + (\pm \frac{i}{2} \text{ arcosh } f)$

$$P(\Delta L) = \frac{\sqrt{f}}{(\Delta L)_{max} \sqrt{2(f^2 - 1)}} \; e^{-2 \Im m\phi}$$

$$-2\Im m\phi = (\Delta L)_{max} \left[\sqrt{f^2 - 1} - f \, \ln \left[f + (f^2 - 1)^{1/2} \right] \right]$$

References

1. A. Bohr and B. R. Mottelson, Nuclear Structure, Vol. II, Benjamin, New York (1965).

2. J. Bardeen, L. Cooper and R. Schrieffer, Phys. Rev. 106 (1957) 102.

3. K. Alder and A. Winther, Electromagnetic Excitations, North-Holland, Amsterdam (1975).

4. R. A. Broglia, O. Hansen and C. Riedel, Adv. in Nucl. Phys., Vol. 6, 287, Plenum Press, New York, 1973.

5. B. Josephson, Phys. Lett. 1 (1962) 251.

6. I. Giaever, Rev. of Mod. Phys. 46 (1974) 245.

7. S. G. Nilsson, Mat. Fys. Medd. Dan. Vid. Selsk. 29, No. 16 (1955).

8. M. Guidry, et al., Nucl. Phys. A361 (1981) 275, and references therein.

9. R. A. Broglia and A. Winther, Heavy-Ion Reactions, Frontiers in Physics, Benajmin, Reading, Massachusetts (1981).

10. C. H. Dasso, T. Dossing, S. Landowne, R. A. Broglia and A. Winther, Nucl. Phys. (in press).

11. R. A. Broglia, G. Pollarolo and A. Winther, Nucl. Phys. A361 (1981).

12. B. R. Mottelson and J. G. Valatin, Phys. Rev. Lett. 5 (1960) 511.

13. F. Stephens and R. Simon, Nucl. Phys. A183 (1972) 257.

14. D. Bes and R. A. Broglia, Nucl. Phys. 80 (1966) 289.

15. R. A. Broglia, D. R. Bes and B. Nilsson, Phys. Lett. 50B
    I. Ragnarsson and R. A. Broglia, Nucl. Phys. A263 (1976) 315.

16. P. G. de Gennes, Superconductivity of metals and Alloys, Frontiers in Physics, Benjamin, Reading (1966).

17. J. D. Garret, et al. (to be published).

18. R. A. Broglia, C. H. Dasso, H. Esbensen, A. Vitturi and A. Winther, Nucl. Phys. A345 (1980) 263.

19. C. Signorini, Procs. of the Workshop on Nuclear Physics, Trieste, Italy, 5-30 October 1981, Eds. C. H. Dasso, R. A. Broglia and A. Winther, North-Holland (1982).

20. P. Bond, et al., Phys. Rev. Lett. 46 (1981) 1565.

21. D. Glassel, et al., Phys. Rev. Lett. 48 (1982) 1089;
    D. V. Harrach, et al., Phys. Rev. Lett. 48 (1982) 1093.

22. D. Brink, Doctoral Thesis, University of Oxford, 1955 (unpublished).

23. J. O. Newton, et al., Phys. Rev. Lett. 46 (1981) 1383.

24. R. A. Broglia, C. H. Dasso, H. Esbensen and A. Winther, Phys. Lett. 104B (1981)

25. R. A. Broglia, C. H. Dasso and A. Winther, Procs. of the International School of Physics, "E. Fermi" on Nuclear Structure and Heavy-Ion Collisions, Varenna, Italy, Eds. R. A. Broglia, C. H. Dasso and R. Ricci, North-Holland, Amsterdam, p. 327 (1981).

26. H. Esbensen, Nucl. Phys. A352 (1981) 147.

27. R. Betts, Procs. of the Workshop on Nuclear Physics, Trieste, Italy, 5-30 October 1981, Eds. C. H. Dasso, R. A. Broglia and A. Winther, North-Holland (1982); F. Becchetti, ibid.

DIFFERENT REGIMES OF DISSIPATIVE COLLISIONS.

A. Gobbi

Gesellschaft für Schwerionenforschung

Darmstadt, W-Germany

The field of dissipative collisions involves a large variety of reaction mechanisms. The most typical characteristics of the process were found to emerge from studies of collisions between very heavy nuclei, at relative velocities of few MeV/u in excess of the Coulomb barrier, where, on the average, the identity of target and projectile remains preserved during interaction times typically of the order of 2-5 10^{-21} s. More recently the investigations were then oriented towards a test of the generality of the phenomenon. New typical domains emerged: the fast fission involving long interaction times, and the projectile splitting for fast violent collisions. In both of these new regimes the identity of target and projectile is not any more preserved until the end of the interaction.

This report shall be subdivided into three sections in accordance with the mentioned domains of interaction times: typical, long and short. They are entitled:

1. Energy dissipation under the influence of the single particle shell structure.
2. Large rearrangement of masses during the fusion process.
3. Dynamical splitting of heavy nuclei (A>86) at incident energies of 12-15 MeV/u.

Since, on this subject, several review articles and detailed reports have, by now, been published, this manuscript is restricted to the general characteristics and to the new developments.

1. Energy dissipation under the influence of the single particle shell structure.

In dissipative collisions the microscopic mechanism of energy dissipation has not been revealed completely. Several, in part totally different, theoretical descriptions were suggested in order to delucidate the most relevant degrees of freedom in which the energy is stored during the dissipative process. One extreme description underlines the role of the collective modes[1] where, as in TDHF-calculations[2], no irreversible dissipation occurs. At the opposite extreme, as in the total level density approach[3], the dissipated energy is immediately converted into heat. The reality may very well lie between these extremes and it still remains a challenge for the theorist to find a general description of the quantum system in evolution towards equilibrium.

Meantime experiments need to be more and more specific and sensitive to the options let open in the model descriptions. In general the distributions of the products observed in dissipative collisions are rather structureless reflecting the quantum fluctuations and the random character inherent to the process. Therefore it has been found that the most sensitive information can be gained from:

(i) the measurement of correlated quantities, in particular the Total Kinetic Energy Loss versus the variance of the element (or mass) distributions, TKEL vs σ_Z^2,

(ii) systematics involving different target-projectile combinations, showing the dependence from mass and charge asymmetry and

(iii) use of colliding nuclei with different structure, in particular, identical nuclei collisions, which simplify the initial conditions.

In the investigations involving a direct comparison between different reactions, great care must be given to the following technical points:

(i) complete cross sections must be measured in order to avoid as much as possible instrumental distortions of the distributions, e.g., in the case of the TKEL vs σ_Z^2 correlation, angle integrated cross sections are needed;

(ii) comparable conditions must be found which are not strongly dependent on trivial differences of the systems, e.g. in the study of the TKEL vs σ_Z^2 correlation, while comparing systems of different asymmetry, it is convenient to take into account the distortion of the distribution due to the asymmetry dependence of the Coulomb repulsion between the nuclei at scission.

Following these considerations, Dakowski[4] was then able to find, on the bases of a model suggested by Nörenberg, a general description reproducing the known data for the TKEL vs σ_Z^2 (σ_A^2) correlation. This is illustrated in figure 1 where, for the 21 reactions investigated, the initial slope of the correlation is compared with the calculated values of average energy-loss per exchanged nucleon. The one to one correspondence between experiment and model becomes meaningfull if one assumes, as in a Brownian motion, that $N_{exch} = \sigma_A^2$.

The physical picture emerging from the model is the following:

1. The single-particle shell-model states, already known from nuclear structure studies, serve as a base for the description of all types of transitions (excitation with and without transfer) during the interaction.

2. The mean energy-loss per transferred nucleon depends, as calculated in a master equation, from the occupation probability of these states, which takes fully into account the Pauli principle.

3. The transition probabilities are governed, like in a direct reaction, by the uncertainty of angular momentum in a single-particle transition, via angular momentum and energy conservation for intrinsic and relative motion.

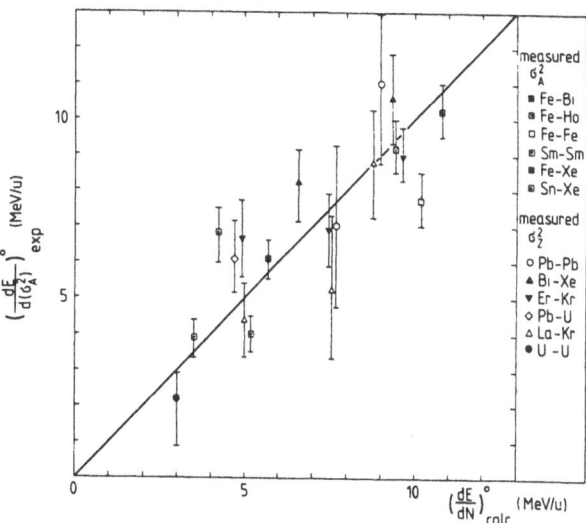

Fig. 1. Correlation between the theoretical $(dE/dN)^0$ values (abscissa) and the $(dE/d(\sigma^2_A))^0$ values extracted from experiments. Both values are obtained from an averaging over the first 25 MeV energy bin of TKEL. The straight line indicates complete agreement between experimental and calculated values. The squares represent systems for which the experimental results were extracted using mass widths. The rest of the data represents those systems, where only charge distributions were experimentally available.

.4. The influence of the single-particle shell structure upon the energy dissipation is most easily understood by considering two magic fragments: it is then seen that with each transition an extra energy roughly equal to the gap energy $\Delta\varepsilon_{shell}$ is needed. Therefore the energy-loss rate for magic fragments should be roughly given by the rate for non magic fragments plus $\Delta\varepsilon_{shell}$. The gap energy is often larger than the energy per nucleon in the relative motion, so that the influence of the shells on the energy dissipation is determinant.

This model description was quite successfull in explaining the observed general trends of the TKEL vs σ_Z^2 correlation, in particular the different behaviour of target projectile combinations like Pb+Pb and U+U. In recent years, improved experimental technics[5] have delivered detailed information about the product distributions in a (N,Z)-plane as a function of energy-loss. In order to describe and understand such observations, Töke[6] has extended the description by Dakowski and performed in a Monte Carlo approach a two dimensional random-walk simulation of stocastic exchanges of individual nucleons between the colliding nuclei. In such an approach the potential energy surface guides the evolution of the system on the projectile-like (N,Z)-plane by determining the available phase space corresponding to the four elementary steps on that plane. In earlier applications of this approach[3,7] the available volume of the phase space, corresponding to a given step, was related to the total level density at the point this step leads to. Recourse to the total level density rather than to the single particle one is, however, not in line with the individual single nucleon exchange (resp.excitation) picture of the kinetic energy dissipation and, indeed, the agreement of such calculations with the experiment has already raised doubts about the applicability of the extreme indipendent nucleon exchange picture[3].

However, the random-walk calculations by Töke were carried out consistently within the single nucleon exchange picture: the available phase space is determined by the single particle level densities (which are kept independent of excitation energy, though different for proton and neutrons) and the relative displacement of the Fermi levels in both partners of the collisions, which is directly related to the potential energy surface.

The quantities which then, in such a random walk simulation, are sensitive to the details of the model (specifically to the response of the branching ratios to the structure of the potential energy surface) are σ_Z^2 values at constant A and σ_A^2 values at constant Z. Results of such calculations are shown in fig. 2 for the $^{56}Fe+^{136}Xe$ reaction. From this it can be concluded that, in spite of its present crudness, the single nucleon exchange model of random-walk agrees well with the experiment.

The observation that, the total level density description is also able to describe most of the features is, at first sight puzzling but, it may be understood as follows: due to the stocastic nature of the process, if the system evolves through few tens of levels or through 10^6 or more levels, the end-result may be very similar, since the single particle spectroscopic strength is distributed over the total number of levels. In the experiment there are almost no observables sensitive to the total density of levels (which may be associated with a temperature); the σ_Z^2 value

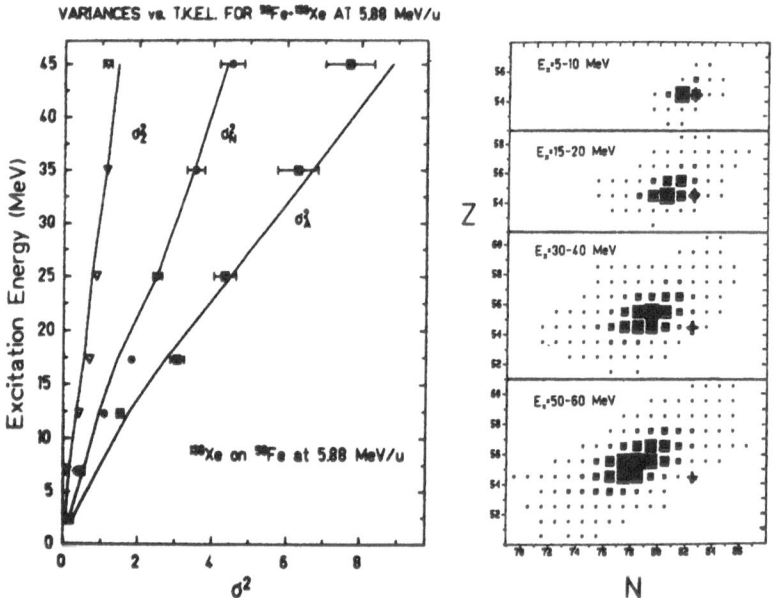

Fig. 2. Results of random-walk simulation for the $^{56}Fe+^{136}Xe$ reaction at 5.88 MeV/u. The experimental results shown in the left part are due to F. Busch et al[7].

at constant A seems to be in this regard the most appropriate quantity, so the models should be accordingly tested.

The described approach is of general applicability and - with a number of refinements introduced - it is expected to be a suitable framework for understanding not only the variances but also the observed drift of the A, Z and N distributions as well.

It is by now well-established that two different time scales govern the drift: a balance of the proton to the neutron abundances in each participant is reached already for small energy losses[8,9], while the trend toward equipartition in mass (or charge) sets in only at the largest energy losses, if ever[9].

These general features correspond qualitatively to the expectations based on liquid-drop potential energy surface calculations: the slope of the potential is steepest along the isospin coordinate. However, calculations performed in the framework of the diffusion model[10] or the fragmentation model[11] show that, under the influence of the liquid drop potential, the element distributions should drift toward symmetry, already for small energy losses. Therefore the absence of a drift in the mass asymmetry coordinate for the non completely damped events is a striking feature of the known data: target and projectile retain, on the average, their identity.

This behaviour has often been ascribed to the stabilizing effect of closed shells which create, in practically all systems studied so far, a dip in the potential energy surface in the vicinity of the injection point: from this point of view, no clear-cut situations has been searched for up to now. The existence of an influence of shell effects on the driving potential between two heavy ions is still a basic problem of actual interest. The ^{184}W+ ^{232}Th reaction, which tests the same (N,Z)-potential energy surface as a collision between two doubly magic ^{208}Pb nuclei (in both cases the same total number of protons and neutrons are involved) was studied by G. Rudolf et al.[12]. It is a system for which the shell effects should manifest themselves more strongly than for any system studied up to now. The strength of the considered shell effect is illustrated in fig. 3b), which shows as a function of the proton and neutron numbers the minimum energy of a dinuclear system calculated on the basis of the shell-corrected liquid drop energy.

However, the experimental data (fig. 3a and c) show that no drift toward the doubly closed ^{208}Pb-shell is observed in the ^{184}W + ^{232}Th reaction.

The absence of drift is suprising. Even if the ^{184}W +^{232}Th system is not situated in the region of steepest gradient, one would expect that the shell effects manifest themselves when the partner have exchanged some nucleons.

In fact, some models have already predicted that the initial identity of the nuclei remains preserved during dissipative collisions: this is the case, for example, when

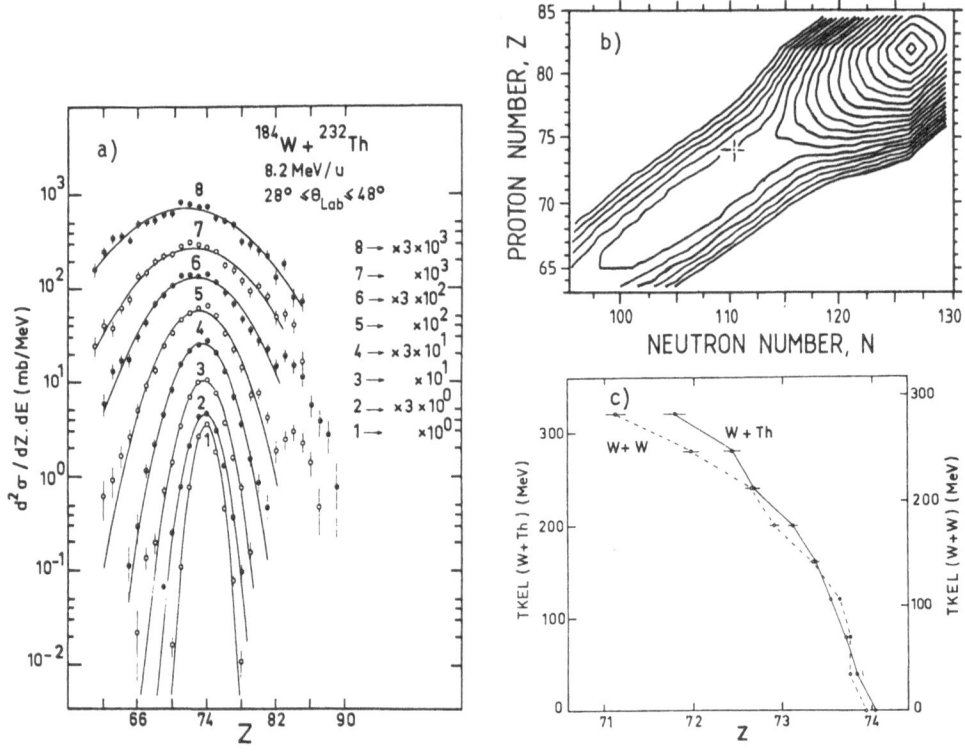

Fig. 3. Investigation of the ^{184}W+^{232}Th-reaction by G. Rudolf et al[12]

a) Element distributions of the projectile-like fragments. Bin number n covers the range TKEL = n x 40+20 MeV.

b) Contour plot of the potential energy surface for a dinuclear system of a total number of 164 protons and 256 neutrons (system ^{208}Pb+^{208}Pb and ^{184}W+^{232}Th).

c) Mean value of the element distributions as a function of TKEL.

excitation via collective modes[1] prevails. From a microscopic point of view, i.e. in a two center shell model, it has been suggested recently that the potential may be diabatic in its nature[13]. This would mean that, at a level crossing, the nucleons do not always follow the state of lowest energy: the system can only undergo small changes of configuration at once. The loss of identity of the two nuclei proceeds only through residual interactions and needs therefore long interaction times, which are related to high excitation energies where shell effects have generally disappeared.

Since the investigated system is located in the vicinity of the strongest possible shell minimum, we are lead to the conclusion that, in general, in deep inelastic collisions the shell structure has no influence on the mean value of the measursed element distributions.

The evidence about the absence of a drift should play, together with other previous specific observations like the influence of shell upon the average energy-loss per exchanged nucleons and upon the σ_A^2/σ_Z^2 values[14], an important role in elucidating the microscopic mechanisms involved in collisions between heavy nuclei. This is of special importance since in the past different descriptions were equally able to describe the general trends of dissipative collisions.

2. Large rearrangement of masses during the fusion process.

Recently considerable effort has been devoted, both experimentally and theoretically, toward a better understanding of fusion as a dynamical process. It has turned out that the study of asymmetric target-projectile combinations, for heavy composite systems at the upper limit of the periodic system, is the most enlightening in this respect. In fact, due to the drastic rearrangement in mass between target and projectile toward an equipartition and due to the instability of the amalgamating system, open to a decay at all stages, the dynamical evolution can be followed through the observation of the distributions in mass and in emission angle of the products. Under these conditions the associated time scale of the process becomes accessible.

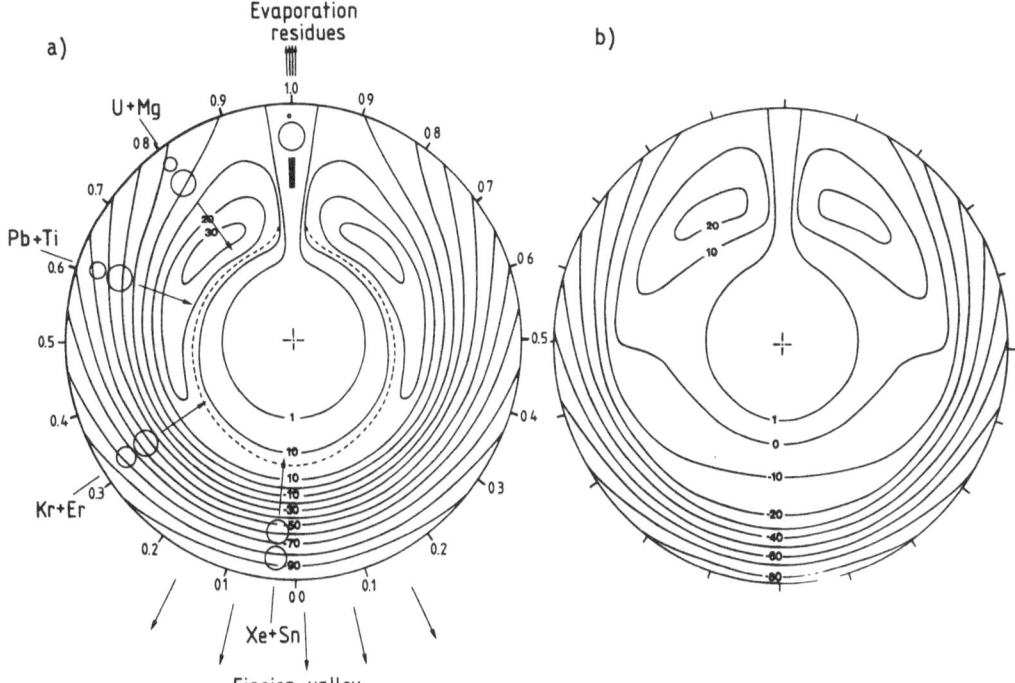

Fig. 4. Potential energy surface for spheres (a) and spheroids (b), represented in polar coordinates.

Figure 4 displays in a contour plot a potential energy surface typical for a composite system at the limits of stability ($^{258}_{104}$X). In the represented polar diagram the radial coordinate is a measure of the distance between the two nuclei while the polar angle describes the change in mass asymmetry. Part a) refers to spherical objects, part b) to aligned elongated spheroids of an axis ratio of 1.5. Both diagrams show that the hight of the conditional saddle responsible, at a given asymmetry, for the molecular binding of the two nuclei is rather low especially around symmetry (where the unconditional saddle is placed). This illustrates the instability of the system.

Such a situation can be seen reflected in the experimental results[15] of figure 5, where the double differential cross section is plotted as a function of product mass and of scattering angle for all the targets bombarded with ^{208}Pb beam at different incident energies. Such a mapping of the transitional region at the limits of stability shows that a real fusion fission occurs only for the light targets, while for heavy targets a symmetric fragmentation is observed only at incident energies in

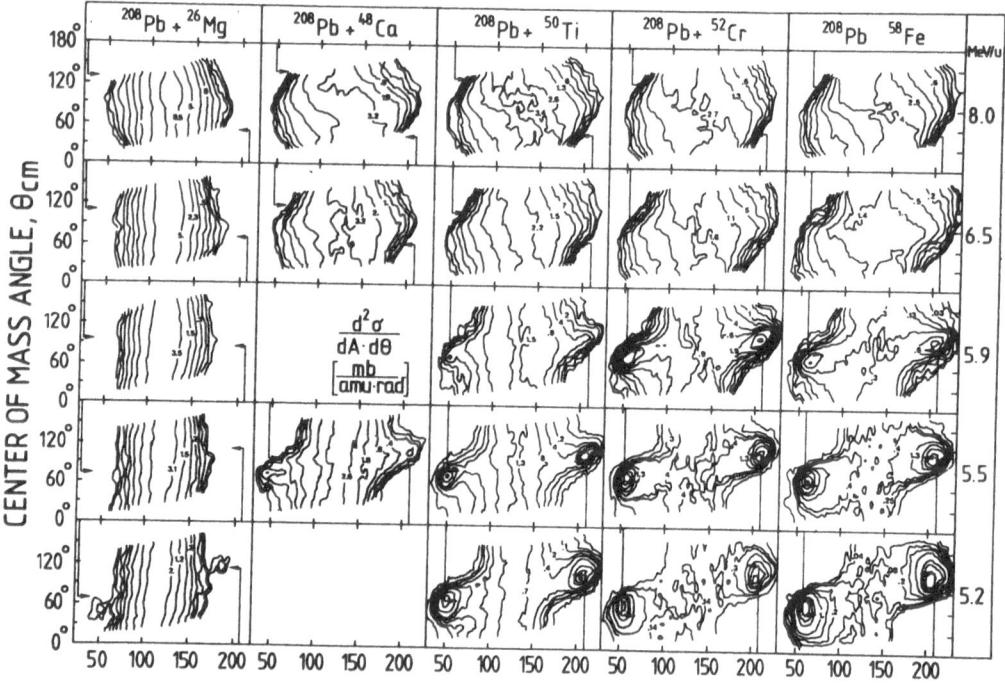

FRAGMENT MASS, A,

Fig. 5. Double differential cross sections $d^2\sigma/(dA.d\theta)$ as a function of the center of mass angle and of the fragment mass. The Rutherford scattering is indicated for the projectile and for the target products by a line up to the grazing angle where an horizontal arrow points in the direction towards symmetric fragmentation. The events were integrated over TKE.

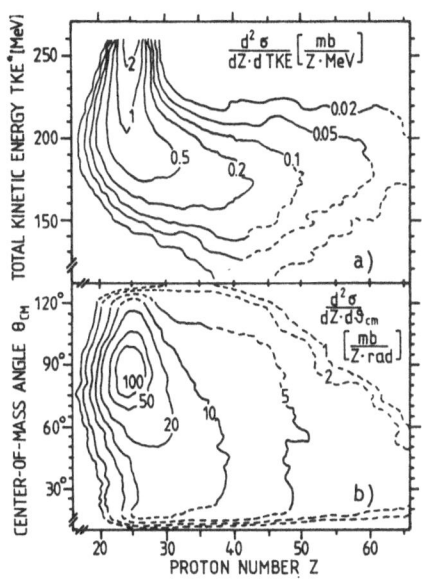

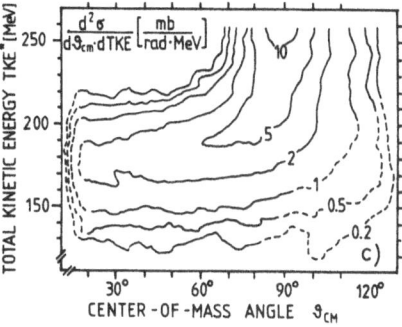

Fig. 6. Results of the $^{56}Fe+^{208}Pb$ reaction at 6 MeV/u.

excess of the barrier and it is not any more separable from the target, projectile-like components. The angular distribution of the symmetric products is in general approximately isotropic, but for the heavy targets and high bombarding energies the forward, backward angles are enhanced by a contribution from the tails of the fast deep inelastic component.

It appears from these measurements that a drift in mass can be observed as a function of scattering angle for the projectile, target-like distributions. Since the described systematic measurement was designed to be selective for the symmetric component and was arbitrarily cutting the deep inelastic component, depending on grazing angle and kinematics, new complementary measurements[16] have been undertaken for the $^{56}Fe+^{208}Pb$-reaction at 5.9 MeV/u: (i) a one particle inclusive measurement of the deep inelastic component, using the position sensitive ionisation chamber and (ii) a radiochemical measurement of the full angular distribution for the symmetric products. The reaction was chosen among the cases of figure 5 so as to observe the continuous evolution in mass from the entrance channel asymmetry until symmetry.

The one particle inclusive data are reported in figure 6. Part a) of the figure displays a Wilczynski diagram typical for a deep inelastic collision. The ridge of the two-dimentional cross section evolves, for increasing energy-loss, toward smaller scattering angles; thus toward the configuration of a long living dinuclear system. Even at the most forward angles the ridge does not reaches the level of a full relaxation in energy (V_{Vi}: Viola-energies of fission).

Fig. 7. Radiochemical measurement of the symmetric fragmentation by Kratz et al[16].

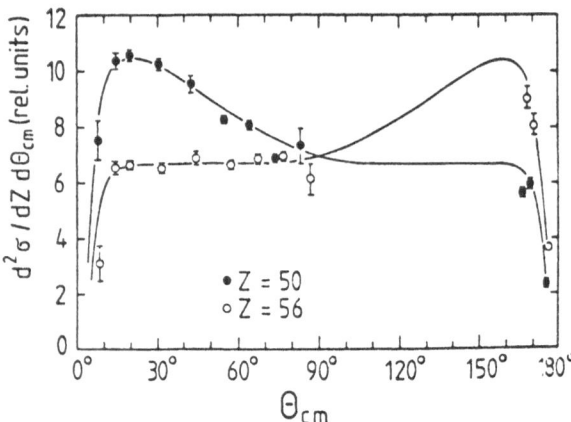

Part b) of figure 14 displays a diffusion diagram. Besides a weak drift toward asymmetry at TKE^*-values $> V_C$ (probably due to a N/Z-equilibration), a drastic drift toward symmetry is clearly seen for $TKE^* > V_C$: the associated large cross section is a striking feature of the Fe+Pb data. The figure shows how the mass drift correlates with a change in the mean total kinetic energy TKE^*. A full relaxation is reached half-way toward symmetry. Large fluctuations in TKE^* are present.

Figure 6c) dispays the correlation between proton number and center of mass scattering angle. In the usual deep inelastic process such a representation was found to be of little relevance since it was observed that the element distribution varies strongly with energy-loss but does not depend on angle, for given energy-loss. However in the regime of a capture process induced by Pb-projectiles on various targets, including Fe, there were good indications[15] that a mass drift can be followed as a function of angle. Indeed the data of figure 6c) show such a trend, for the first time unambiguously, It demonstrates that the mass drift occurs within a time corresponding to a fraction of one rotation of the dinuclear system. Thus with the help of a model calculation it should be possible to extract a rather reliable time scale for the mass drift.

A preliminary estimation gives a value of about 8.10^{-21}s for the mass drift toward symmetry. The possibility of extracting a time scale is of great interest and makes of heavy ion collision a process which, in this respect, is preferable to fission.

In addition to the inclusive data of figure 6, angular distributions were gained with an off-line X-ray activation technic, between 5° and 178° in the center of mass system, at E/B= 1.1, 1.2, 1.3 and 1.6, for product atomic numbers 38<Z<83. For all values of Z (except for $Z_{tot}/2$) angular distributions asymmetric around 90° are observed. As an example figure 7 shows angular distributions for Z=50 and Z=56 at E/B=1.2. The forward backward asymmetry is most pronounced at the lowest bombarding energy and decreases gradually with increasing bombarding energy.

All these measurements show that a good mapping in angle (and of course in mass) is essential in order to distinguish among processes as deep inelastic, fast fission and fusion fission although a continous transition can not be excluded. Additional measurements are needed.

3. Dynamical splitting of heavy nuclei (A>86) at incident energies of 12-15 MeV/u.

The extention of dissipative collision studies to higher relative velocities at 10-20 MeV/u is of great interest for several reasons:

(i) as opposed to the investigations at barrier energies, there is a strong enhancement of the coherent motion of the nucleons of the colliding nuclei, which becomes comparable in magnitude to the random component of the Fermi motion. This situation favours the collective excitation modes. For a given energy-loss the energy is dissipated into fewer degrees of freedom, the interaction time is shorter, damping and residual interaction are smaller. The neck snaps-in faster. Collective states may be able to decay not only into the complicated deep inelastic states but also into the preequilibrium channels.

(ii) Although total excitation energies of 500-1000 Mev can easily be reached, the mean excitation energy per nucleon is, in these heavy systems, still below the binding energy per nucleon so that the mean field still prevents the large bulk of nucleons from escaping. However it is possible at these energies, that few nucleons are promoted into the continuum from where they undergo a preequilibrium decay, so that it may becomes possible to observe more directly the transition energy of the exchanged nucleons.

(iii) Since the collisions are more violent larger deformations are involved. Part of the initial coherence may appear as an internal coherent motion ($v_{rel} \simeq v_{sound} \simeq 18$ MeV). During the collision products more deformed than the saddle shape configuration undergo a prompt decay: this is called splitting, fragmentation or instantaneous fission. Threshold effects as a function of bombarding energy are expected. Through preequilibrium decay and proximity effects, a direct probe of the interacting zone, crucial for our understanding, can in principle be gained.

In a first experiment at 12 MeV/u, using the ^{86}Kr+^{166}Er reaction, we surprisingly discovered that the projectile had a high probability to undergo a fragmentation into two heavy products[17]. The large probability, much higher than expected from compound nucleus reaction studies, induced to postulate the presence of a dynamical process called splitting. Detailed angular and mass distributions of this process have been recently obtained in the case of the Xe+Sn reaction by Harrach, Glässel et al[18].

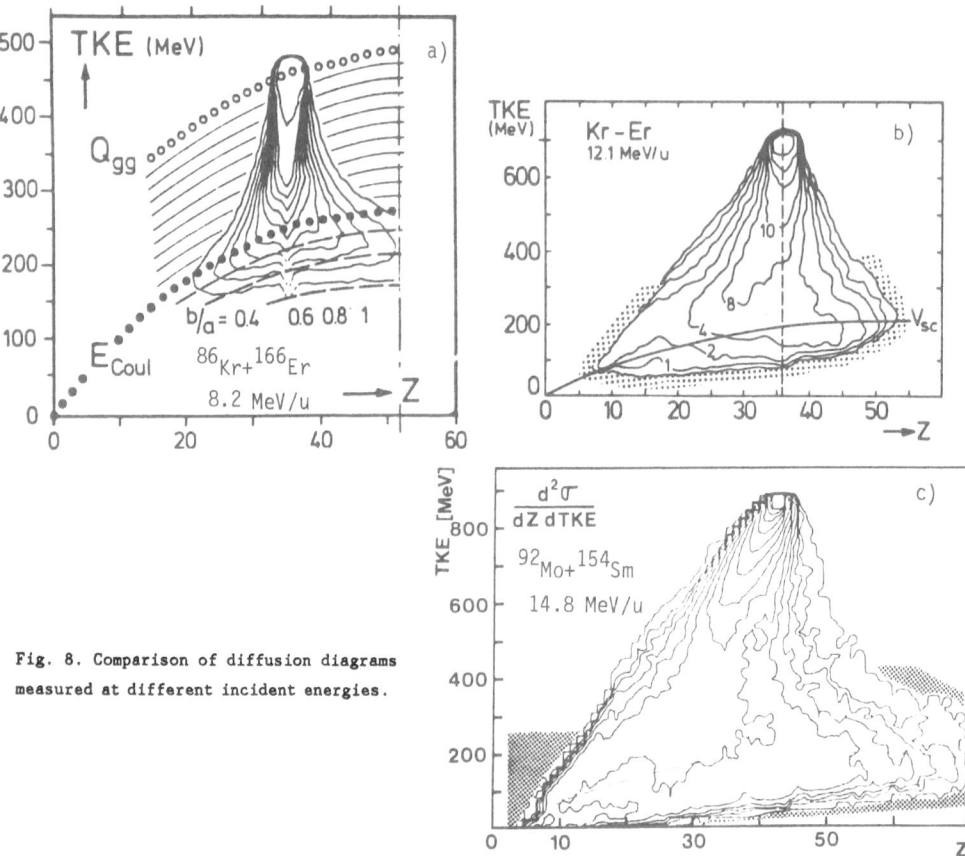

Fig. 8. Comparison of diffusion diagrams measured at different incident energies.

Since the beginning of this year energies in excess of 12 MeV/u are available at the UNILAC-accelerator. Few month ago we have performed an experiment[19] at 14.8 MeV/u with beams of ^{98}Mo and ^{92}Mo on targets of ^{92}Mo, ^{98}Mo, ^{147}Sm, ^{154}Sm and ^{238}U. A new set-up, especially conceived for the detection of three and four-body reactions, was used[20].

The new measurement has confirmed the trends observed earlier as a function of total kinetic energy in the element distributions at 12 MeV/u. This is illustrated in fig-ure 8 which compares results obtained at incident energies of 8, 12 and 14.8 MeV/u. The enhanced production of elements lighter than the projectile does not set-in, for the 3 cases investigated, at the same excitation energy (same energy-loss) but clearly depends on the relative velocity: at the higher bombarding energy it sets-in at lower excitation energies. This shows that we deal here not with decay properties but with dynamical properties of the system. The other interesting observation which emerges from the inclusive data of figure 8 is that the ridge in cross section of elements lighter than the projectile, observed at 14.8 MeV/u, lies definitly outside the range predicted from the systematics of TKEL vs σ_Z^2, well established from the

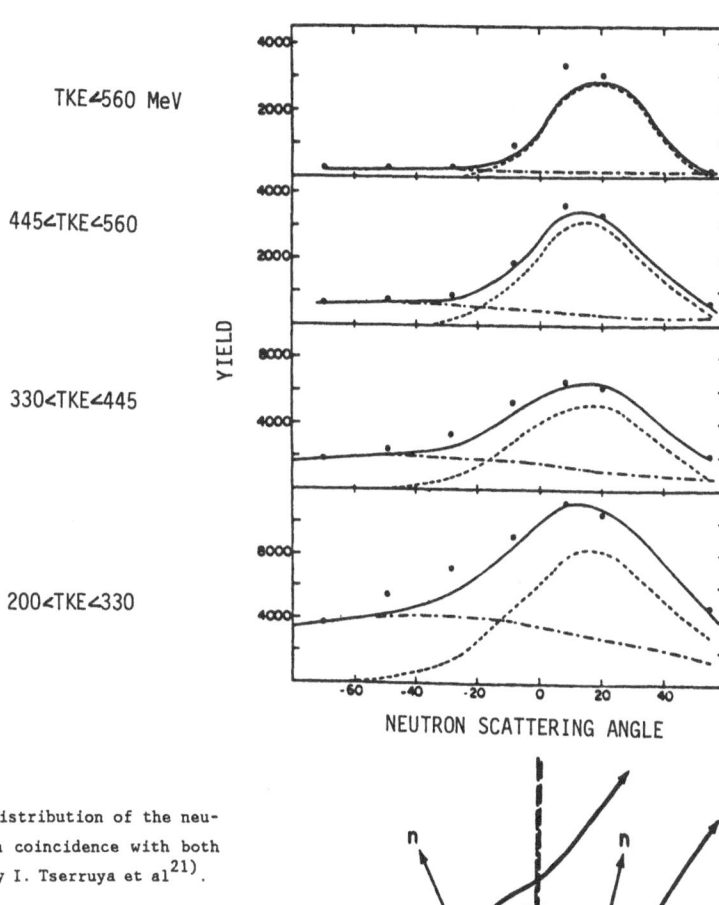

TKE ≤560 MeV

445 < TKE < 560

330 < TKE < 445

200 < TKE < 330

YIELD

NEUTRON SCATTERING ANGLE

Fig. 9. Angular distribution of the neutrons measured in coincidence with both binary products by I. Tserruya et al[21].

low bomarding energy measurements. These trends of the enhanced production of light elements appear also for the symmetric target-projectile combinations we have investigated, showing that part of the total charge is missing if we consider the process as binary.

At the present stage of the investigation it is not possible to say quantitatively to what extent the missing charge is only due to the emission of fragments (splitting) or if a preequilibrium emission of light charged particles (α,p) is also of importance.

Up to now direct evidences for preequilibrium light particle emission in collisions between heavy nuclei (A>80) are very scarce. This is due, on one side, to the present capability of accelerators but also to the complexity of the necessary exclusive experiments. A good evidence was founded by Tserruya et al[21] for the ^{86}Kr+^{166}Er reaction at 11.9 MeV/u. This is illustrated in figure 9 which shows the presence of a weak preequilibrium component of neutrons emitted, in quasielastic events mainly at forwards angles on the side of the projectile-like fragments and for fully damped events, more towards the target-like component. The qualitative features of the data could be reproduced by a simple model[21], which assumes that 10% of the neutrons are knocked-out from the interaction-zone, along the direction of projectile motion. The measured neutron multiplicities were found in good agreement with the predictions of evaporation model calculations.

The results on light charged particle emission are even more scarce than the one on neutrons. Up to now no evidence for a preequilibrium emission could be proved.

A quite striking feature of the high energy data is the presence of a continous distribution of the light fragments extending all the way between protons, alphas and the heavier fragments expected from a fission of the projectile-like products. This feature is demonstrated from figure 10 which shows a scatter plot with the results obtained from the ionisation chamber: the proton number of such light products, which possess to a large extend a high speed and are not stopped within the gas volume of the ionisation chamber, can be identified on the bases of the dispayed diagram, energy-loss versus velocity.

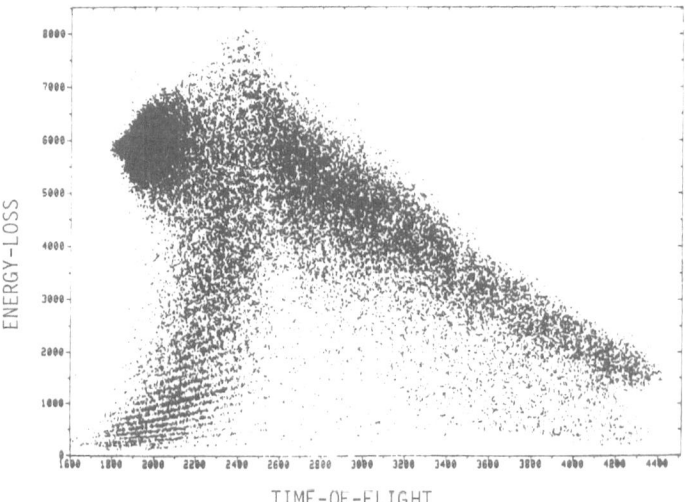

Fig. 10. Products measured in the ionisation chamber for the ^{92}Mo+^{92}Mo reaction at 14.8 MeV/u.

At the moment we know the following, about the light fragments: (i) they are emitted in three or more body reactions, (ii) they have a broad velocity distribution with maximum yield at about a value which corresponds to the Coulomb repulsion energy of a binary reaction, (iii) the angular distribution, although forward peaked, is rather broad, (iv) they are neutron reach. It is still premature to speculate about there origine. A complete kinematic reconstruction is in progress.

Nevertheless the observations so far made suggest that, at high incident energies, a considerable fraction of the kinetic energy is transmitted to a few degrees of freedom also of collective nature (possibly even mainly of collective nature). Such excitation modes can be seen as "doorway" states which then couple either to the more complicated excitation modes, or directly to outgoing channels such as fast particle emission or projectile splitting.

Until now only a few pilot experiments with energetic very heavy ions were performed and therefore the obtained evidences, although quite promising, have in many aspects still a preliminary character.

Acknowledgements: I am very much indebed to R. Bock, S. Bjørnholm, Y.T. Chu, M. Dakowski, S. Gralla, K.D. Hildenbrand, U. Lynen, A. Olmi, W.F.J. Müller, M. Petrovici, G. Rudolf, H. Stelzer and J. Töke for contributing many ideas and hard work in the realisation of the results I have presented here.

References

1) R. Broglia, this Summer School

2) R.J. Cusson, J.A. Maruhn, H. Stöcker, Z. Physik A294 (1980) 257.

3) J.J.,Griffin et.al., Nucl. Phys. A 369 (1981) 181.

4) M. Dakowski, A. Gobbi and W. Nörenberg, Nucl. Phys. A378 (1982) 189.

5) D. Schüll, W.C. Shen, H. Freiesleben, R. Bock, F. Busch, D. Bangert, W. Pfeffer, and F. Pühlhofer, Phys. Lett. 102B (1981) 116.

6) J. Töke and A. Gobbi, GSI-Annual-Report (1982).

7) F. Busch et.al. in Proc. Int. Workshop, IX Hirschegg Kleinwalsertal (1981) 112.

8) B. Gatty, D. Guerreau, M. Lefort, J. Pouthas, X. Tarrago, J. Galin, B. Gauvin, J. Girard, and H. Nifenecker, Z. Physik A273 (1975) 65 and Nucl. Phys. A253 (1975) 511.

   J.V. Kratz, W. Brüchle, H. Gäggeler, M. Schädel, K. Sümmerer, and G. Wirth, Z. Physik A296 (1980) 141.

9) G. Rudolf, A. Gobbi, H. Stelzer, U. Lynen, A. Olmi, H. Sann, R. Stokstad, and D. Pelte, Nucl. Phys. A330 (1979) 243.

10) G. Wolschin, Preprint MPI, Heidelberg (1979) 28. U. Brosa and S. Grossmann, Proceedings of the XIVth Summer School on Nuclear Physics, Mikolajki, Poland (1981).

11) J.A. Maruhn, J. Hahn, H.J. Lustig, K.-H. Ziegenhain, and W. Greiner, Progress in Particle and Nuclear Physics, vol. 4, Oxford (1980) p. 257.

12) G. Rudolf, J.C. Adloff, D. Disdier, V. Rauch, F. Scheibling, M. Dakowski, A. Gobbi, K.D. Hildenbrand, W.F.J. Müller, A. Olmi, submitted for publication to PL.

13) W. Nörenberg, Phys. Lett. 104B (1981) 107.

14) E.C. Wu, K.D. Hildenbrand, H. Freiesleben, A. Gobbi, A. Olmi, H. Sann, and U. Lynen, Phys. Rev. Lett. 47 (1981) 1974.

15) H. Sann, R. Bock, Y.T. Chu, A. Gobbi, A. Olmi, U. Lynen, W. Müller, S. Bjornholm and H. Esbensen, Phys. Rev. Lett. 47 (1981) 1248.
R. Bock, Y.T. Chu, M. Dakowski, A. Gobbi, E. Grosse, A. Olmi, H. Sann, D. Schwalm, U. Lynen, W. Müller, S. Bjornholm, H. Esbensen, W. Wölfli, and E. Morenzoni, to appear in Nucl. Phys. 1982.

16) J.V. Kratz, K. Sümmerer, G. Wirth, C. Gregoire, R. Lucas, J. Poiton, S. Bjornholm, A. Gobbi, G. Guarino, A. Olmi, Proc. Int. Workshop on gross properties of nuclei and nuclear excitations, Hirschegg, 1982.

17) A. Olmi, U. Lynen, J.B. Natowitz, M. Dakowski, P. Doll, A. Gobbi, H. Sann, H. Stelzer, R. Bock, and D. Pelte, Phys. Rev. Lett. 44 (1980) 383.

18) P. Glässel, D.v. Harrach, L. Grodzins, and H.J. Specht, Phys. Rev. Lett. 48 (1982) 1089
D.v. Harrach, P. Glässel, L. Grodzins, S.S. Kapoor, H.J. Specht, Phys. Rev. Lett. 48 (1982) 1093.

19) M. Petrovici, ICOSAHIR Conf. Saclay May 3-7 (1982) to appear in Nucl. Phys. A.

20) G. Augustinski, R. Bock, H. Daues, A. Gobbi, S. Gralla, K.D. Hildenbrand, M. Ludwig, W.F.J. Müller, A. Olmi, M. Petrovici, W. Quick, H. Sann, H. Stelzer, J. Töke, Ann. Rep. GSI, Darmstadt.

21) I. Tserruya, A. Breskin, R. Chechik, Y. Fraenkel, S. Wald, N. Zwang, R. Bock, M. Dakowski, A. Gobbi, H. Sann, R. Bass, G. Kreyling, R. Renfordt, K. Stelzer, and U. Arlt, Phys. Rev. Lett. 47 (1981) 16.
I. Tserruya et.al., to be published.

22) R.Y. Cusson, J.A. Maruhn, W. Greiner and H. Stöcker, "TDHF study of angular momentum transfer in 6 ans 12 MeV/u heavy ion reactions of Kr on Er", preprint.

Charge Equilibration in Deep-Inelastic Peripheral Collisions

D.H.E. Gross and K.M. Hartmann

Bereich Kern- und Strahlenphysik, Hahn-Meitner-Institut für Kernforschung
1000 Berlin 39

Deep-inelastic collisions between heavy ions are being used to study the relaxation of various macroscopic degrees of freedom. These include mass, charge, scattering angle, angular momentum and shape. The energy loss undergone during the reaction is usually employed as the time scale.

It was quickly realized that, already for small energy losses, the charge-to-mass ratio of the detected projectile-like fragment is related to the charge-to-mass ratio of the combined projectile-plus-target system. The rapid equilibration of this degree of freedom gave rise to two opposing theoretical interpretations of the data:

(I) A giant dipole-like mode of the combined projectile-plus-target system is excited in which neutrons and protons are set in a quantized collective motion against each other[1,2]. It's zero point fluctuations determine the width of the charge distribution at constant mass.

(II) Stochastic motion on the potential energy surface (PES) of the combined system determines the neutron and proton numbers of the projectile-like fragment. The system evolves on the PES in the direction of maximum phase space i.e. it evolves so as to minimize its potential energy and thereby maximize the excitation energy of the fragments and thus the number of available states [3-6]. In contrast to (I) the width of the charge distribution at constant mass is of purely statistical origin and proportional to T/C, where T is the temperature and C the curvative of the PES.

No theory exists as yet which combines the quantal fluctuations in (I) with the stochastic ones in (II). We develop a model, appropriate for peripheral collisions, which treats rigorously, at a microscopic level, both quantal and classical contributions to the fluctuation. To this end we assume that the projectile and target move on classical trajectories given by the Newton equation of motion with Gross-Kalinowski friction[7]. The neutrons and protons are pictured as moving in their respective single particle potentials formed from the combined projectile-plus-target system as shown in fig.1. As the projectile and target pass each other on their classical trajectories an approximately parabolic single particle potential barrier is formed between them through which neutrons and protons are able to pass from

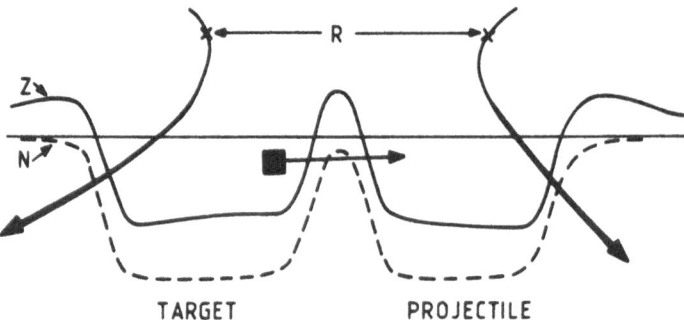

Fig.1: Single particle proton and neutron potentials for the combined projectile-
plus-target system.

one nucleus into the other. The following qualitative consequences, to be borne
out by calculations below, follow immediately: (i) The net nucleon flow is from
the nucleus with the higher to the one with the lower Fermi level. The Coulomb
force acting between a __strongly asymmetric__ projectile-target system is such that
at close separations the protonFermi level of the lighter partner is higher than
that of the heavier partner. The net proton flow is thus away from symmetry as is
clearly evident in the Xe+Fe reaction of fig.2. (ii) There is no net nucleon flow
for __identical__ nuclei since the Fermi levels in projectile and target remain always
equal. However the fluctuations in the number of neutrons and protons transferred
is different. Proton flow is hindered by the higher barrier through which the protons must tunnel, see fig.1. This "neutron-rich flow" is clearly evident for the
near symmetric Xe+Sn system shown in fig.3.

The above are typical phenomena expected for peripheral quasi-elastic collisions
where the neutron and proton single particle barriers are still present. Our theory
aims at accounting for such data. In more central, deep-inelastic collisions these
barriers disappear quickly and other degrees of freedom like neck formation become
more important.

The probability $P(N,Z,t)$ that the projectile-like fragment contains at time t,N
neutrons and Z protons is assumed to satisfy the Fokker-Planck equation

$$\frac{\partial P}{\partial t} = \left(-\frac{\partial}{\partial N} v_N + \frac{\partial^2}{\partial N^2} D_{NN} - \frac{\partial}{\partial Z} v_Z + \frac{\partial^2}{\partial Z^2} D_{ZZ} \right) P \quad . \tag{1}$$

The drift (v_N, v_Z) and diffusion (D_{NN}, D_{ZZ}) coefficients are obtained from the above-described microscopic model where neutrons and protons move in their re-

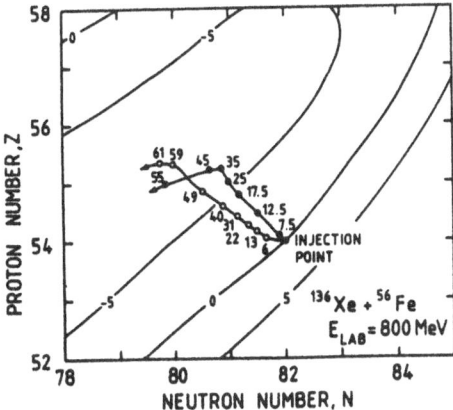

Fig.2: Evolution of the mean neutron and proton numbers on the PES as described in text. Full circles are the experimental[9] and open circles the calculated mean values at various energy losses.

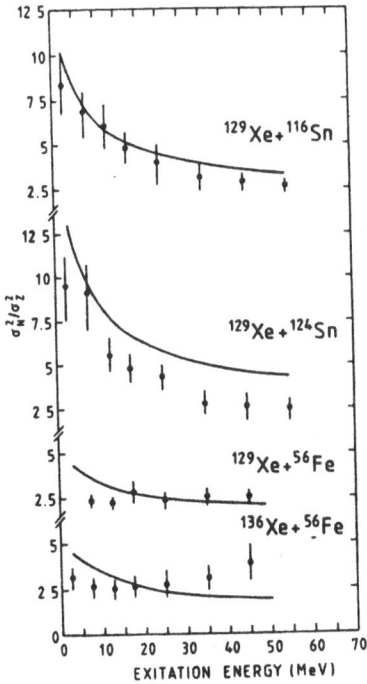

Fig.3: Ratio of neutron to proton variance as function of excitation energy for two near-symmetrical and two asymmetrical systems. Data from ref.10).

spective single particle potentials. The probability that a single phase space cell makes a transfer from target to projectile (say) is the product of the probabilities that (i) the phase space cell in the target is occupied (ii) the corresponding phase space cell in the projectile is empty and that (iii) the phase space cell tunnels through the barrier. The first two of these contributions lead to <u>classical</u> phase space fluctuations while the last contribution gives rise to <u>quantal</u> fluctuations.

The probability that some fraction of the total number of phase space cells in the target transfer to the projectile is given by a generalized binomial distribution formed from the probabilities to transfer single phase space cells. The drift and diffusion coefficients are calculated respectively from the time derivatives of the first and second moments of this distribution and the corresponding distribution for transfer in the opposite direction (projectile to target) [6].

The resulting expression for the neutron (proton) drift velocity is the product of the flux of neutrons (protons) through the barrier and of the driving force. The term describing the flux contains the opening and closing of a window (or the lowering and heightening of the single particle barriers) as the projectile and target pass each other. The driving force, being the difference in the projectile and target neutron (proton) Fermi levels, is the slope of the PES. The diffusion coefficient is, for small energy losses, given by the product of the flux and the magnitude of the driving force and, for large energy losses, by the product of the flux and the nuclear temperature.

In fig.4 we show the dynamical behaviour of several quantities for a trajectory leading to a small energy loss (left) and one leading to a large energy loss (right) in the reaction $^{136}Xe \rightarrow ^{56}Fe$ at E_{lab} = 800 MeV. (i) The proton separation energy S_Z decreases substantially as the nuclei pass each other while S_N remains essentially at its asymtotic value. This enhances proton transfer thereby countering the effect of a higher proton barrier (see fig.1). (ii) However as is shown in frames (c) and (d) of fig.4, the higher proton barrier is more important so that the proton tunneling probability P_Z along the line joining the centres of the fragments as well as the proton window radius ρ_Z remain smaller than the corresponding quantities P_N and ρ_N for the neutrons. (iii) The opposite signs for the driving forces F_Z and F_N indicate that protons and neutrons experience forces in opposite directions - the protons from Xe to Fe while the neutrons from Fe to Xe. (iv) Of critical importance for nucleon diffusion is the relative velocity $\dot{R}$ of the two fragments. It enhances diffusions during the approach phase ($\dot{R}<o$) by helping nucleons to transfer above the Fermi sea. In the exit channel ($\dot{R}>o$) it hinders diffusion by pushing the transferred nucleons below the Fermi sea. This gives rise to strongly entrance channel/exit channel asymmetric diffusion coefficients D_{NN} and D_{ZZ}. (v) Note also that these diffusion coefficients are largest during the time that the tunneling probabilities P_N and P_Z lie between zero and unity. There is thus an important quan-

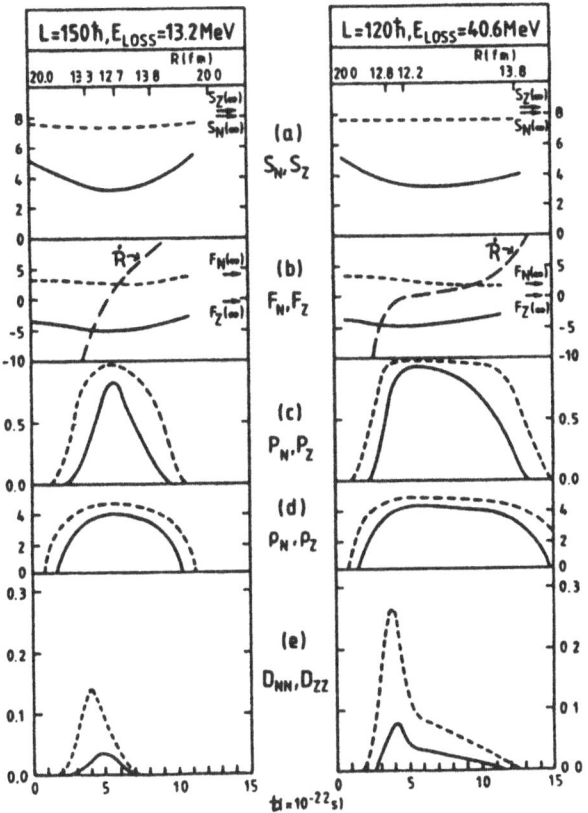

Fig.4: Dynamical behaviour of various quantities described in the text for the reaction $^{136}Xe + ^{56}Fe$ at E_{lab} = 800 MeV. Full lines for protons, dashed lines
for neutrons. The symbols are explained in the text.

tal contribution to D_{NN} and D_{ZZ}. This is in contrast to a classical window where,
since the nucleons either <u>definitely</u> do not (P=0) or do (P=1) pass from one nucleus
to the other, there is no fluctuation caused by the tunneling. In addition to this
quantal fluctuation there is the fluctuation caused by the finite temperature soft-
ening of the Fermi distributions for nucleons and holes.

The asymptotic neutron-proton distribution $P(N,Z,t\rightarrow\infty)$ is most conveniently described by its first and second moments i.e. by the mean values $\langle N \rangle$ and $\langle Z \rangle$, the
variances σ_N^2, σ_Z^2 and the covariance σ_{NZ}^2. These moments satisfy the following
coupled first order differential equations in time

$$\frac{d\langle N \rangle}{dt} = v_N \qquad , \qquad \frac{d\langle Z \rangle}{dt} = v_Z \qquad\qquad (2)$$

$$\frac{d\sigma_N^2}{dt} = 2\left\{\sigma_N^2 \frac{d}{dN} + \sigma_{NZ}^2 \frac{d}{dZ}\right\}v_N + 2D_{NN} \qquad , \qquad (3)$$

$$\frac{d\sigma_Z^2}{dt} = 2\left\{\sigma_Z^2 \frac{d}{dZ} + \sigma_{NZ}^2 \frac{d}{dN}\right\}v_Z + 2D_{ZZ} \qquad , \qquad (4)$$

$$\frac{d\sigma_{NZ}^2}{dt} = \left\{\sigma_N^2 \frac{d}{dN} + \sigma_{NZ}^2 \frac{d}{dZ}\right\}v_Z + \left\{\sigma_Z^2 \frac{d}{dZ} + \sigma_{NZ}^2 \frac{d}{dN}\right\}v_N \qquad . \qquad (5)$$

Since eqs.(3) and (4) contain the source terms D_{NN} and D_{ZZ}, σ_N^2 and σ_Z^2 grow initially linearly with the time. In contrast eq.(5) contains no source term (since we have at the outset excluded cluster transfer of nucleons i.e. short-range correlations). The covariance σ_{NZ}^2 thus begins to increase more slowly (quadratically with the time). Neutron-proton correlation does however build-up because of the long-range correlation induced by the driving force (motion on the PES). A convenient measure for the neutron-proton correlation is given by

$$\chi = \frac{\sigma_{NZ}^2}{\sqrt{\sigma_N^2 \sigma_Z^2}} \qquad . \qquad , \qquad (6)$$

which in a statistical model has values between zero (no correlation) and unity (full correlation).

Fig.2 shows the experimental [9] (solid dots) and calculated (open circles) $\langle N\rangle$ and $\langle Z\rangle$ values at successively larger energy losses for the Xe like nucleus in the ^{136}Xe $+^{56}$Fe reaction at E_{lab} = 800 MeV, superimposed on the PES appropriate for a near grazing collision at the distance of closest approach (contour lines). The system first rolls down the steepest slope of the PES in the (N-Z) direction (charge equilibration) followed by evolution along the (N+Z) direction towards symmetry.

The variation with excitation energy of the ratio σ_N^2/σ_Z^2 is compared in fig.3 for near symmetric (Xe+Sn) and strongly asymmetric (Xe+Fe) systems. The increase of this ratio for all the systems at low excitation energies ("neutron-rich flow") is attributed to a hindrance of proton flow due to the higher proton single particle barrier [8]. However for the asymmetric systems proton diffusion becomes easier and hence σ_N^2/σ_Z^2 smaller for the following reason: the Coulomb field of the heavier nucleus decreases the separation energy of protons in the lighter nucleus more than the Coulomb field of the lighter nucleus decreases the separation energy of protons in the heavier nucleus. This results in an increase of the driving force for asymmetric systems and thus a net proton flow from lighter to heavier system (see Fig.2) and also an increase in $\sigma_z{}^2$ (recall that D_{ZZ} is proportional to the driving force for small temperatures).

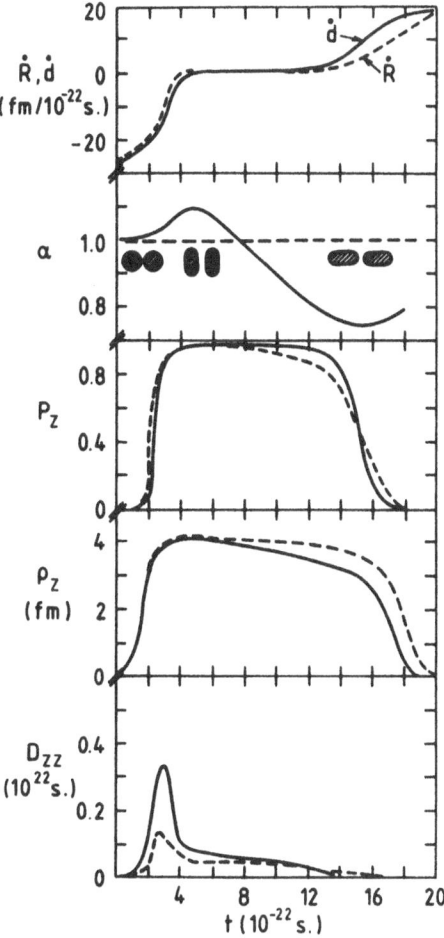

Fig.5: Comparison of the dynamical behaviour of various quantities in a spherical (dashed curve) and deformed (solid curve) calculation for the ^{40}Ar → ^{100}Mo reaction at E_{lab} = 270 MeV and incident angular momentum ℓ = 100.

In order to account for the observed energy loss in deep-inelastic collisions and also for recent fusion cross-section data, Gross and Satpathy[11] have modified the Gross-Kalinowski friction model to include dynamical deformations. The friction force is assumed to depend on the distance d between the nuclear surfaces rather than on the distance R between their centres. The further assumption of nose-to-nose symmetry and the same deformation α (= ratio of axes) in projectile and target simplify the equations of motion considerably. The nuclei become oblate deformed ($\alpha > 1$) in the entrance channel and prolate deformed ($\alpha < 1$) in the exit channel as is shown in fig.5 for one trajectory in the ^{40}Ar+^{100}Mo reaction at 270 MeV. Here for example the proton tunneling probability P_z and window radius ρ_z are only slightly modified as compared with the spherical calculation. Due to the similar behaviour of the relative velocity of the centres $\dot{R}$ in the spherical calculation and of the relative velocity $\dot{d}$ of the surfaces in the deformed calculation the proton diffusion coefficient D_{zz} is similarly enhanced in the entrance channel and hindered in the exit channel (see fig.5).

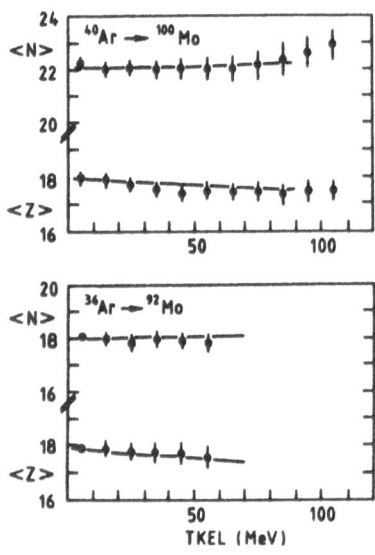

Fig.6: Energy loss dependence of $\langle N\rangle$ and $\langle Z\rangle$ for the $^{40}Ar \to {}^{100}Mo$ and $^{36}Ar \to {}^{92}Mo$ reactions at $E_{lab} = 270$ MeV. Data from ref.[12].

The $^{40}Ar + {}^{100}Mo$ and $^{36}Ar + {}^{92}Mo$ systems at $E_{lab} = 270$ MeV have recently been studied by Bohne et al. at the HMI[12]. These authors have carefully reconstructed the primary from the observed distributions. Fig.6 compares the experimental and the calculated mean neutron and proton values. The slight proton drift away from symmetry is nicely reproduced.

The ratio of the neutron and proton variances σ_N^2 / σ_Z^2 are compared for these two systems in fig.7. This ratio is smaller for the ^{36}Ar induced reaction both because there are fewer neutrons in this system and because ^{92}Mo has a closed neutron shell at N=50.

Fig.7: Energy loss dependence of σ_N^2 / σ_Z^2 for the same systems as in fig.6.

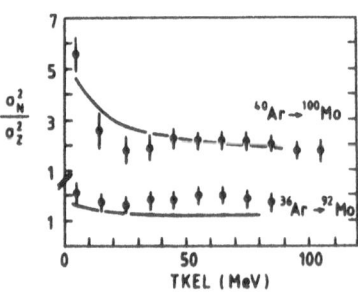

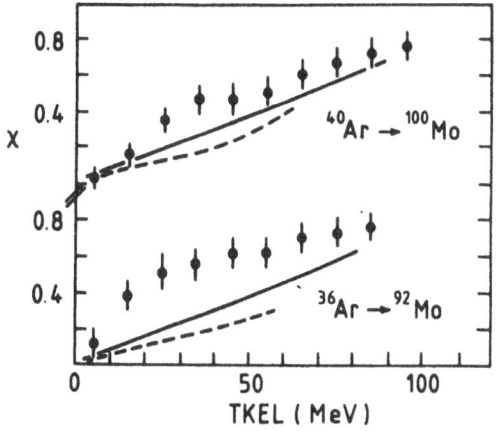

Fig.8: Energy loss dependence of χ for the systems in fig.6. For comparison the spherical calculation is also shown (dashed line).

Finally we compare the proton-neutron correlation coefficient χ (eq.(6)) for these two reactions in fig.8. For low energy losses (TKEL < 50 MeV) the observed N-Z correlation for the ^{36}Ar induced reaction builds up much more rapidly than would be expected from independent particle motion on the PES. We speculate that this may be an indication for additional correlation induced by alpha particle transfer from ^{36}Ar.

Alpha-particle flow, in addition to neutron and proton flow, may be described by adding the terms

$$\left(-\frac{\partial}{\partial\alpha}\,v_\alpha + \frac{\partial^2}{\partial\alpha^2}\,D_{\alpha\alpha}\right)P(N,Z,\alpha,t) \qquad ,$$

to the Fokker-Planck equation (1), which may again be solved by moment expansion. Since the experiment measures only the neutron and proton numbers $\hat{N} = N + 1/2\alpha$ and $\hat{Z} = Z + 1/2\alpha$ in the projectile-like fragment, the distribution $P(\hat{N},\hat{Z},t)$ must be constructed from $P(N,Z,\alpha,t)$. This leads to the appearance of the source term $1/2\,D_{\alpha\alpha}$ in the equation for $d\sigma_{\hat{N}\hat{Z}}/dt$, see eq.(5). Consequently $\sigma_{\hat{N}\hat{Z}}$ grows more rapidly (linearly instead of quadratically) with time i.e. energy loss.

Summarizing, we have developed a well-defined independent particle model for proton and neutron exchange between colliding heavy ions. The model should be specially reliable for quasi-elastic reactions where projectile and target overlap little thus allowing fairly well-defined single particle neutron and proton barriers to form. In this energy loss region the model may be used as a background model for predicting new phenomena e.g. cluster transfer. At higher energy losses other degrees of freedom, like neck formation, may become important. It may then be necessary to include the inertia terms in the Fokker-Planck equation (1)[13].

References:

1) H. Hofmann, C. Gregoire, R. Lucas and C. Ngo, Z. Phys. A293 (1979) 229

2) U. Brosa and D.H.E. Gross, Z.Phys. A294 (1980) 217
   E.S. Hernandez, W.D. Myers, J. Randrup and B. Remaud, Nucl. Phys. A361 (1981) 483

3) W.U. Schröder, J.R. Huizenga and J. Randrup, Phys. Lett. 98B (1981) 355

4) A.C. Merchant and W. Nörenberg, Phys. Lett. 104B (1981) 15

5) J.J. Griffin, Y. Boneh, K.K. Kan and M. Dworzecka, Nucl. Phys. A369 (1981) 181

6) D.H.E. Gross and K.M. Hartmann, Phys. Rev. C24 (1981) 2526

7) D.H.E. Gross and H. Kalinowski, Phys. Rep. 45 (1978) 175

8) U. Brosa and D.H.E. Gross, Z.Phys. A298 (1980) 91

9) D. Schüll, W.C. Shen, H. Freiesleben, R. Bock, F. Busch, D.Bangert, W.Pfeffer and F. Pülhofer, Phys. Lett. 102B (1981) 116

10) D. Schüll, W.C. Shen, W.F.W. Schneider, H. Freiesleben, D. Bangert, F. Busch and F. Pülhofer, GSI Scientific Report 1980, page 18.

11) D.H.E. Gross, R.C. Nayak and L. Satpathy, Z. Phys. A299 (1981) 63
    D.H.E. Gross and L. Satpathy, Phys. Lett. 110B (1982) 31

12) W. Bohne, private communication

13) S. Grossmann and H.J. Krappe, Z. Phys. A298 (1980) 41

NEUTRON-PROTON ASYMMETRY AND FAST FISSION : TWO EXTREME TIME EVOLUTIONS IN DISSIPATIVE HEAVY ION REACTIONS

Christian NGÔ

DPh-N/MF, CEN Saclay, 91191 Gif-sur-Yvette Cedex, France

Abstract

We review two different phenomena occuring in dissipative heavy ion collisions : neutron-proton asymmetry and fast fission. The first one is very fast and exhibits quantum fluctuations. The second one is very slow and dominated by statistical fluctuations. A dynamical model treating both these processes is also described.

Introduction

A nucleus is a complex object which is already very difficulty to understand. Nuclear reactions where we shoot one heavy ion on another should be even much more complicated. When the bombarding energy $E \leq 10$ MeV/A, fortunately, it turns out that despite this complexity simple features can be observed. These features can be understood in terms of a few macroscopic parameters. This is so because they result from a coherent motion of many nucleons. In other words they are of collective nature. The experimental data have strongly suggested to introduce new concepts borrowed from other fields of physics especially to the physics of liquid solid and gases. These concepts had to be adapted to nuclei, because, despite their resemblance with macroscopic objects there are nevertheless some specific differences. To illustrate this point let me just quote two examples :

1. The number of nucleons in a nucleus is large compared to unity but it still remains small (≤ 250) compared to the case of macroscopic systems where there is an enormous number of elements ($\sim 10^{23}$). Therefore nuclear physics provides a rather unique example of microsystems and care should be sometimes exercized when applying for instance methods of statistical mechanics. For example, a macroscopic system at equilibrium can be equally well described using a microcanonical ensemble or a canonical ensemble : both give the same result as far as we do not look at the energy fluctuations of the total system. Therefore a canonical distribution is preferentially used because of its greater simplicity. However for systems, like nuclei, there exist differences which can amount to a few percent in some particular cases [1].

Nuclear physics, in the low energy domain, is dominated by the mean field and the mean free path of the nucleons is large. If we consider a nucleus as a gas, it is quite different from an ordinary gas where the mean free path is small compared to the dimensions of the container where it is enclosed. For instance for O_2 at normal conditions of pressure and temperature the mean free path $\Lambda \sim 500$ Å $= 5 \times 10^{-8}$m. For nuclei the mean free path is of the order or larger than the dimension of the nucleus. This means that most of the collisions that a nucleon suffers occur with the mean field. For an ordinary gas the mean free path can of course be increased by decreasing the pressure because it is inversely proportional to the number of molecules per unit volume. If the pressure becomes 10^{-9} atmosphere the mean free path of the oxygene molecules will become $\Lambda \sim 50$ m. In this case most of the collisions will occur with the walls of the container. Such a gas is usually called a Knudsen gas. However it should also be kept in mind that a nucleus is a degenerate Fermi gas so that quantum effects will also be present.

In these lectures I would like to concentrate on two specific aspects of heavy ion collisions : charge equilibration and fast fission. I would like to apologize to the specialists in this field for remaining as simple as possible and skipping most of the technical details. I will mainly restrict myself to the physical aspects of these two subjects.

Reference

[1] H. Hofmann, Fizika, 9 (1977) 441.

NEUTRON-PROTON ASYMMETRY

1. Introduction

Due to the nature of the nucleon-nucleon interaction there is a strong correlation
between neutrons and protons in isolated nuclei. This is expressed, in the binding
energy of nuclei, by the existence of a symmetry energy term in liquid drop formulas. What happens to this correlation when another nucleus its brought in close contact? This can be investigated in deep inelastic reactions because of the following
reasons :
- A broad range of interaction times are involved in deep inelastic reactions and
allow to investigate dynamical changes of this correlation.
- A lot of excitation energy is transformed from the relative motion to the intrinsic degrees. It is interesting to know the possible influence of such an excitation
energy on the neutron-proton correlation.
The first study of this correlation was done by Gatty et al.[1] . They have shown
that it varies very rapidly during the course of the reaction and reaches an equilibrium value. This equilibrium value is a property of the composite system and does
not depend on the value it has initially in the projectile or target nucleus. Various names have been used in the litterature : charge equilibration,N/Z equilibration,isospin transfer [2] . The equilibrium value that the neutron-proton asymmetry
reaches in a deep inelastic process can be understood in terms of a simple model
where we assume that we form a composite system consisting of two liquid drops in
contact. The potential energy of such a system is equal to :

$$\mathcal{E} = E_{LD}(Z_1,A_1) + E_{LD}(Z_2,A_2) + V_N(R) + \frac{Z_1 Z_2 e^2}{R} + \frac{\ell(\ell+1)\hbar^2}{2\mu R^2} \qquad (1)$$

where the two first terms are the binding energy of the two nuclei (Z_1,A_1) and $(Z_2,$
$A_2)$, and the three last terms respectively the nuclear, Coulomb and centrifugal interaction between these nuclei separated by a distance R. Because the total number
of neutrons and protons is kept constant,$\mathcal{E}$ only depends on Z_1,A_1 and ℓ.
If we want to investigate neutron-proton asymmetry we have to measure simultaneously the atomic number and the mass of the fragments. Such an experimental investigation provides information about two macroscopic variables : mass and neutron-proton asymmetry. The first one is well known in deep inelastic reactions [2] . It corresponds to a transfer of nucleons from one nucleus to the other. The direction of
the transfer is such that the potential energy of the composite system $\mathcal{E}$ decreases

when nucleons are exchanged. Mass asymmetry is also known to have a very slow evolution towards equilibrium. Typical relaxation times of the order of $5\text{-}10 \times 10^{-21}$s can be estimated. On the other hand neutron-proton asymmetry is very fast, with a relaxation time which can be estimated to be of the order of a few 10^{-22}s. In the long run, mass asymmetry and neutron-proton asymmetry correspond to an exchange of nucleons. How can we get such a big difference between the relaxation times? This can be roughly understood with the following picture :

Assume that we form a composite system. Then there will be a flow of nucleons from the projectile to the target and vice-versa. The flow in one direction can be very large and in this case many nucleons will go in both directions. The evolution of mass asymmetry corresponds to the difference between these two flows. If they are almost equal we can imagine that mass asymmetry changes very slowly. On the contrary neutron-proton asymmetry is directly related to one of these two flows. Consequently it can be much faster.

Because of the large difference between the relaxation times of mass and neutron proton asymmetry, we expect that as soon as mass transfer is large we will observe the equilibrium value for neutron-proton asymmetry. This means that we can only follow a dynamical evolution of neutron-proton asymmetry to equilibrium when mass asymmetry is close to the initial one. However to get this dynamical information we need a quantity which is related in some way to the time. Because the energy loss in the relative motion is also very fast (its relaxation time is of the order of $3\text{-}5 \times 10^{-22}$s) it can be used as a clock. Therefore the investigation of neutron-proton asymmetry for small mass transfer as a function of the energy loss can provide some indication about its relaxation to equilibrium. This kind of experiment has been performed the first time on the ^{86}Kr + ^{92,98}Mo at GSI [3]. The measure of the mass of one fragment is directly related to mass asymmetry whereas the measure of the atomic number for a given mass (isobaric-distribution) provides information on neutron-proton asymmetry. The study of the isobaric distribution will thus give us all information concerning the neutron-proton asymmetry.

2. Experimental results

At equilibrium the mean value of the isobaric distribution can be easily obtained if we look for the minimum of the potential energy $\mathcal{E}$ for a fixed mass of one fragments. The result obtained in this way is in very good agreement with the experimental data [1,3].

Additional information can be obtained if we look at the evolution of the FWHM of the isobaric distribution. Fig. 1 shows its evolution for several masses, as a function of the total kinetic energy of the products, for the ^{86}Kr + ^{92}Mo system. The FWHM increases strongly as the total kinetic energy of the fragments decreases. A

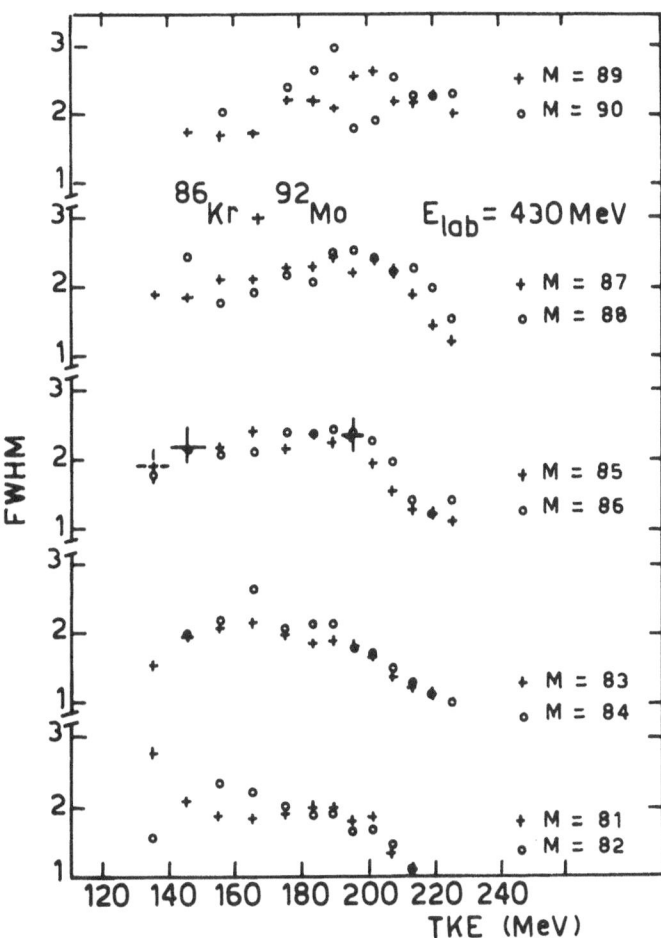

Fig. 1. FWHM of the isobaric distributions for several masses as a function of the total kinetic energy. From ref.[3].

plateau seems to be reached when the energy loss is larger than about 30 MeV. Such an evolution shows the relaxation of the neutron-proton asymmetry until equilibrium is reached. It also means that the neutron-proton asymmetry relaxes to equilibrium in a time which is shorter than the energy damping. Indeed around 100 MeV energy loss are necessary for the kinetic energy in the relative motion to be completely relaxed. Other experiments, on different systems, have shown a similar behavior [4a] except one, concerning the Xe + Au system [4b]. For this latter system the FWHM seems not to reach a plateau for large energy losses and the absolute value of the FWHM is too small compared to what we can extrapolate from the Kr + Mo system.

3. Static interpretation of neutron-proton asymmetry

The equilibrium isobaric distribution can be understood in terms of simple potential
energy considerations. As we have already discussed above, the mean equilibrium value is obtained from the condition :

$$\frac{d\mathscr{E}}{dZ_1}\bigg|_{A_1 = Cte} = 0 .$$ (2)

As far as the second moment of this distribution is concerned, the problem reduces
to a harmonic oscillator coupled to a heat bath at temperature T. The heat bath consists of the intrinsic degrees. In this case we know from elementary quantum mechanics that the variance σ at equilibrium is given by :

$$\sigma^2 = \frac{1}{B\Omega^2} \left[\frac{1}{2} \hbar\Omega + \frac{\hbar\Omega}{e^{\frac{\hbar\Omega}{T}} - 1} \right]$$ (3)

where B is the inertia and Ω the frequency of the motion. These two quantities are
related to the stiffness coefficient C by :

$$C = B\Omega^2 .$$ (4)

Equation (3) has two limiting cases of particular interest :
1. When the phonon energy $\hbar\Omega$ is much smaller than the temperature : $\hbar\Omega \ll T$ eq.(3)
reduces to :

$$\sigma^2 = \frac{T}{B\Omega^2} = \frac{T}{C}$$ (5)

in such a case we have statistical fluctuations. This means that,at equilibrium,the
FWHM of the isobaric distribution should increase like $\sqrt{T}$ i.e. with increasing energy loss. In the case where statistical fluctuations can be observed, the harmonic
oscillator is excited as the temperature of the heat bath increases. The equilibrium
distribution is just the Boltzmann distribution.
2. When the phonon energy $\hbar\Omega$ is much larger than the temperature, $\hbar\Omega \gg T$, the variance is given by :

$$\sigma^2 = \frac{\hbar}{2B\Omega} = \frac{\hbar\Omega}{2C}.$$ (6)

This means that, at equilibrium, the FWHM of the isobaric distribution is independent of the temperature, i.e. of the energy loss. We observe quantum fluctuations.
These fluctuations result from the zero point motion.
The preceding considerations concerning the equilibrium distribution in a harmonic
oscillator coupled to a heat bath are summarized in the schematic picture shown in
fig. 2. In both cases the heat bath has a temperature T=2 MeV and the stiffness
coefficient is the same. The inertia is chosen in such a way that on the left hand

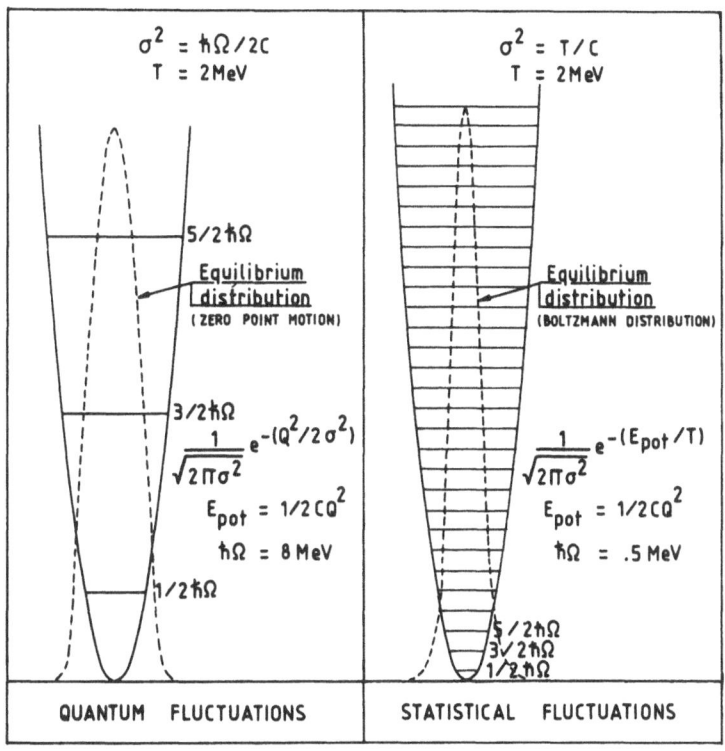

$\sigma^2 = \hbar\Omega/2C$
T = 2 MeV

$\sigma^2 = T/C$
T = 2 MeV

5/2 $\hbar\Omega$

Equilibrium distribution
(ZERO POINT MOTION)

Equilibrium distribution
(BOLTZMANN DISTRIBUTION)

3/2 $\hbar\Omega$

$\dfrac{1}{\sqrt{2\pi\sigma^2}} e^{-(Q^2/2\sigma^2)}$

$E_{pot} = 1/2 \, CQ^2$

$\hbar\Omega = 8 \, MeV$

1/2 $\hbar\Omega$

$\dfrac{1}{\sqrt{2\pi\sigma^2}} e^{-(E_{pot}/T)}$

$E_{pot} = 1/2 \, CQ^2$

$\hbar\Omega = .5 \, MeV$

5/2 $\hbar\Omega$
3/2 $\hbar\Omega$
1/2 $\hbar\Omega$

QUANTUM FLUCTUATIONS

STATISTICAL FLUCTUATIONS

Fig. 2. Schematic presentation of the difference between statistical and quantum fluctuations (see text).

side the phonon energy is equal to 8 MeV whereas on the right hand side it is equal to 0.5 MeV. This means that the levels on the left are spaced by 8 MeV whereas they are spaced by 0.5 MeV on the right. As a consequence the heat bath does not excite the left hand side harmonic oscillator and we observe the zero point motion with an equilibrium distribution which is represented by the dashed line. On the contrary, we observe on the right hand side that the heat bath excites the harmonic oscillator. The equilibrium distribution is the Boltzmann distribution which is also a gaussian represented by the dashed line. To summarize, on the left hand side, the collective variable Q exhibits quantum fluctuations whereas on the right hand side, it exhibits statistical fluctuations.

It should be noted that, in the discussion above, we have assumed the inertia B and the stiffness coefficient C to be constant during the process. Of course this is an oversimplified picture because these transport coefficients are expected to change during the course of the reaction, especialy the inertia [5]. However even this very simplified picture gives us a rather good understanding of the experimental results. From the above discussion we can conclude that, for the $^{86}Kr + ^{92}Mo$ system for which we observe a plateau for the FWHM, we have likely to deal with quantum fluctuations.

In this case we have to apply eq.(6). From the potential energy $\mathcal{E}$ of the composite system we can easily calculate the stiffness coefficient C :

$$C = \frac{d^2\mathcal{E}}{dZ_1^2}\bigg|_{A_1} .$$ (7)

If we extract from the experimental data the variance σ at equilibrium, we can estimate the phonon energy associated to the mode using liquid drop formula of ref. [6]. In this case we find about 8 MeV for the ^{86}Kr + ^{92}Mo system. This gives us a hint concerning the possible origin of the collective mode which is responsible of charge equilibration. Indeed it is natural to relate the isospin mode to the longitudinal component of the giant dipole resonance of the composite system. In such a case the phonon energy would be of the order of :

$$\hbar\Omega = \frac{78}{A_1^{1/3} + A_2^{1/3}} \text{ (MeV) .}$$ (8)

If we apply this formula to the ^{86}Kr + ^{92}Mo system we indeed find a phonon energy of the order of 8 MeV which is in good agreement with the one deduced from eq.(6). Such a high value of the phonon energy is also consistent with the fact that we observe quantum fluctuations. For other systems a similar agreement can be found except for the Xe + Au system which can only be understood in terms of statistical fluctuations i.e. using eq.(5).

4. Time dependent Hartree-Fock calculation of neutron-proton asymmetry

We have assumed that charge equilibration is a collective effect. This is by no means obvious and was an *a priori* assumption. Charge equilibration could, for instance, result of single particle effects. However a very careful calculation treating the nucleon exchanges as a random walk process and taking into account the Pauli principle [7] has shown that it was not possible to reproduce the results concerning systems for which quantum fluctuations are observed. This might indicate that if quantum effects are present they could be due to a collective effect. It is interesting to study the very beginning of a dissipative heavy ion collision like deep inelastic or compound nucleus formation in a microscopic way where no collective degree of freedom has been introduced *a priori*. Such an investigation can be performed, using time dependent Hartree-Fock,if we look at the dynamical evolution the neutron-proton asymmetry at the very beginning of a dissipative heavy ion collision. Such a calculation has been performed in ref.[8] on two systems : ^{16}O + ^{40}Ca and ^{16}O + ^{48}Ca. Information concerning neutron-proton asymmetry has been obtained by looking at the evolution of the center of mass of neutrons with respect to the center of mass of protons of the whole system. For the 84 MeV ^{16}O + ^{40}Ca and ℓ=46, fig. 3 displays the

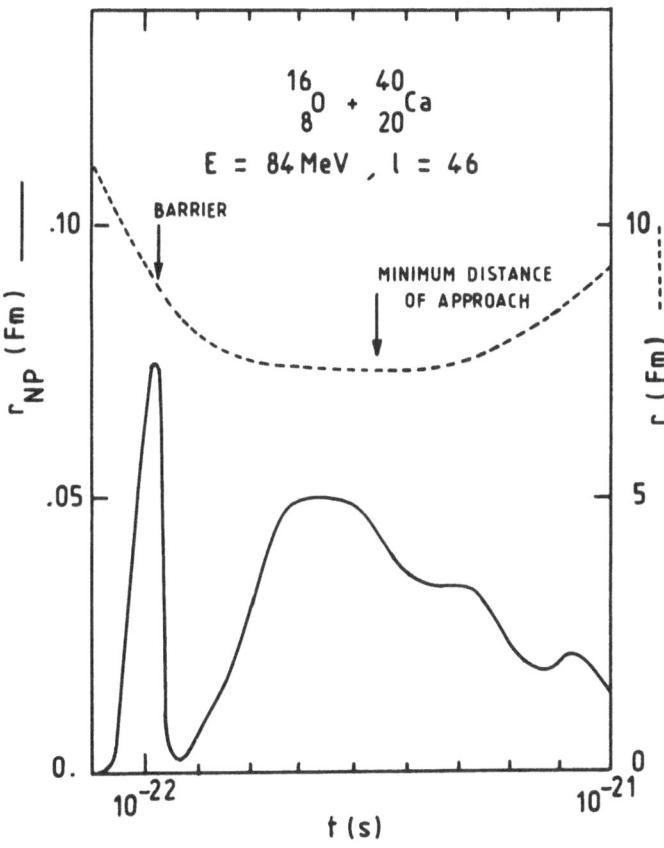

Fig. 3. Distance r_{np} between the centers of mass of the protons and neutrons of the composite system plotted as a function of time (full line). The dotted line corresponds to the distance between the centers of mass of the fragments as a function of time. From ref.[8].

modulus r_{np} of the distance between the center of mass of neutrons and protons of the composite system (full line) as a function of time. This trajectory corresponds to a deep inelastic collision. The origin of time has been taken when the two ions are at a distance of 11 Fm. Two peaks can be observed : one around 10^{-22}s, and a second one later. This means that we have very likely two collective modes corresponding to different frequencies. By looking at the projection $\vec{r}_{np} \vec{i}$ of the vector $\vec{r}_{np}$ joining the centers of mass of the neutrons and protons on the principal axis of inertia unit vector, $\vec{i}$, of the composite system, we will extract the longitudinal component of the motion. This is shown in fig. 4 where we observe that the first peak has dissapeared whereas the broad one remains. This means that two modes are excited at the very beginning of the collision : first a transversal one when $t \sim 10^{-22}$s, and then the longitudinal one when $t \sim 3 \times 10^{-22}$s. This is what indeed could be expected when we extrapolate the splitting of the giant dipole resonance of

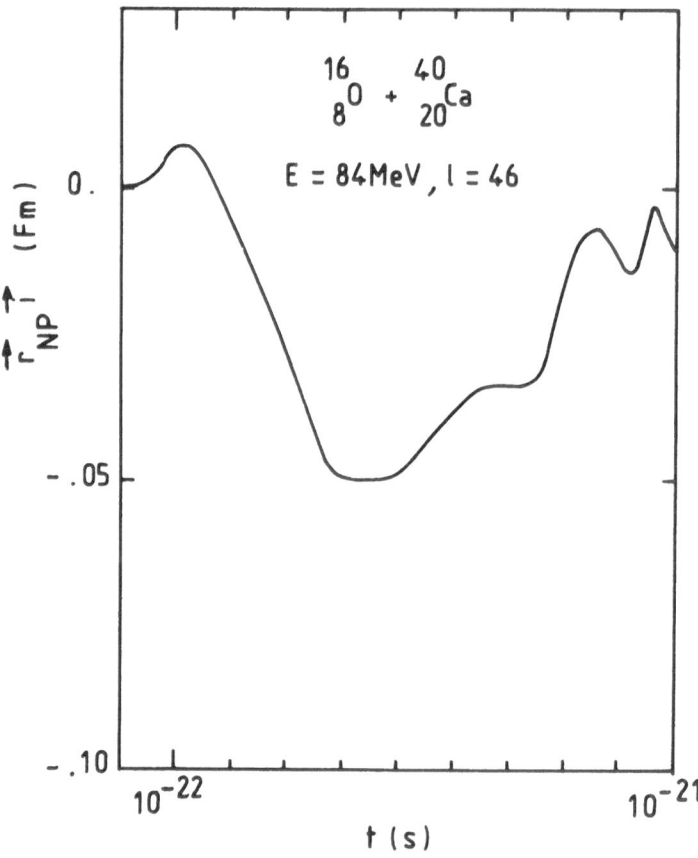

Fig. 4. Projection $\vec{r}_{np} \cdot \vec{i}$ of the distance between the center of mass of the protons and neutrons on the principal axis of inertia of the composite system plotted as a function of time. From ref.[8].

a spherical nucleus which is stretched. Furthermore we see that the transversal mode is excited on top of the fusion barrier whereas the longitudinal mode is excited in the way between the fusion barrier and the minimum distance of approach. The evolution of the neutron-proton asymmetry occurs on this latter time scale indicating that it is directly connected to the longitudinal motion. The frequency of the longitudinal motion can be extracted from the oscillations in fig. 5, for the same system but for $\ell=44$, which corresponds to a fusion trajectory. When the longitudinal mode is excited it oscillates with a period of the order of 3×10^{-22}s. The corresponding phonon energy is about 13-14 MeV. This value is in surprisingly good agreement with eq.(8).

The relaxation time of neutron-proton asymmetry can also be estimated from the TDHF calculation by looking at the $^{16}O + {}^{48}Ca$ system which corresponds to an initial neutron excess out of equilibrium. Fig. 6 shows $\vec{r}_{np} \vec{i}$ as a function of time. We see

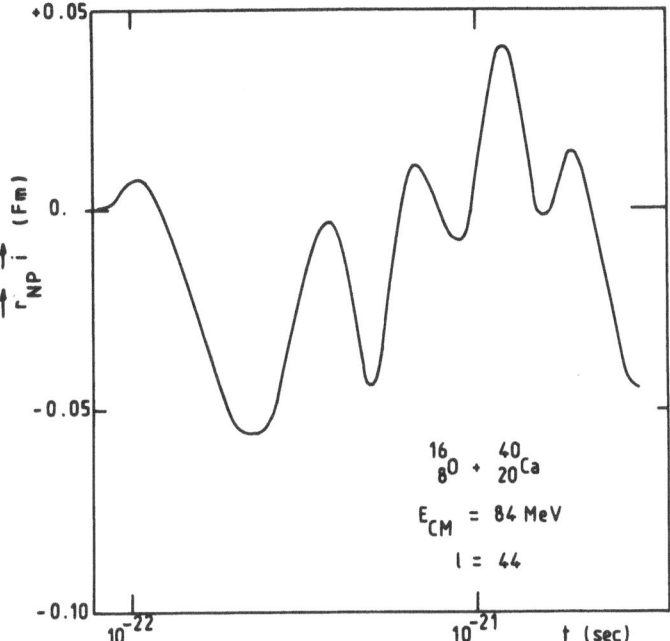

Fig. 5. Same as fig. 4 for ℓ=44, which corresponds to a fusion reaction. From ref.[8].

that we start with a very large negative value of $\vec{r}_{np}\,\vec{i}$ which falls rapidly to almost zero. This evolution arises because of two effects : first there is a pure kinematical effect due to the fact that the two ions move with respect to each other
and second there is a change due to neutron-proton asymmetry relaxation. The first
effect can be calculated and subtracted. The result is that neutron-proton asymmetry
relaxes to equilibrium in a time of the order of 2×10^{-22}s. This relaxation occurs
before the minimum distance of approach has been reached but starts after the top of
the fusion barrier has been overcomed.
To summarize, it seems now very likely that the neutron-proton asymmetry degree of
freedom is of collective nature and that it can be related to the longitudinal component of the giant dipole resonance of the composite system.

5. Dynamical treatment of neutron-proton asymmetry

We shall now use a dynamical model which will be described in the next lecture, to understand the evolution of charge equilibration to equilibrium. This model describes
explicitly the relative motion of the two ions during the collision, together with
mass and the neutron-proton asymmetry transfer [9]. In this lecture we will restrict
ourselves to this latter characteristic of the model. The model basically describes

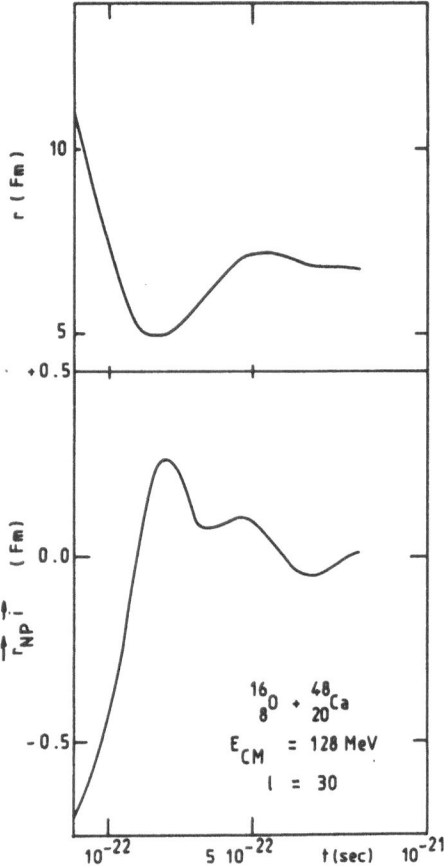

Fig. 6. Bottom : same as fig. 4 for the 128 MeV $^{16}O + ^{48}Ca$ reaction with ℓ=30. Top : distance between the center of mass of the fragments as a function of time. From ref. [8].

the dynamical evolution of the set of collective degrees using a transport equation. As far as neutron-proton asymmetry mode is concerned, quantum effects are explicitely taken into account. To simplify the presentation, let us assume that we have to only deal with one collective degree Q which will be in this case the neutron-proton asymmetry. P will be the conjugate momentum. Then the Wigner transform f of the collective density operator satisfies the following transport equation [10]

$$\frac{\partial f}{\partial t} = - \frac{P}{B} \frac{\partial f}{\partial Q} + B\Omega^2 Q \frac{\partial f}{\partial P} + \gamma \frac{\partial}{\partial P} \frac{Pf}{B} + D \frac{\partial^2 f}{\partial P^2} \, . \tag{9}$$

This equation can be deduced using the approach of Hofmann and Siemens [11] based on linear response theory. In this equation, γ is the friction and D the diffusion coefficient. The diffusion coefficient D is related to γ by the fluctuation dissipation theorem which can be expressed as a generalized Einstein relation :

$$D = \gamma \, T^* \qquad (10)$$

where T^* is the mean energy of the oscillator put into the heat bath at temperature T :

$$T^* = \frac{\hbar\Omega}{2} \, \text{Cotgh} \, \frac{\hbar\Omega}{2T} . \qquad (11)$$

The usual Einstein relation :

$$D = \gamma \, T \qquad (12)$$

is only obtained in the high temperature limit, i.e. when T is much greater than $\hbar\Omega/2$. It should be noted that the transport eq.(9) can describe the low temperature limit where quantum effects are dominating but it can also describe the high temperature limit where statistical fluctuations appear as well as the transition between these two regimes.

The solution of this equation is a gaussian [10,11] and it is sufficient to know the first and second moments of the macroscopic variables. The first moments satisfy a classical Newton equation with a friction force proportional to the collective velocity :

$$\frac{d<P>}{dt} = - B\Omega^2 <Q> - \gamma \frac{<P>}{B} \qquad (13)$$

where the brackets in eq.(13) denote the average value of what is inside. The second moments satisfy a coupled set of first order linear differential equations and give the fluctuations around the mean values. For time going to infinity we get an equilibrium solution which is exactly the same as the one which we have discussed in section 3. In particular the equilibrium variance is given by :

$$\sigma^2 = \frac{T^*}{B\Omega^2} = \frac{T^*}{C} \qquad (14)$$

which is identical to eq.(6) because $T^* = \frac{\hbar\Omega}{2}$ in the low temperature limit.

We would like now to discuss a little bit the physical meaning of the transport equation. This can be done with the help of fig. 7. The transport eq.(9) can be rewritten using a Poisson bracket :

$$\{H,f\} = \frac{\partial H}{\partial Q} \frac{\partial f}{\partial P} - \frac{\partial f}{\partial Q} \frac{\partial H}{\partial P} \qquad (15)$$

as :

$$\frac{\partial f}{\partial t} = \{H,f\} + \gamma \frac{\partial}{\partial P} \frac{P}{B} f + D \frac{\partial^2 f}{\partial P^2}. \qquad (16)$$

If we would remove the two last terms on the left of eq.(16), then it is just the Liouville equation in collective space. This would be the case if there is no dissipation ($\gamma=0$). The trajectory followed by the system in the phase space is shown

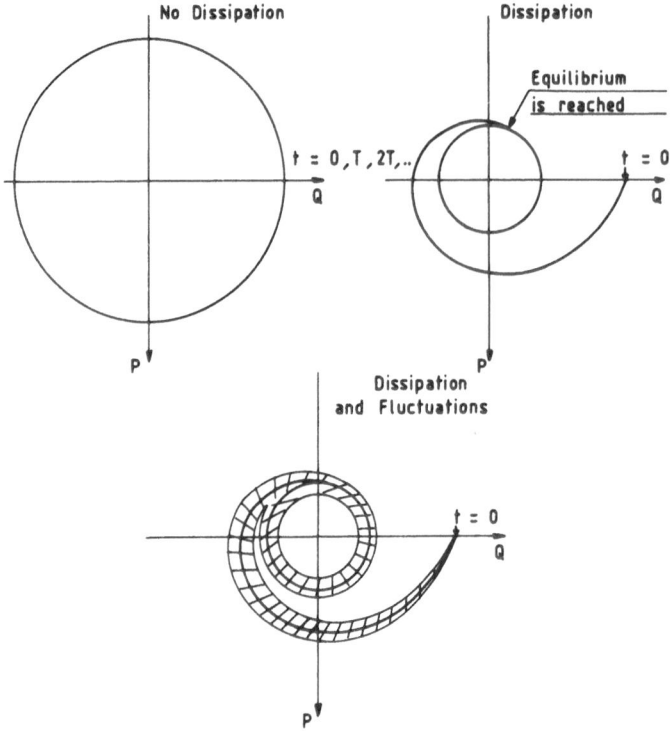

Fig. 7. Schematic presentation of different types of trajectories in phase space when there is no friction (top left), when there is friction (top right) and when there is friction and fluctuations (bottom). See text.

on the top left side of fig. 7 (the units have been chosen in such a way that it is a circle). The second term on the left of eq.(16), proportional to γ, corresponds to a dissipative term. It will be responsible for a drift of the mean value towards equilibrium. We will then observe a trajectory in phase space which is similar to the one schematized on top of the right side in fig. 7. In this picture, Q will decrease, become negative and then become positive again before it reaches equilibrium which is a circle corresponding to the zero point motion. However, we know that each time we have dissipation, we have also fluctuations. The effect of these fluctuations is given by the last term of eq.(16). The physical effect of this term is to spread the density distribution in phase space. In other words we have a diffusion, around the mean value. The distribution will broaden when the time increases until it reaches equilibrium. On the bottom of fig. 7 is shown a schematic trajectory in phase space including the diffusion term. The dashed area is supposed to correspond, for instance, to around 90 % of the total density distribution at a given time. This discussion for one collective degree can be easily generalized to several collective degrees as it has been done for the dynamical model described in the next lecture. If we apply this model to the ^{86}Kr + ^{92}Mo system, we can now calculate the evolution of the variance of the isobaric distribution as a function of the energy loss. The results are shown in fig. 8 where the full line corresponds to the calculation.

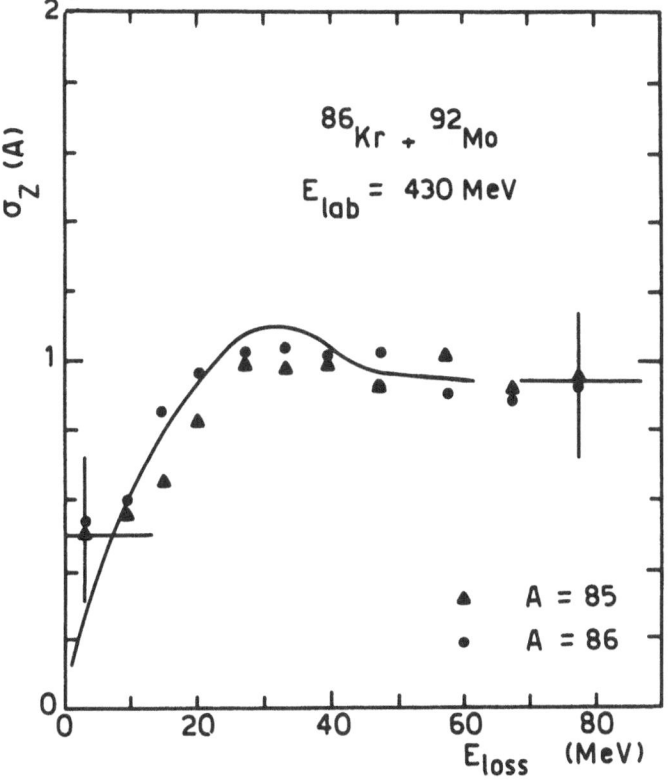

Fig. 8. Calculation of the variance of the isobaric distribution for A=86 as a function of the energy loss (full line) compared to the experimental data. From ref.[9].

6. Conclusion

To our present stage of knowledge it seems that in most of the cases we can associate neutron-proton asymmetry to the longitudinal component of the giant dipole resonance of the composite system. As far as the experimental results are concerned some care should sometimes be exercized. Indeed it is the secondary products which are detected after the collision and we want information on the primary products. The deexcitation of the primary products can change a lot the primary distributions. This is especially true for the light systems and has been investigated in great details by Cauvin [12].

We have seen that the relaxation time of neutron-proton asymmetry is very short. However it should also be kept in mind that this mode seems to only be excited after the system has passed the fusion barrier. This results from the TDHF calculation of ref.[8] but also from more phenomenological approaches [9].

Finally, due to this short relaxation time, which becomes of the order of the relaxation time for the intrinsic degrees, we might also expect memory effects in the sense that the relaxation process is non Markovian. Linear response theory can also provide a treatment of such processes [14].

References

[1] B. Gatty, D. Guerreau, M. Lefort, J. Pouthas, X. Tarrago, J. Galin, B. Cauvin, J. Girard and H. Nifenecker, Z. Phys. A273 (1975) and Nucl. Phys. A253 (1975) 511.

[2] C. Ngô, Lectures given at the international summer school, Piana Brasov, Roumania (1980) p. 395.

[3] M. Berlanger, A. Gobbi, F. Hanappe, U. Lynen, C. Ngô, A. Olmi, H. Sann, H. Stelzer, H. Richel and M.F. Rivet, Z. Phys. A291 (1979) 133.

[4] a A.C. Mignerey, V.E. Viola, H. Bvener, K.L. Wolf, B.G. Glagola, J.R. Birkelund, D. Hilscher, J.R. Huizenga, W.U. Schröder and W.W. Wilcke, Phys. Rev. Lett. 45 (1980) 508.
  b J. Poitou, R. Lucas, J.V. Kratz, W. Brüchle, H. Gäggeler, M. Schädel and G. Wirth, Phys. Lett. 88B (1979) 69.

[5] U. Brosa and D.H.E. Gross, Z. Phys. A294 (1980) 217.

[6] A.E.S. Green and N.A. Engler, Phys. Rev. 91 (1953) 40.

[7] D.H.E. Gross and K.M. Hartmann, Phys. Rev. C24 (1981) 2526.

[8] P. Bonche and C. Ngô, Phys. Lett. 105B (1981) 17.

[9] C. Grégoire, C. Ngô and B. Remaud, Phys. Lett. 99B (1981) 17.
  C. Ngô, C. Grégoire and B. Remaud, 3rd Adriatic Europhysics Study Conference on the dynamics of heavy ion collisions, Hvar (Yugoslavia) (1981), North Holland Company, p. 211.
  C. Grégoire, C. Ngô and B. Remaud, Nucl. Phys. in press.

[10] H. Hofmann, C. Grégoire, R. Lucas and C. Ngô, Z. Phys. A293 (1979) 229.

[11] H. Hofmann and P.J. Siemens, Nucl. Phys. A257 (1976) 165 and Nucl. Phys. A275 (1977) 464.

[12] B. Cauvin, Thèse de Doctorat d'Etat, Orsay (1980).

[13] E. Tomasi, C. Grégoire, C. Ngô and B. Remaud, J. Phys. Lett. 43 (1982) L115.

[14] H. Hofmann, A. Jensen, C. Ngô and P.J. Siemens, to be published.

FAST FISSION PHENOMENON

In this lecture we shall consider the case of dissipative heavy ion collisions occurring on a large time scale.

1. Experimental definition of fusion

Assume that we form a compound nucleus (Z,A) by bombarding a target nucleus (Z_2,A_2) with an incident projectile (Z_1,A_1). Then it will be formed with some excitation energy, E^*, and angular momentum ℓ. It will de-excite to more stable nuclei either by emitting light particles and γ rays, or by fissioning in two fragments. In the first case a distribution of residual nuclei (called also evaporation residues) with a mass close, but smaller than A, will be observed. In the second case a distribution of nuclei with masses centered around $\frac{A}{2}$ will be detected. These latter products form what we can call a symmetric fragmentation mass distribution. A schematic mass spectrum is shown in fig. 1. In addition to the two above components, deep inelastic products are also detected with masses close to the one of the projectile and of the target. The complete fusion cross section, σ_{CF}, is defined as the sum of two terms :

$$\sigma_{CF} = \sigma_{FL} + \sigma_{ER} \qquad (1)$$

the fission like cross section σ_{FL} (associated to the symmetric fragmentation component), and the evaporation residues cross section σ_{ER}.

The fission like cross section is usually obtained by measuring $d^2\sigma_{FL}/dAd\Omega$ at a laboratory angle which roughly corresponds to a fission at $\sim 90°$ in the center of mass system. σ_{FL} is then obtained assuming a $\frac{1}{\sin\theta}$ angular distribution for the fission like fragments which seems to be the case to a good approximation.

The evaporation residues cross section σ_{ER} is obtained after measuring the angular distribution of the residual nuclei which is forwards peaked.

We already see that there are some cases where it is difficult to obtain the fusion cross section :

1) for a symmetric system deep inelastic products are mixed with fission like products ;

2) for very asymmetric systems (small A_1 and large A_2) at very high bombarding energy it is difficult to separate evaporation residues and heavy deep inelastic products.

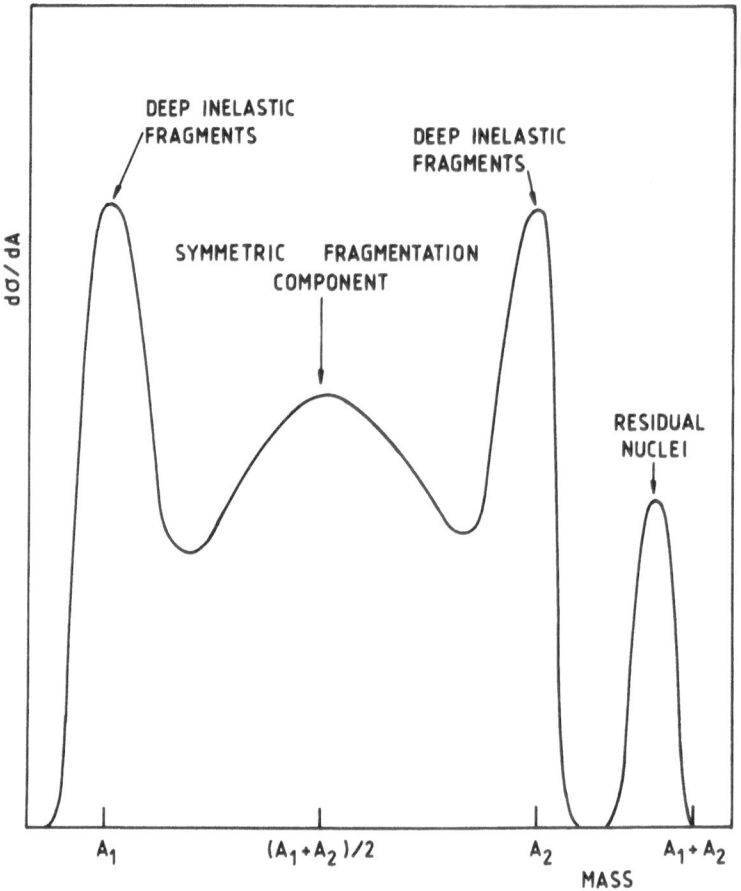

Fig. 1. Schematic mass distribution in heavy ion collisions.

2. Critical angular momentum for fusion

The results on complete fusion cross section are often expressed in terms of a critical angular momentum, ℓ_{CR}, which is defined as follows :

It is assumed that the lowest ℓ values or impact parameters go to fusion and that the sharp cut off approximation is valid (see fig. 2). With this hypothesis, the critical angular momentum is the largest ℓ value which fuses and is defined by the relation :

$$\sigma_{CF} = \frac{\pi}{k^2} \sum_{\ell=0}^{\ell_{CR}} (2\ell+1) = \frac{\pi}{k^2} (\ell_{CR}+1)^2 \tag{2}$$

where k is the wave number. The critical angular momentum depends on the system and on the bombarding energy. It is a property of the entrance channel but not of the compound nucleus [1].

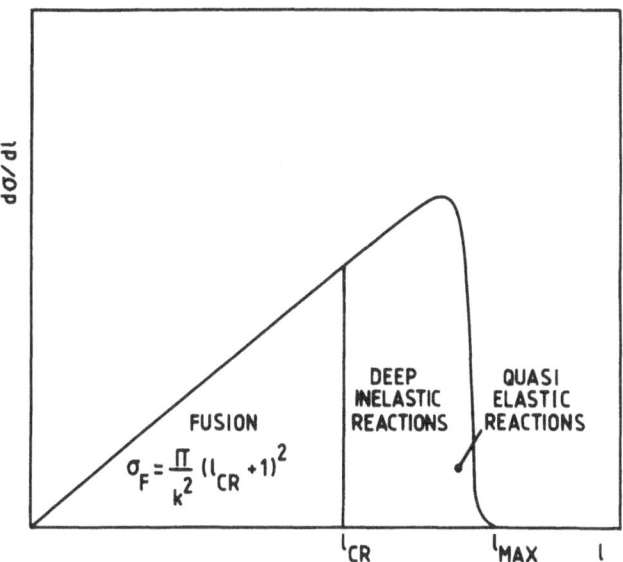

Fig. 2. Definition of the critical angular momentum ℓ_{CR}.

3. Disappearance of the fission barrier due to angular momentum

Compound nuclei with a high fissility parameter have a large probability to decay by
fission. For a given compound nucleus the major effect on the fission probability
comes from the angular momentum. This effect has been investigated in great details
by Cohen, Plasil and Swiatecki [2] within the rotating liquid drop model. It is schematically explained in fig. 3 where the potential energy of the fissioning nucleus
is plotted as a function of the elongation. For $\ell=0$ the fission barrier, $B_f(\ell=0)$ is
the energy difference between the saddle and the compound nucleus configurations.
When we bring angular momentum in the compound nucleus, we have to add to this potential energy curve rotational energy which is inversely proportional to the moment
of inertia of the configuration. Because the moment of inertia increases with the
elongation of the fissioning system, the effective barrier against fission decreases.
This can be seen in fig. 3 for the curve labelled ℓ. For some value, ℓ_{B_f}, this effective barrier vanishes. This means that when $\ell > \ell_{B_f}$ the compound nucleus becomes unstable with respect to fission. Consequently we can imagine that it is not possible
to form a compound nucleus with an angular momentum larger than ℓ_{B_f}. Indeed such an
entity would be so unstable that it would disintegrates before it has lived long
enough to forget its formation. In such a case it would not be a real compound nucleus.

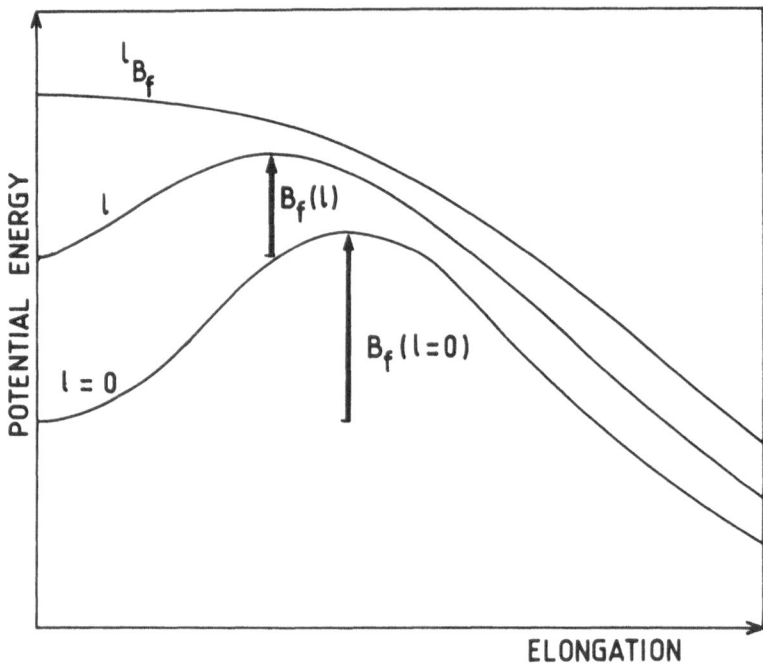

Fig. 3. Schematic evolution of the potential energy of the fissioning system with angular momentum. For $\ell = \ell_{B_f}$ the fission barrier vanishes.

4. Is fusion identical to compound nucleus formation?

When light ions are used to bombard light or medium targets, fusion can be identified
to compound nucleus formation. The critical angular momentum remains always smaller
than ℓ_{B_f}. However experiments using heavier projectiles have set several problems
which justify the use of the word fusion instead of compound nucleus formation. For
medium systems the critical angular momentum can be larger than ℓ_{B_f}. Several examples
of this are indicated in fig. 4. ℓ_{B_f} was calculated according to the rotating liquid
drop model of ref.[2]. This figure seems to indicate that fusion is not identical to
compound nucleus formation and we should try to understand why.

5. Why very heavy systems do not fuse?

Another interesting question concerns systems for which $Z_1 Z_2 \gtrsim 2500-3000$ because they
cannot fuse anymore [3]. This experimental fact can be understood in terms of the potential energy between the two heavy ions. It turns out that fusion data can be rather well explained using sudden potentials. A sudden potential is calculated assuming that the densities of the two ions remain frozen during the collision. This

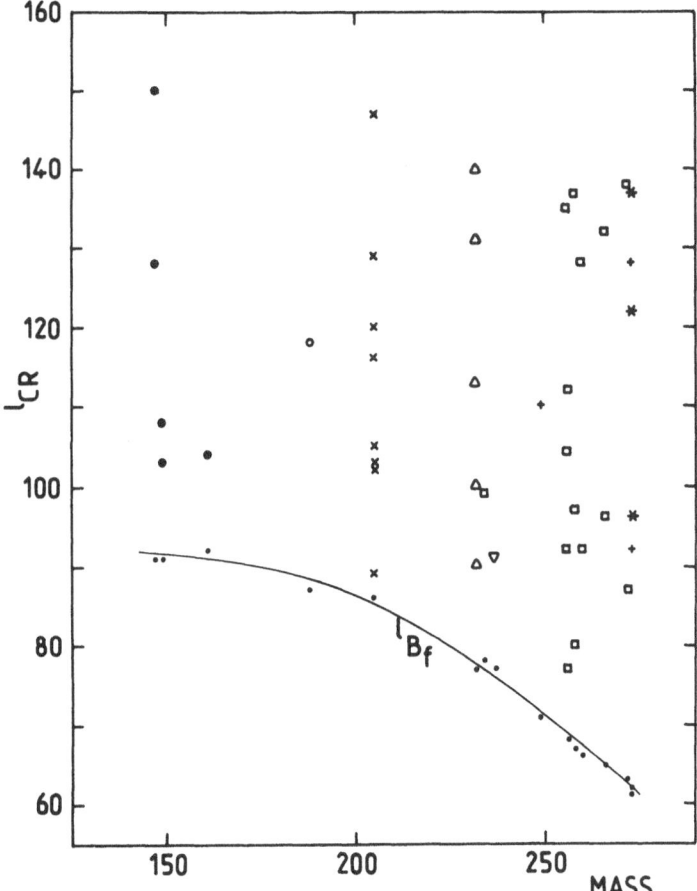

Fig. 4. Values of the critical angular momentum (symbols) obtained for different reactions. The abscissa is the mass of the compound nucleus which is expected to be obtained. The small dots correspond to ℓ_{B_f} calculated according to ref.[2] (a full line is drawn among them to guide the eyes).● ref.[16], o ref.[17], x ref.[7], Δ ref.[8], ⊓ ref.[18], ∇ ref.[19], + ref.[20], * ref.[21].

implies that the decision of whether the system will fuse or not will be taken at the early stage of the collision, probably before the two ions have reached the closest distance of approach. Indeed otherwise deformations should then play a very important role which in fact they do not,except for small corrections.

In fig. 5 is shown the total interaction potential $V(R)$, the nuclear part $V_N(R)$ and the Coulomb part $V_C(R)$, as a function of the distance R separating the center of mass of the two ions, for the $^{40}Ar + ^{197}Au$ system. This potential has been calculated for a head-on collision using the energy density formalism of ref.[4]. We see that $V(R)$ exhibits a pocket. This pocket is necessary to trap the colliding system. When the system is trapped, due to friction forces, we have fusion. Fusion is therefore strongly correlated to the existence of a pocket. Because of the increasing Coulomb repulsion with Z_1Z_2, it turns out that the pocket disappears when $Z_1Z_2 \gtrsim 2500\text{-}3000$. It is also when this inequality is satisfied that we do not observe fusion anymore. An example of a potential energy curve where there is no more a pocket is shown in fig. 6 for the $^{136}\text{Xe} + ^{197}\text{Au}$ system. For this system there should not be fusion anymore.

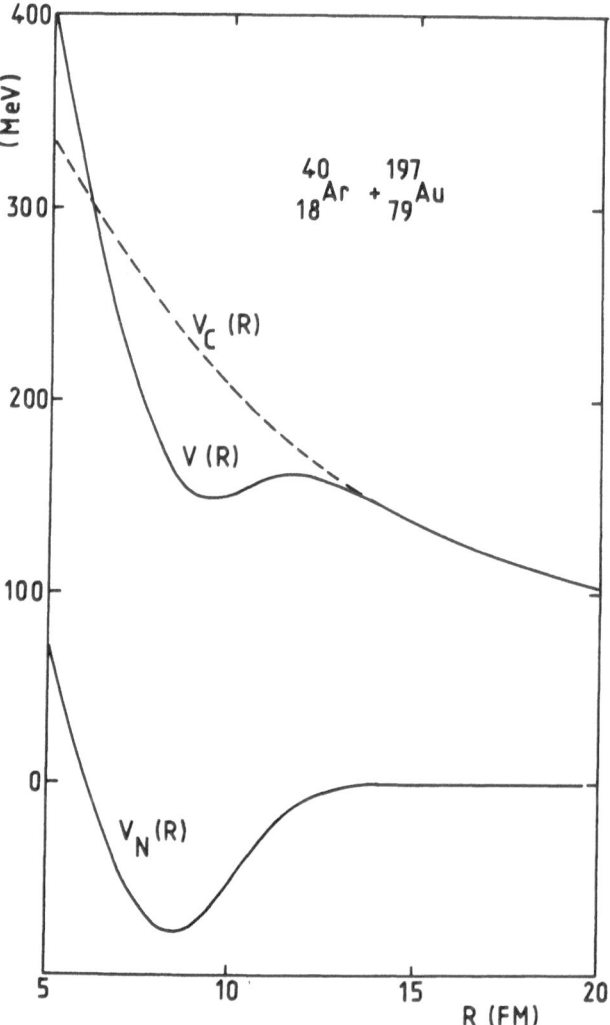

Fig. 5. Interaction potential V(R) versus R. V_N is the nuclear part and V_C the Coulomb one. From ref.[4].

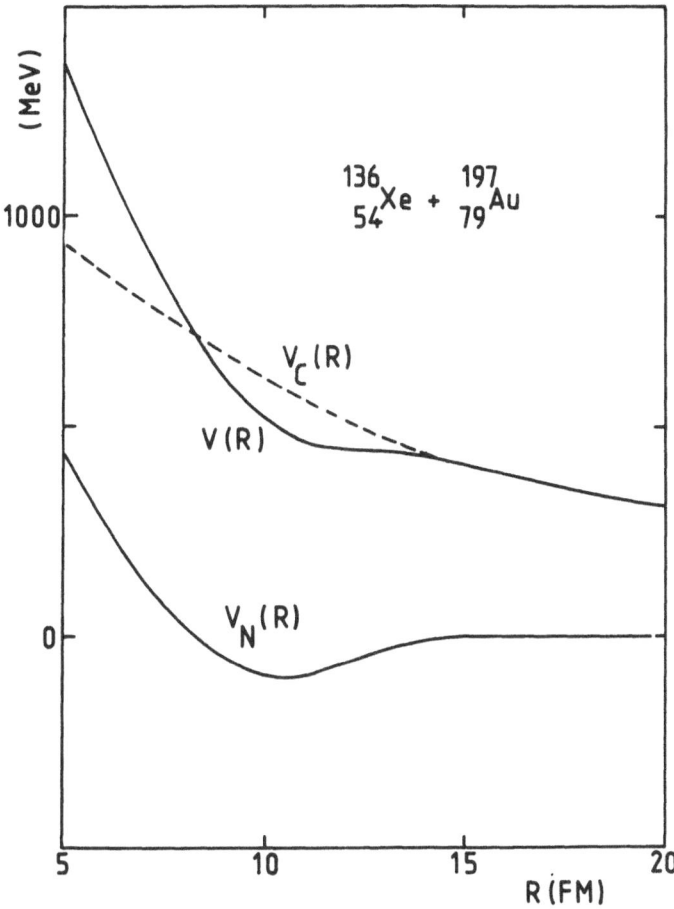

Fig. 6. Same as fig. 5. From ref.[4].

6. Why can ℓ_{CR} be larger than ℓ_{B_f}?

For medium systems ℓ_{CR} can be larger than ℓ_{B_f}. We could think of several possibilities to explain this surprising result :

1) ℓ_{B_f} has been calculated using the rotating liquid drop model. We could think that it has been underestimated. Indeed, at temperature zero, ℓ_{B_f} can be calculated to be much larger in some cases due to shell effects [5]. We are however in a situation where the temperature is high ($\sim$ 2-3 MeV) so that shell effects are very likely wiped out. Furthermore it should be noted that the rotating liquid drop model has been used at temperature zero. Finite temperature effects have a tendency to decrease the height of the fission barrier [5,6]. Consequently we would expect that ℓ_{B_f} also decreases. These remarks indicate that there are no serious reasons to suspect that ℓ_{CR} cannot be larger than ℓ_{B_f} especially if we consider the large differences between ℓ_{CR} and ℓ_{B_f} which can be reached.

2) We can imagine that fusion not only includes compound nucleus formation, but that it also includes incomplete fusion. Incomplete fusion corresponds to the case where a fast particle is emitted at the very beginning of the reaction before the two remaining partners fuse together. Because the light particle removes angular momentum, the critical angular momentum can be larger than ℓ_{B_f}, but we still form a compound nucleus.

3) We can also imagine that when $\ell > \ell_{B_f}$ we do not form a compound nucleus but a composite system which decays in two fragments. If the life time of this system is large enough, it has time to relax its mass asymmetry to equilibrium and then fissions. The mass distribution of the products will look similar to those of fission following compound nucleus formation. However the life time will be smaller than in the case of compound nucleus fission because this latter has not been formed. For this reasons such a phenomenon can be called fast fission. Fig. 7 shows a schematic representation of what could happen.

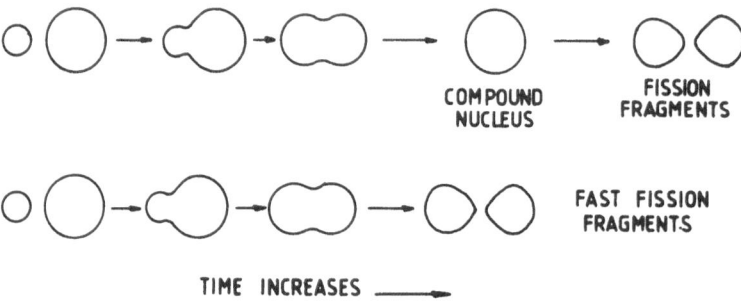

Fig. 7. Schematic picture of compound nucleus fission and fast fission.

7. Mass distribution of the fission like products

The total kinetic energy distribution of the fission like products does not show any possible existence of two different components. The mass distribution of the fission like products is however characterized by an important increase of the full width half maximum (FWHM) when the bombarding energy - or excitation energy - increases. This has been extensively investigated on the ^{40}Ar + ^{165}Ho system [7]. In fig. 8 the FWHM is plotted as a function of the excitation energy E^* of the compound nucleus. We observe a strong increase of the FWHM with the excitation energy. We expect the FWHM to vary because of three reasons :

1) when the excitation energy increases, the statistical fluctuation increases. In the case of a harmonic oscillator we have seen in the preceding lecture that the FWHM should increase like $\sqrt{T}$. However the experimental evolution is much more important than the preceding one.

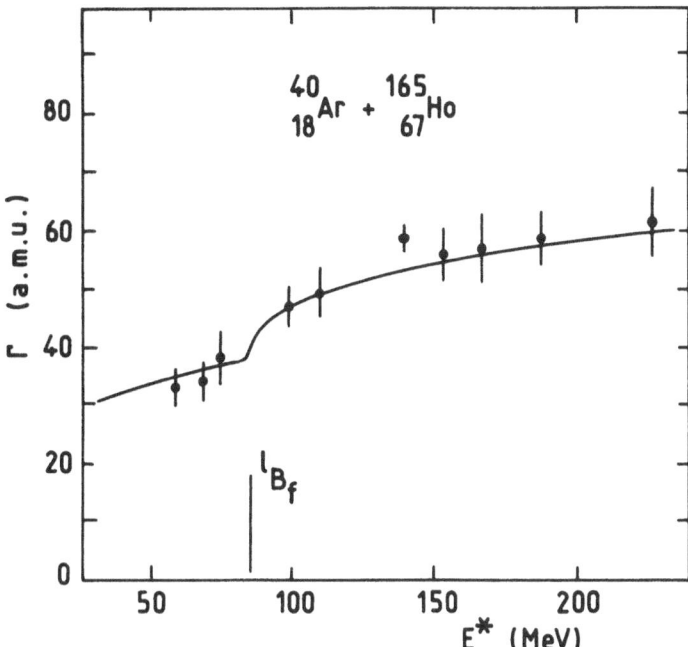

Fig. 8. FWHM of the fission like mass distribution as a function of the excitation energy. The dots are the experimental points of ref.[7]. The full curve is the result of a calculation discussed in section 11.

2) A contribution to the fission like mass distribution of incomplete fusion - fission events can also increase the FWHM. Indeed the fission like mass distribution would result of the superposition of several slightly shiffed distributions. As a consequence the total distribution would be broader than each individual one. Nevertheless incomplete fusion cross sections are much too low in order to explain the increase of the FWHM with E^*. This has been checked for example on the Cl + Au system by detecting α particles emitted in coincidence with the fission like fragments.

3) The potential energy surface governing the process depends on angular momentum. This means that the stiffness along the mass asymmetry coordinate will change. To account for the experimental observation it should decrease when the angular momentum increases. Caculating potential energy surfaces for fission as a function of angular momentum, Faber has shown that the stiffness along mass asymmetry decreases when ℓ increases. But if we look at a given configuration, for instance at the saddle point, it does not decrease enough in order to explain the experimental data.

To summarize, it turns out that the FWHM of the fission like mass distribution increases a lot when the excitation energy increases. For the Ar + Ho system this is particularly apparent when $E^* \sim 80\text{-}100$ MeV. It is in this region that the critical angular momentum becomes larger than ℓ_{B_f}. The experimental data seem to indicate a strong dependence of the stiffness along the mass asymmetry coordinate as a function of angular momentum. This might be an indication that for $\ell > \ell_{B_f}$ some other mechanism like the fast fission suggested above comes into play. However it should be noted that these experiments are by no means a proof for the existence of such a mechanism.

8. Dynamical model for dissipative heavy ion collisions

We shall now give the philosophy of a dynamical model [10] which is able to describe deep inelastic collisions together with compound nucleus cross section. Furthermore, under certain conditions which have to be satisfied, a new type of dissipative mechanism occurs : fast fission. We will not describe in details the model but refer the reader to ref.[10] where extensive descriptions are done. We will restrict ourselves to the basic ideas under which the model is based.

The aim is to describe the heavy ion collision in terms of a few collective degrees. Four are explicitly taken into account : the distance r between the center of mass of the two moving ions, θ, the polar angle, $x=(A_2-A_1)/(A_1+A_2)$ the mass asymmetry defined from the masses A_1 and A_2 of the fragments and $y = N_1-Z_1$ the neutron excess of the light fragment. The density distribution which describes the dynamical evolution of the dissipative heavy ion collision will satisfy a transport equation obtained within the harmonic approximation using the linear response theory of Hofmann and Siemens [11]. This equation is a generalization to several degrees of freedom of the transport equation discussed in the previous lecture.

The experiment tells us that the deformation degrees of freedom are very important and that they should be treated in some way in order to correctly reproduce the energy loss observed in deep inelastic reactions. In the model described here these deformations will be simulated by making a transition between an entrance and an exit potential.

The entrance potential is calculated in the sudden approximation where the densities of the two ions remain frozen. We will use the energy density potential developed in ref.[12]. This will allow us to define the fusion landscape.

During the reaction, if the overlap and the time are large enough, there will be a reorganization of the densities of the two incident ions. For strong dissipative collisions, the situation in the exit channel is better described using an adiabatic potential. We are in a situation similar to fission and the system reseparates in the fission landscape. We will use the adiabatic potential calculated according to the model of Ledergerber and Pauli based on the liquid drop approximation applied to a sequence of shapes describing the fission process.

The trick of the model is to allow a time dependent transition between the entrance and the exit potential. This transition will be complete only if the overlap between the two nuclei is large enough. This will be done by taking for the total interaction potential V the following phenomenological form :

$$V(t) = V_s \chi(t) + V_a(1 - \chi(t)). \tag{3}$$

The transition between the sudden and the adiabatic potential will be done by a re-organization parameter $\chi(t)$. $\chi(t)$ is assumed to satisfy *a priori* the following differential equation :

$$\frac{d\chi(t)}{dt} = - \frac{1}{\tau_{SA}} \Phi(r,\chi) \chi(t), \tag{4}$$

with the following boundary condition :

$$\chi(0) = 1. \tag{5}$$

τ_{SA} has the dimension of a time. It corresponds to the relaxation time associated with the reorganization of the densities (transition from the sudden to the adiabatic potential). Φ is a form factor which is proportional to the density overlap between the two ions. It allows a transition between the sudden and the adiabatic potential to only occur when the two ions interact strongly enough.

This idea of a transition between a fusion and a fission landscape was already pointed out by Swiatecki in 1972 [13]. Of course the way we describe it might be a very rough estimate of what is really going on. However we believe that such a description will give us the main physical ideas of what occurs in dissipative heavy ion collisions.

An important parameter in this description is the reorganization time τ_{SA}. It turns out that the value $\tau_{SA} = 10^{-21}$s gives an overall description of the experimental data concerning most of the systems. This value corresponds to what we could estimate by looking at the typical time to excite shape vibrations. As far as the other transport coefficients, inertia and friction, are concerned we refer the reader to ref.[10].

The solution of the transport equation provides a density distribution function from which it is possible to calculate mean values of macroscopic variables but also the distribution around these mean values. We shall now discuss some of the results which are obtained from the model.

9. How fast fission phenomenon occurs

We shall consider the 340 MeV Ar + Ho system for which the critical angular momentum is larger than ℓ_{B_f}.

In fig. 9 is shown the mean value of x as a function of the mean value of r for several values of the initial orbital angular momentum. These curves correspond to mean trajectories followed by the system in the (r,x) plane.

For $\ell=150$ the interaction between the two ions is rather deep. Some mass is exchanged between the two nuclei and the total kinetic energy in the relative motion is completely damped. Such a trajectory is typical of deep inelastic reactions.

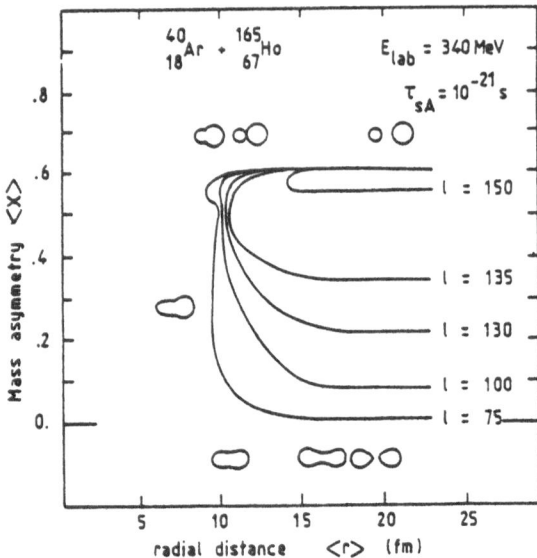

Fig. 9. Typical trajectories in the r,x plane for the 340 MeV ^{40}Ar + ^{165}Ho system (see text). From ref.[10].

For ℓ=75, 100, 130 and 135 the system is trapped in the pocket of the entrance potential. Then a lot of mass is transferred from the heavy nucleus to the light one. For ℓ=75 we see that mass asymmetry reaches equilibrium. But, because for these ℓ values the compound nucleus has no more a fission barrier, the system escapes again by fissioning in two fragments. Because broad statistical fluctuations exist around the mean mass we will observe a broad mass distribution. This kind of trajectory corresponds to fast fission. The interaction time for this process ranges from 10^{-21} to 10^{-20}s. It is larger than the one of a deep inelastic reaction. In fact this kind of process is a delayed deep inelastic reaction. The delay arises because the system is trapped into the pocket of the interaction potential. It can only escape when this pocket has dissapeared due to the reorganization of the densities. Such an effect can be observed when the compound nucleus has no fission barrier. Finally, for ℓ values smaller than ℓ=72,the compound nucleus has a nonvanishing fission barrier. The system which is trapped into the pocket cannot escape before a compound nucleus is formed. This compound nucleus will later on fission in two fragments.

In other words we can say that fast fission corresponds to the case where we go directly from the fusion valley to the fission valley without passing by the compound nucleus configuration. This is schematically represented in fig. 10.

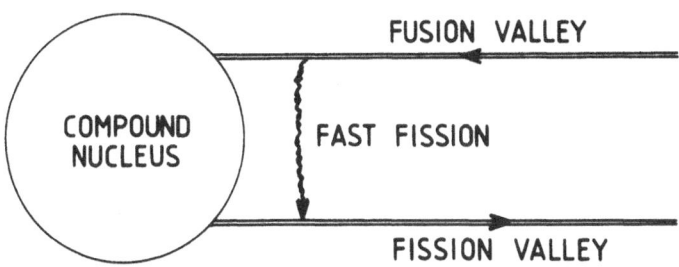

Fig. 10. Schematic representation of fast fission as a mechanism which allows to go directly from the fusion to the fission valley.

10) Compound nucleus and fast fission cross sections

The model gives the compound nucleus cross section σ_{CN} and also the fast fission cross section σ_{FF}. The complete fusion cross section σ_{CF} as it is defined experimentally is :

$$\sigma_{CF} = \sigma_{CN} + \sigma_{FF}. \tag{6}$$

From σ_{CF} we can deduce ℓ_{CR}. When $\ell_{CR} \leq \ell_{B_f}$, $\sigma_{FF} = 0$ and $\sigma_{CF} = \sigma_{CN}$ but when $\ell_{CR} > \ell_{B_f}$, $\sigma_{FF} \neq 0$ and all ℓ values in between ℓ_{B_f} and ℓ_{CR} give fast fission. This is the case for instance for all ℓ values between 72 and 135 for the 340 MeV Ar + Ho system whereas all ℓ values between 0 and 72 give a compound nucleus.

In fig. 11 is shown the excitation function of the Ar + Ho system as a function of $1/E_{CM}$, the inverse of the center of mass energy. The dots correspond to the experimental measurements whereas the full curve corresponds to the calculation. The agreement is qualitatively correct. The full curve is the sum of σ_{CN} (long dashed curve) and σ_{FF} (short dashed curve). For small values of E_{CM} we only form a compound nucleus ($\ell_{CR} < \ell_{B_f}$). For bombarding energies larger than 170 MeV, ℓ_{CR} becomes larger than ℓ_{B_f} and fast fission appears. In this model compound nucleus formation is bounded by ℓ_{B_f} and it is not possible to form a compound nucleus with an angular momentum larger than ℓ_{B_f}. An important consequence of the model is that the critical angular momentum is bounded at large bombarding energies. The reason is that the pocket which is necessary, for the system to be trapped in the entrance channel, disappears because of the angular momentum.

11) Fluctuations around the mean values

Because the model treats explicitly fluctuations around the mean values, it is possible to calculate the FWHM of the fast fission mass distribution. If turns out that

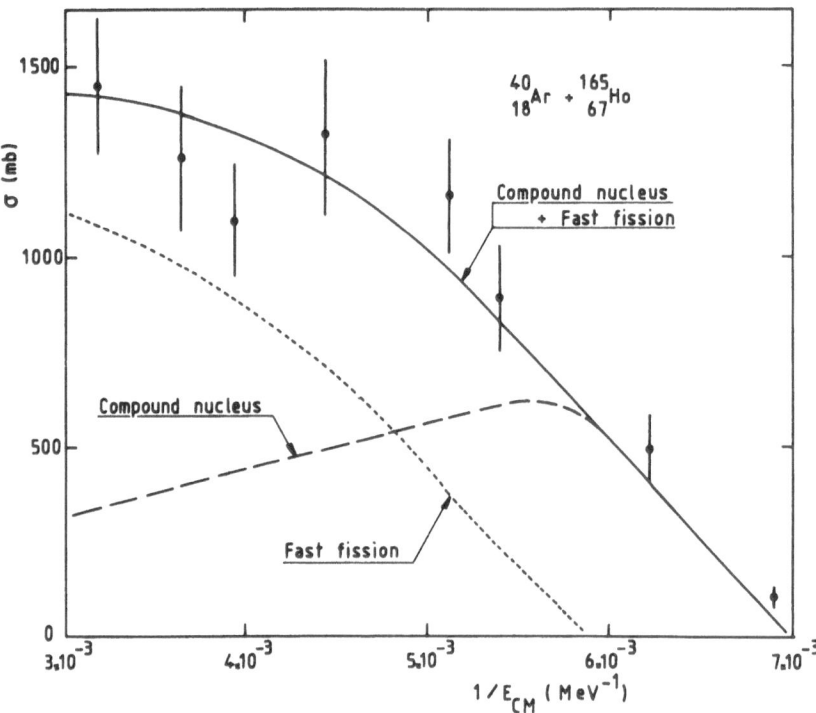

Fig. 11. Experimental fusion cross sections (dots) from ref.[7] as a function of $1/E_{CM}$ are compared with a theoretical calculation (see text) from ref.[10].

these fluctuations are of statistical nature due to the small phonon energy of the mass asymmetry degree of freedom compared to the temperature of the intrinsic system. Nevertheless this model is not able to calculate the FWHM of the mass distribution of the fission products following compound nucleus formation. This calculation was done separately in ref.[14] by solving a similar transport equation using propagators on gaussian bundles. Adding this contribution to fast fission gives the full curve in fig. 8. When $E^* \leq 80$ MeV we have only compound nucleus fission but when $E^* \geq 80$ MeV fast fission starts to contribute to the fission like mass distribution. This latter contribution is calculated using the dynamical model described above. We observe in fig. 8 an overall agreement with the experimental results. The main difference between compound nucleus fission and fast fission can roughly be explained as follows : for compound nucleus fission the mass distribution is almost given by an equilibrium distribution at scission whereas for fast fission it is almost an equilibrium distribution at the turning point. Because the stiffness are different for these two configurations it explains the jump which is observed around 80-100 MeV excitation energy.

Finally fig. 12 shows $d^2\sigma/dAdE$ for the 340 MeV Ar + Ho system as a function of A and E. A is the mass of the fragment and E the total kinetic energy of both fragments.

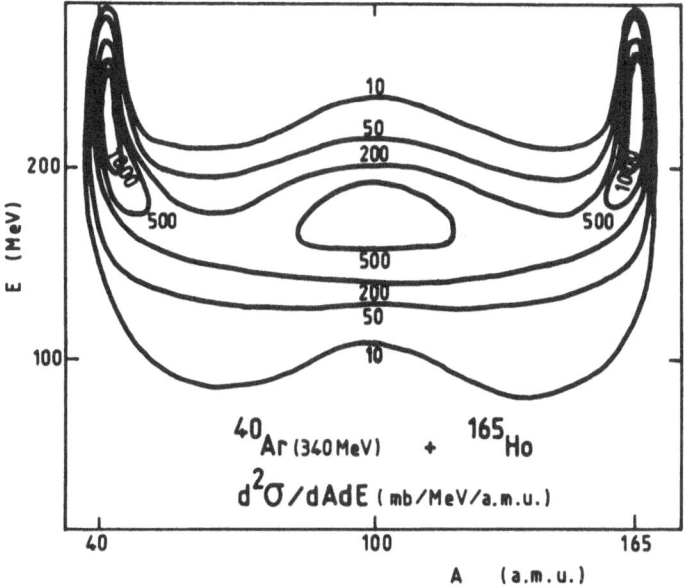

Fig. 12. $d^2\sigma/dAdE$ (μb/MeV/a.m.u) versus the mass A and the total kinetic energy E of the fragments for fast fission and deep inelastic products. From ref.[10].

This two-dimensional contour plot only contains fast fission and deep inelastic products but does not contain fission following compound nucleus formation fragments. The striking feature is that fast fission looks almost similar to fission following compound nucleus formation.

12) Case of heavy systems : quasifission

Systems with $2000 \lesssim Z_1 Z_2 \lesssim 2500$ show an interesting behavior. Because the compound nucleus has a high fissility parameter, the saddle point configuration is more compact than the configuration corresponding to the bottom of the pocket of the entrance potential. Therefore, even if the system is trapped into the pocket, the system cannot give a compound nucleus but immediately fissions. This occurs although the fission barrier is non zero. This special aspect of fast fission has been predicted by Swiatecki [15] and was called quasifission. In the model of ref.[15], in order to get fusion for these heavy systems, an extra energy (extra push) is needed to bring the system to a more compact configuration. It should be noted that this extra push notion does not exist in our model. In order to get it,we would be obliged to considerably reduce the value of τ_{SA}.

13) Conditions under which fast fission can be observed

We shall now summarize the conditions under which fast fission can be observed. The situation is schematically presented in fig. 13.

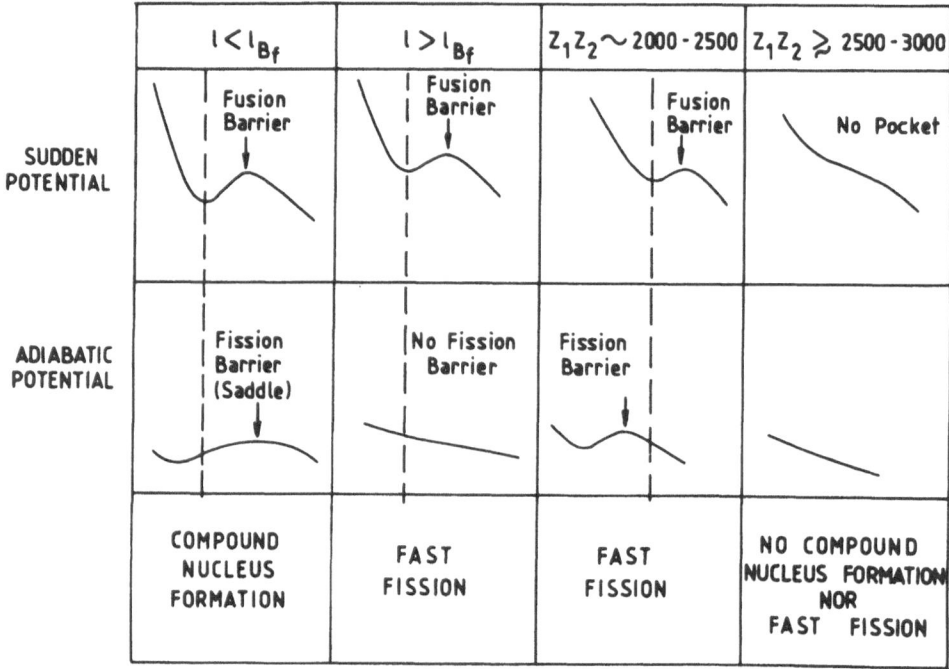

Fig. 13. Schematic description of the possibilities for observing fast fission.

When $Z_1Z_2 \lesssim 2000$ the saddle configuration is not too compact and is located outside the bottom of the pocket of the entrance potential. If $\ell \leqslant \ell_{B_f}$, and if the system is trapped, we will obtain a compound nucleus. If $\ell > \ell_{B_f}$ the adiabatic potential has no more a fission barrier. If the system is trapped in the entrance potential, it will subsequently escape after it has relaxed mass asymmetry : we will observe fast fission.

For systems with $2000 \lesssim Z_1Z_2 \lesssim 2500$, the saddle configuration is more compact and located inside the position of the bottom of the pocket of the entrance potential. If the system is trapped, it will lead to fast fission even if the fission barrier is non zero.

Finally when $Z_1Z_2 \gtrsim 2500$-3000 there is no more a pocket in the entrance potential. The system cannot be trapped and we cannot observe neither a compound nucleus, nor fast fission.

References

[1] A.M. Zebelman and J.M. Miller, Phys. Rev. Lett. 30 (1973) 27.
[2] S. Cohen, F. Plasil and W.J. Swiatecki, Ann. Phys. 82 (1974) 557.
[3] M. Lefort, C. Ngô, J. Péter and B. Tamain, Nucl. Phys. A216 (1973) 166.
[4] H. Ngô and C. Ngô, Nucl. Phys. A348 (1980) 140.
[5] M. Diebel, K. Albrecht and R. Hasse, Nucl. Phys. A355 (1981) 66.
[6] R.W. Hasse and W. Stocker, Phys. Lett. 44B (1973) 26.
[7] C. Lebrun, F. Hanappe, J.F. Lecolley, F. Lefebvres, C. Ngô, J. Péter and
    B. Tamain, Nucl. Phys. A321 (1979) 207.
    B. Borderie, M. Berlanger, D. Gardes, F. Hanappe, L. Nowicki, J. Péter, B.
    Tamain, S. Agarwal, J. Girard, C. Grégoire, J. Matuszek and C. Ngô, Z. Phys.
    A299 (1981) 263.
[8] V. Bernard, C. Grégoire, C. Mazur, C. Ngô, M. Ribrag, G. Fan, P. Gonthier,
    H. Ho, W. Kühn and P. Wurm, Nucl. Phys. in press.
[9] M. Faber, Report AIAU-8023, Wien (1980).
[10] C. Grégoire, C. Ngô and B. Remaud, Phys. Lett. 99B (1981) 17, Nucl. Phys. (in
     press) and   Dynamics of heavy-ion collisions proceedings of the 3rd Adriatic
     europhysics study conference (1981) (North Holland) p. 211.
     E. Tomasi, C. Grégoire, C. Ngô and B. Remaud, J. Phys. Lett. 43 (1982) L115.
     C. Grégoire, C. Ngô, E. Tomasi, B. Remaud and F. Scheuter, ICOSAHIR (Saclay,
     mai 1982) to appear in Nucl. Phys.
[11] H. Hofmann and P.J. Siemens, Nucl. Phys. A257 (1976) 165 and Nucl. Phys.
     A275 (1977) 464.
[12] C. Ngô, B. Tamain, M. Beiner, R. Lombard, D. Mas and H. Deubler, Nucl. Phys.
     A252 (1975) 237.
[13] W.J. Swiatecki, J. Phys. C5 (1972) 45.
[14] C. Grégoire and F. Scheuter, Z. Phys. A303 (1981) 337.
[15] W.J. Swiatecki, Phys. Script. 24 (1981) 113 and Nucl. Phys. A376 (1982) 275.
[16] H.C. Britt, B.H. Erkkila, R.H. Stockes, H.H. Gutbrod, F. Plasil, R.L. Fer-
     gusson and M. Blann, Phys. Rev. C13 (1976) 1483.
[17] B. Heusch, C. Volant, H. Freiesleben, R.P. Chesnut, K.D. Hildenbrand,
     F. Pühlhofer, W. Schneider, B. Kohlmeyer and W. Pfeffer, Z. Phys. A288
     (1978) 391.
[18] H. Sann, R. Bock, Y. Chu, A. Olmi, U. Lynen, N. Müller, S. Bjornholm and
     H. Esbensen, Phys. Rev. Lett. 47 (1981) 1248.
[19] C. Ngô, J. Péter, B. Tamain, M. Berlanger and F. Hanappe, Z. Phys. A283
     (1977) 161.
[20] F. Hanappe, C. Ngô, J. Peter and B. Tamain, Proc. 3rd IAEA Symposium of phy-
     sics and chemistry of fission, Rochester (1973) p. 289.
[21] S. Leray, G. Fan, C. Grégoire, H. Ho, C. Mazur, C. Ngô, A. Pfoh, M. Ribrag,
     N. Sanderson, L. Schad, E. Tomasi, R. Wolski and P. Wurm, ICOSAHIR (Saclay,
     mai 1982) p. 88.

Conclusion

In these lectures we have described two different phenomena occuring in dissipative heavy ion collisions : neutron-proton asymmetry and fast fission.
Neutron-proton asymmetry has provided us with an example of a fast collective motion. As a consequence quantum fluctuations can be observed. The observation of quantum or statistical fluctuations is directly connected to the comparison between the phonon energy and the temperature of the intrinsic system. This means that this mode might also provide a good example for the investigation of the transition between quantum and statistical fluctuations which might occur when the bombarding energy is raised above 10 MeV/A. However it is by no means sure that in this energy domain enough excitation energy can be put into the system in order to reach such high temperatures over the all system. The other interest in investigating neutron-proton asymmetry above 10 MeV/A is that the interaction time between the two incident nuclei will decrease. Consequently, if some collective motion should still be observed, it will be one of the last which can be seen.
Fast fission corresponds on the contrary to long interaction times. The experimental indications are still rather weak and mainly consist of experimental data which cannot be understood in the framework of standard dissipative models. We have seen that a model which can describe both the entrance and the exit configuration gives this mechanism in a natural way and that the experimental data can, to a good extend, be explained. The nicest thing is probably that our old understanding of dissipative heavy ion collisions is not changed at all except for the problems that can now be understood in terms of fast fission. Nevertheless this area desserve further studies, especially on the experimental side to be sure that the consistent picture which we have on dissipative heavy ion collisions still remain coherent in the future.

Acknowledgments

These lectures have been supported by the University of Sevilla and the French ministère des relations extérieures. I would like to thank both of them and especially Prof. Gonzalo Madurga.
The experimental results presented in these lectures have been gathered during the last few years. I would like to thank all my colleagues from GSI, Orsay and Saclay for the fruitful collaborations and discussions. As far as the interpretations are concerned, they result from a collaboration between Munich, Nantes and Saclay where Helmut Hofmann has played the role of a strong driving force. I would like to particularly thank Michel Berlanger, Paul Bonche, Adriano Gobbi, Christian Grégoire, Francis Hanappe, Helmut Hofmann, Marc Lefort, Renée Lucas, Alessandro Olmi, Bernard Remaud, Hervé Richel, Marie France Rivet, Bernard Tamain and Eglé Tomasi who have contributed to the work contained in these lectures.

Finally I would like to thank Mrs P. Gugenberger and E. Thureau for preparing the manuscript and Mr J. Matuszek for drawing the figures.

<u>PRE-EQUILIBRIUM PROCESSES IN NUCLEAR REACTIONS</u>

P.E. Hodgson
Nuclear Physics Laboratory
Oxford, U.K.

1. Introduction

The energy distribution of particles emitted from a nuclear interaction at moderate
energies often shows many of the features illustrated in Fig.1. There is a broad
maximum at lower energies, a rather featureless continuum at intermediate energies,
and some more or less well resolved peaks at higher energies. If we investigate how
the intensities of these features vary with emissions angle, we find that the intensities of the peaks vary quite strongly with angle, with a marked forward excess and
a pronounced oscillatory structure. The continuum at intermediate energies usually
shows a forward excess, but little structure, while the broad peak at low energies is
symmetrical about 90°, usually with little angular variation.

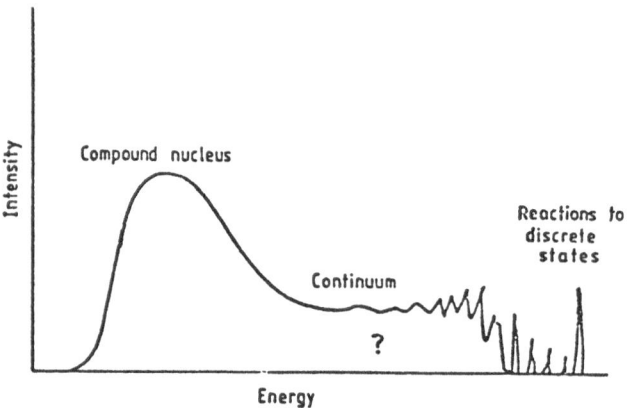

Fig. 1 Typical energy spectrum of particles from intermediate energy reactions.

The physical processes responsible for the peaks are generally quite well understood.
The high energy peaks are due to direct reactions to particular states of the residual
nucleus. These take place in times comparable with the projectile transit time, and
usually involve rather few target nucleons. As the excitation energy of the residual
nucleus increases the energy differences between successive states become less and
their widths increase, until they are no longer resolved individually and merge into
a continuum. Some of the cross-section is due to direct reactions, but at the lower
emergent energies it is more and more likely that the interaction has taken place in

two or more stages. Finally, the broad peak at low energies corresponds to the particles emitted from the final equilibrated nucleus in which the excitation energy is shared statistically among all the nucleons.

Some of this qualitative understanding has developed to the stage at which quite accurate and reliable calculations of the cross-section can be made. The distorted wave theories of inelastic scattering and transfer reactions can generally account quite well for the direct reactions to individual nuclear states, and the comparison with experimental data often allows information on nuclear structure to be obtained. The statistical theories of compound nucleus decay likewise account for the broad peak at low energies. In some cases, particularly at lower incident energies, the cross-sections of the reactions to individual final states also have a compound nucleus component, and the cross-section can then be understood as the incoherent superposition of direct and compound nucleus processes (Hodgson, 1971).

The continuum at intermediate energies is much less well understood, and indeed is the subject of many detailed investigations at the present time (Hodgson, 1980). This continuum is partly due to direct processes, but it also contains a component due to particles emitted after the direct process but before the final attainment of statistical equilibrium. These pre-equilibrium or pre-compound particles, and the processes responsible for them, are the subject of this review.

It is important to distinguish between the multistep compound process, in which all the particles are bound throughout the interaction, and the multistep direct process in which at least one particle is in the continuum. The former predominates at lower energies while the latter becomes increasingly important as the energy increases.

The particles emitted in the direct and compound nucleus stages may be recognised by their energy and angular distributions. Direct particles are emitted predominantly in the same direction as the incident particle, with similar energies, while compound nucleus particles are emitted in all directions, with an angular distribution symmetric about 90°, and when the final states merge into a continuum have the characteristic Maxwellian energy distribution.

Pre-equilibrium particles are generally more energetic than those from the compound nucleus, and those from the multistep direct process are predominantly emitted in the forward direction. Those from the multistep compound process are emitted symmetrically about 90° and a more specific test of their presence is provided by an analysis of the fluctuations in the cross-section as a function of energy. As described in Section 2, these provide a measure of the interaction time, and show that there are processes taking place in times intermediate between those taken by the direct process (the transit time, 10^{-22}-10^{-23} secs) and by the decay of the compound nucleus (around 10^{-16} secs). This applies only to the pre-equilibrium particles emitted by the multistep compound processes which show fluctuations with correlation widths determined by the lifetimes of the intermediate states. The multistep direct process on the other hand, takes place in a time very similar to the incident particle transit time, and so the particles emitted show no energy fluctuations.

Many theories have been developed to account for the pre-equilibrium processes, differing greatly in their basic approach, their assumptions and their success in accounting for the experimental data. In this review the semi-classical theories, particularly the exciton model and its variations, are discussed in Section 3, and the quantum-mechanical theories in Section 4.

2.   Analyses of Fluctuations

The process of nuclear excitation takes place in a number of stages, as indicated schematically in Fig.2. The first nucleon-nucleon interaction excites a particle-hole

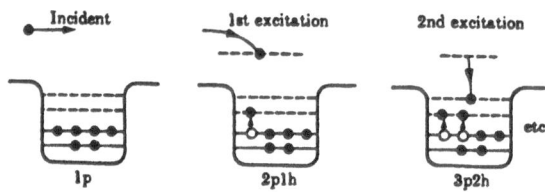

Fig. 2   Succesive stages in nuclear excitation.

pair, and these particles and holes propagate through the nucleus exciting further particle-hole pairs and so on until the excitation is spread through the whole nucleus. The nuclear level density increases very rapidly with excitation energy, so the number of states that can be excited at any one stage is much greater than the number excited at the previous stage. At each stage the process is most likely to go on to the next stage, but there is a small probability that it returns to an earlier stage of lower excitation, and also it is possible for particles to be emitted into the continuum; these are the pre-equilibrium particles.

The excitation function gives a measure of the time-scale of the process. The direct reaction cross-sections have a smooth energy variation with energy whereas the compound nucleus processes give rapidly fluctuating cross-sections like that shown in Fig.3. The signature of a pre-equilibrium process is thus an excitation funtion with structure intermediate between the smooth energy variation of the direct process and the rapid fluctuations of the compound nucleus process. Since the compound nucleus process is certainly present, we look for compound nucleus fluctuations superposed on a structure varying more slowly with energy as shown in Fig.4. This example, incidentally, shows that it is possible for the pre-equilibrium processes to contribute to reactions to discrete states of the final nucleus.

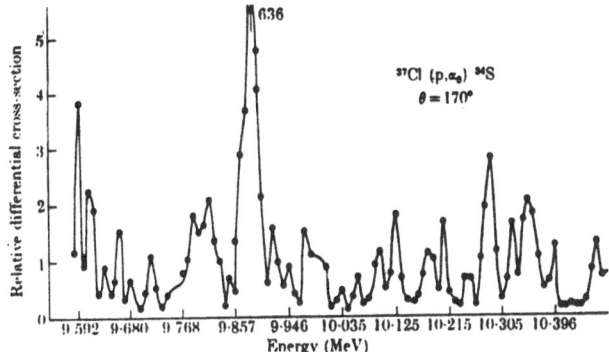

Fig.3　Excitation function for the reaction $^{37}Cl(p,\alpha_o)^{34}S$ from 9.59 to 10.4 MeV, showing strong fluctuations (Von Brentano et al 1964).

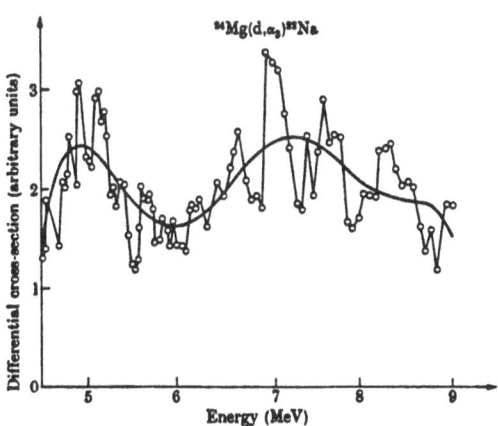

Fig.4　Experimental excitation function of $^{24}Mg(d,\alpha_3)^{22}Na$ at $\Theta_{lab} = 85°$. The solid line is the experimental average excitation function. (Levi et al 1966).

A convenient way of analysing these fluctuations is provided by the energy autocorrelation function

$$C(\varepsilon) = \frac{<\sigma(E)\ \sigma(E+\varepsilon)>}{<\sigma(E)><\sigma(E+\varepsilon)>} - 1 \qquad (2.1)$$

If the fluctuations are characterised by a unique correlation width Γ, the autocorrelation function

$$C(\varepsilon) = \left| \frac{\sigma}{1 + i\varepsilon/\Gamma} \right|^2 \qquad (2.2)$$

so that

$$\frac{C(\varepsilon)}{C(0)} = \frac{\Gamma^2}{\Gamma^2 + \varepsilon^2} \tag{2.3}$$

The presence of compound nucleus fluctuations is thus shown by plotting $C(\varepsilon)/C(0)$ as a function of ε and seeing if it has the expected Lorentzian form. An example of such a plot is shown in Fig.5.

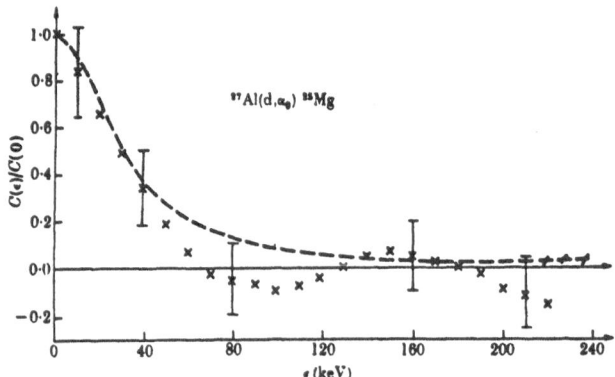

Fig.5   Normalized correlation function for the ^{27}Al$(d,\alpha_o)^{25}$Mg reaction from 1.4 to 2.3 MeV at an angle of 150°. The dashed curve is the theoretical Lorentzian expression (2.3) with $\Gamma = 30$ keV. The curve follows the data for $\varepsilon \lesssim \Gamma$, but thereafter the data continues to fluctuate while the curve goes asymptotically to zero. The fluctuations are due to statistical uncertainties associated with the finite range of the experimental data. (Gadioli et al, 1965).

If however there are also present pre-equilibrium processes with characteristic correlation widths Γ_n appreciably greater than that of the compound nucleus process, then it will not be possible to fit the autocorrelation function with a single Lorentzian, and indeed, a superposition is required (Friedman et al, 1981).

$$C_N(\varepsilon) = \left| \sum_{n=1}^{N} \frac{\sigma_n}{1 + i\varepsilon/\Gamma_n} \right|^2 \tag{2.4}$$

An analysis of the excitation functions of the ^{27}Al$(h,p)^{29}$Si reaction by Bonetti et al (1980ab) gave autocorrelation functions that cannot be fitted by a single Lorentzian, as shown in Fig.6. Subsequent analysis with the expression (2.4) corresponding to two correlation widths enabled the data to be fitted, as shown in Fig.7. The values of the largest and smallest correlation widths found in this analysis are ≈ 200 keV and ≈ 50 keV respectively. A similar analysis of fluctuations in the ^{15}N$(^{12}$C$,\alpha)^{23}$Na reaction by Hussein (1981) gave coherence widths of 400 keV and 70 keV. These enable the interactions times to be estimated, since a width of 100 keV corresponds to a time of 6×10^{-21} secs, and smaller widths to proportionately longer times.

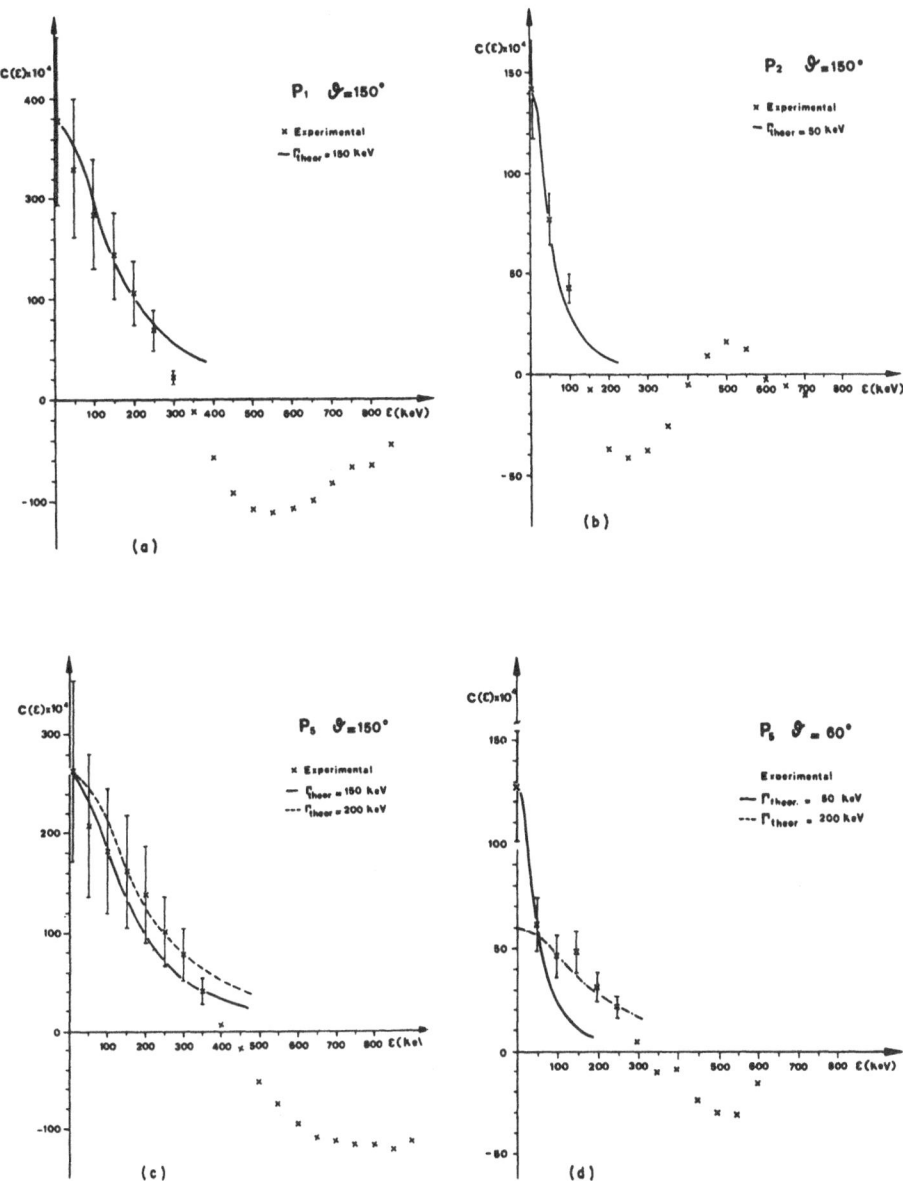

Fig. 6 Autocorrelation data for the $^{27}Al(h,p)^{29}Si$ reaction compared with Lorentzian curves corresponding to a single coherence width (Bonetti et al, 1980a).

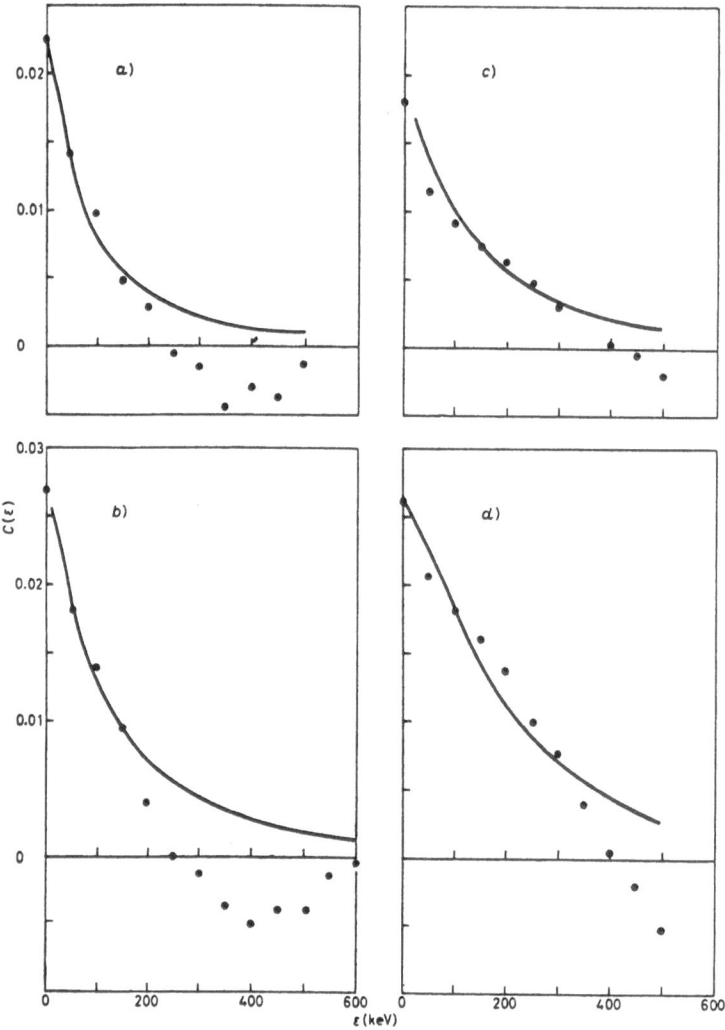

Fig. 7 Autocorrelation data for the ^{27}Al(h,p)^{29}Si reaction compared
with the expression (2.4) corresponding to two correlation
widths (Bonetti et al, 1980b).

Analyses like these show the presence of pre-equilibrium particles emitted from early
stages in the nuclear excitation process. It is however, difficult to analyse pre-
equilibrium processes in detail by fluctuation analyses partly because the measurements demand very high resolution and are extremely tedious and partly because their
accuracy is inherently limited by the amount of data available in any energy interval
small compared with the energy itself.

The analysis of pre-equilibrium processes thus requires a detailed theory of the interaction that permits the energy and angular distributions of the emitted particles

to be calculated. Some of the presently available theories are discussed in the following sections.

3. The Exciton Model

There are many varieties of exciton model, of varying degrees of sophistication, and these have been extensively applied to calculate the cross-section of reactions with contributions from pre-equilibrium processes (Griffin, 1966, 1967; Williams, 1970; Blann, 1975; Gadioli and Gadioli Erba, 1980). The earlier theories gave only the total cross-sections, but later developments enabled the angular distributions to be calculated as well (Akkermans et al, 1979, 1980).

The essential feature of these exciton models is the description of the successive excitation of the nucleus by a series of stages characterized by the number of excitons, or particle-hole pairs. In the first stage, the projectile interacts with a nucleon and excites a particle-hole pair. The particles and holes from the first stage initiate a cascade of nucleon-nucleon interactions that spreads the incident energy among all the particles of the compound system. At each stage in this cascade there are probabilities, determined by a two-body transition matrix element and a level density function, for a transition either to a more complicated state with a higher exciton number or back to one of lower exciton number, or directly in to the continuum. The latter case corresponds to pre-equilibrium emission.

The first particles emitted during this cascade retain to some extent the direction of the projectile, giving a characteristically forward-peaked angular distribution. The degree of forward peaking decreases down the cascade as the particles are scattered at each stage. Eventually, the incident energy is fully shared among the target nucleons and statistical equilibrium is established. The subsequent emission of particles then follows the statistical theory for the decay of the compound nucleus, and the angular distribution becomes symetrical about 90° with respect to the incident direction.

The varieties of the model differ in the details of the calculation and in the expressions used for the transition matrix element and the level density distribution.

These exciton models have been used to calculate the cross-sections of a large number of reactions. Their relative simplicity makes them very suitable for obtaining the cross-sections of rather complicated processes in which many particles are emitted. The usual method of comparison with the experimental data is to used one set to fix the values of the parameters of the model, in particular the transition matrix element, and then to use this value in subsequent calculations. The success of the model is assessed by its ability to account for a wide range of data with the same values of these parameters.

A Monte Carlo procedure is frequently used to evaluate the cross-sections. At each stage of the cascade a random number determines the decay mode, taking account of the known probabilities of the various processes that can occur. A second random number determines the energy of the emitted particle. A large number of such calculations are made so that statistically meaningful results are obtained.

A particularly extensive series of calculations with the exciton model has been made by Gadioli and colleagues (Gadioli et al, 1977ab; Hogan et al, 1979). They studied the interaction of 10-100 MeV protons with a wide range of nuclei, and calculated the total cross-sections of many reactions involving the emission of several neutrons, protons and alpha-particles. The cross-sections are measured radiochemically, so it is not possible to distinguish between different reactions (like (p,pn) and (p,d) for example) leading to the emission of the same numbers of neutrons and protons. In such cases the sum of the calculated cross-sections for the two reactions is compared with the experimental data. For the lighter target nuclei, the energy range of the incident protons spans both the region where the compound nucleus reactions predominate, and the higher regions where the pre-equilibrium processes predominate.

The emission of alpha-particles is particularly favoured for the interactions with the lighter nuclei, and comparison with the data in these cases enables the ability of the exciton model to give alpha emission probabilities to be assessed. In order to calculate the alpha emission probability it is assumed that the alpha-particles are present in the nucleus with a pre-formation density ϕ_α, a momentum distribution proportional to p^6 and are distributed in single alpha-particle states.

Some typical excitation functions for total cross-sections calculated with the exciton model are compared with the experimental data in Figs. 8-11. Figs. 8 and 9 refer to (p,n) and (p, 2pn) reactions respectively, and the overall agreement is satisfactory. In each case, the peak at lower energies is due to the compound nucleus process and the tail extending to higher energies is mainly composed of the pre-equilibrium particles. Fig. 10 refers to the (p, 2p3n) reaction and shows the effect of two competing processes. The peak at lower energies is due to the (p,αn) reaction, and that higher energies to the (p, 2p3n) reaction with the nucleons emitted individually. The next figure shows a similar comparison for the ^{50}Ti(p, 2p2n)^{47}Sc reaction, in which the lower energy peak is attributed to the (p,α) reaction. The dashed line shows the result of a calculation that does not include the emission of pre-equilibrium alpha-particles. Comparison with the data shows that the evaporated alpha-particles dominate in the region of the low-energy peak, while the pre-equilibrium alpha-particles dominate at higher energies.

There have been many calculations of pre-equilibrium processes in neutron induced reactions using the exciton model, and these have been compared with the data on (n, n'), (n, p), (n,α) and (n, xnyp) reactions (Gadioli and Gadioli Erba, 1980). Some data on (n, p) reactions has been analysed by Braga-Marcazzan et al (1972), who found that the exciton model calculations are in good overall accord with the measured proton energy spectra.

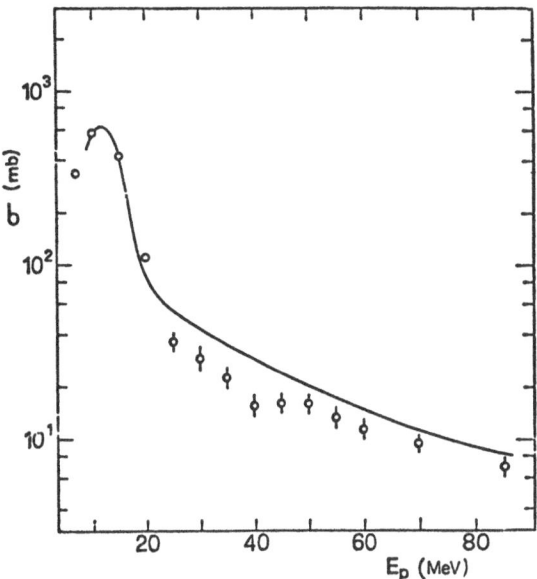

Fig. 8 Excitation function for the ^{48}Ti$(p,n)^{48}$V reaction compared with exciton model calculations (Gadioli et al 1981).

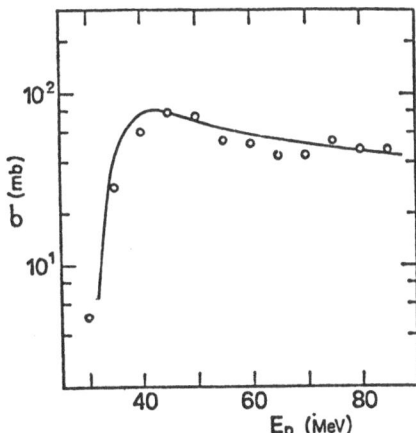

Fig. 9 Excitation function for the ^{48}Ti$(p,2pn)^{46}$Sc reaction compared with exciton model calculations (Gadioli et al 1981)

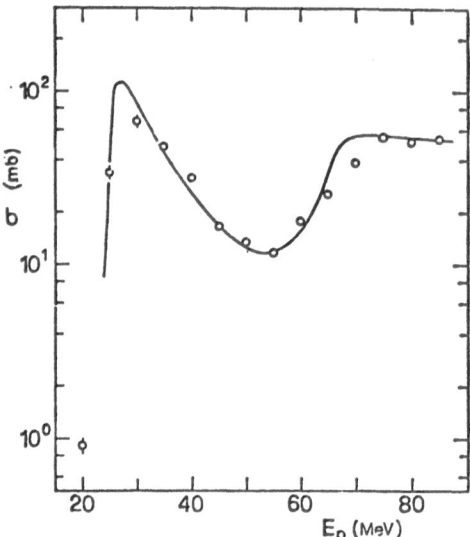

Fig. 10 Excitation function for the ^{48}Ti(p,2p3n)^{44}Sc reaction compared
with exciton model calculations (Gadioli et al 1981).

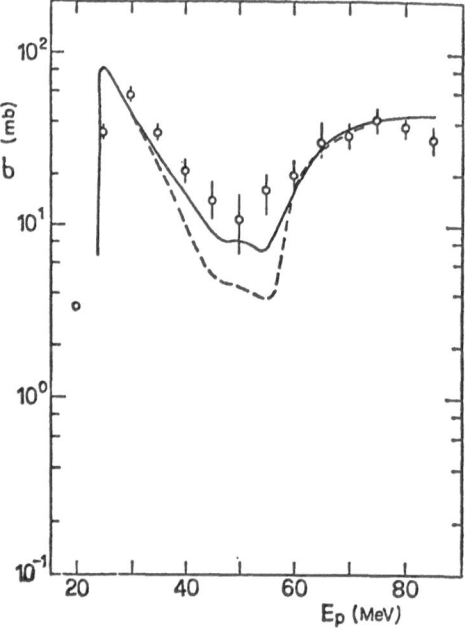

Fig. 11 Excitation function for the ^{50}Ti(p,2p2n)^{47}Sc reaction compared
with exciton model calculations. The dashed line shows the
effect of omitting the pre-equilibrium alpha-particles (Gadioli
et al 1981)

The contribution of pre-equilibrium processes to a series of reactions with the common compound nucleus ^{63}Cu has recently been studied by Parker et al (1982). They bombarded targets of ^{59}Co, ^{56}Fe and ^{51}V with alpha-particles, ^{7}Li and ^{12}C nuclei respectively over a range of energies using the stacked foil technique. The total cross-sections of the reactions leading to seven of the radioisotopes formed in these reactions are shown by the histograms in Fig. 12. Calculations of the cross-sections were made using the statistical model for the decay of the compound nucleus, and the results are also included in the Figure. Comparison between the calculations and the experimental results permits several conclusions to be drawn concerning the reaction mechanism. In particular, the good overall fit to the reactions initiated by ^{12}C indicates that these reactions proceed mainly through the compound nucleus process. In the case of the reactions initiated by alpha-particles and ^{7}Li there are significant deviations very similar to those already discussed for proton-induced reactions that indicate the presence of pre-equilibrium processes.

This conclusion has been further tested by measurements of the energy distribution of the recoil nuclei. In these measurements, a stack of aluminium catcher foils was placed behind the target, and measurements of the radioactivities of the foils enabled the projected range distribution of the recoil nuclei to be found. The projected range distributions were calculated for purely statistical emission and are compared with the results of the measurements in Figs. 13 and 14.

There is once again good agreement for the reactions initiated by ^{12}C, but there is a notable excess of short range recoils in the case of the reactions initiated by ^{7}Li. Short range recoils are expected to result from pre-equilibrium emission, since this process takes a greater proportion of the momentum of the incident particle than the compound nucleus formation, so that there is less available for the recoil.

These data are presently being analysed using the exciton model, and detailed comparisons with the experimental data should allow the mechanism of the pre-equilibrium processes to be explored in more detail.

3. Quantum-Mechanical Theories

More recently, several quantum-mechanical theories of pre-equilibrium reactions have been developed. Essentially, these are based on detailed expressions for the microscopic processes occuring in the nucleus, and the statistical averaging is introduced to make the calculations tractable. So far, only those of Tamura and collaborators (1977, 1978, 1979) and of Feshbach, Kerman and Koonin (1980) have been developed to the stage of numerical comparison with experiment, so attention will be concentrated on these.

(a) Theory of Tamura

The problem of extending the theory in a praticable way to more highly excited states

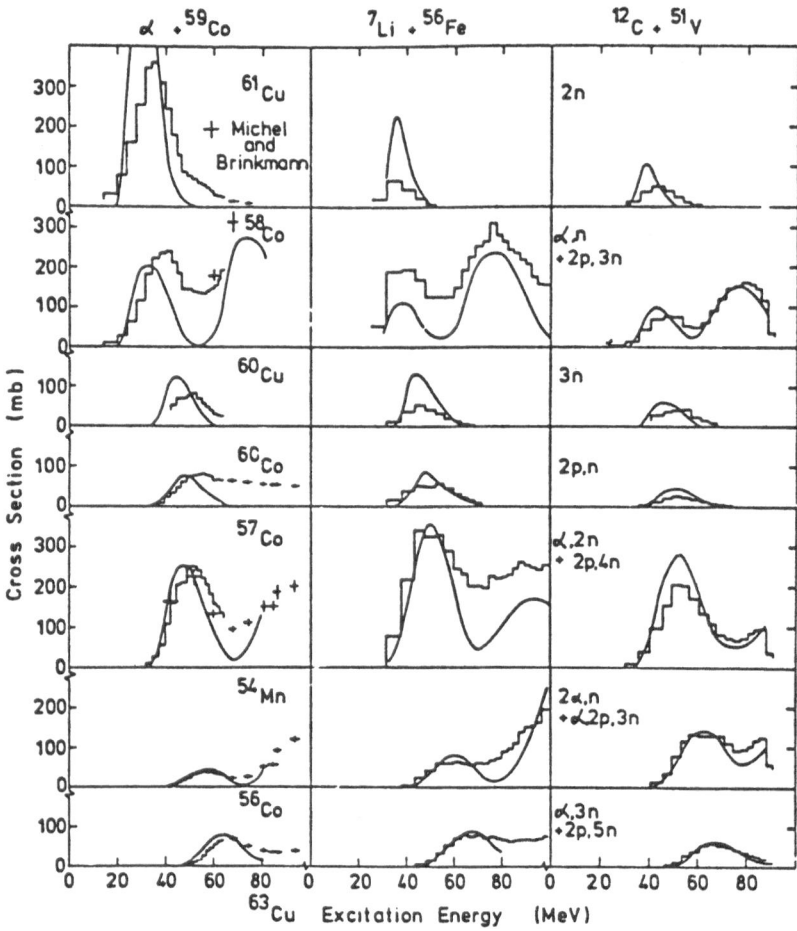

Fig. 12 Cross-sections for the production of seven radioisotopes from the entrance channels α+^{59}Co, ^{7}Li+^{58}Fe and ^{12}C+^{51}V as a function of compound nuclear excitation. The histograms show the experimental data and the curves the results of statistical evaporation calculations (Parker et al, 1982).

and into the continuum has been tackled by Tamura et al (1977, 1978, 1979). They realized that the very complexity of reactions to the continuum permits important simplifications to be made in the calculations. Since the aim is to obtain averaged cross-sections corresponding to the excitation of a large number of states of different configurations within a range of energies it is permissible to ignore the interference contributions. It is thus a good approximation to use the single-particle shell model for the final states. The one-step contributions to the cross-section in energy bands are evaluated using the DWBA theory with the corresponding spectroscopic densities and two-step contributions by appropriate integrations over the possible intermediate states.

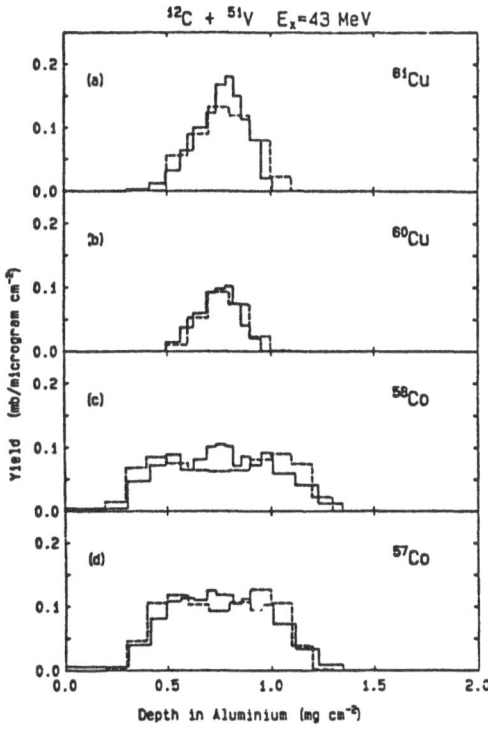

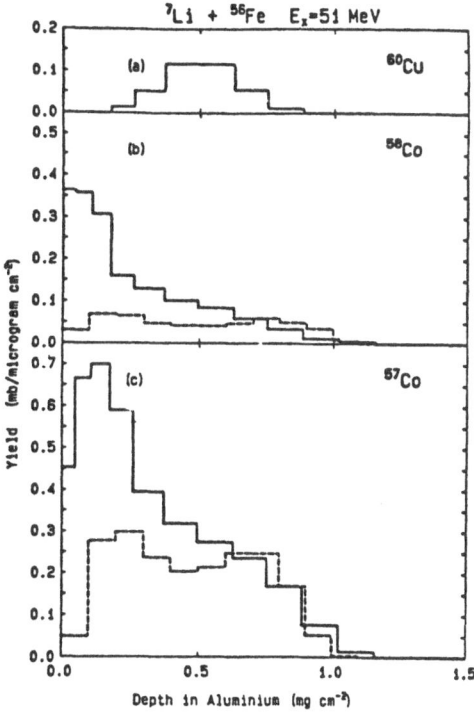

Fig. 13 Projected recoil range distributions for ^{61}Cu, ^{60}Cu, ^{58}Co
and ^{57}Co produced by the reaction of ^{12}C
with ^{51}V at an equivalent ^{63}Cu excitation
energy of 43 MeV. The solid histograms
are measured distributions and the dashed
histograms show the results of the statistical evaporation calculations (Parker et
al, 1982).

Fig. 14

Projected range distributions for ^{60}Cu,
^{58}Co and ^{57}Co produced by the interaction of ^{7}Li with ^{56}Fe at an equivalent
excitation energy of 51 MeV (Parker et
al, 1982)

Initially they applied their theory to calculate the cross-section for the inelastic
scattering of protons by ^{27}Al and ^{209}Bi, and subsequently they extended it to (p,α)
reactions (Tamura and Udagawa, 1977, 1978) and to deep inelastic collisions between
heavy ions (Udagawa et al, 1979). In this study they examined the cross-sections for
the continuum spectra in the $^{27}Al(^{20}Ne,^{16}O)$ and $^{27}Al(^{20}Ne,^{12}C)$ reactions at 120 MeV
and assumed that they take place by one-step and two-step alpha transfer respectively. Since it is too time-consuming to calculate all the DWBA matrix elements for
heavy-ion reactions, the matrix elements were parametrised by simple yet accurate
expressions and the subsequent integrations were carried out numerically. The values
of the parameters specifying the matrix elements were determined by a least squares
fit to a large number of exact finite-range DWBA calculations (Udagawa et al, 1979).

The computing time is substantially reduced by the Q-window, which severly restricts
the range of integration. Penetrability effects in the region of the Coulomb barrier
were taken into account by a Coulomb reduction factor, and a simple functional form
was used for the level density.

In their work on proton inelastic scattering, Tamura et al (1977) used the distorted
wawe theory to describe the one-step process and summed over all possible particle-
hole pairs consistent with energy conservation. Their results for (p,p') at 62 MeV
on ^{27}Al and ^{209}Bi are shown in Fig. 15, and it is apparent that the solid lines,
which represent the sum of the one-step and two-step reaction cross-sections, are in
good agreement with the experimental angular distributions. Since all the parameters
of the calculation are obtained from other work the absolute value of the cross-sec
tion is obtained.

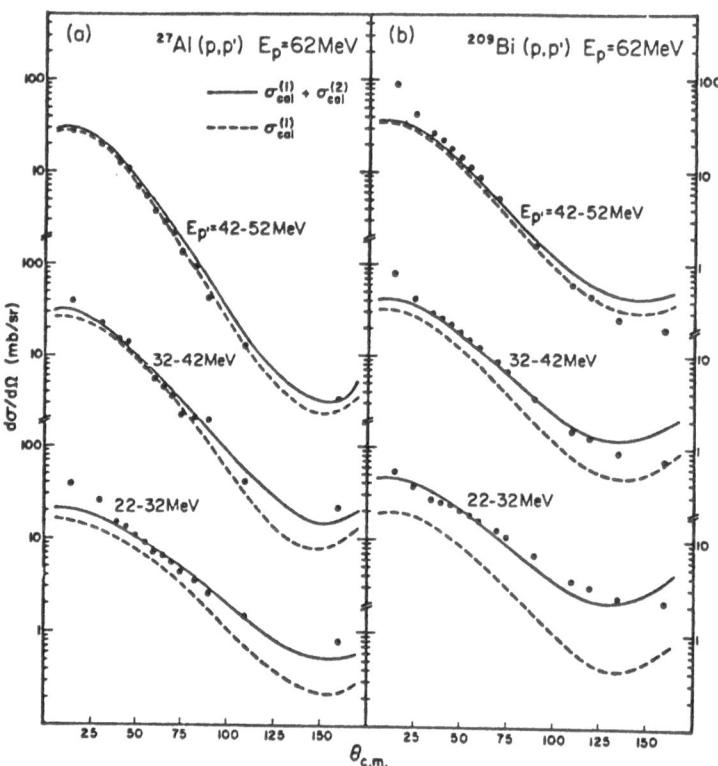

Fig. 15 Comparison of the calculated (p,p') cross sections with experiment. Before comparison, both the experimental and theoretical cross section were inegrated over 10 MeV energy bins
as shown in the figure. The solid lines represent the sum of
one- and two-step cross sections, while the dotted lines represent only the one-step cross section (Tamura et al 1977).

b) <u>Theory of Feshbach, Kerman and Koonin</u>

As in the exciton model, this theory assumes that the reaction takes place in a number of stages of increasing complexity. The projectile enters the nucleus and collides with a nucleon, producing a two-particle one-hole excitation. The secondary particles can themselves interact, producing three-particle two-hole excitations and so on. At each stage there is a finite probability that the compound system will make a transition to the final state, and this gives the pre-equilibrium cross-section. To make the computation practicable, it is assumed that transitions within the chain of excitations can only be made one step at a time; this is called the "chaining hypothesis". This means that from the nth stage we can only go to the $(n\pm1)^{th}$ stage, or on to be the final state; it is not possible to miss a stage and go from the nth state to the $(n\pm2)^{th}$ stage, for example. A distinction is also made between the multistep direct processes, in which at least one of the particles is in the continuum, and the multistep compound in which all the particles are bound at each stage.

A further assumption is that the relative phases of certain matrix elements are random, but this is done in a different way for the two processes. In the case of the multistep compound reaction, matrix elements with different quantum numbers are assumed to have random phases, and this ensures that no interference terms remain after averaging, so that the cross-section is symetric about 90°. In the case of the multistep direct process, however, it is assumed that only those matrix elements that interfere constructively on averaging are those involving the same change in momentum of the particle in the continuum. This ensures that the memory of the initial direction is preserved, giving an anisotropic angular distribution. The calculations described here are for the multistep direct process only.

As in the exciton model, the excited nuclear states are characterised by their number of excitons, or particle-hole excitations and the class os states with $(2n+1)$ excitons is labelled P_n. The incident kinetic nergy is spread through the nucleus by the two-body residual interaction, which is responsible for the transitions along the chain. The total cross-section for a reaction from a state with particle momentum k_i to one of particle momentum k_f is the sum of the cross-sections of reactions of all possible numbers of stages. It is convenient to separate the single-step calculation and write

$$(\frac{d^2\sigma_{if}}{dUd\Omega}) = (\frac{d^2\sigma_{if}}{dUd\Omega})_{one-step} + (\frac{d^2\sigma_{if}}{dUd\Omega})_{multi-step} \equiv S_1 + S_M \qquad (4.1)$$

We now introduce the cross-section for a transition from the nth to the $(n+1)^{th}$ stage, and denote it by

$$W_{n,n-1} \equiv \frac{d^2W_{n,n-1}(\underline{k}_n,\underline{k}_{n-1})}{dU_n \, d\Omega_n} \qquad (4.2)$$

Now consider the probability of a two-step reaction. This can be made in one way as shown in Fig. 16 the cross-section is

$$S_2 = W_{21} S_1 \tag{4.3}$$

The three-step reaction can be made in two ways, as in the figure. The sum of their cross-sections is

$$S_3 = W_{32}W_{21}S_1 + W_{12}W_{21}S_1 \tag{4.4}$$

Similarly the probability of a four-step reaction is

$$S_4 = W_{43}W_{32}W_{21}S_1 + W_{23}W_{32}W_{21}S_1 + W_{21}W_{12}W_{21}S_1 \tag{4.5}$$

The level density increases rapidly with the number of excitons, and so terms with retrograde steps are much smaller than those without retrograde steps. For the same reason terms with retrograde steps towards the beginning of the excitation chain are much smaller than terms with similar retrograde steps further down the chain. The terms in the above expressions are thus arranged in order of decreasing

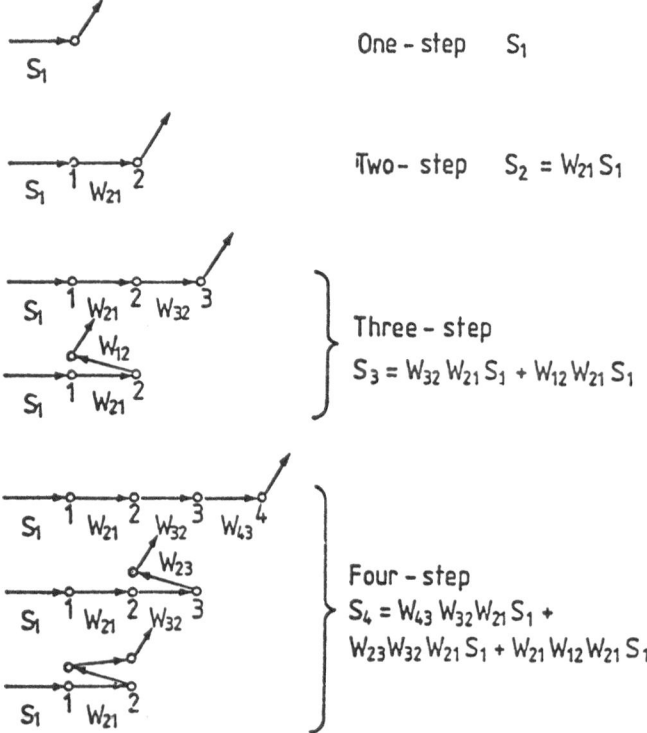

Fig. 16 Schematic diagram illustrating the various processes contributing to multistep reaction cross-sections.

magnitude, and we retain only the first two terms. The number of retrograde terms omitted in this way rises rapidly with the number of steps; there are two for five-step processes and seven for six-step processes. Their inclusion would thus greatly increase the complexity of the calculation without appreciably affecting the result.

Omitting these terms, the total multistep cross-section becomes

$$S_M = \sum_n S_n = \sum_n \sum_{m=n-1}^{n+1} W_{mn} W_{n,n-1} \cdots \cdots W_{21} S_1 \qquad (4.6)$$

At each stage we must integrate over all angles and momenta so that the full expression for the multistep cross-section is

$$\left(\frac{d^2\sigma_{if}}{dUd\Omega}\right)_{multistep} = \sum_n \sum_{m=n-1}^{n+1} \int \frac{dk_1}{(2\pi)^3} \int \frac{dk_2}{(2\pi)^3} \cdots \int \frac{dk_n}{(2\pi)^3} \times \qquad (4.7)$$

$$\times \frac{d^2 W_{mn}(k_f, k_n)}{dU_f d\Omega_f} \frac{d^2 W_{n,n-1}(k_n, k_{n-1})}{dU_n d\Omega_n} \cdots \frac{d^2 W_{2,1}(k_2, k_1)}{dU_2 d\Omega_2} \left(\frac{d^2\sigma_{if}}{dU_1 d\Omega_1}\right)_{single\ step}$$

The transition matrix element

$$\frac{d^2 W_{n,n-1}(k_n, k_{n-1})}{dU_n d\Omega_n} = 2\pi^2 \rho(k_n) \rho_n(U) < |v_{n,n-1}(k_n, k_{n-1})|^2 > \qquad (4.8)$$

where $\rho(k_n) = mk/(2\pi)^3 \hbar^3$ is the density of particle states in the continuum, $\rho_n(U)$ the level density of the residual nucleus at excitation energy U, and $v_{n,n-1}(k_n, k_{n-1})$ is the matrix element describing the transition from a state (n-1) to a state n when the particle in the continuum changes its momentum from k_{n-1} to k_n. This matrix element is given by the distorted wave Born approximation expression

$$v_{a,b}(k_i, k_f) = \int \chi_a^{(-)*} <\psi_f|V(r)|\psi_i > \chi_b^{(+)} dr \qquad (4.9)$$

where V(r) is the effective interaction for the transition, $\chi_a^{(-)}$ and $\chi_b^{(+)}$ the incoming and outgoing distorted waves and ψ_i and ψ_f the initial and final nuclear states. To include all the transition strength, the spectroscopic factors are always taken to be unity.

In the expression (4.8) for the transition matrix element, the angular brackets indicate an appropriate averaging procedure. When the transition probability is averaged over many final states, the interference terms cancel and the orbital angular momenta contribute incoherently, so that the averaged value of the squared matrix element becomes

$$<|v(k_i, k_f)|^2> = \sum_L (2L+1) < |v_L(k_i, k_f)|^2 > R(L) \qquad (4.10)$$

where R(L) is the spin distribution function of the residual nucleus levels that is normalised by $\sum_{L}(2L+1)R(L)=1$.

Similarly, the averaged single-step cross-section is given by

$$\left(\frac{d^2\sigma_{if}}{dUd\Omega}\right)_{single-step} = \sum_{L}(2L+1)\rho_2(U)R_2(L) \left<\left(\frac{d\gamma}{d\Omega}\right)_L\right> \qquad (4.11)$$

where the suffix 2 refers to the number of excitons in the residual nucleus after the first interaction.

This formalism has been used by Bonetti et al (1981) to calculate the cross-sections of several (p,n) reactions from 25 to 45 MeV. In this work the inelastic cross-section was calculated microscopically for each transferred L-value as a function of angle and for a series of values of the incoming and outgoing energies and for all possible pairs of initial and final bound states consistent with energy conservation, using a single-particle shell model to describe the states. The average value of these cross-sections gives $<d\sigma/d\Omega>_L$ for each value of L.

In the evaluation of the matrix element (4.9) a Yukawa potential of range 1 fm was used for $V(r)$ and the incoming and outgoing distorted waves were calculated using the potentials of Becchetti and Greenlees (1969). The bound state wave functions were calculated in a Saxon-Woods potential of standard parameters. The results of some typical calculations for $\Delta L=1$ transitions in ^{48}Ca at 45 MeV are shown in Fig. 17.

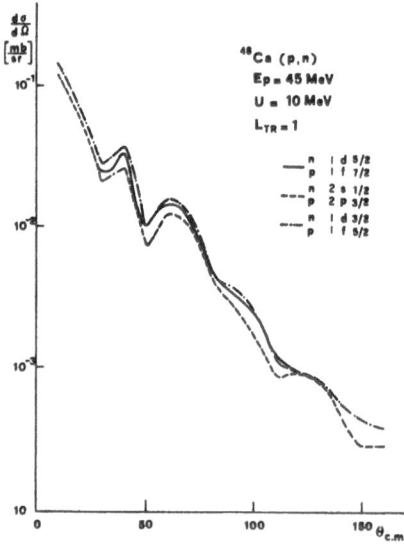

Fig. 17 Calculated differential cross-sections for some typical transitions in ^{48}Ca at 45 MeV between shell-model states corresponding to $\Delta L=1$, showing their overall similarity.

To obtain the single-step cross-section (4.11), the Ericson expression was used for the n-exciton state level density

$$\rho_n(U) = g(gU)^{n-1}/p!h!(n-1)!$$ (4.12)

where n=p+h, and the spin distribution function

$$R_n(L) = \frac{2L + 1}{\pi^{1/2}n^{3/2}\sigma^3} \exp - \frac{(L+1/2)^2}{n\sigma^2}$$ (4.13)

where σ is the spin cut-off parameter.

These expressions are also used to calculate the transition probabilities (4.8) and hence the multi-step contribution. The level density $\rho_n(U)$ describes the final states of the interaction when a particle in the continuum with momentum $\underline{k}_{n-1}$ collides with a bound nucleon, changing its momentum to $\underline{k}_n$ and creating a particle-hole pair. The final state density is that of a particle-hole pair, that is $\rho_2(U)$ for all stages of the chain.

In the multistep calculation the averaged expression (4.11) is inserted in (4.7) and then integrated over all intermediate energies and angles. The calculation of the subsequent steps then follows because the source term of each step comes from the results of the previous steps. It requires the evaluation of v_{ab} at each value of the incoming and outgoing energies compatible with energy conservation and for each angle and transferred L-value. In practice this was carried out for a number of discrete energy steps and the distribution between neutrons and protons was ignored for the intermediate stages as this was found to make little difference to the final result. An averaged value of the interaction strength was used, taking account of Austin's result that it is about four times as strong between like as between unlike particles.

Calculations were made of the energy and angular distributions of (p,n) reactions on [48]Ca, [90]Zr and [208]Pb at 45 MeV, and on [120]Sn at 25, 35 and 45 MeV, and some of the results are shown in Figs. 18 to 22. Essentially all the parameters were taken from previous work, except the strength parameter V_0 which was adjusted to optimise the fit in each case. The resulting values of V_0 were found to be remarkably constant and are consistent with that found by Austin from analyses of particular inelastic interactions to discrete final states. Since V_0 operates at each step, the magnitude of the final m-step cross-section is proportional to V_0^{2m}. The value of V_0 thus determines the relative contributions of the different steps as well as the absolute values of the cross-sections, and hence the energy and angular distributions. The overall agreement with experiment found with an essentially constant value of V_0 thus confirms the validity of the theory.

These calculations show the importance of multistep processes in the Feshbach chain. At 45 MeV they account for about one-half of the total (p,n) cross-section. The mul-

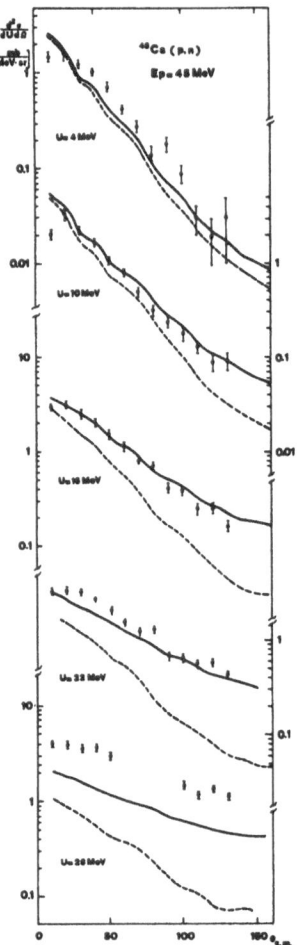

Fig. 18 Comparison between calculated and experimental differential
cross-sections for the ^{48}Ca(p,n) reaction at several excitation energies of the residual nucleus. At the highest excitation energy (u=28 MeV), the experimental cross-section
exceeds the calculated value due to the presence of multiple
particle emission.
- - - single-step contribution; —— total.

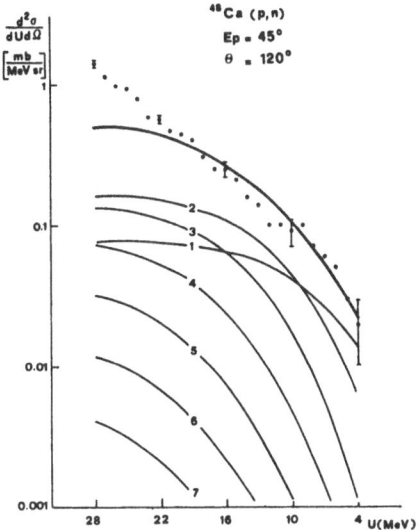

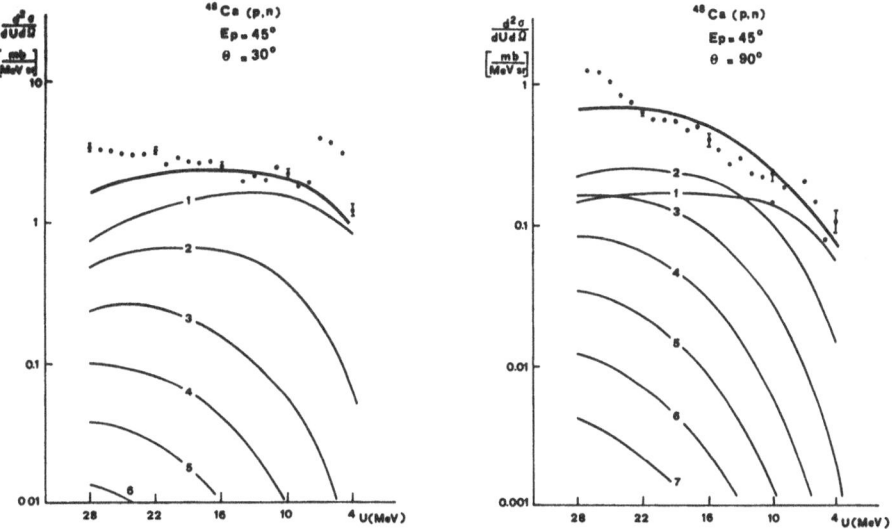

Fig. 19 (a,b,c). Comparison between calculated and experimental energy
spectra for ^{48}Ca(p,n) at several angles showing the contributions
of various steps. The short fall at the higher excitation energies is attributed to multiparticle emission. It is notable that
in the backward direction the two-step and even the three-step
cross-section is sometimes greater than the one-step cross-section.

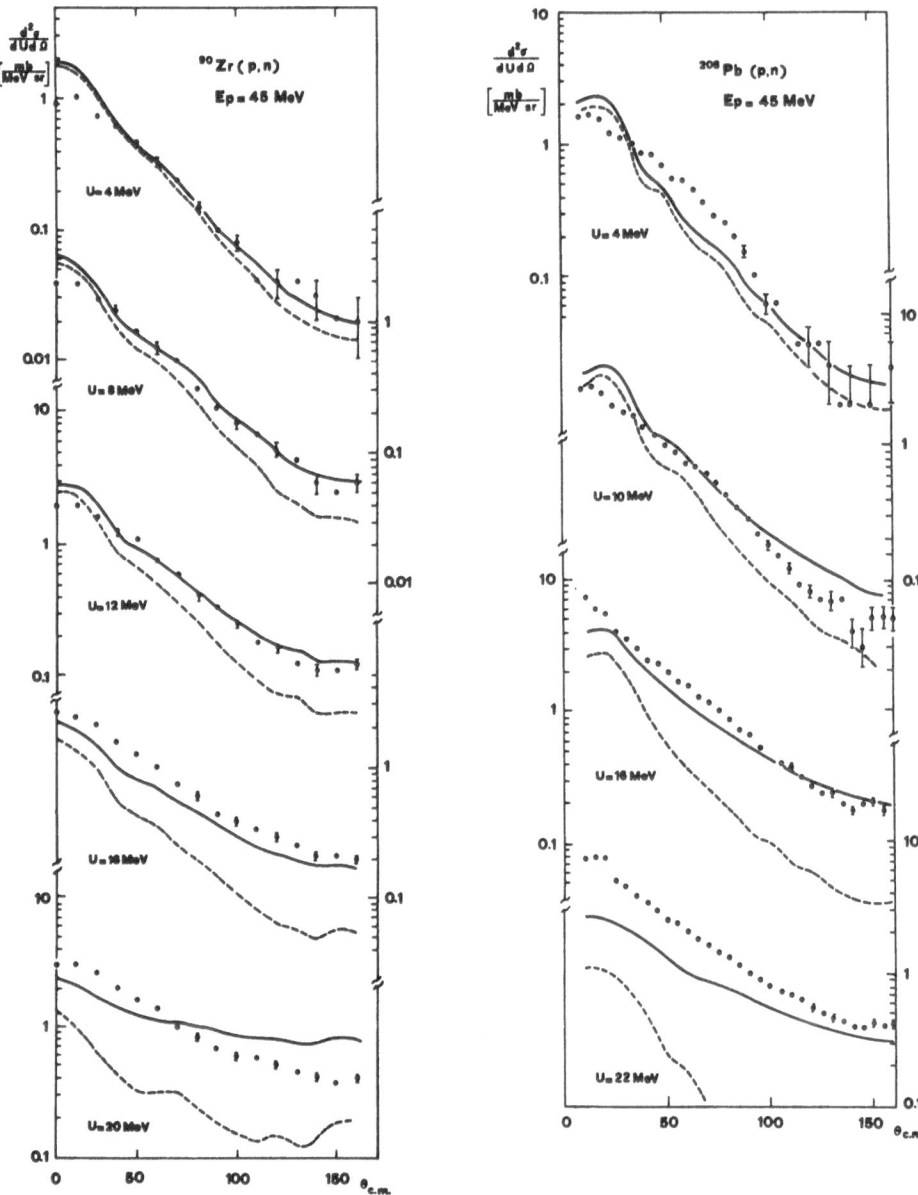

Fig. 20 Comparison between calculated
         and experimental differential
         cross-sections for the
         ^{90}Zr(p,n) reaction.
         - - - single step contribution;
         ——— total

Fig. 21 Comparison between calcula-
         ted and experimental diffe-
         rential cross-section for
         the ^{208}Pb(p,n) reaction.
         - - - single step contribution;
         ——— total

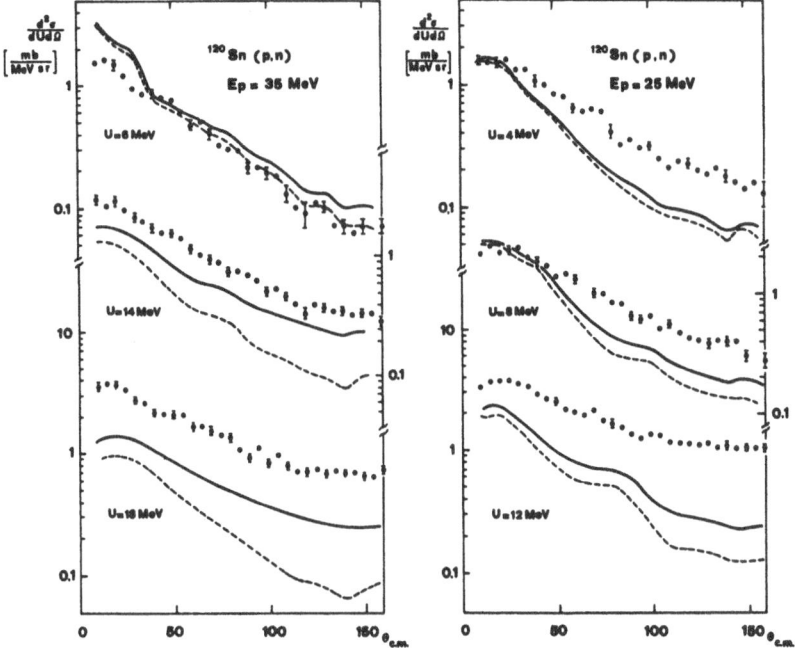

Fig. 22   Comparison between calculated and experimental differential
cross-section for the ^{120}Sn(p,n) reaction at 35 MeV and 25
MeV incident proton energy.

tistep contribution increases quite strongly with the emission angle and with the
residual excitation energy.

There are some disagreements between the calculations and the data, such as the missing strength at high excitation energy. This may be due to contributions from multiple emission, which is included in the measurements but not in the calculations.
At lower incident energies, such as the ^{120}Sn reaction at 25 MeV, the spectrum is
not well fitted, probably because the statistical multistep compound emission process is not included in the calculations.

This analysis has recently been extended to include the analysing powers, and calculations have been made for the ^{58}Ni($\vec{p}$,p') reaction at 65 MeV (Bonetti et al, 1982).
Data depending on the polarisation are always more sensitive than the differential
cross-section to the details of the interaction mechanism and thus provide a severe
test of any theory of the reaction.

The analysing power is very likely to depend on the number of steps in the reaction:
as the number increases the magnitude of the analysing power is expected to decrease

as the memory of the initial polarisation is gradually lost, until finally it becomes zero for the evaporated particles. The $(\vec{p},p')$ measurements of Sakai et al (1980) show however that the analysing power in the continuum is large and positive at backward angles, where the contribution of the one-step direct process is expected to be small. This at once indicates that the effects of multistep processes are likely to be important in this angular range.

The analysing power is defined by

$$A_y = \frac{\sigma_L - \sigma_R}{\sigma_L + \sigma_R} \qquad (4.14)$$

where σ_L and σ_R are the left and right cross-section respectively. These can each be written as a sum of single-step and multistep cross-sections.

$$\sigma_{L,R} = \sigma^*_{L,R} \text{ (single step)} + \sigma_{L,R}(\text{multistep}) \qquad (4.15)$$

The multistep cross-section is given by (4.7), with the integrations over the intermediate angles restricted to the values of θ for which each left and right cross-section is defined.

The transition matrix elements for the inelastic scattering process were evaluated using a complete microscopic model that includes the non-central components of the two-body residual interaction, particularly the $\underset{\sim}{L}.\underset{\sim}{S}$ term. Such an interaction has been found able to given the analysing powers measured for reactions to describe final states (Satchler, 1967; Greaves et al, 1972; Escudié et al, 1974; Hosono et al, 1978). The effective two-body interaction has the form

$$V(r) = V_c(r) + V_0(r) + V_\sigma(r)(\underset{\sim}{\sigma}_1 \cdot \underset{\sim}{\sigma}_2) + V_\tau(r)(\underset{\sim}{\tau}_1 \cdot \underset{\sim}{\tau}_2) + V_{\sigma\tau}(\underset{\sim}{\sigma}_1 \cdot \underset{\sim}{\sigma}_2)(\underset{\sim}{\tau}_1 \cdot \underset{\sim}{\tau}_2) +$$
$$+ \left[V_{LS}(r) + V_{LST}(r)(\underset{\sim}{\tau}_1 \cdot \underset{\sim}{\tau}_2)\right] \underset{\sim}{L}.\underset{\sim}{S} + \left[(V_T(r) + V_{T\tau}(\underset{\sim}{\tau}_1 \cdot \underset{\sim}{\tau}_2)\right] S_{12}. \qquad (4.16)$$

The radial dependence has the Yukawa form $V_{exp}(-r/\mu)/(r/\mu)$.

In the calculation, the parameter of Sakaguchi (1979) for polarised 65 MeV protons on ^{58}Ni were used for the entrance channel, and the global parameters of Menet et al (1971) were used for the other channels. The parameters of Austin (1980) were used for the central components of the two-body effective interaction, while the non-central components (tensor and spin-orbit) were fixed by simultaneously fitting the analysing powers of the discrete states of the reaction and of a part of the continuum at very low excitation energy where the higher order effects should be negligible. The values obtained in this way are the same as those of Hosono et al (1978) for the tensor components, while those of the spin-orbit components were adjusted to fit the data for the reaction to the 2^+ state obtained by Fricke et al (1967) at 40 MeV, and by Kocker et al (1976) at 60.2 MeV.

Calculations were made for four steps of the statistical multistep direct excitation chain, to excitation energies of up to 22 MeV in the residual nucleus. The results given in Fig. 23 show the quality of the fits to the differential cross-sections is even better than that previously obtained for the (p,n) reaction at 45 MeV. The calculations also reproduce the overall features of the analysing powers (Fig. 24) for residual nucleus excitation energies up to 18 MeV, and in particular the magnitude in the backward direction is correctly given together with the rather uniform variation with angle.

These results show that the statistical multistep direct theory is able to account for the overall features of both the differential cross-sections and the analysing powers of nuclear reactions to continuum states.

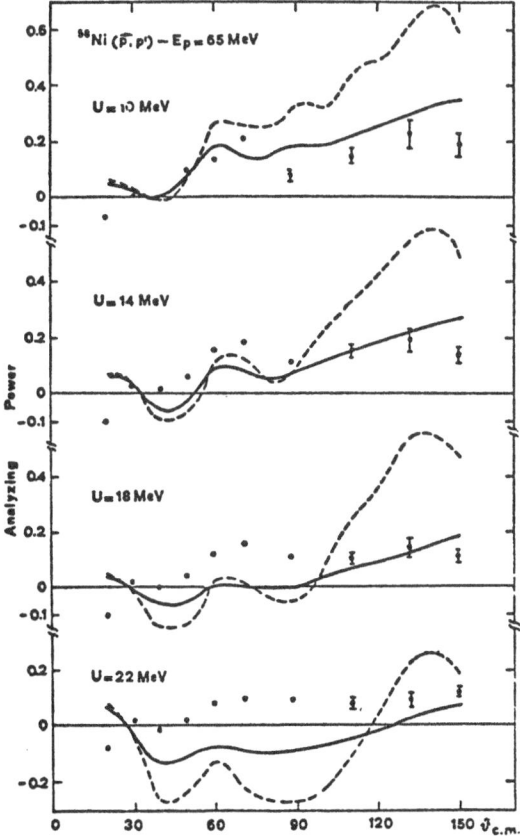

Fig. 23 Comparison between calculated and experimental differential cross-sections for the inelastic scattering of 65 MeV protons by ^{58}Ni.

- - - single-step contribution; ——total.

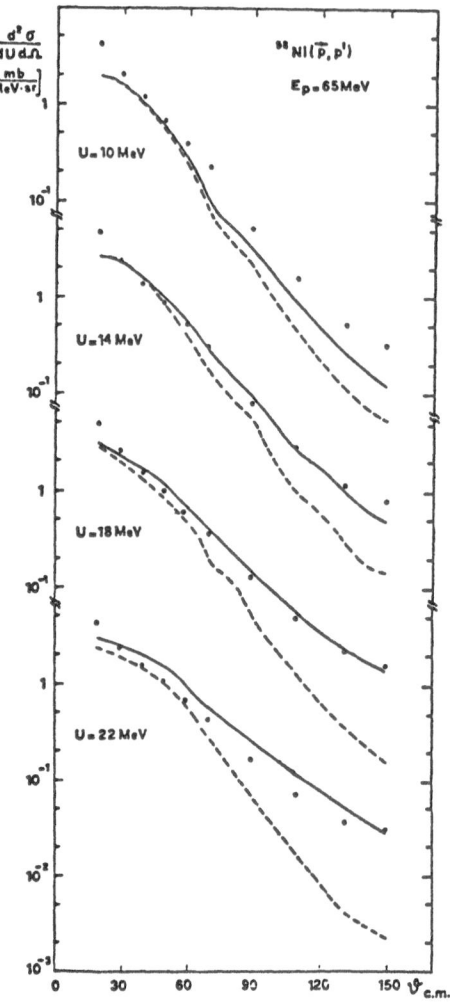

Fig. 24 Comparison between calculated and experimental analysing powers
for the inelastic scattering of 65 MeV protons by ^{58}Ni.
- - - single-step contribution; —— total.

References

J.M.Akkermans, H.Gruppelaar and G.Reffo, Phys.Rev. C22.73.1980.

J.M.Akkermans, Phys.Let. B82.20.1979; Z.Physik A292.57.1979.

S.M.Austin, Telluride Conference MSCUP-302.1979.

M.Blann, Ann.Rev.Nucl.Phys. 25.123.1975.

R.Bonetti, L.Colli Milazzo, A.de Rosa. G.Inglima, E.Perillo, M.Sandoli and F.Shahin,
     Phys.Rev. C21.816.1980a.

R.Bonetti, L.Colli Milazzo and A.Garegnani, Lett. Nuovo Cim. 29.496.1980b.

R.Bonetti, M.Camnasio, L.Colli-Milazzo and P.E.Hodgson, Phys.Rev. C24.71.1981.

R.Bonetti, L.Colli-Milazzo, I.Doda and P.E.Hodgson, Phys.Rev. (In press) 1982.

G.M.Braga-Marcazzan, E.Gadioli Erba, L.Colli-Milazzo and P.G.Sona, Phys.Rev. C6.1398.
    1972.
P.Von Brentano, J.Ernst, O.Häusser, T.Mayer-Kuckuk, A.Richter and W.Von Witsch,
    Phys.Lett. 9.48.1964.
T.Ericson, Adv.Phys. 9.425.1960.
J.L.Escudié, R.Lombard, M.Pignanelli, F.Resmini and A.Tarrats, Phys.Rev. C10.1645.
    1974.
H.Feshbach, A.Kerman and S.Koonin, Ann.Phys. (New York) 125.429.1980.
M.P.Fricke, E.E.Gross and A.Zucker, Phys.Rev. 163.1153.1967.
W.A.Friedman, M.S.Hussein, K.W.McVoy and P.A.Mello, Phys.Lett. 99B.179.1981.
E.Gadioli and E.Gadioli-Erba, IAEA Technical Report, 1980.
E.Gadioli, E.Gadioli-Erba, J.J.Hogan and K.I.Burns, Private communication, 1981.
E.Gadioli, E.Gadioli-Erba, J.J.Hogan and K.I.Burns, (In press) 1982.
E.Gadioli, E.Gadioli-Erba and J.J. Hogan, Phys.Rev. C16.1404.1977a; Nuovo Cim. 40.383.
    1977b.
E.Gadioli, I.Iori, M.Mangialaio and G.Pappalardo, Nuovo Cim. 38.1105.1965.
P.D.Greaves, V.Hnizdo, J.Lowe and O.Karban, Nucl.Phys. A179.1.1972.
J.J.Griffin, Phys.Rev.Lett 17.478.1966; Phys.Lett. 24B.5.1967.
P.E.Hodgson, Nuclear Reactions and Nuclear Structure (Oxford University Press 1971)
    Ch.11.
P.E.Hodgson, Continuum Processes in Heavy-Ion Reactions. Theoretical Summary of the
    Conference on Continuum Spectra of Heavy-Ion Reactions. Nuclear Science Research
    Conference Series, Vol.2. Edited by T.Tamura, J.Natowitz and D.H.Youngblood
    (Harwood Academic Publishers, 1980) p.445.
J.J.Hogan, E.Gadioli, E.Gadioli-Erba and C.Chung, Phys.Rev. C20.1831.1979.
H.Hosono, H.Kondo. T.Saito, N.Natsuoka, S.Nagamachi, S.Kato, K.Ogino, Y.Kadata and
    T.Noro, Phys.Rev.Lett. 41.621.1978.
M.S.Hussein, Contributed paper to Versailles Conference, 1981.
D.G.Kocker, F.E.Bertrand, B.E.Gross and E.Newman, Phys.Rev. C14.1392.1976.
C.Levi, M.Mermaz and L.Papineau, Phys.Lett. 22.483.1966.
J.J.Menet, E.E.Gross, J.J.Malanify and A.Zucker, Phys.Rev. C4.1114.1971.
D.J.Parker, J.Asher, T.W.Sonlon and I.Naqib (In press) 1982.
H.Sakaguchi, Proceedings of the Tsukuba Conference on Polarisation Phenomena, 1979.
H.Sakai, K.Hosono, N.Matsuoka, S.Nagamachi, K.Okada, K.Maeda and H.Shimizu, Phys.
    Rev.Lett. 44.1193.1980.
G.R.Satchler, Nucl.Phys. A95.1.1967.
T.Tamura and T.Udagawa, Phys.Lett. 71B.273.1977; 78B.189.1978.
T.Tamura, T.Udagawa, D.H.Feng and K.-K.Kan, Phys.Lett. 66B.109.1977.
T.Udagawa, T.Tamura and B.T.Kim, Phys.Lett. 82B.349.1979.
F.G.Williams, Phys.Lett. 31B.184.1970.

LIGHT NUCLEI FAR FROM STABILITY

A. Dobado   and   A. Poves

Departamento de Física Teórica

Universidad Autónoma de Madrid

Canto Blanco, Madrid -34 Spain

1.- Introduction

The production of new isotopes far from the stability by means of the Isotope Separators on Line has provided new and very interesting data relevant to the study of nuclear structure. We shall comment here on the results for light nuclei obtained by several collaborations at ISOLDE[1]. The important point to justify focusing on light nuclei is the feasibility of refined theoretical calculations. Moreover the domain of exotic nuclei brings in new situations of major interest, among them let us quote:

- The appearance of new regions of deformation. This is the case when filling a shell with neutrons, in the presence of a few protons. The expected neutron-shell closure does not appear because the large isovector core polarization produces a level crossing between different major shells. This is the situation for ^{31}Na and ^{32}Mg (ref.2)), where the lowering of the $1f_{7/2}$ orbit plays a fundamental role. There are theoretical approaches in the framework of the Hartree-Fock theory (ref. 3)) and a more recent program of extended shell model calculations by the Glasgow group (ref.4)) giving support to the interpretations quoted above.

- The experimental determination of isobaric multiplets by means of the study of proton rich nuclei. These data are very important in testing the charge independence of nuclear forces. Recently, the measure of the ^{32}Cl (0^+T=2) mass[5] has allowed a determination of the value of the cubic term in the mass formula of the multiplet, d=0.35±1.09 KeV, in complete agreement with the charge symmetry hypothesis[6].

- The decay of very exotic nuclei feeds states of nuclei, also far from stability or in other cases new types of states of more familiar nuclei. This is the case, for instance, of the study being carried out at ISOLDE on the heavy potassium isotopes[7]. They have been produced up to mass 54 (^{54}K has a lifetime of 10±5 ms), and up to mass 52 its decays have permitted by means of beta and neutron spectroscopy, the study of particle-hole states of Ca- isotopes. In the heavier cases even the Ca-ground states were not known, as the ^{52}Ca.

We shall be concerned mainly with the last domain of the study of light nuclei far from stability. The interest of doing a theoretical spectroscopy of these nuclei rests on several facts: a) it can have a fruitful interaction with experiments in its present situation; b) there are complete calculations[8] for the low part of the fp- shell that have fixed a very important part of the interaction to be used;

c) the study of non-natural parity states allows to explore the interaction between orbits belonging to different major shells.

In this paper we shall try to describe theoretically the spectra of heavy potassium isotopes (A=47 to 50) and the corresponding particle-hole spectra of Calcium isotopes, within a shell model approach.

2.- General Comments on the Calculation, Model Space. Interaction.

Before doing a shell model calculation there are several "ab initio" choices, ours were clearly stated in ref. 8). Those relevant to this calculation are: 1) try to keep close to realistic interactions, and 2) make, if needed, only monopole changes.

The model space suitable for the study of the nuclei we are interested in, includes the hole orbits $2s_{1/2}$ and $1d_{3/2}$, and the particle orbits $1f_{7/2}$ $2p_{3/2}$ $1f_{5/2}$ $1p_{1/2}$. Obviously this space is far bigger than what is needed and tractable. We shall describe later the truncations we make. As starting interaction, we take the Lee-Kahama-Scott[9], resulting from a G-matrix calculation using the Lee interaction, The monopole changes of the fp- part of the interaction are known from ref. 8), and shall be taken into account from the beginning. Nothing is known on the behaviour of the (sd)-(fp)- part of the interaction. The experience of p, (sd) and (fp)- complete calculations shows that realistic interactions are excellent except for their monopole behaviour. There are theoretical arguments for that to be so; let us comment that the monopole part of the interaction carries on almost all the binding energy of the nuclei - it, so to speak, its Hartree-Fock part - and, it is well-known that Brueckner calculations up to the order used in the calculation of realistic interactions for finite nuclei, fail to give the right saturation.

The first step of the calculation is then to fix the monopole centroids of the interaction between particles and holes.

3.- Fit of the Monopolar Parameters.

The monopole (isoscalar and isovector) hamiltonian can be written as:

$$H_m = \sum_r \varepsilon_r \hat{n}_r + \sum_r a_{rr} \hat{n}_r^2 + \sum_{r<s} a_{rs} \hat{n}_r \cdot \hat{n}_s + \sum_r b_{rr} \hat{T}_r + \sum_{r<s} b_{rs} \hat{T}_r \cdot \hat{T}_s$$

$$(1)$$

where n and T are operators and a's and b's are the corresponding averages of matrix element (for the definitions and details on the monopole hamiltonian, see ref. 10)). We are interested in using the ^{40}Ca as an inert core; after the particle hole transformation, formula (1) reads:

$$. \; H_{mm} = E(^{40}Ca) + \bar{\varepsilon}_h \, \hat{n}_h + \sum_r \hat{\varepsilon}_r \, \hat{n}_r + a_{hh} \, \hat{n}_h^2 + \sum_r a_{rr} \, \hat{n}_r^2 + \sum_{r<s} a_{rs} \, \hat{n}_r \, \hat{n}_s$$

$$- \sum_r a_{hr} \, \hat{n}_h \hat{n}_r + b_{hh} \, \hat{T}_h^2 + \sum_r b_{rr} \, \hat{T}_r^2 + \sum_{r<s} b_{rs} \, \hat{T}_r \cdot \hat{T}_s \tag{2}$$

if there is just one hole orbit. The $\bar{\varepsilon}$ are the single particle or hole energies referred to the new vacuum, and $\hat{n}_h$, $\hat{T}_h$ are the number and isospin operators for the
holes.

The basis states in our model have the structure

$$|\Psi\rangle = |\, (sd)^{-1} (1f_{7/2})_{T_1}^{n_1} (T_i) \, (2p_{3/2})_{T_2}^{n_2} (T_j) (1f_{5/2})_{T_3}^{n_3} (T_f)\rangle \tag{3}$$

(The $2p_{1/2}$ is omitted for reasons to be explained later).

The monopolar energy of these states is given by $E(\psi) = \langle\psi|H_m|\psi\rangle$ which
is a function of $(n_1 n_2 n_3)$ $(T_1, T_2, T_3, T_i, T_f)$. We do not write the explicit formula
(it is rather lengthy and results in a generalization of the Bansal-French formula
- case of two shells - with a few Racahs added), but we shall use it to fit the
particle hole centroids to the experimental data. The most important parameters are,
a_{sf}, a_{df}, b_{sf}, b_{sp} that we shall determine using data in ^{47}K (whose experimental
spectrum is presented in figure 1). As no calculation of the Calcium ground state
will be made, we need a prescription to locate the particle hole states, in order
to do that we take the Q_β of each particular process and the Coulomb energy difference $(^ACa-^ASc)$, from the experiment. These two data allow the determination of the
excitation energy of a particle hole state in calcium as

$$. \quad \Delta E_{ph} = Q_\beta + m_e - \Delta m_{np} + \Delta E^{coul} - \Delta E \left(K_{gs} - CA_{ph} \right)$$

This introduces incertitudes in the calculation through a) the Q_β error
bars and b) the Coulomb energy extraction, for that quantity we shall take a constant $E^{Coul} = +7.0$ MeV which is the approximate value for A≈44. Calculations of ref.
11) show that for A≈47 this may have decreased to $E^{Coul} = +6.8$ MeV but we shall
keep the above value. In any case modifications in Q_β and E^{Coul} would lead only to
global shifts of the spectra.

We calculate then $\Delta E(gs(K) - ph(Ca))$ and using the monopole formula for
the first p-h states in ^{47}Ca, we find:

$$\tfrac{1}{2} \, b_{sf} = 1.10 \pm 0.15 \qquad \tfrac{1}{2} \, b_{df} = 1.10 \pm 0.15 \qquad (MeV)$$

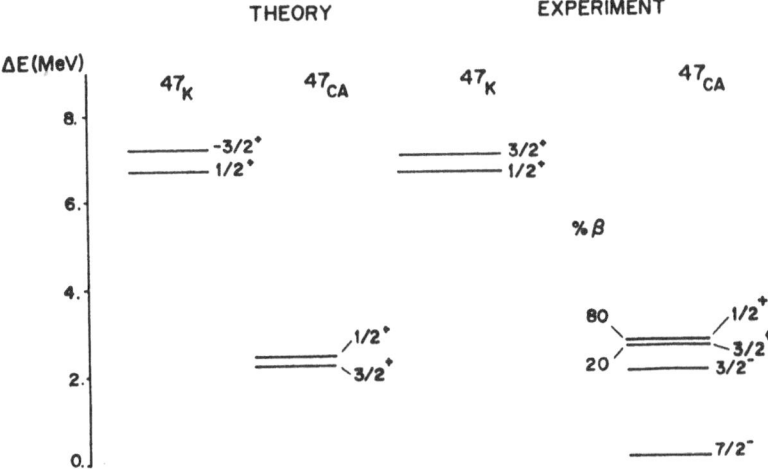

Figure 1.   Experimental and computed levels for A = 47

The variation of the difference $E(1/2^+) - E(3/2^+)$ in the known potassium isotopes (figure 2) allows, in the same scheme, the determination of the difference $a_{sf} - a_{df} = 0.3 \pm 0.1$ MeV. Finally one can estimate $a_{sf} \approx 0.2$ via the binding energies referred to ^{40}Ca. The results of the fit do not deviate drastically from the LKS values for the crucial quantities such as: $(\bar{V}^1_{sf} - \bar{V}^1_{df}) = +0.5$ MeV with LKS, and +0.32 MeV with the fitted values. This quantity governs the structure of the K-isotopes g.s.. The good behaviour of $\Delta \bar{V}^1$ justifies doing a first trial, keeping the a_{sp}, a_{dp}, b_{sp}, b_{dp} values of LKS.

Calculations with the monopolar formula for A=48, 49, 50 locate well the calculated centroids on the regions of β- strength, and allow us to use the interaction as described above, for the shell model calculation.

4.- Shell Model Results.

Once the interaction is fixed, we are faced with the problem of the dimensions of the space. Even with the very high isospins involved, a complete calculation is unfeasible. We have chosen the following truncation:
- the maximum number of holes is one;
- the $(1f_{7/2})$ is allowed to have 8 or 9 particles;
- the $(2p_{1/2})$ orbit is always empty.

The last part of the truncation may seem arbitrary, but up to A=50 the $2p_{1/2}$ orbit excitations will couple weakly with the dominant components, while $1f_{5/2}$ excitations are fundamental. This restriction allows us to test the influence of

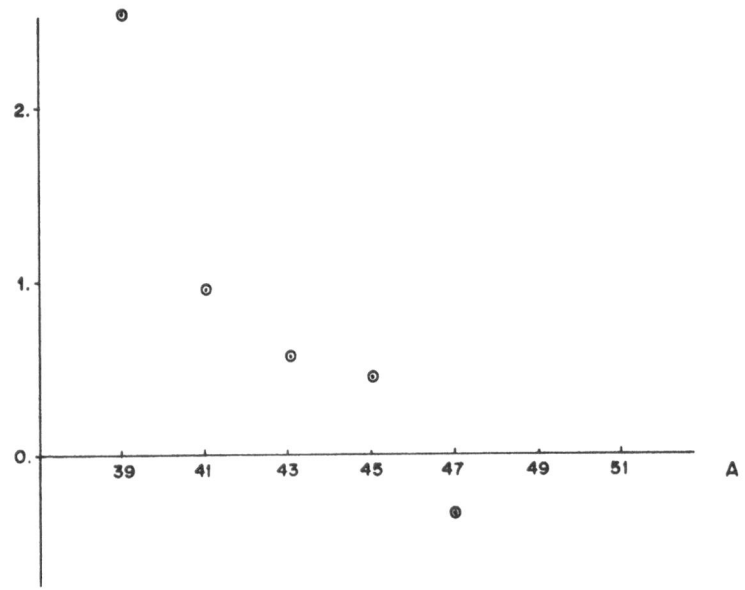

Figure 2. Variation of the difference $E(1/2^+) - E(3/2^+)$ in the known K-isotopes

$(1f_{5/2})^{2,3}$ configurations. Nevertheless, we plan another set of calculations, including the $(2p_{1/2})$ orbit but limiting the number of particles on $(2p_{1/2}\ 1f_{5/2})$ to one on each. Obviously, all that paraphernalia is just due to computer limitations.

Diagonalizing in ^{47}K-^{47}Ca introduces cosmetic modifications in the values of the parameters which take finally the values:

$$a_{sf} = -0.10 \qquad \tfrac{1}{2}\,b_{sf} = 1.05 \qquad a_{df} = -0.40 \qquad \tfrac{1}{2}\,b_{df} = 1.17 \ (MeV)$$

The results of the calculations are presented in figures 3, 4, 5. We have proceeded as follows. In each case a calculation of the potassium is made to determine the spin of the ground state, subsequently only those spins of the Calciums attainable by Gamow-Teller β- decay are computed.

In fig. 1 we give the results for A=47 where the fit has been made.

In fig. 3 the A=48 results are shown, deserving some comments:
- The g.s. of ^{48}K is predicted to be $J^{\pi} = 2^-$.
- The level grouping and sequency is well understood in general. Notice the right placement of the two low lying 3^- states.
- Experimental levels having a black dot are strongly fed by the β- decay. On the

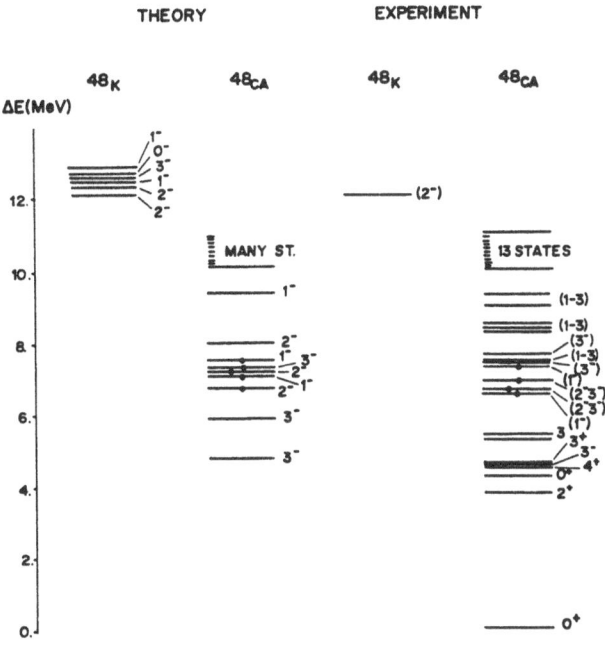

Figure 3. Experimental and computed levels for A = 48

calculated Calcium levels we put a black dot when their structure is the same as that of the ^{A}K g.s.. Although this is just a guide, the correspondence is quite nice.

- The structure of the ^{48}K g.s. is 50% $|d_{3/2}^{-1} f_{7/2}^8 p_{3/2}^1>$ + 50% $|s_{1/2}^{-1} f_{7/2}^8 p_{3/2}^1>$, indicating level crossing.

In fig. 4 we present the comparison between theory and experiment for mass 49. Our results can be summarized as follows:

- We predict a 3/2$^+$ as spin of the g.s. of ^{49}Ca. This is the second crossing of the $2s_{1/2}$ and $1d_{3/2}$ in the potassium chain, the first occurring in mass ∿46, 47.
- An open circle on a calculated level means it cannot be connected with the ^{49}K g.s. by the G.T. operator.
- The ^{49}K g.s. is almost pure $1d_{3/2}^{-1} f_{7/2}^8 p_{3/2}^2$ configuration.
- The global trends of the experimental spectrum are reproduced. Nevertheless, the study of the β- transitions appears to be very important in this case to decide on the quality of the agreement.
- The $(2s_{1/2})$- quasiparticle appears at 1.0 MeV of excitation energy in ^{49}K.

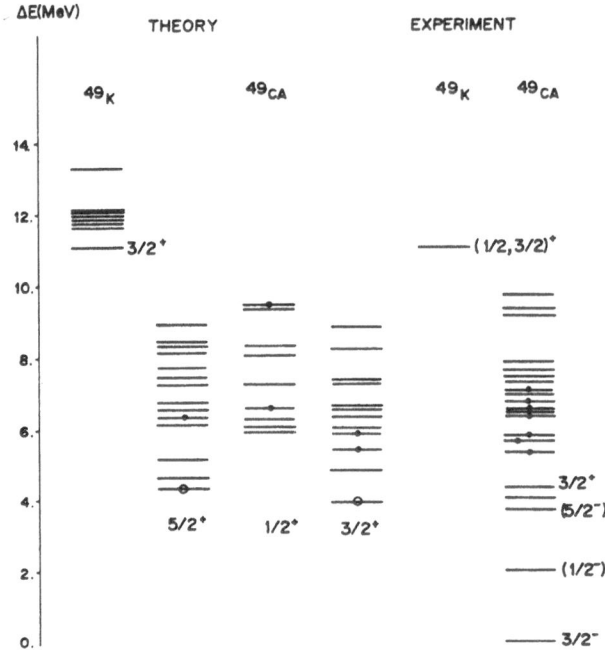

Figure 4. Experimental and computed levels for A = 49

Finally in fig. 5 we present the results for A=50.
- In that case the prediction of 0^- as g.s. of ^{50}K simplifies a lot the interpreta-
tion of the data because only 1^- states of ^{50}Ca can be fed.
- The structure of the 0^- is almost pure $d_{3/2}^{-1}$ $f_{7/2}^8$ $p_{3/2}^3$.
- The calculated spectra reproduce quite well the experimental trends.
- The first calculated 1^- state lies at an excitation energy of 7.0 MeV and should
correspond to the experimental state at 6.51 MeV of excitation energy. The relevance
that this state might be 1^- for astrophysical reasons has been recently argued[12].

In conclusion, we have seen how it is possible to describe the spectroscopy of potassium and calcium (ph) states in the framework of extended shell model.
A first set of results - obtained with a realistic interaction plus monopole corrections - shows a global agreement with the experimental data. Several predictions are
made. New calculations are going on with a double goal: a) to test the truncation
and b) to reach masses 51 and 52.

The authors want to thank the Strasbourg group at ISOLDE for its kind
collaboration and for sending us their results prior to publication.

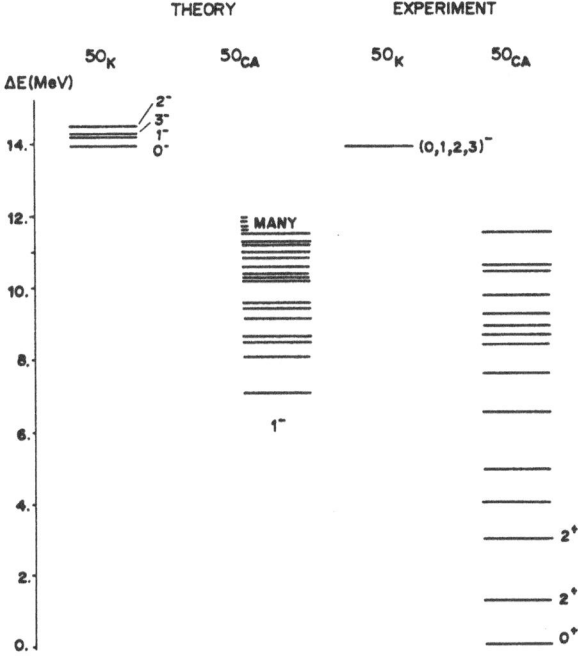

Figure 5.   Experimental and computed levels for   A = 50

References

1) ISOLDE refs. in "4th. International Conference on Nuclei far from Stability",
    Proc. CERN 81-09 (1981).
2) C. Detraz in 1), p. 361;   C. Thibault in 1), p. 47.
3) X. Campi et al., Nucl. Phys. A251 (1975) 193.
4) A. Watt et al., J. Phys. G. 7 (1981) L145.
5) E. Magberg et al., Phys. Rev. Lett. 39 (1977) 792.
6) M.S. Antony et al., Rapport d'Activité CRN 80-01. Strasbourg 1980.
7) L.C. Carraz et al., Phys. Lett. 109B (1982) 419.
8) A. Poves and A. Zuker, Phys. Reports 70 (1981) 235.
9) S. Kahana, H. Lee and K. Scott, Phys. Rev. 180 (1969) 956.
10) E. Pasquini, Ph.D. Thesis, Strasbourg 1976.
11) E. Caurier and A. Poves, Nucl. Phys. in press.
12) W. Hillebrandt et al., ISOLDE proposal sub. to PSCC (1982).

DIRECT PROTON DECAY OF ^{147}Tm

D. Schardt

GSI Darmstadt
6100 Darmstadt, Federal Republic of Germany

Abstract

Using on-line mass separation of evaporation residues from the reaction ^{58}Ni + ^{92}Mo
→ ^{150}Yb*, a proton line of 1055 ± 6 keV energy and 0.56 ± 0.04 s half-life was observed.
The activity was assigned to the direct proton decay of ^{147}Tm. Beta-delayed protons
registered at the same mass position were found to arise from the decays of
^{147}Er($T_{1/2}$ = 2.5 ± 0.2 s) and ^{147m}Dy($T_{1/2}$ = 57 ± 4 s).

1. Introduction

With increasing distance from the line of stability the proton separation energies of
neutron-deficient isotopes decrease steadily due to the Coulomb repulsion. Beyond the
"proton-drip-line" proton decay from the ground-state becomes energetically possible,
in direct analogy to alpha decay. Since proton radioactivity (also named Coulomb-
delayed or self-delayed proton decay) was first proposed by Marsden {1,2} (1914) and
Rutherford {3} (1919) many efforts have been made to detect this decay mode (for a
historical review see {4}). Beside the production of nuclei beyond the proton-drip-
line, the main experimental problem is the small "observation-window" {5}, defined by
the lowest detectable decay branch on one side and by the shortest detectable half-
life on the other side. Because of the very strong dependance of the decay width on
the energy and angular momentum carried away by the emitted proton, this window is
much narrower than for alpha decay. Most favorable experimental conditions are offered by heavy-ion-based on-line systems, making use of the high yields for extremely
neutron-deficient isotopes obtained from heavy-ion reactions. The first successful
experiment, leading to the discovery of proton-radioactivity of ^{151}Lu (see Fig. 1),
was recently reported by Hofmann and co-workers {4}. The 1231 keV proton activity
was produced in ^{58}Ni +^{96}Ru reactions and investigated at the velocity filter SHIP.

Here we want to present new results on a second case of proton radioactivity, which
was studied at the GSI on-line mass separator. In an earlier experiment {6}, a 1055
keV proton line has been observed in ^{58}Ni +^{92}Mo reactions and preliminary assigned
to the direct proton decay of ^{147}Tm. This interpretation was confirmed by a new experiment, including a negative result from a positron-proton coincidence measurement
and a cross bombardment.

In parallel to our studies of direct proton decay, we investigated beta-delayed pro-

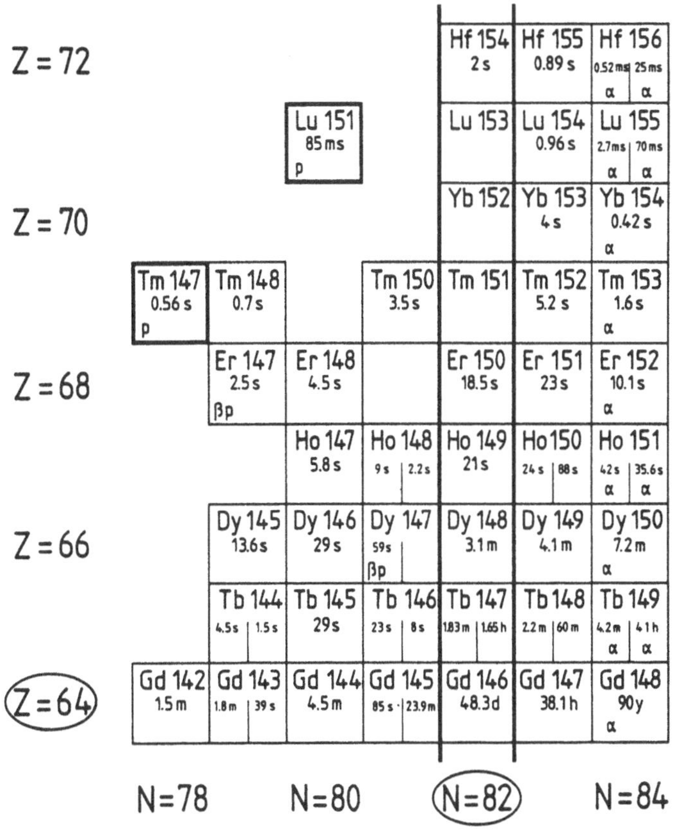

Fig. 1   Part of the chart of nuclides showing neutron-deficient nuclei
in the rare-earth region around N = 82. The proton emitters
^{151}Lu and ^{147}Tm, which were recently discovered at GSI Darmstadt,
are marked by heavily outlined boxes. Both proton emitters were
produced as p2n evaporation residues from reactions of ^{58}Ni ions
from the UNILAC with enriched targets of ^{96}Ru and ^{92}Mo, respectively.

ton emission of A = 147 precursors. As reported in our earlier work, the measured
proton energy distribution shows a remarkable peak structure. New spectroscopic information obtained from proton-gamma coincidence and half-life measurements revealed,
that the peak structure arises from the decay of ^{147m}Dy, while contrastingly the beta-delayed proton spectrum of ^{147}Er is structureless. In the first case, the protons
are emitted from highly excited states in ^{147}Tb to final states in ^{146}Gd. In view of
the growing evidence for a doubly closed shell structure of ^{146}Gd, the observed proton decays are of particular interest.

2. Experimental Techniques

The experiments were performed at the GSI on-line mass separator {7} connected to the heavy-ion accelerator UNILAC. Very neutron-deficient rare-earth isotopes were produced as evaporation residues from reactions of 5.0 MeV/u ^{58}Ni ions with targets of ^{92}Mo (2.9 mg/cm^2). The recoiling products were stopped in a tantalum catcher inside a thermal ion source. This is a 0.3 cm^3 tungsten cavity heated from the bottom by electron bombardment. A 1.9 mg/cm^2 thick tungsten foil served as ion source entrance window. The targets were protected by a 0.7 mg/cm^2 tantalum and three 20 µg/cm^2 carbon heat shields. A detailed description of the target-ion-source system is given in {8}. Overall efficiencies in the percent range were achieved for ^{153}Er, ^{153}Tm, ^{154}Yb and ^{157}Lu.

The re-ionized reaction products were mass separated and investigated by decay spectroscopy. Various detection systems were placed into three separate beam lines: for "in-beam" measurements the mass separated ion beams were implanted into carbon foils of 12 µg/cm^2 thickness behind which a telescope of surface barrier detectors was placed (25.3 µm thick ∆E-detector, 700 µm thick E-detector). Such a telescope detector allows selective detection of low-energy protons in the presence of intense beta radiation. In the ∆E-detector protons up to 1.3 MeV, in the E-detector up to 10 MeV are completely stopped if entering perpendicular to the detector surface. For positron-proton coincidence measurements a 2 inch diameter, thin plastic scintillator was placed behind the telescope. A cooled single detector of 500 µm thickness was used for high-resolution proton energy measurements.

At the central mass position a fast tape transport system was used, in order to reduce background radiation from longer-lived isotopes and to measure half-lives. The activity was collected on aluminized mylar tape and periodically moved into a counting position. This was equipped with a Ge(Li) and an intrinsic Germanium detector in a 180° geometry. For studying proton-gamma coincidences, a silicon detector telescope was placed in between the tape and the Ge(Li) detector.

Linear signals from ∆E and E-detectors were stored event-by-event on magnetic tape, using the GSI data acquisition system GOLDA2 {9}. The off-line analysis was performed by using the program system SATAN {9}.

3. Results

3.1. Direct proton decay of ^{147}Tm

The most probable candidates for direct proton decay, which may be reached by the reaction ^{58}Ni +^{92}Mo, are very light isotopes of the odd-Z elements thulium and holmium. The partial alpha decay half-lives for these isotopes are expected to be much longer than those for beta decay, which are predicted to be of the order 1 s {10}. The compound nucleus ^{150}Yb has three neutrons less than the lightest known ytterbium isotope. Although such neutron-deficient fusion product are predominantly deexcited by the emission of protons and alpha particles, one could hope to reach very light thulium isotopes beyond the proton-drip-line.

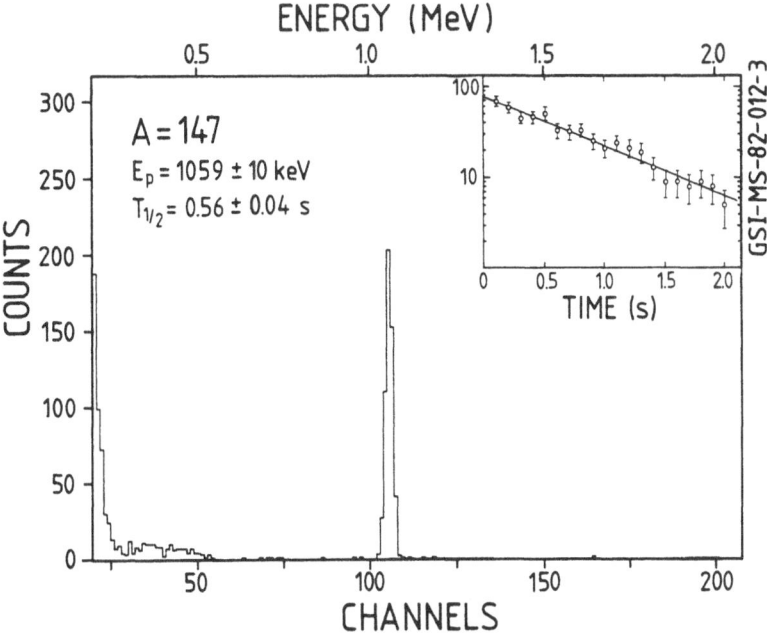

Fig. 2   Energy spectrum obtained at mass 147 from the ΔE-detector in anti-coincidence with the E-detector of one of the silicon telescopes. The activity was collected on aluminized mylar tape and periodically moved into a counting position with a cycle time of 2 s. Each event was tagged by the time elapsed since end of the last transport, using a time-to-digital converter. The resulting decay curve for the proton line is shown as an inset.

Fig. 2 shows the energy spectrum recorded at mass number 147 from the ΔE-detector of one of the telescopes. A single proton line was observed at an energy of 1059 ± 10 keV (width 30 keV FWHM). The half-life was remeasured with higher precision to be 0.56 ± 0.04 s. In a separate experiment at the velocity filter SHIP {4}, the low-energy proton line in the reaction ^{58}Ni + ^{92}Mo was detected independently. The energy of 1055 ± 6 keV from this measurement agrees very well with the value given above and is preferred in view of its higher accuracy. At optimal ion source conditions the counting rate was 7 protons/min. normalized to a ^{58}Ni beam intensity of 10 particle-nA, which is an imporvement of a factor of 5 compared to our earlier experiment {6}. The proton character of the new activity was confirmed by an energy loss measurement (Fig. 2), using a 280 μg/cm^2 carbon foil. The proton line was not observed, when the ^{92}Mo target was replaced by a ^{93}Nb target.

In order to prove that the protons are emitted directly and not beta-delayed, a positron-proton coincidence measurement was carried out. The result is shown in Fig. 3. The distribution between channels 50 and 150 arises from the energy loss of beta-de layed protons of higher energy (see section 3.2). They are emitted quasiprompt follo-wing a preceeding beta decay and provide a good test of the coincidence set-up. Al-though the coincidence efficiency was rather small, it can be clearly seen that the 1055 keV protons are not beta-coincident. From their intensity in the single spectrum one would expect 22 coincident events if the proton emission would follow beta decay, whereas only 1 (random) event was observed in the coincident spectrum.

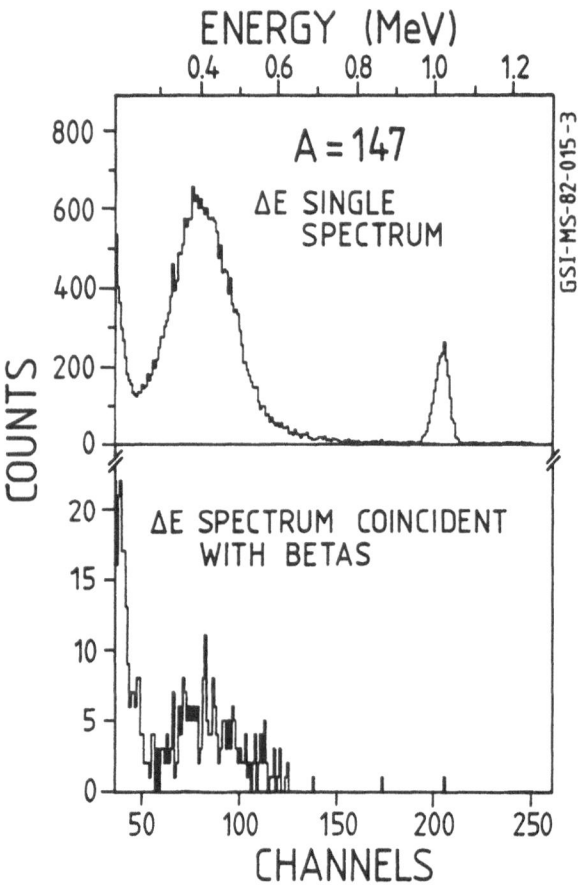

Fig. 3 Single and beta-coincident proton spectrum taken with a 22 μm thick silicon detector at mass 147. The lower spectrum was obtained by operating the silicon detector in coincidence with a thin plastic scintillator. Beta-delayed protons, characterized by the energy loss distribution between channels 50 to 150, are recorded in coinciden-ce, whereas the 1055 keV proton line is not beta-coincident (see al so section 3.1).

In the following, the assignment of the proton line to ^{147}Tm and a comparison of the measured half-life with penetrability calculations will be discussed.

Since the mass number was unambiguously determined by magnetic separation to be 147, the proton emitter has to be one of the very neutron-deficient reaction products within this isobaric chain. Those produced directly by evaporation of three particles are ^{147}Yb, ^{147}Tm, ^{147}Er and ^{147}Ho with estimated ratios of production rates of 1: 7×10^2 : 7×10^4 : 5×10^5 (calculated with the evaporation code HIVAP {11}). With a small calculated cross section of 0.2 μb for the 3n-channel, ^{147}Yb is not likely to be the origin of this proton line. The element assignment can be further based on the systematics of known proton separation energies S_p and extrapolation by means of current mass formulae {12,13}. For the comparison shown in Fig. 4 we have chosen three mass formulae with very different approaches. The predictions of Liran-Zeldes and of mass formulae and based on the Garvey-Kelson relations (not shown in Fig. 4) generally agree fairly well with the experimental values. Larger discrepancies occur between experimental and theoretical values when applying the mass formula of Myers.

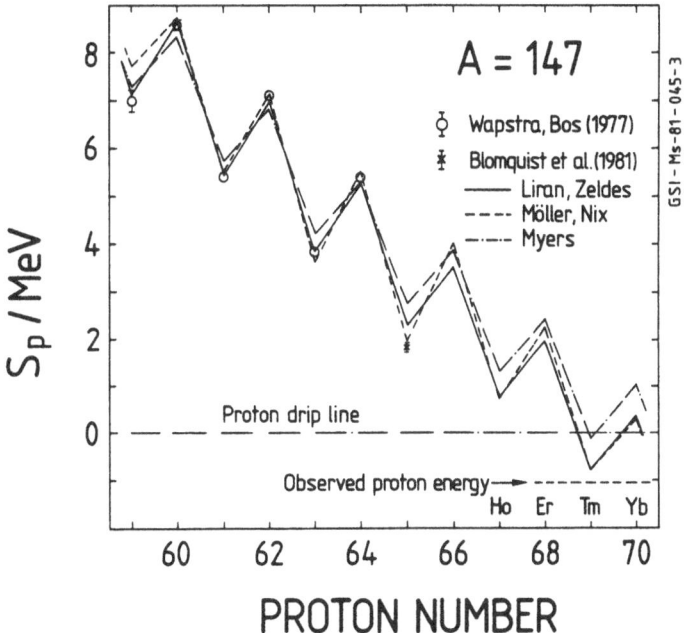

Fig. 4   Proton separation energies for A = 147 nuclei with predictions from different mass formulae {12,13}. The experimental values are taken from the 1977 mass evaluation of Wapstra et al {14} and from a shell model analysis of high-spin states of nuclei above ^{146}Gd {15}.

In particular for odd-Z, even-N isotopes the predicted S_p values are too high. As was discussed by Hofmann et al {16}, Myers' droplet formula apparently underestimates both the proton-proton and the proton-neutron pairing energy. According to the predicted separation energies shown in Fig. 4 and those of other current mass formulae {12}, ^{147}Yb, ^{147}Er, ^{147}Ho and its beta decay daughters are proton bound. The S_p values taken from Liran-Zeldes are 0.40, 2.0 and 0.80 MeV, respectively. The odd-Z ^{147}Tm nucleus, however, is expected to be unstable against proton emission from the ground-state. The predicted S_p values vary between -0.12 MeV (Myers) and -0.98 MeV (Jaenecke-Eynon). Liran-Zeldes and Möller-Nix predict about -0.80 MeV.

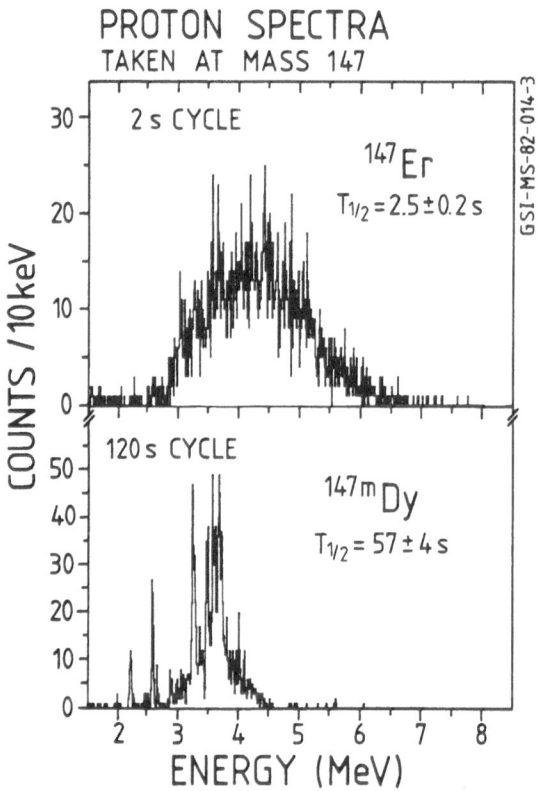

PROTON SPECTRA
TAKEN AT MASS 147

2 s CYCLE

^{147}Er

$T_{1/2} = 2.5 \pm 0.2$ s

120 s CYCLE

^{147m}Dy

$T_{1/2} = 57 \pm 4$ s

COUNTS / 10 keV

ENERGY (MeV)

GSI-MS-82-014-3

Fig. 5 Energy spectra of beta-delayed protons recorded at mass 147 with a silicon detector telescope, which was placed in the counting position of a tape transport system. The spectra were generated by summing coincident signals of the ΔE- and E-detectors.

Similar trends as seen for A = 147 isobars (Fig. 5) are obtained by comparing S_p values of thulium isotopes with mass formula predictions. The lightest thulium isotope with known S_p value (2.17 + 0.08 MeV {14}) is ^{156}Tm. The value predicted by Myers (2.55 MeV) is about 400 keV higher, those taken from Comay-Kelson (2.16 MeV), Jaenecke-Eynon (2.00 MeV), Liran-Zeldes (2.04 MeV) and Möller-Nix (1.98 MeV) are closer to the experimental value. Guided by the mass formulae of Liran-Zeldes, we can now extrapolate the S_p values of both the A=147 nuclei and the thulium isotopes to ^{147}Tm and then obtain a proton decay energy between 0.5 and 1.1 MeV. We therefore conclude that out of the ensemble ^{147}Yb, ^{147}Tm, ^{147}Er, ^{147}Ho the nucleus ^{147}Tm is the only candidate for ground-state proton decay and that the observed 1055 keV proton activi_ty can be well explained as the direct proton decay of its ground-state or a low-lying isomeric state.

We have estimated the partial proton half-life for ^{147}Tm by calculating the decay constant $\lambda_p = T_{j1} \times f$ as product of the transmission coefficient T_{j1} for a proton of to_tal spin j and orbital angular momentum l and the "frequency factor" $f = v_F/2R = 6.8 \times 10^{21}$, where v_F is the Fermi velocity and R is the nuclear radius. The transmission coeffi-cient was calculated with a WKB method using the real part of the optical model po-tential given by Becchetti and Greenless {17}. The resulting half-lives for $s_{1/2}$ $d_{3/2}$ and $h_{11/2}$ single particle proton states, which are expected to be very close to the Fermi surface, are 65 μs, 0.6 ms and 2.0 s, respectively. A direct comparison with the experimentally determined half-life of 0.56 s is not possible, since the proton branching ratio is unknown. However, it can be seen that a partial proton half-life $T_{1/2} > 0.56$ s requires a hindrance factor corresponding to at least 4-5 units of ℏ ca-rried away by the emitted proton. One can speculate that, similar as in the case of ^{151}Lu {4}, this would be in accordance with a transition between an $11/2^-$ state in ^{147}Tm and the 0^+ ground-state of ^{146}Er.

3.2. Beta-delayed proton decay of ^{147}Er and ^{147m}Dy

A complex spectrum of beta-delayed protons, extending from 2 to 8 MeV, was measured simultaneously with the ^{147}Tm proton line at mass number 147. It is characterized by a remarkable peak structure in the lower energy part up to about 4 MeV. Starting from predicted ($Q_{EC} - S_p$) values and calculated production rates, we considered ^{147}Er, ^{147}Ho and ^{147}Dy as probable precursors. As in our first experiment {6} more detailed spec-troscopic information was not available, a definite assignment was not posible at that time. For further investigations the tape station described in section 2 was used, allowing half-life and coincidence measurements. It was operated with collection-coun_ting periods of 2, 12 and 120 s. The earlier measured proton spectrum shown in {6} turned out to be a superposition of essentially two components (Fig. 5). At a cycle time of 2 s a structureless, bell shaped spectrum typical for medium and heavy mass precursors was observed. Based on the measured coincidence relations (Fig. 6) with the known {18} $2^+ \rightarrow 0^+$, $(4^+) \rightarrow 2^+$ transitions in ^{146}Dy, the 2.5 s proton activity is assigned to the new precursor ^{147}Er. The endpoint energy of about 8 MeV agrees fair-ly well with the value of 8.6 MeV taken from the Liran-Zeldes mass table. At the lon_gest cycle time of 120 s a spectrum with a distinct peak structure, extending up to about 4.5 MeV was observed. From the measured half-life of 57 ± 4 s we assign it to the known {19} 59-s ^{147m}Dy as precursor. The endpoint energy of 4.6 MeV predicted by Liran-Zeldes is in good agreement with the measured one. In the beta-delayed proton

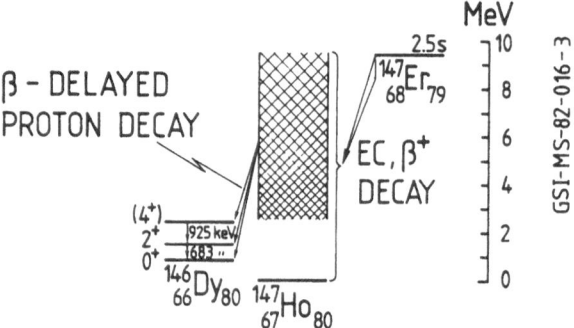

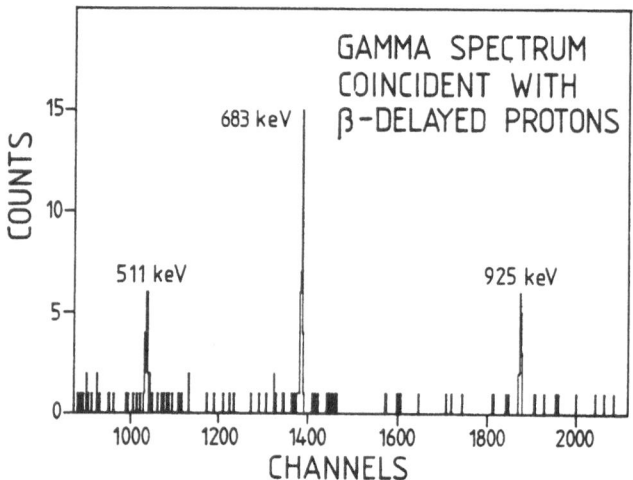

Fig. 6   Identification of beta-delayed protons of the precursor ^{147}Er by
measured proton-gamma coincidence relations. The gamma transitions
in ^{146}Dy are known from the work of Nolte et al {18}

decay of ^{147m}Dy the protons are emitted from excited states in ^{147}Tb. The non-observa-
tion of proton-gamma coincidences in this decay suggests that only the ground-state
of the final nucleus ^{146}Gd is fed by the proton decays. The proton decaying levels
are lying at about 4.3 - 6.8 MeV excitation energy, assuming a proton separation energy of 2.3 MeV (Liran-Zeldes) for ^{147}Tb.

Both assignments were confirmed by a cross bombardment, using a target of ^{93}Nb instead of ^{92}Mo. The spectrum of ^{147m}Dy was measured with about the same intensity as be-

fore, now being produced through a 3pn-reaction. The ^{147}Er spectrum disappeared, because of the much smaller production through the p3n-channel compared to the 2pn-channel in the ^{92}Mo reaction.

4. Outlook

The identification of the proton emitters ^{151}Lu and ^{147}Tm has shown that direct proton decay of extremely neutron-deficient isotopes beyond the proton-drip-line is indeed observable under favourable experimental conditions. The measured decay energies may serve as first fixpoints, allowing more reliable predictions of the properties of other candidates in this region. A search for proton radioactivity among promethium, europium, terbium and holmiun isotopes seems to be promising, whereas for higher-Z elements the competition from alpha decay poses a severe problem. For a further understanding of the observed proton decays, more detailed spectroscopic information is required. In case of ^{147}Tm, a determination of the partial proton decay width is important, because the measured total half-life is of the order of the predicted beta half-life. In view of the improved source strength of about 1/s such a measurement seems to be feasible. Since the spectroscopy of neighbouring nuclei is still completely unknown, it will be difficult, however, to obtain more experimental information about the nature of the proton decaying states.

References

1. Marsden, E.: Phil. Mag. 27,824 (1914)

2. Marsden, E.: Phil. Mag. 30,240 (1915)

3. Rutherford, E.: Phil. Mag. 37,537 (1919)

4. Hofmann, S., Reisdorf, W., Münzenberg, G., Heßberger, F.P., Schneider, J.R.H., Armbruster, P.: Z. Physik A 305,111 (1982)

5. Bogdanov, D.D., Bochin, V.P., Karnaukhov, V.A., Petrov, L.A.: Soviet J. Nucl. Phys. 16,491 (1973)

6. Klepper, O., Batsch, T., Hofmann, S., Kirchner, R., Kurcewicz, W., Reisdorf, W., Roeckl, E., Schardt, D., Nyman, G.: Z. Physik A 305,125 (1982)

7. Bruske, C., Burkard, K.H., Hüller, W., Kirchner, R., Klepper, O., Roeckl, E.,: Nucl. Instr. Meth. 186,61 (1981)

8. Kirchner, R., Burkard, K.H., Hüller, W., Klepper, O.: Nucl. Instr. Meth. 186, 295 (1981)

9. Chestnut, R.P., Grasmück, B., Hadsell, R.W., Lebershausen, W., Lowsky, J., Plappert, W., Richter, M., Rother, H.P., Siart, O., Steitz, M., Winkelmann, K.: EDAS, (GOLDA + SATAN) User's Guide, GSI Darmstadt, (1981)

10. Takahashi, K., Yamada, M. Kondoh, T.: Atomic Data and Nuclear Data Tables 12, 102 (1973)

11. Reisdorf, W.: Z. Phys. A 300,227 (1981) and to be published

12. Maripuu, S. (ed.): 1975 Mass Predictions, Atomic Data and Nuclear Data Tables 17, Nos. 5-6,476 (1976)

13. Möller, P., Nix, J.R.: Los Alamos Preprint LA-UR-80-1996 (1980)

14. Wapstra, A.H., Bos, K.: Atomic Data and Nuclear Data Tables 19, No. 3 (1977)

15. Blomquist, J., Kleinheinz, P., Broda, R., Daly, P.J.: Proc. Int. Conf. on
     Nuclei Far From Stability, Helsingor, CERN 81-09,545 (1981)

16. Hofmann, S., Münzenberg, G., Faust, W., Heßberger, F.P., Reisdorf, W.,
     Schneider, J.R.H., Ambruster, P., Güttner, K., Thuma, B.: Proc. Int. Conf. on
     Nuclei Far From Stability, Helsingor, CERN 81-09 (1981)

17. Becchetti, Jr., F.D., Greenless, G.W.: Phys. Rev. 182,1190 (1969)

18. Nolte, E.,Korschinek, G., Hick, H., Colombo, G., Komninos, P., Gui, S.Z.,
     Schollmeier, W., Kubik, P., Geier, R., Heim, U., Morinaga, H.: Proc. Int. Conf.
     on Nuclei Far From Stability, Helsingor, CERN 81-09,253 (1981)

19. Toth, K.S., Rainis, A.E., Bingham, C.R., Newman, E., Carter, H.K., Schmidt-Ott,
     W.D.: Phys. Lett. 56 B,29 (1975).

DESCRIPTION OF HIGH SPIN STATES[*]

Amand Faessler

Universität Tübingen

Institut für Theoretische Physik

D-7400 Tübingen, W.-Germany

Abstract:

In this talk mainly three points will be discussed: (i) The nature of the second backbending (bb) in the rare earth nuclei will be theoretically investigated. It will be confirmed that 2. bb is due to the alignment of a $h_{11/2}$-proton pair. The oscillating behaviour of the 2. bb can explain the variation of the second anomaly in the N=90 isotones. Strong variations of the 2. bb with the neutron number can be explained by the change of the deformation β. (ii) The smooth increase of the moment of inertia in the actinide nuclei is explained mainly by the alignment of a $i_{13/2}$ proton pair. This is confirmed by measurements of the g-factors and the blocking effect in odd proton nuclei where the odd proton is in a $i_{13/2}$ level. In ^{248}Cm, where one has a smaller deformation, one finds a roughly equal role for $i_{13/2}$ and $j_{15/2}$ neutrons for the increase of the moment of inertia. (iii) In this third part we study high spin spectra in the even mass Hg-isotopes by coupling two quasi-particles to an interacting boson core. This calculation can reproduce the anomalous behaviour of the yrast states around angular momentum 8, 10 and 12. The agreement for the yrast state is similar in an earlier calculation where two quasi-particles have been coupled to an asymmetric rotor. The g-factors of these states measured in Bonn and the reduced quadrupole transition probabilities are also nicely reproduced in both models.

[*]Lectures given at the La Rábida Summer School on Heavy Ion Collisions in La Rábida, Spain from June 7-19, 1982.

1. Introduction

In these lectures on the description of high spin states I shall concentrate on three topics:

(i) First we shall discuss the nature of the second backbending. This second anomaly[1] of the moment of inertia has been predicted by Faessler and Ploszajczak[1] and found by the Berkeley Group[2]. The nature of this anomaly has been explained in detail by Faessler and Ploszajczak[3] as an alignment of a $h_{11/2}$ proton pair. Since the protons are responsible for this second anomaly we expect that the strength of the second backbending is varying with the proton number. This is in analogy to the oscillation of the strength of the backbending due to the $i_{13/2}$ neutrons as a function of the neutron number in the rare earth nuclei[4]. By this oscillation we are able to explain the variation of the 2. bb in the N=90 iso-tones ^{156}Dy, ^{158}Er and ^{160}Yb. But on the other side one finds also that the se-cond anomaly is varying very strongly with the neutron number. ^{156}Er shows 2. bb, ^{158}Er shows upbending and ^{160}Er shows no second anomaly. According to the ex-planation as an alignment of a $h_{11/2}$ proton pair the strength of the 2. bb should not depend on the neutron number. But we shall show that the neutron number is affecting the deformation β of the nucleus and by changing the deformation one can cross different $h_{11/2}$ proton levels in the same way as one crosses by in-creasing the proton Fermi surface those levels. This yields again the same oscil-lations and can explain the variation of the second anomaly in the Er isotopes.

(ii) In the third chapter we will discuss the increase of the moment of inertia in the actinide region. We shall show that this increase is mainly due to the alignment of a $i_{13/2}$ proton pair. This explanation is supported by measurements of the g-factors at high spin states. Häusser et al.[5] found an increase at high spin states over the value $g_R=Z/A$. We shall see that one gets also support from the spectra in odd proton actinide nuclei where the odd nucleon is blocking the $i_{13/2}$ proton level. In such bands one does not find an excessive increase of the moment of inertia over the variable moment of inertia model. On the other side we shall see that in ^{248}Cm both proton $i_{13/2}$ and neutron $j_{15/2}$ pairs con-tribute by their alignment to the increase of the moment of inertia.

(iii) In chapter 4 we shall extend the increasing boson approximation[6] to couple two quasi-particles which can then also align their angular momenta to such an IBA core. With this model we will then study the spectra, transition probabili-ties and g-factors in the even (and also odd) mass mercury isotopes. These even mass isotopes show along the yrast line a strange anomaly: The 8^+, 10^+ and 12^+ yrast levels are almost degenerate to a triplett. In earlier work we have already shown that those states can be explained by coupling either $i_{13/2}$ neutrons or $h_{11/2}$ protons to a triaxial rotor[7,8,9].

Before we discuss the 2. bb I would like to remember a few facts which we need from the first backbending which is due to the alignment of a $i_{13/2}$ neutron pair. The best way of bringing a large amount of angular momentum into the nucleus is still a heavy ion fusion reaction with the subsequent evaporation of a few neutrons. After this reaction one measures the deexcitation of the fast rotating nucleus by γ-rays. From the energies of the electric quadrupole transition γ-rays one can re-construct the ground state band. In the rare earth nuclei one finds normally a spec-trum close to a rigid rotor $E_I=(\hbar^2/2\theta_J)J(J+1)$. The deviations from a pure rotational spectrum are described by a moment of inertia θ_J which depends on the total angular momentum J. This varying moment of inertia θ_J is determined by the energy of neigh-bouring angular momenta J and J-2. One often plots this moment of inertia as a func-tion of the classical rotational frequency of the nucleus which is determined in lowest order from the same energy differences $\hbar\omega=(E_J-E_{J-2})/2$. Figure 1 shows the energy as a function of the square of the angular momentum J(J-1) and the backben-ding plot of the same two nuclei ^{158}Er and ^{174}Hf where twice the moment of inertia is plotted against the square of the rotational frequency. Figure 2 shows the ori-gin of the anomaly of the moment of inertia as seen for example in ^{158}Er in the se-cond half of Figure 1. It is explained by the intersection of two bands[10]. The se-cond half of Figure 2 shows that the intersection of two rotational bands with different moments of inertia can reproduce the backbending anomaly seen in the plot of the moment of inertia against the rotational frequencies squared.

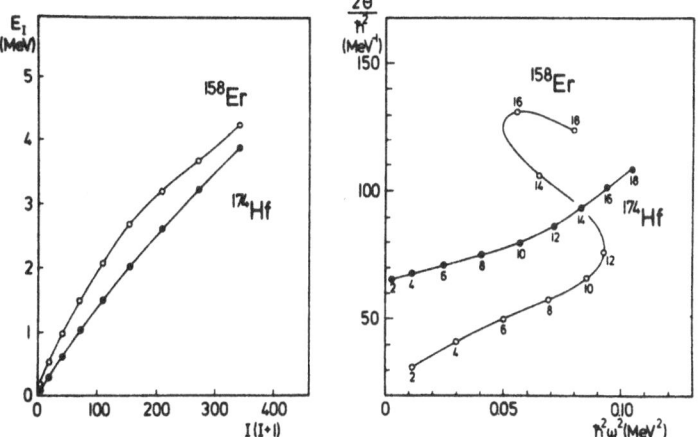

Fig. 1:
On the left hand side the plot of excitation energy vs I(I+1) for the gsb in ^{158}Er
and ^{174}Hf, and on the right hand side the plot of twice the moment of inertia against
the square of the rotational frequency of the nucleus. The moment of inertia is defined by the excitation energy differences $(2\theta/\hbar^2) = (4I-2)/E_I - E_{I-2}$. The rotational
frequency is given by half the transition energy.

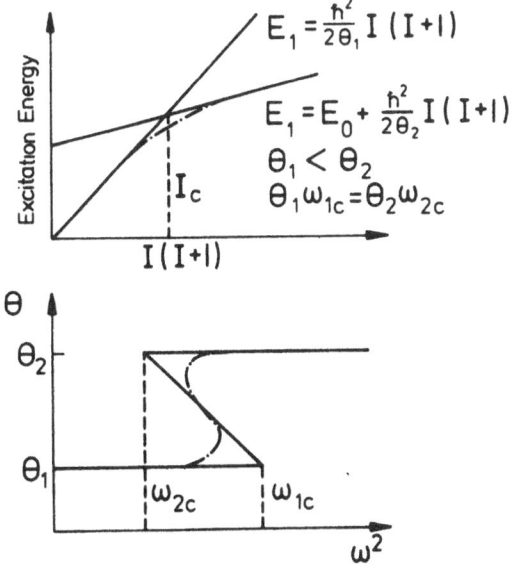

$$E_1 = \frac{\hbar^2}{2\theta_1} I (I+1)$$

$$E_1 = E_0 + \frac{\hbar^2}{2\theta_2} I (I+1)$$

$$\theta_1 < \theta_2$$

$$\theta_1 \omega_{1c} = \theta_2 \omega_{2c}$$

Fig. 2:
The upper part shows the intersection of two bands in the diagram
excitation energy vs, the square
of the angular momentum I(I+1).
The dashed-dotted line indicates
the yrast energy if one assumes an
interaction between the two crossing bands. The lower part gives
the backbending plot for two intersecting bands without any interaction (solid line) and with
interaction (dashed-dotted line).

From Figure 2 one sees that if we have no interaction between the two bands we get an extremely strong backbending which is described in the backbending plot by a mirrored Z. If we have a small interaction V between the two bands we still get a strong backbending. But if one increases the interaction $|V|$ the backbending disappears and one finds only an upbending. So we should keep in mind the fact that a weak interaction $|V|$ between the two bands means strong backbending while a strong interaction $|V|$ indicates no backbending.

The intersection of two bands with different moments of inertia Θ is naturally not yet an explanation of the nature of backbending. We have to understand the configuration of the upper band and why it has such a large moment of inertia. Stephens and Simon explained the nature of the upper band by an alignment of a $i_{13/2}$ neutron two quasi-particle state along the total angular momentum. This alignment is due to the Coriolis force.

$$V_{Coriolis}(i) = -\frac{\hbar^2}{\Theta} j(i) \cdot I$$

$$<i_{13/2} \; \Omega | j_x | i_{13/2} \; \Omega \pm 1> = [j(j+1) - \Omega(\Omega \pm 1)]^{1/2}$$

(1)

One sees from eq. (1) that the Coriolis force acts strongest on the largest single-particle angular momentum j. Near the Fermi surface this is the $i_{13/2}$ neutron. One sees also that at the beginning of the shell where the projection of the single-particle angular momentum Ω is 1/2 one finds a much stronger Coriolis force than at the end of the shell where $\Omega = j$. Since the neutron Fermi surface is at the beginning of the rare earth nuclei at the $i_{13/2}$ Nilsson level with $\Omega = \frac{1}{2}$ and at the end near the level with $\Omega = 1\frac{3}{2}$ we expect at the beginning of the rare earth nuclei a strong and at the end no backbending at all. This can be read of eq. (1) since the Coriolis force disappears for $\Omega = 1\frac{3}{2}$. Indeed one finds more backbending nuclei in the beginning of the rare earth region than at the end. But on the other side exist very strong backbenders in the W and Os region. Inspection of the data[12] shows that the strength of backbending for example represented by the absolute value of the interaction between the two intersecting bands is oscillating[4]. This is qualitatively shown in Figure 4.

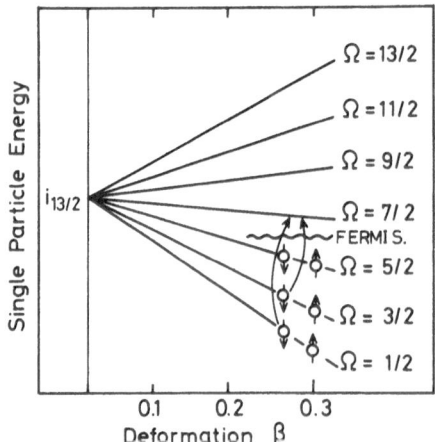

Fig. 3.

Fig. 3:
Sketch of the single-particle energies of the $i_{13/2}$ neutron shell as a function of the deformation β. The Coriolis force tries to align a neutron pair along the total angular momentum perpendicular to the symmetry axis. If the Fermi surface is roughly in the middle of the $i_{13/2}$-shell the aligned states are two holes mainly concentrated in the $\Omega = \frac{1}{2}$ and $\Omega = \frac{3}{2}$ levels.

Hartree-Fock-Bogoliubov calculations of the backbending anomaly were able to reproduce qualitatively correct this oscillations of the backbending strength for the
different rare earth nuclei[13,14]. Experimentalists interpreting their data have for
example assumed that the backbending found in the W, Os isotopes is due to alignment
of a $h_{9/2}$ proton pair[15]. We found that opposite to this explanation the backbending
anomaly even in the Os isotopes must be explained by the alignment of an $i_{13/2}$ neutron pair[14].

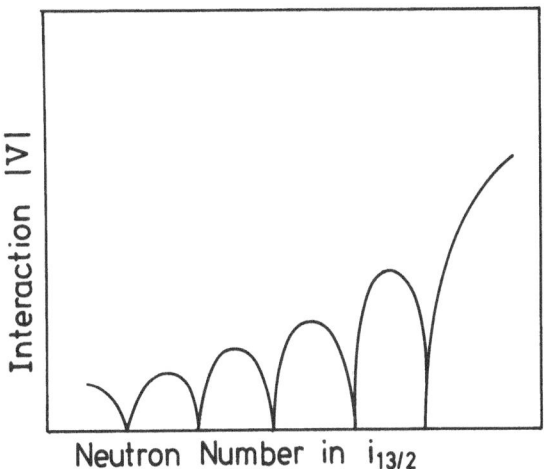

Fig. 4:
Absolute value of the interaction strength $|V|$ between the two intersecting bands as
a function of the number of neutrons in the $i_{13/2}$ neutron shell for the rare earth
nuclei. Zero interaction V means extremely strong backbending while a large inter-
action $|V|$ indicates only upbending.

The physical nature of these strange oscillations of the backbending strength has
been explained by Grümmer, Schmid and Faessler[16]. The backbending oscillations as
a function of the neutron number because the nucleus saves energy. The Coriolis
force is only able to scatter during the alignment neutrons into higher $i_{13/2}$
levels. But between the Fermi surface and the first empty $i_{13/2}$ Nilsson level are
many free levels of the core. It would cost less energy to scatter two nucleons
which leave behind the two aligned hole states shown in Figure 3 into these core
levels. The Coriolis force is not able to do this. But apart of the Coriolis force
we have also a pairing force which is able to scatter pairs of particles coupled
to angular momentum 0 from the $i_{13/2}$ valence shell into the core. Since the core
nucleons are all paired to angular momentum 0 and since the $i_{13/2}$ neutron shell is
a positive parity intruder state into an area of only negative parity levels the
pairing force is the only way for bringing pairs of nucleons from this valence
shell into the core. Since it is cheaper in energy to scatter the two particles of
which the holes get aligned into the core the upper aligned band will have roughly
two neutrons less in the $i_{13/2}$ valence shell. If one considers separately the lower
band with N particles in the valence shell and the upper band with N-2 particles in
the valence shell, the expression for the Coriolis force (1) tells us that an $i_{13/2}$
neutron pair will align already at lower angular momentum in the upper band compared to the lower one. Since the interaction between the two bands is due to the

pairing force which can only scatter pairs of particles between states which have the same number of depaired nucleons, the interaction between the two bands will be zero in the angular momentum region where the upper band has already an aligned two quasi-particle states and the lower band is still fully paired. If one goes through the rare earth nuclei by increasing the Fermi surface for the neutrons one finds such a zero between N=2 and 4, 4 and 6, 6 and 8, 8 and 10, 10 and 12 in the $i_{13/2}$ neutron valence shell. Thus one expects 5 $(j-\frac{3}{2})$ zeros in the interaction between the ground and the upper band for the $i_{13/2}$ shell. This has been found experimentally[17] and theoretically[4,14,16].

Let me summarize: We believe today that backbending is due to the intersection of two rotational bands where the upper band is due to the alignment of two $i_{13/2}$ neutrons. The quasi-particle states which are aligned are mainly two hole-states and the nucleons corresponding to those two holes are scattered by the pairing force into the core. So roughly speaking the ground state band has N-nucleons in the $i_{13/2}$ neutron shell while the upper band has N-2 nucleons. The oscillation of backbending comes from the interplay between the pairing and the Coriolis force[16].

2. Second Backbending

The second anomaly of the moment of inertia has been predicted theoretically[1] and has been found by the Berkeley Group[2] in 1977 in ^{158}Er. In the meantime several groups[2,18,19] have measured second anomalies in different nuclei. The data are mainly from the groups in Berkeley[2], in Strasbourg and Cracow[18] and Copenhagen[19]. The 2. bb has been explained by Faessler and Ploszajczak[3] as the alignment of a proton $h_{11/2}$ pair. Thus one expects in the spirit discussed in the introduction os-cillations of the strength of the second anomaly as a function of the proton number. It is therefore interesting to study the N=90 isotones where $^{156}_{66}$Dy shows no second anomaly, $^{158}_{68}$Er shows a second upbending and $^{160}_{70}$Yb shows a strong 2. bb. To study this behaviour as an oscillation of the strength of the 2. bb we performed Hartree-Fock-Bogoliubov[13,14] calculations for these nuclei.

The Hartree-Fock-Bogoliubov approach for the first backbending has been discussed extensively in the literature[13,14,20,21,22]. We thus give here only an outline of the main ideas used here.

The Hartree-Fock-Bogoliubov approach combines the degrees of freedom of the Har-tree-Fock (HF) and the pairing (BCS) approaches. The ground state expectation value of the many-body Hamiltonian is minimized with respect to the HF and BCS degress of freedom at the same time. The ground state wave function is a quasi-particle Slater determinant built of all quasi-particle anihilation operators applied to the par-ticle vacuum.

$$|\text{HFB}> = \prod_{\text{all } i} \alpha_i |0>$$

$$\alpha_i^+ = \sum_a \{c_a^+ A_{ai} + c_a B_{ai}\}$$

(2)

We produced in two steps: In the first one we build a trial HFB wave function taking the HFB solution of the following Hamiltonian.

$$H_t = \sum_{\tilde{a}\tilde{b}} <\tilde{a}|h|\tilde{b}> c_{\tilde{a}}^+ c_{\tilde{b}} - \sum_{\tilde{a}>0} \Delta_\tau (c_{\tilde{a}}^+ c_{-\tilde{a}}^+ + c_{-\tilde{a}} c_{\tilde{a}})$$

$$- \omega_I \sum_{\tilde{a}\tilde{b}} <\tilde{a}|j_x|\tilde{b}> c_{\tilde{a}}^+ c_{\tilde{b}}$$

$$|\tilde{a}|h|\tilde{b}> = \varepsilon_{\tilde{a}} \delta_{\tilde{a},\tilde{b}}$$

(3)

$$-\alpha_\tau \hbar\omega_0 <\tilde{a}|r^2[\beta \cos \gamma Y_{20} + \frac{\beta \sin \gamma}{\sqrt{2}} (Y_{22}+Y_{2-2})]$$

$$+ r^2\beta_4 Y_{40}|\tilde{b}>$$

with: $$|\pm\tilde{a}> = N\{|a> \pm e^{i\pi j_x}|a>\}$$

In the second step we take this trial wave function which is constrained to an averaged angular momentum $[I_x] = [I(I+1)]^{1/2}$ and calculate the particle number pro-jected expectation value of the many-body Hamiltonian $\hat{H}$ and minimize this expression to the deformation parameters β, γ and β_4 and with respect to the pairing properties described by the pairing gaps Δ_p for the protons and Δ_n for the neutrons.

$$E_I(\beta,\gamma,\beta_4\;;\Delta_p,\Delta_n) = \frac{<\psi_t(I)|\hat{H}\hat{P}_p\hat{P}_n|\psi_t(I)>}{<\psi_t(I)|\hat{P}_p\hat{P}_n|\psi_t(I)>}$$

(4)

$$\frac{<\psi_t|\hat{I}_x\hat{P}_p\hat{P}_n|\psi_t>}{<\psi_t|\hat{P}_p\hat{P}_n|\psi_t>} = \sqrt{I(I+1)}$$

The second equation (4) guarantees that the particle number projected average angular momentum has the right average value already before the minimization of the energy given in the first part of eq. (4). The Lagrange multiplier ω_I which is classically the rotational frequency is determined by this equation. The energies obtained by minimizing the first part of eq. (4) are then compared with the experimental data. The single-particle basis and the pairing plus quadrupole plus hexadecapole force Hamiltonian is taken from Kumar and Baranger[23] but supplemented with a hexadecapole force[24].

To minimize the calculation for the N=90 isotones ^{156}Dy, ^{158}Er and ^{170}Yb we vary in the minimization of the energy expectation value (4) only the proton gap parameter Δ_p for each angular momentum the other parameters are kept fixed to the value obtained in the intrinsic system for the ground state ($\beta=0.26$; $\gamma=0$; $\beta_4=0.05$; $\Delta_n=0.9$ MeV). The parameters of the Hamiltonian are: $A\cdot G_p=25$ MeV; $A\cdot G_n=20$ MeV; the quadrupole force constant $\chi=72\cdot A^{-1\cdot4}$ MeV. To obtain the right result we are forced to modify slightly the single-particle energies[23] according to the prescription of Chasman[25]. This means that according to the analysis of the single-particle proton energy near the magic shell Z=64' one has to shift the single-particle $h_{11/2}$ proton level 900 keV up to get the energy gap at Z=64[26].

Figure 5 shows the results for the three N=90 isotones. A quantitative agreement for the first backbending is not too good since we have kept the neutron gap parameter fiexed to 0.9 MeV. But we see that the theory (dashed line) nicely reproduces the fact that one has for Dy no second anomaly, for Er an upbending and for Yb a strong 2. bb. This correct behaviour is only obtained since we shifted the $h_{11/2}$ proton level[23] according to the prescription of Chasman[25] and since we included the hexadecapole deformation. The reason for the variation of the strength of the 2. bb is the oscillating behaviour of the interaction between the aligned $i_{13/2}$ neutron band and the aligned $h_{11/2}$ proton band. This interaction is shown in Figure 6. The interaction strength V reaches its maximum inbetween the $i_{13/2}$ neutron levels for the first backbending (see Fig. 3) and inbetween the $h_{11/2}$ proton Nilsson levels for the second backbending. The minima are obtained when the Fermi surface crosses one of the inner Nilsson levels ($\Omega=\frac{3}{2}$ up to $\Omega=11\frac{1}{2}$ for $i_{13/2}$ and $\Omega=\frac{3}{2}$ up to $\Omega=\frac{9}{2}$ for $h_{11/2}$). From what we have said above we would expect that first backbending is not changing with varying the proton number and the second backbend is not modified when the neutron number is changed. But this contradicts the fact that ^{156}Er shows upbending while ^{160}Er shows no second anomaly. Figure 7 shows the interaction strength in units of the oscillator energy ($\hbar\omega_0 = 7.6$ MeV). One sees that one just gets the experimental result: ^{156}Er has a very small interaction strength V while the interaction strength for ^{158}Er is appreciably larger and the interaction strength for ^{160}Er is so large that one finds no second anomaly. This is obtained by the variation of the deformation β which in Figure 7 is selfconsistently determined for each Er isotopes in the HFB approach. Looking to Figure 3 one sees immediately that also by keeping the Fermi surface fixed but increasing the deformation β one can cross the $h_{11/2}$ proton Nilsson levels and therefore produce oscillations in the same way as by increasing the proton Fermi surface.

Concluding this chapter we can say that the present investigation strongly supports the explanation of the second anomaly as an alignment of a $h_{11/2}$ proton pair.

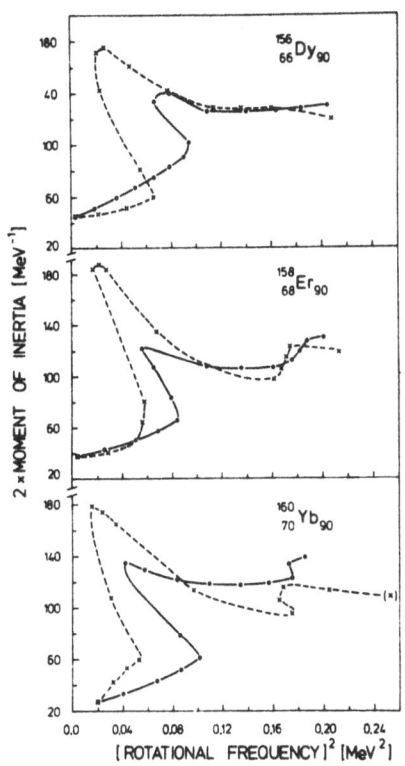

Fig. 5:
Backbending plots for the N=90 iso-
tones $^{156}_{66}$Dy, $^{158}_{68}$Er and $^{160}_{70}$Yb. The
solid line gives the experi-
ment[2,18,19] and the dashed line the
theory[27]. The theoretical results
are calculated in the HFB approach.

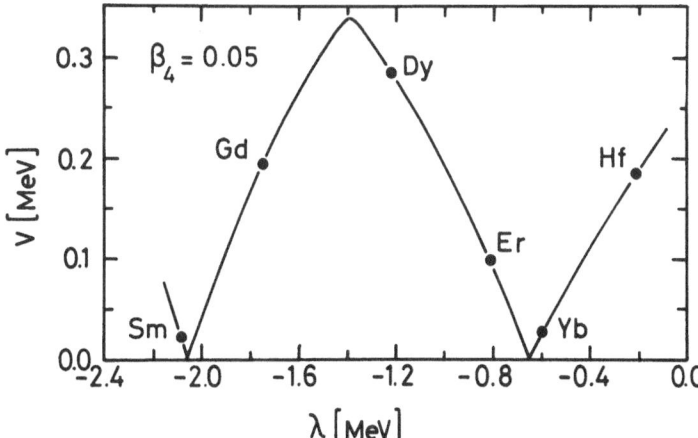

Fig. 6:
Interaction V at the
crossing of the pro-
ton 2 q.p. $h_{11/2}$ band
as a function of the
proton Fermi surface
λ for $\beta_4 = 0.05$. The
position of the dif-
ferent N=90 isotones
are indicated. One
sees that Dy has a
large interaction V
and thus no second
anomaly. Er shows a
medium interaction V
and therefore shows
upbending and Yb has
a very small inter-
action V and there-
fore shows a second
anomaly.

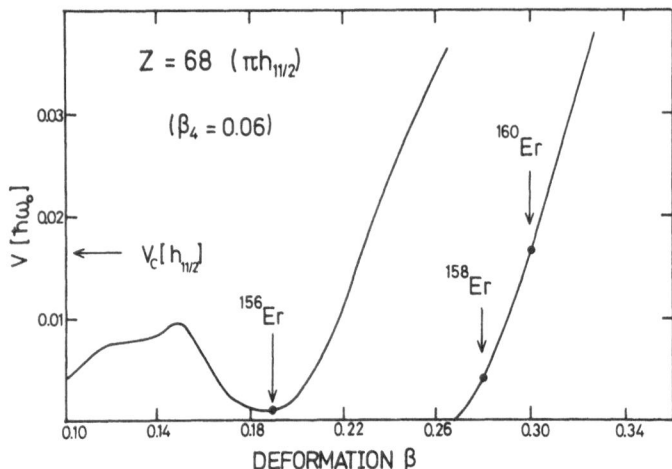

Fig. 7:
The interaction strength $|V|$ in units of the oscillator energy $\hbar\omega_0 = 41/A^{1/3}$ [MeV] as a function of the deformation β. The deformation of the three Er isotopes (Z=68) are determined selfconsistently by the HFB approach. The interaction strength is extracted from the quasi-particle energies of the cranked Hartree-Fock-Bogoliubov approach. A jump in the interaction strength V at the deformation β=0.27 is due to a crossing of the $h_{11/2}$ quasi-particle state by an other level with different quantum numbers.

3. The moment of inertia in the actinides

The anomalous increases of the moment of inertia called backbending are in the rare earth region connected with the alignment of an $i_{13/2}$ proton pair and $h_{11/2}$ neutron pair. The two large corresponding single-particle states for the actinides are the $j_{15/2}$ neutron level and the $i_{13/2}$ proton Nilsson states. The proton $i_{13/2}$ levels are practically all empty while the $h_{11/2}$ proton states are more occupied than the $i_{13/2}$ neutron states in the rare earth nuclei. At the beginning of the rare earth region one finds that as expected from the larger angular momentum the $i_{13/2}$ neutron level is responsible for the first backbending and the $h_{11/2}$ proton level produces the second anomaly. In the transuranic nuclei the situation is slightly different: At the beginning of the actinides the $i_{13/2}$ proton level is practically empty while the $j_{15/2}$ neutron level is already partially filled. This might lead to the expectation that in the actinides the $i_{13/2}$ protons and the $j_{15/2}$ neutrons are equally contributing to the increase of the moment of inertia. In addition the moment of inertia is already a factor 2 larger in the actinides than in the rare earth nuclei even at low angular momenta. Thus we have a weaker Coriolis force and expect therefore not such a rapid alignment already at quite low angular momenta as in the rare earth nuclei. Indeed we know experimentally that the actinides show no backbending but a continuous increase of the moment of inertia by almost a factor 2 from angular momentum zero to about angular momentum 30.

To study the cause of this increase of the moment of inertia we performed Hartree-Fock-Bogoliubov calculations using the same method as presented in chapter 2[28]. Figure 8 shows the backbending plot calculated for ^{236}U compared with Coulomb excitation data from GSI and Frankfurt[29,30]. Figure 9 shows the alignment plot for the same ^{236}U nucleus. It gives the contribution to the angular momentum as a function of the angular momentum coming from different pairs of nucleons. One sees clearly that the $i_{13/2}$ proton pair is the one which is mainly contributing and shows the strongest alignment. This is also confirmed experimentally.

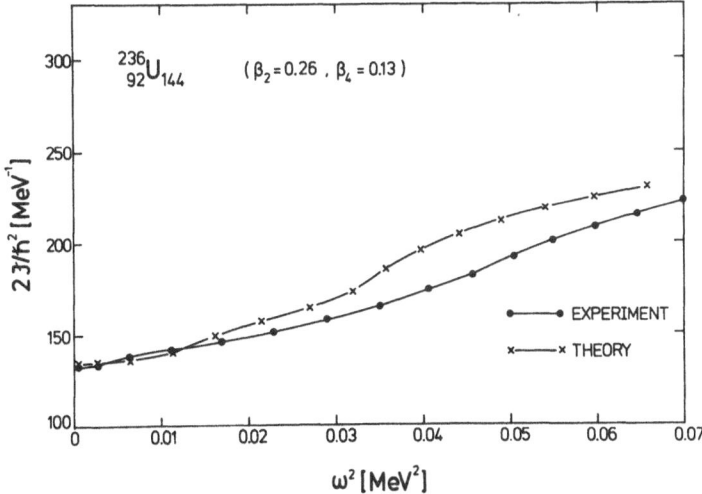

Fig. 8:
Backbending plot of twice the moment of inertia versus the rotational frequencies squared for ^{236}U. The curve with the dotts are the experimental data[29,30] and the crossed curve ar the theoretical results of a cranked Hartree-Fock-Bogoliubov calculation. The deformation is kept constant in the Hartree-Fock-Bogoliubov variation to the selfconsistent value obtained in the intrinsic system for the ground states.

The above result is supported by two experimental facts: (i) Simon et al.[31] have measured rotational bands in the neighbouring nuclei of ^{236}U which differ by a $j_{15/2}$ neutron hole (^{235}U) and by a $i_{13/2}$ proton particle (^{237}Np). Figure 10 shows the aligned single-particle angular momentum i[π] as a function of the rotational frequency for the favoured and unfavoured bands (^{235}U: crosses favoured, dots unfavoured; ^{237}Np: dots favoured, crosses unfavoured). One sees that the bands in ^{235}U show a similar alignment at around $\hbar\omega$ = 0.2 MeV while the alignment in ^{237}Np is blocked by the odd $i_{13/2}$ proton. This indicates that the alignment of the core is due to $i_{13/2}$ protons.

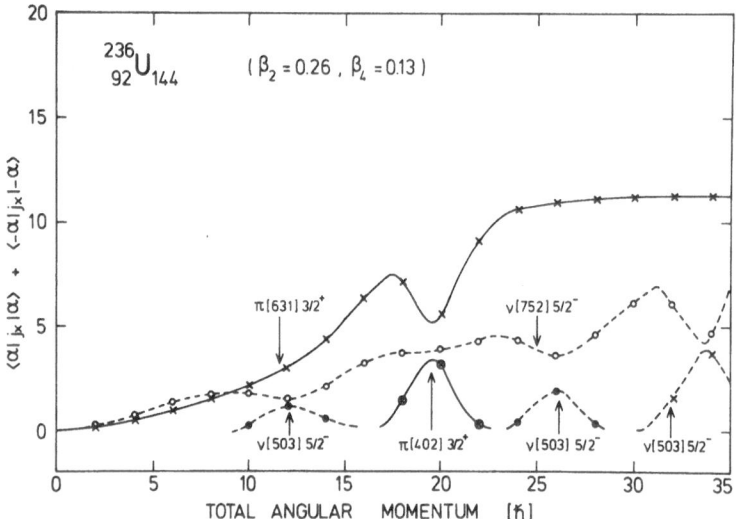

Fig. 9:
Rotational alignment plot of the contributions to the total angular momentum of different pairs of nucleons as a function of the total angular momentum for ^{236}U. One sees that the $i_{13/2}$ proton pair is mainly contributing.

(ii) Second experimental indication that the increase of the moment of inertia in the actinides is mainly due to the alignment of an $i_{13/2}$ proton is found by g-factor measurements[32]. Häusser et al.[32] performed Coulomb excitation in ^{232}Th and ^{238}U with ^{127}I and ^{142}Nd. In Coulomb excitation with ^{142}Nd higher angular momenta are reached. A fit to the rotation of the angular distribution in strong magnetic fields yields unsatisfactory results assuming a constant g-factor. A g-factor increasing with angular momentum yields a good agreement with the data. These results indicate especially for ^{238}U that the alignment of a $i_{13/2}$ proton pair is stronger than the alignment of a $j_{15/2}$ neutron pair.

The situation seems to be a little bit different in the ^{248}Cm region. The Coulomb excitation data[33] have been nicely explained by Faessler and Ploszajczak[33] using the Hartree-Fock-Bogoliubov approach described here. From the backbending plot Figure 11 and especially from the alignment plot in Figure 12 for ^{248}Cm one sees that in this nucleus at the end of the actinide nuclei with the small deformation β=0.2 the alignment of the $i_{13/2}$ protons and $j_{15/2}$ neutrons are competing with each other at least according to our theoretical analysis.

Essential for the qualitative and almost quantitative understanding of the actinide nuclei by a cranked Hartree-Fock-Bogoliubov approach is the particle number projection before the variation. We performed also calculations without particle number projection and found that pairing is collapsing at a too low angular momentum. This produces a drastic increase of the moment of inertia which is not found experimentally. Such results are for example found in these nuclei by Frauendorf et

al. who used the so-called cranked shell model approach where no projection on good
particle number is performed[34].

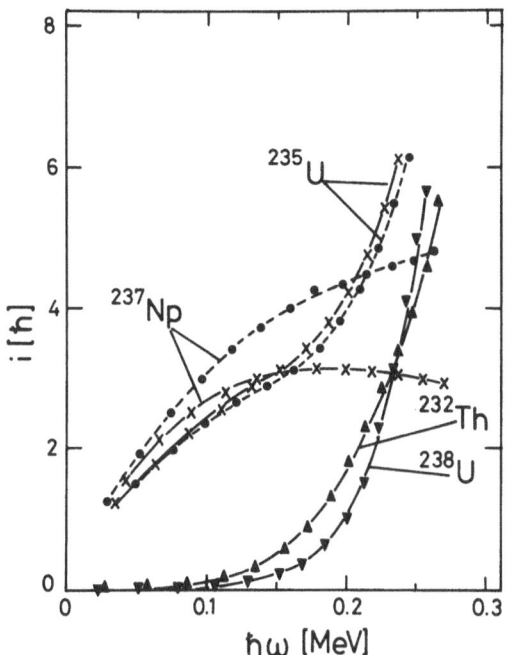

Fig. 10:
Alignment in the odd mass nuclei
^{235}U and ^{237}Np compared to the alignment in even mass nuclei ^{232}Th and
^{238}U. i is the angular momentum of a
nucleon of the aligned pair extracted from the total angular momentum
after the angular momentum of the
variable moment of inertia is subtracted for a given rotational frequency ω. In the odd mass nuclei the
results are plotted for the favoured
and the unfavoured bands (^{235}U:
crossed favoured, dotted unfavoured;
^{237}Np: dotted favoured, crossed unfavoured). The alignment of the core
is blocked in ^{237}Np an odd $i_{13/2}$
proton. The alignment should therefore be due to a $i_{13/2}$ proton pair.
Data are taken from Simon et al.[31].

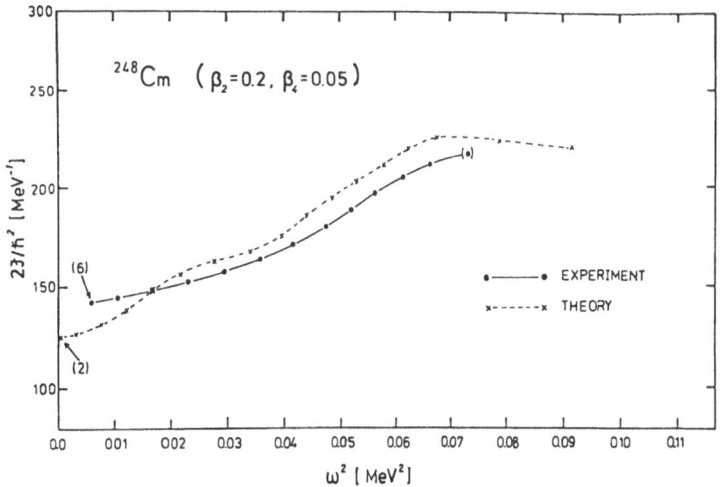

Fig. 11:
Backbending plot for ^{248}Cm. Twice the moment of inertia is plotted against the ro-
tation frequencies squared for the experimental data[33] and for a cranked Hartree-
Fock-Bogoliubov theory[33].

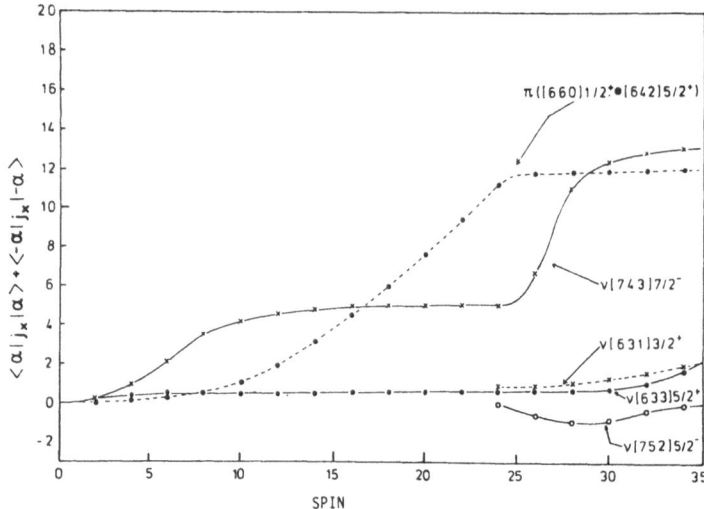

Fig. 12:
Rotational alignment plot of the angular momentum of a pair of nucleons in ^{248}*Cm as*
a function of the total angular momentum. One sees that first a $i_{13/2}$ *proton pair*
is fully aligned and only at angular momentum 25 to 30 a $j_{15/2}$ *neutron pair gets*
aligned completely.

are not orthogonal to the pure boson states. The orthogonalization[35] yields quasi-particle states a^+_{jm} which are connected with the bosons and the particle states c^+_{jm} in the following way:

$$c^+_{jm} = u_j\, a^+_{jm} - v_j\, \frac{1}{\sqrt{N}}\, (s^+ \tilde{a}_j)_{jm}$$

$$-u_j\, \sum_{j'}\, \beta_{j'j}\, \frac{\sqrt{10}}{\hat{j}}\, (d^+ \tilde{a}_{j'})_{jm} \tag{9}$$

$$-v_j\, \sqrt{\frac{10}{N}}\, \sum_{j'}\, \beta_{j'j}\, \frac{1}{\hat{j}}\, s^+ (d a^+_{j'})_{jm}$$

with: $\hat{j} = \sqrt{2j+1}$

This expression corresponds to the usual quasi-particle transformation. But it conserves the particle number. If one introduces transformation (9) into the fermion part of the Hamiltonian (5) one obtains from a one-body fermion operator 16 terms and from a two-body fermion operator already 256 terms. In using the transformation (9) we retain for simplicity only one boson operators, sd, (s$^+$d)... and make approximations where appropriate to low seniority states (e.g. $s^+s \sim \hat{N} \sim N$). More details are described in the work of Morrison, Faessler and Lima[36]. The two parameters (7) of the coupling of the odd nucleons to the quadrupole moment of the core are fitted in the odd mass nucleus ^{199}Hg.

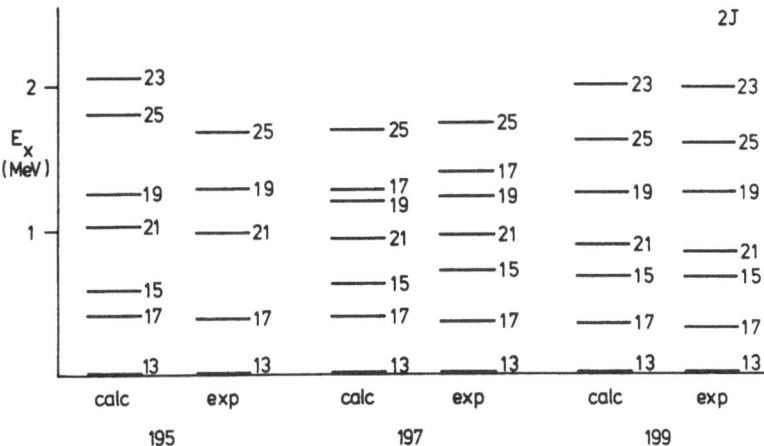

Fig. 13:
Odd mass Hg nuclei described by coupling an odd nucleon to an IBA core[36]. The two parameters of coupling of an odd nucleon to the quadrupole moment of the core are adjusted in ^{199}Hg. The results for ^{195}Hg and ^{197}Hg are parameterfree predictions. The angular momenta in these odd mass nuclei are given as 2I.

Figure 14 and 15 contain the results of coupling two quasi-particles to an IBA core and to a triaxial rotor[40] for ^{196}Hg. To get no double counting we allow that the quadrupole-quadrupole interaction acts only between nucleons of different charge while the residual interaction represented by the surface delta interaction (the

4. Two quasi-particles coupled to an interacting boson and a rigid rotor core

In this chapter we want to extend the interacting boson approximation[6] in such a way that we can couple two quasi-particles (2 q.p.) to it. We shall see that this is possible in a natural way without introducing more parameters than needed for the description of odd mass nuclei[35]. This extension has been done by Morrison, Faessler and Lima[36]. We shall apply this model to the even mass Hg nuclei for describing the positive pairty states. These nuclei show around angular momentum 8^+, 10^+ and 12^+ a strange anomaly along the yrast line, which has been explained by Yadav, Toki and Faessler as the intersection of an aligned $i_{13/2}$ neutron or an aligned $h_{11/2}$ proton and with the ground state band[7]. Recently Kroth[37] et al. found that in the lighter Hg even mass isotope the 12^+ yrast state has been overlooked. Thus there is no need for proton $h_{11/2}$ alignment from the experimental point of view. This is also supported by the measuremnt of the g-factors of the 10^+ in ^{194}H and ^{196}H[38]. These measurements[38] yield negative g-factors $g(^{194}$Hg$) = -0.24$ (4) and $g(^{196}$Hg$) = -0.18$ (9) in agreement with the values expected for an $i_{13/2}$ neutron structure. One therefore has to assume that in the Hg nuclei the 2 q.p.'s of $h_{11/2}$ proton nature lie farther away from the Fermi surface than the $i_{13/2}$ neutron states. In the present calculation we include therefore only neutron quasi-particle states. The purpose of this chapter is, to compare the results of the IBA + 2 q.p.'s with calculations performed in the asymmmetric rotor + 2 q.p.'s model[7] and the data.

The Hamiltonian of the IBA + 2 q.p.'s consists out of 4 parts.

$$\hat{H} = \hat{H}_{IBA} + \hat{H}_{o\ part} + V_{SDI\ part}$$

$$+ \hat{H}_{IBA,\ part} \tag{5}$$

The first part $\hat{H}_{IBA}$ describes the core in the interacting boson approximation[6]. The Hg nuclei are well described[39] in the O(6) limit[6].

$$H_{IBA} = AP_6 + BC_5 + CR^2 \tag{6}$$

with the parametrization[36] A=470 keV, B=600 keV, C=0 keV. P_6 is the Casimir operator of O(6), C_5 the quadratic Casimir operator of O(5) and R^2 is the well-known Casimir operator of O(3) and represents the square of the angular momentum of the core. $\hat{H}_{o\ part}$ describes the unperturbed spherical shell model Hamiltonian for the two valence nucleons. $V_{SDI\ part}$ is the residual interaction between the two nucleons and at the same time the monopole part describes the pairing interaction between all the nucleons. The last part $\hat{H}_{IBA,part}$ describes the interaction of the two nucleons with the quadrupole moment of the core.

$$\hat{H}_{IBA,part} = \sum_{ii'} Q_2 \cdot \hat{q}_{jj'2}$$

$$\hat{q}_{jj'2m} = q_{jj'} |c_j^+ \tilde{c}_j|_{2m}$$

$$Q_{2m} = T_o \{ (s^+ \tilde{d} + d^+ \tilde{s})_{2m} + \chi (d^+ \tilde{d})_{2m} \} \tag{7}$$

with: $\tilde{c}_{jm} = (-)^{j+m} c_{j-m}$; $\tilde{d}_{2m} = (-)^m d_{2-m}$

The boson-fermion states

$$\psi = N(s^+)^m (d^+)^n (c^+)^k |0> \tag{8}$$

monopole particle-particle part is the pairing force) acts between particles of the same charge. If the second excited zero plus state of the core (pairing vibrational state) is not included in the basis one obtains for the strength of the surface delta interaction k=0.18 MeV in rough agreement with the BCS pairing strength $G_n \approx 37/A$ MeV[7].

$$V_{SDI, part} = - 4\pi \sum_{\gamma\mu} (k/2 \sum_{abcd} <a|Y_{\gamma\mu}|c>$$

$$|b|Y^*_{\gamma\mu}|d> N(c^+_a c^+_b c_d c_c) \tag{10}$$

When boson pairing vibration states were included in the basis the SDI strength k was reduced to 0.15 MeV to retain approximate agreement for the excitation energy of the second excited 0^+ state.

The amplitude u_i, v_i were calculated using the usual BCS approach. For the IBA plus 2 q.p.'s we used the single-particle basis by $p_{1/2}$, $1f_{5/2}$, $2p_{3/2}$, $0i_{13/2}$, $0h_{9/2}$, $1f_{7/2}$. For the triaxial rotor + 2 q.p.'s only the $0i_{13/2}$ neutron state was included. We needed therefore a larger quenching for the g-factors in the rotor than in the IBA model. Due to the larger single-particle basis one obtains agreement for the g-factor of the 10^+ state in ^{196}Hg with the usual quenching factor of 0.6 while the result for the triaxial rotor + 2 q.p.'s gets agreement with the data for a quenching of 0.4 for the spin g-factor of the neutron. Results are summarized in Figure 14 and 15. The comparison of the results obtained in the IBA + 2 q.p. and the rotor + 2 q.p. model show a close agreement between the two models and the data. Naturally the IBA + 2 q.p. model is able to describe excited zero plus states which are not contained in a triaxial rotor description. To obtain those states one would need to allow for beta vibrational states. Coupling of two quasiparticle states to a full Bohr-Mottelson Hamiltonian is in progress[41].

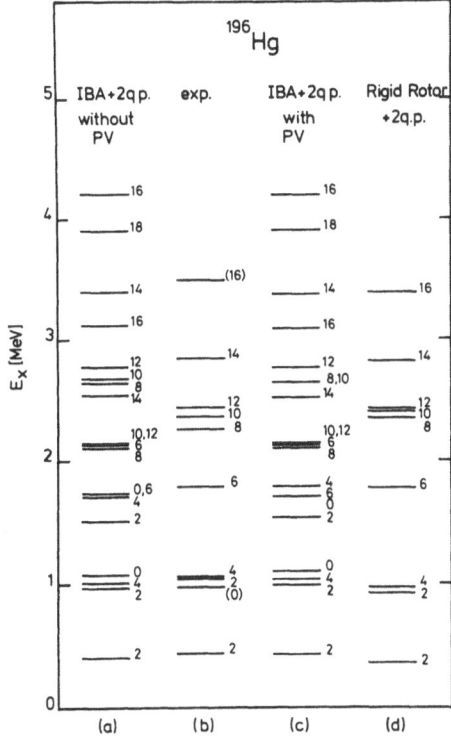

Fig. 14:
Experimental and theoretical energy spectra for ^{196}Hg. The left column is the result of an IBA + 2 q.p.'s calculation without inclusion of the second excited zero plus ("pairing vibrational") state. The second column gives the experimental data. The third column shows the calculation given in the first column with the inclusion of the pairing vibrational state. We have the second excited zero plus state at the same energy as the strength of the surface delta interaction is reduced from 0.18 MeV to 0.15 MeV. The last column gives the result of the rotor + q.p.'s calculation[7].

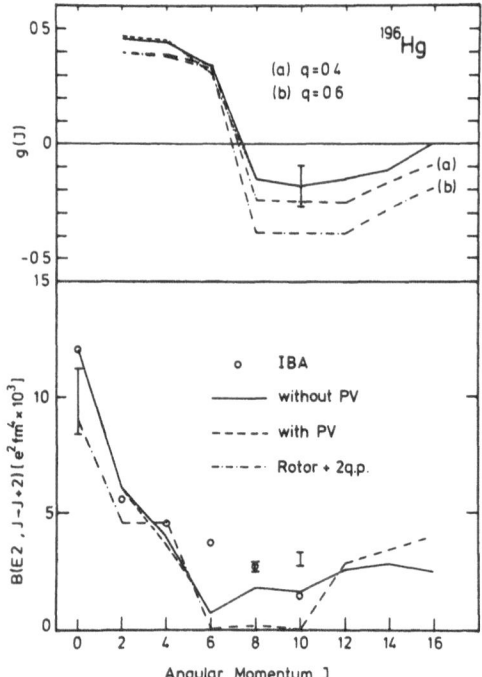

Fig. 15:
g-factors and reduced quadrupole transition probabilities for ^{196}Hg. The upper part shows the g-factors in the IBA + 1 q.p.'s model using a quenching factor of 0.6 for the neutron spin. Here we included 6 neutron single-particle shells. In the rotor + 2.p. model only the $0i_{13/2}$ neutron shell is included and thus we need a quenching factor of 0.4 to obtain agreement with the data. The dashed line shows the result with and the solid line without the inclusion of the second excited zero plus state from the boson basis in the IBA + 2 q.p. calculation. In the lower part circles indicate the result of the pure IBA model. Since the IBA model contains only states up to 12^+ these points stop at total angular momentum 10 in the above Figure.

5. Conclusions

In these lectures we have first discussed the second backbending which is due to the alignment of a $h_{11/2}$ proton pair. We have seen that the 2. bb in the N=90 isotones can be naturally explained by the oscillation of backbending with increasing proton number. $^{156}_{66}Dy$ shows no second anomaly, $^{158}_{68}Er$ shows upbending and $^{160}_{70}Yb$ shows a strong second anomaly. This variation can be explained by the oscillation of the strength of backbending. Backbending is strong if the interaction between the two intersecting bands is small and it is weak or does not exist at all if the interaction V between the intersecting bands is large. Since the interaction V oscillates in such a way that the absolute value has a maximum if the Fermi surface lies inbetween the Nilsson levels of the $h_{11/2}$ protons and it has a minimum if the Fermi surface is on top of one of the inner $h_{11/2}$ proton Nilsson levels. This explanation suggests that adding neutrons does not modify the nature of 2.bb. But this contradicts the experimental data which gives a strong second anomaly for ^{156}Er, an upbending for ^{158}Er and no second anomaly for ^{160}Er. But here we showed by selfconsistent cranked Hartree-Fock-Bogoliubov calculations that the deformation is drastically increasing from about β=0.18 to β=0.29. In this way the Fermi surface is again crossing the $h_{11/2}$ Nilsson levels but due to the change of the deformation and this yields oscillations in the same way as if the Fermi surface is crossing those Nilsson levels by increasing the proton number.

In the second part we discussed the reason for the increase of the moment of inertia in the actinide nuclei. We showed that in the uranium region this increase is mainly due to the alignment of an $i_{13/2}$ proton pair. This result of selfconsistent cranked Hartree-Fock-Bogoliubov calculations with particle number projection before the variation agree with experimental data in neighbouring odd mass nuclei: If an odd proton $i_{13/2}$ is blocking a level, the alignment of the core is drastically reduced while this is not the case for an odd $j_{15/2}$ neutron. The important role of the $i_{13/2}$ protons for the increase of the moment of inertia in the actinides is supported by g-factor measurements which show an increase at high spin states. In ^{248}Cm the situation seems to be different according to our self-consistent calculations. There $i_{13/2}$ protons and $j_{15/2}$ neutrons seem to play a similar important role for the increase of the moment of inertia. In all the calculations in the actinides it is essential that one performs a particle number projection before the variation to obtain even qualitatively reasonable results.

In the last part of this talk we coupled two quasi-particles to an IBA and an asymmetric rotor core. With these models we described positive parity spectra in the even mass nuclei of Hg isotopes. We especially explained the anomaly of the 8^+, 10^+, 12^+ yrast states which are due to an intersection of the ground state band with an aligned $(i_{13/2})^{-2}$ neutron band. Both models are able to explain this anomaly in agreement with the data.

At the end I would like to thank my collaborators M. Ploszajczak, F. Grümmer, K.W. Schmid, I. Morrison and C. Lima with whom the above work has been done.

References

1. M. Ploszajczak, A. Faessler, Z. Phys. A283 (1977) 349.
2. T.Y. Lee et al., Phys. Rev. Lett. 38 (1977) 1454.
3. A. Faessler, M. Ploszajczak, Phys. Lett. 76B (1978) 1.
4. R. Bengtsson, I. Hamamoto, B. Mottelson, Phys. Lett. 73B (1978) 259.
5. O. Häusser et al., Phys. Rev. Lett. 48 (1982) 383.
6. A. Arima, F. Iachello, Ann. Phys. 99 (1976) 253 and 111 (1978) 201 and
   Phys. Rev. Lett. 40 (1978) 385.
7. H.L. Yadav, H. Toki, A. Faessler, Phys. Rev. Lett. 39 (1977) 1128 and more
   recent unpublished calculations of A. Faessler and C. Lima.
8. H.L. Yadav, H. Toki, A. Faessler, Phys. Lett. 76B (1978) 144 and 89B (1980) 307.
9. A. Faessler, Proceedings of the Gull Lake Conference on Nuclear Spectroscopy,
   Gull Lake, Michigan 1979, Springer Verlag 1980.
10. A. Faessler, W. Greiner, R.K. Sheline, Nucl. Phys. 62 (1965) 241.
11. F.S. Stephens, R.S. Simon, Nucl. Phys. A183 (1972) 257.
12. R.M. Lieder, H. Ryde, Adv. in Nucl. Phys. 10 (1978) 1, edited by M. Baranger
    and E. Vogt.
13. A. Faessler, K.R. Sandhya Devi, F. Grümmer, K.W. Schmid, R. Hilton, Nucl. Phys.
    A256 (1976) 106.
14. A. Faessler, M. Ploszajczak, K.R. Sandhya Devi, Nucl. Phys. A301 (1978) 382.
15. A. Neskakis, R.M. Lieder, M. Müller-Veggian, H. Beuscher, W.F. Davidson,
    C. Mayer-Böricke, Nucl. Phys. A261 (1976) 189.
16. F. Grümmer, K.W. Schmid, A. Faessler, Nucl. Phys. A326 (1979) 1.
17. R. Bengtsson, S. Frauendorf, Nucl. Phys. A314 (1979) 27 and A327 (1979) 139.
18. F.A. Beck et al., Phys. Rev. Lett. 42 (1979) 493.
19. L.L. Riedinger et al., Phys. Rev. Lett. 44 (1980) 568.
20. H.J. Mang, Phys. Rep. 18 (1975) 325.
21. A. Goodman, Adv. Nucl. Phys. 11 (1979) 263.
22. A. Faessler, M. Ploszajczak, K.W. Schmid, Progress in Particle and Nuclear
    Physics, 5 (1980) 79.
23. M. Baranger, K. Kumar, Nucl. Phys. 110 (1968) 490, 529.
24. M. Ploszajczak, A. Faessler, G. Leander, S.G. Nilsson, Nucl. Phys. A301
    (1978) 477.
25. R.R. Chasman, Phys. Lett. 102B (1981) 299.
26. P. Kleinheinz, Proc. Symp. on High Spin Phenomena in Nuclei (Argonne National
    Laboratory, 1979).
27. A. Faessler, M. Ploszajczak, Phys. Rev. C22 (1980) 2609.
28. M. Ploszajczak, A. Faessler, J. Phys. G8 (1982) 709.
29. H. Ower, Th.W. Elze, J. Idzko, K. Stelzer, P. Fuchs, E. Grosse, H. Emling,
    D. Schwalm, H.J. Wollersheim, N. Kaffrell, N. Trautmann, Contribution to the
    Int. Conf. on Nucl. Behaviour at High Angular Momentum (Strasbourg), April
    22-24, 1980, p. 119.
30. P. Fuchs, H. Emling, F. Folkmann, E. Grosse, D. Schwalm, R.S. Simon, H.J.
    Wollersheim, J. Idzko, D. Pelte, 1977 GSI-J-1-78 (GSI Annual Report).
31. R.S. Simon, F. Folkmann, Ch. Briancon, J. Libert, J.P. Thibaud, R.J. Walen,
    S. Frauendorf, Z. Physik A298 (1981) 121.
32. H. Häusser et al., Phys. Rev. Lett. 48 (1982) 383.
33. R.B. Piercey, J.H. Hamilton, A.V. Ramayya, H. Emling, P. Fuchs, E. Grosse,
    D. Schwalm, H.J. Wollersheim, N. Trautmann, A. Faessler, M. Ploszajczak,
    Phys. Rev. Lett. 46 (1981) 415.
34. S. Frauendorf et al., private communication.
35. O. Scholten, Thesis, Groningen, 1980.
36. I. Morrison, A. Faessler, C. Lima, Nucl. Phys. A372 (1981) 13.
37. R. Kroth, K. Hardt, M. Guttormsen, G. Mikus, J. Recht, W. Vilter, H. Hübel,
    C. Günther, Phys. Lett. 99B (1981) 209.
38. R. Kroth, S.K. Bhattarcherjee, C. Günther, M. Guttormsen, K. Hardt, H. Hübel,
    A. Kleinrahm, Phys. Lett. 97B (1980) 197.
39. I. Morrison, R.H. Spear, Phys. Rev. C23 (1981) 932.
40. A.S. Davydov, G.F. Filippov, Nucl. Phys. 8 (1958) 237.
41. A. Faessler, C. Lima, H. Müther, work in progress.

SHAPE COEXISTENCE AND A NEW REGION OF STRONG DEFORMATION
IN NUCLEI FAR FROM STABILITY

J. H. Hamilton
Physics Department, Vanderbilt University
Nashville, TN 37235/USA

I. INTRODUCTION

The discoveries of the coexistence of overlapping low energy states with well deformed and near spherical shapes in both the light mercury nuclei far off stability[1,2] and in ^{72}Se (refs. 2-4) helped break down the idea that a given nucleus
had one fixed shape. More recently other types of shape coexistence where changes
in the shape as one goes to somewhat higher spins in near spherical transitional
nuclei[5] and to very high spins in deformed nuclei also have become the subject of
interest[6] (The latter is discussed in the third lecture[7]). The A = 70 region has become an important new testing ground for many types of nuclear structure models because of the richness of different collective motions and structures which are found
in this region and the rapid changes seen in some structures with the addition of only
two protons or two neutrons. The variety of different structures and the rapidity of
their changes can be seen in an earlier survey of the moments of inertia as a function of the rotational frequency $(\hbar\omega)^2$ for the yrast states in nuclei in this region
(Fig. 1). The first nucleus studied to high spin was ^{72}Se (ref. 3) and its strong
forward bend of $\mathcal{J}$, very like the forward bends of $\mathcal{J}$ seen[1,2] in $^{184-188}$Hg, was

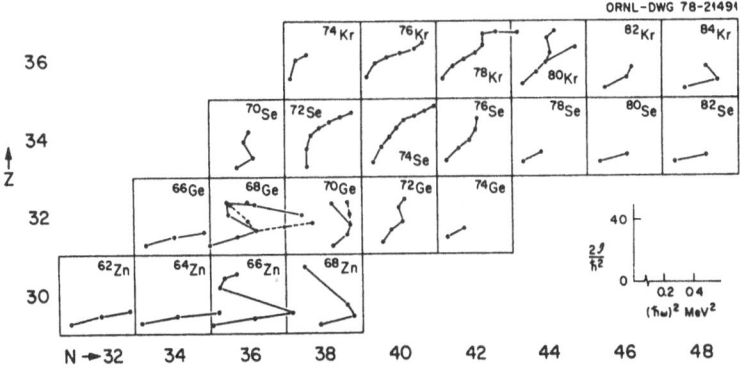

ORNL-DWG 78-21491

Figure 1. Moments of inertia of yrast cascades.[4]

similarly interpreted in terms of shape coexistence[2-4]. A somewhat similar forward
bend was observed in ^{74}Se but not in ^{70,72}Ge. The latter two nuclei also have very
low energy, first excited 0_2^+ states like ^{72,74}Se. There is definite experimental

evidence[8,9] supported by theoretical calculations[8,10] that the 0_2^+ states in $^{70,72}_{32}$Ge indeed have much larger deformations than their ground states as in $^{72,74}_{34}$Se. However, there are no well developed rotational bands built on the 0_2^+ states in ^{70,72}Ge. This was considered by some as a serious problem for the shape coexistence picture. Finally, while in $^{184-188}$Hg and ^{72}Se the overlapping levels with different deformation could be reproduced by potentials with second minima at large deformation (for examples see refs. 11,2,3), it is important to understand the origin of the shape coexistence in the A = 70 region in terms of the single particle orbitals.

Recent studies[12] of ^{74,76}Kr have illuminated the origin of the shape coexistence in this region and given evidence for super deformation of the ground states where both N and Z are at or near 38. The $2_1^+ \rightarrow 0_1^+$ energy does not indicate super deformation in $^{74}_{36}$Kr$_{38}$. The origin of the masking of the strong deformation now is understood in terms of the interaction of two close lying 0^+ states[12]. A similar masking is now understood to occur in the light Pt isotopes. This paper will describe the discovery of super deformation in the region of Z and N both at or near 38. Simultaneously, Möller and Nix[13] carried out calculations that predicted that nuclei in this region of N $\approx$ Z $\approx$ 38 should have the strongest ground state deformation of any nuclei with $\beta \approx 0.4$. New studies in the Sr nuclei[14,15] support this conclusion. Indeed, we have found experimentally and now predicted theoretically a new region of super deformation. The understanding of the origin of this deformation gives insights into nuclear shape coexistence in a wide range of different masses.

II. SHAPE COEXISTENCE AND SUPER DEFORMATION WHEN N $\approx$ Z $\approx$ 38

The clearest evidence[3] for shape coexistence in this region has been $^{72}_{34}$Se$_{38}$. Our studies of the light Kr isotopes provide the clues to understanding this region. The B(E2;2-0) for ^{78,80,82,84,86}Kr were measured by Sakamoto et al.[16]. They found nearly, a factor of 10 increase in B(E2) strength in going from N = 50, $^{86}_{36}$Kr, B(E2; 2-0)$_{spu}$=6.5 to ^{78}Kr where it is 51.8 spu. This suggests the onset of large collective effects possibly associated with deformation. However, the 2-0 energies (455-617 keV) do not suggest large deformation. On the other hand, the B(E2) values up to the 10^+ level in ^{78}Kr show large collective strength.[17]

We investigated the levels in ^{76}Kr and our results are shown in Fig. 2 (ref. 12). Note the dominance of rotational-like band structures. We measured the B(E2) strengths of the transitions from the 4^+ to 10^+ levels by Doppler-shift, line-shape analysis.[12,18] Both singles and coincidence spectra line shapes were analyzed. The results are given in Table I. The B(E2)'s show surprisingly large collective strength--the largest of any nucleus in this region. For comparison the B(E2)$_{exp}$/ B(E2)$_{sp}$ for the 2-0 and 4-2 transitions in $^{68-72}$Ge are the order of 10-20 spu. These data provide strong evidence for large collective effects associated with

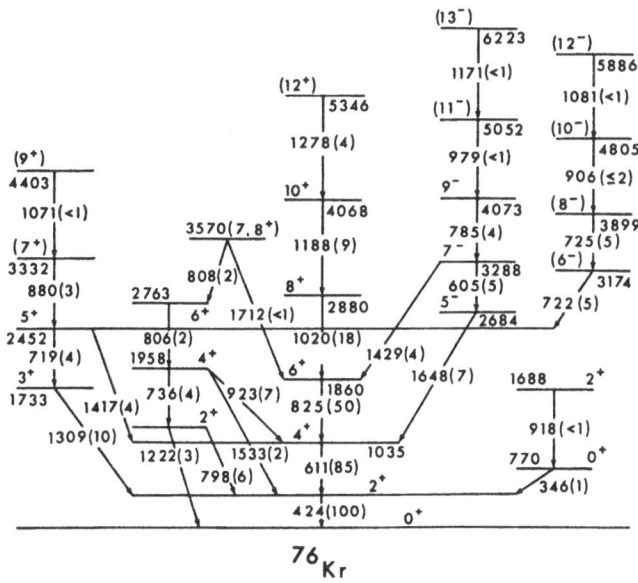

Figure 2. Energy levels of ^{76}Kr from heavy ion, in-beam studies.[12]

Table I. Measured mean lives and extracted B(E2) values[12,18] for transitions in ^{76}Kr

E_γ (keV)	$I_i \rightarrow I_f$	τ_{mean}(ps)	B(E2)/B(E2)$_{sp}$	B(E2)/B(E2)$_{rot}$[ψ]
424	$2^+ \rightarrow 0^+$	53(7)[†]	59(7)[†]	0.75(9)
611	$4^+ \rightarrow 2^+$	6.5	$77(^{23}_{14})$[††]	$0.70(^{21}_{14})$
825	$6^+ \rightarrow 4^+$	1.25(12)	89(8)	0.72(6)
1020	$8^+ \rightarrow 6^+$	0.30(3)	129(13)	1.0(10)
1188	$10^+ \rightarrow 8^+$	0.14(2)	129(19)	0.98(15)
1278	$(12^+) \rightarrow 10^+$	0.24(5)*	52(11)*	0.38(9)*

*Composite state plus feeding lifetime and composite B(E2) compared to single particle values.

†Nolte et al., (Ref. 19).

††Based on an error weighted average (τ = 5.0(20)) of our data[12] and that of Nolte et al. (τ = 8.2 ± 2.3).

ψIntrinsic quadrupole moment normalized to the 8 → 6 transition.

nuclear deformation. From the 2^+ to 10^+ level, the B(E2) values generally follow the gradual increase expected in a rotational model in sharp contrast to the rapid increase in B(E2) in a vibrational model.

Note in ^{76}Kr that the energy of the first excited, 0_2^+, level has continued to drop sharp relative to the 0_1^+ ground state as N decreases analogous to the similar sharp drops seen in Ge and Se nuclei around N = 40. But there is no rotational band built on it as in ^{72}Se. The only feeding to the 0_2^+ level is from a 2^+ level at 1688 keV. Note the 2^+-0_2^+ energy of 917 keV is more than twice the 2_1^+-0_1^+ energy of 424 keV. This is in sharp contrast to ^{72}Se where these two energies are the reverse, eg., the 2^+-0_2^+ energy is low compared to the 2_1^+-0_1^+ energy. A (2^+)-2^+-0_2^+ cascade (882-917 keV γ rays) is seen in UNISOR studies[20] of the decay of ^{76}Rb. These data indicate that something different and unusual is happening in ^{76}Kr.

We carried out a similar study of the energy levels in ^{74}Kr (ref. 12). The moments of inertia for $^{74-80}$Kr are shown in Fig. 3. One can see from Fig. 3, that at low

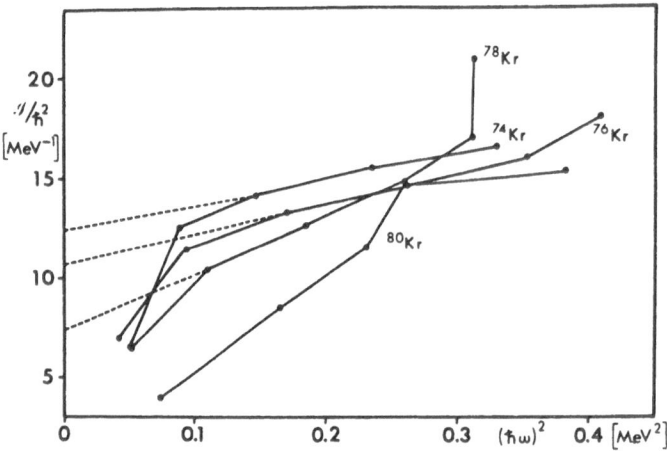

Figure 3. Moments of inertia for $^{74-80}$Kr.

spin $\mathscr{J}$ of each band becomes larger when going from N = 44 to N = 38, except for ^{74}Kr, where the point corresponding to the 2 → 0 energy in ^{74}Kr strongly deviates. This tendency also is seen in ^{76}Kr to a lesser degree. These nuclei exhibit forward bends in $\mathscr{J}$ above the 2_1^+ states analogous to those in ^{72,74}Se$_{38,40}$ which were interpreted (refs. 2, 3) in a shape coexistence picture with bands built on the ground and 0_2^+ states of quite different deformations. All the above data led us to suggest that the ground state of ^{76}Kr is well deformed and the 0_2^+ level is associated with a near-spherical shape[12], in contrast to the reverse situation in ^{72,74}Se. A similar situation should be occurring in ^{74}Kr.

How can we understand large ground state deformation but relatively small $\mathscr{J}$'s extracted from the relatively large 2 → 0 energies in $^{74-76}$Kr? The relatively large 2 → 0 energies and correspondingly small $\mathscr{J}$ in ^{74,76}Kr, which make these nuclei look less deformed than they really are, can arise from an interaction between the 0_1^+ deformed ground states and higher states such as the 0_2^+ states to push down the 0_1^+ energies. The origin of shape coexistence for N $\approx$ 38 nuclei is related to the number of protons which delicately controls whether a deformed shape or near-spherical shape is lowest in this region. Our ^{74,76}Kr data give evidence that their ground states have remarkably large deformation.

The origin of strong deformation and shape coexistence in this region can be attributed to the gaps in the single-particle spectrum seen in Fig. 4 at N (or Z) = 40, $\delta \approx 0$, and N(Z) = 38, $\delta \approx 0.3$, that stabilize the nuclear shape. Evidence for the spherical subshell closure around N = 40 is found when Z is close or equal to 28 or 50, or around Z = 40 when N $\approx$ 28 or 50 because then the protons (neutrons) prefer a spherical shape, as seen for example in $^{66}_{28}$Ni$_{38}$($^{90}_{40}$Zr$_{50}$). However, as Z moves away from 28 or 50 the level density for a spherical shape becomes very high and the minimum of the proton deformation energy moves to deformed shapes and similarly for the neutrons which have almost identical single-particle levels. Away from Z(N) = 28 and 50 closed shells, maximal deformation is expected at N(Z) $\approx$ 38. However, the deformed state can coexist with a nearly spherical configuration in a delicate balance. Which one is lower depends on the proton number. For $^{70,72}_{32}$Ge$_{38,40}$ and $^{72,74}_{34}$Se$_{38,40}$ the coexistence of nearly spherical ground states with deformed 0^+_2 states has been reported[2,4,8] as noted earlier. In ^{72,74}Se, the deformed band becomes yrast at I $\approx$ 2-4 because of its lower rotational energy. For the Ge isotopes, the bands built on the two different shapes are not well developed and so not seen because of the smaller deformation (two protons less than Se). In the Kr isotopes, the 36 protons favor deformation even more.

From Fig. 4 one can see that for N = 38, the deformed structure is generated from

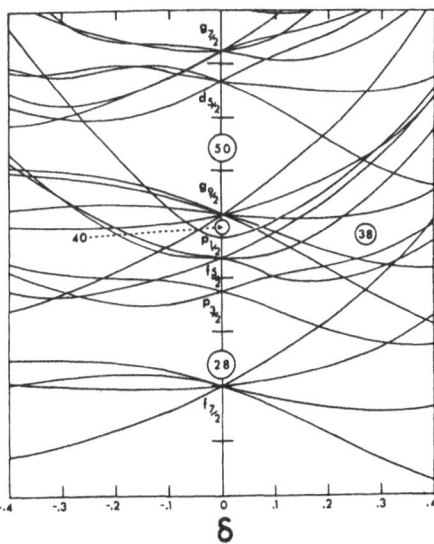

Figure 4. The Nilsson single particle spectrum for neutrons. Note the gap at N=38.

the spherical one by transferring two pairs of neutrons from the $f_{5/2}$ into the $g_{9/2}$ shell. Correlations of the pairing type contain the multiple scattering of pairs from the $f_{5/2}$ to the $g_{9/2}$ shell (and vice-versa) and may be an important source for the coupling of different structures.

To quantify our interpretation, we analyzed[12] the Kr yrast bands in a two-band mixing model. For $I \gtrsim 6$ h, where $\mathcal{J}$ is nearly linear, we considered the yrast levels to be purely deformed. The up bends above 8^+-10^+ (Fig. 3) are related to the alignment of $g_{9/2}$ proton pair. The position of the unperturbed deformed levels were determined by extrapolating the linear part of $\mathcal{J}(\omega^2)$ down to $\omega = 0$. This corresponds to a Harris or variable moment of inertia (VMI) parametrization of the deformed g bands $\hbar\omega = E_\gamma/2$; $\mathcal{J}\hbar = (I + 1/2)/\omega$; $\mathcal{J} = \mathcal{J}_0 + \omega^2 \mathcal{J}_1$ (see Bengtsson and Frauendorf[21]). The deviation between the extrapolated and measured levels (Fig. 3) is much larger in ^{76}Kr than in ^{78}Kr in accordance with the higher 0_2^+ energy in ^{78}Kr.

The shifts $\delta E = E(0_1^+)^0 - E(0_1^+)$ were extracted[12] where $E(0_1^+)^0$ is the extrapolated and $E(0_1^+)$ the measured ground-state energy. The difference between the unperturbed levels $\Delta E_0 = E(0_2^+)^0 - E(0_1^+)^0$ is equal to $E(0_2^+) - E(0_1^+) - 2\delta E$. The interaction[21], V, is equal to $1/2(\Delta E^2 - \Delta E_0^2)^{\frac{1}{2}}$, where $\Delta E = E(0_2^+) - E(0_1^+)$. The close values of V for ^{76}Kr and ^{78}Kr indicate that the smaller energy perturbations in ^{78}Kr are related to the higher energy of the (unperturbed) 0_2^+ state. Assuming that the interaction V in ^{74}Kr has a value similar to tht of ^{76}Kr, 0.33 MeV, we predict the 0_2^+ energy at 0.68 MeV in ^{74}Kr. The extracted unperturbed 2 → 0 energies in the deformed ground bands are 200 and 237 keV in ^{74,76}Kr, respectively. By scaling the unperturbed 2 → 0 energy by $A^{5/3}$, one may compare the deformation of ^{74}Kr to that of ^{240}Pu, which has one of the largest ground state deformations known. The 200-keV transition in ^{74}Kr would correspond to 28 keV in ^{240}Pu compared with its actual value of 43 keV. This is an unusually large ground-state deformation, slightly larger than the "super deformation" recently reported for ^{100}Sr with its scaled 30-keV 2_1^+ energy.[22] Such interaction of two 0^+ levels and their splitting may have masked strong ground-state deformation in other regions. Indeed, recently analysis indicate that this is clearly happening in the light platinum isotopes, too.

These data extend our understanding of the coexistence of different nuclear shapes first proposed in ^{72}Se. However, in ^{74,76}Kr the role of the near-spherical and deformed minima are reversed with the ground states well deformed and the excited 0_2^+ states associated with the near-spherical minima. The present data give evidence for large ground-state deformation in these light Kr isotopes.[12] This interpretation for the N = 38 and 40 Kr nuclei supports the expectation that at these neutron numbers as the proton number approaches the middle between Z = 28 and 50 closed shells, the protons can drive a nucleus with a pair of $g_{9/2}$ neutrons toward deformation. Recently Möller and Nix[13] calculated nucleon masses and ground state shapes for 4023 nuclei from ^{16}O to ^{279}112 with a Yukawa-Plus-Exponential Macroscopic

Model and a folded Yukawa single-particle potential. Their calculations[13] predict that nuclei with both N and Z at or near 38 should be among the most strongly deformed ones in nature with $\beta \simeq 0.4$.

Further support for the importance of deformation when N(Z) = 38 and Z(N) is well removed from a closed shell come from the lightest and heaviest Sr nuclei far from stability.[14,15,22] In Fig. 5 is shown the yrast levels of $^{80}_{38}Sr_{42}$ obtained from an analysis of 10% of the data from a recent in-beam study (ref. 14) at the new Holifield tandem. The reaction $^{32}S + ^{51}V$ was carried out with terminal voltages of 16-19 MeV. One sees that the 2-0 energy continues to drop and is the lowest known for any nucleus in the A = 80 region. This sharp drop in the 2-0 energies indicates that the Sr nuclei are moving toward large ground-state deformation in $^{76}_{38}Sr_{38}$ as N decreases toward N = 38 with the 38 protons strongly supporting deformation, too. We also have bombarded ^{58}Ni target with ^{24}Mg at the Holifield tandem to search for ^{78}Sr (ref. 15). We employed the neutron multiplicity technique to pick out the (2p,2n) reaction to ^{78}Sr from other four particle reaction channels which were identified (4p),(3p,n), (α,3p). The data are still under analysis but an approximately 278-504 keV cascade is observed which would fit as the 0-2-4 cascade. Further analysis and other cross experiments are needed to establish these γ-rays as belonging to ^{78}Kr.

The sudden onset of strong deformation was reported in ^{98}Sr (refs. 23,24) and "super" deformation subsequently in ^{100}Sr (ref. 24). The origin of this deformation was related[22] to a gap at N = 60 in the neutron Nilsson levels. The potential energy surfaces calculated for the Sr nuclei show minimum at large deformation for both prolate and oblate shapes for N = 60 and 62 which are not present for N < 60. The experimental data favor prolate deforma-

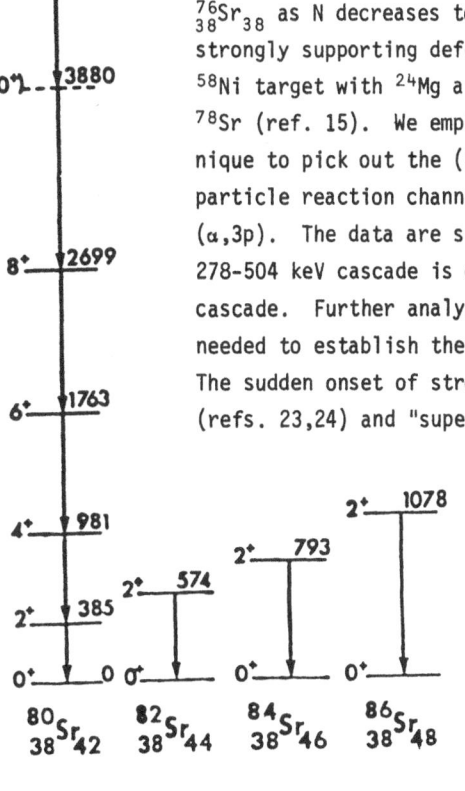

Figure 5. Levels in ^{80}Sr seen in beam.[14]

tion. This N = 60 gap gets strongly reinforced by the gap at Z = 38 at large prolate deformation. The reinforcement of these two gaps at similar large prolate deformation is undoubtedly the cause of the sudden large deformation in $^{98,100}Sr$ just as when both N and Z are near 38 they reinforce each other to give unusually large deformation in the ^{74,76}Kr ground states. The importance of the gap at Z = 38 in deformation in this region is seen in that as one goes away from Z = 38 the deformation begins to decrease for the N = 60 and 62 nuclei, eg. the 2_1-0_1 energies are 129.2, 151.9, and 192.2 keV for N = 62 $^{100}_{38}$Sr, $^{102}_{40}$Zr, and $^{104}_{42}$Mo (refs. 22, 25,26).

The question of shape coexistence in the heavy Zr isotopes was considered earlier[27,28] to account for the spherical ground states and low energy 0_2^+ states in ^{96,98}Zr and ground state deformation in ^{100}Zr (ref. 29). However, we already noted that the work of Flynn et al.[30] indicates that the low energy 0_2^+ state in ^{96}Zr is not a deformed band head. The onset of deformation in ^{100}Zr with N = 60 is sudden, E_{4^+}/E_{2^+} goes from 1.6, 2.65 and 3.14 in ^{98,100,102}Zr, respectively. The lowest energies known for the 0_2^+ state in even-even nuclei are at the remarkably low energies of 215.5 and 331.3 keV in ^{98}Sr and ^{100}Zr, respectively.[24,31] These two unusual 0_2^+ states and the deformed ground states of ^{98}Sr and ^{100}Zr have been discussed in terms of shape coexistence.[24,31] In ^{98}Sr a high energy 2^+ state, characteristic of the vibrational 2^+ state of a near-spherical nucleus, is seen on top of the 0_2^+ state to support shape coexistence. At Helsingnør, Ragnarsson[32] noted that their calculations[33] also indicate that in ^{100}Zr with N = 60 there are two coexisting close lying 0^+ states and that for ^{98,102}Zr with N = 58 and 62 one expects coexisting 0^+ states with spacings of 1.0-1.5 MeV. However, farther away from N = 60 the energy differences rise very rapidly and it becomes difficult to observe any coexistence.[33] Their work supports the interpretation in ^{96}Zr that the low energy 0_2^+ state is probably not a deformed state. A similar sharp crossing of spherical and deformed shapes with shape coexistence observed in a narrow range of neutron numbers around N = 60 in the Sr nuclei also is indicated.

III. SHAPE COEXISTENCE IN OR NEAR CLOSED SHELL NUCLEI

The $^{184,188}_{80}$Hg nuclei already are considered classic textbook examples of shape coexistence.[2] In this section we will concentrate on further examples of shape coexistence in nuclei with Z at or near a closed shell. First, let us look at the light platinum isotopes where behavior similar to that in ^{74,76}Kr is indicated; namely, the interaction of two close lying 0^+ states to push down the 0_1^+ and enlarge the 2_1^+-0_1^+ energies to mask large ground state deformation. Shape coexistence in the light platinum nuclei, $^{180-186}$Pt, was suggested by the calculations of Frauendorf et al.[11] which gave potential energy surfaces with oblate and prolate minima at similar deformations. They suggested[11] the low lying 0_2^+ states in $^{182-186}$Pt and the 2^+, 4^+ states built on them form a sequence similar to the shape-isomeric band in the adjacent Hg isotopes.

Hagberg et al.[34] reported a large odd-even mass staggering of the moments of inertia of these platinum nuclei. The odd mass nuclei exhibit spectra and $\mathscr{J}$ characteristic of well-deformed shapes, eg., $\mathscr{J}$ is similar to that of the deformed Yb nuclei (Fig. 6), but the even-even Pt neighbors show much smaller $\mathscr{J}$'s based on their 2-0 energies. Wood[35], however, has pointed out that if one looks at the higher spin states in the yrast cascades of the Pt isotopes that by the 8^+ states they are characteristic of well-deformed nuclei. In Fig. 7 is shown the rotational parameter $\hbar^2/2\mathscr{J}$ for each spin state of the yrast band for the N = 100 isotones, from ^{176}Yb and ^{178}Hf with well-deformed ground states to ^{186}Hg with a near-spherical ground state. Note

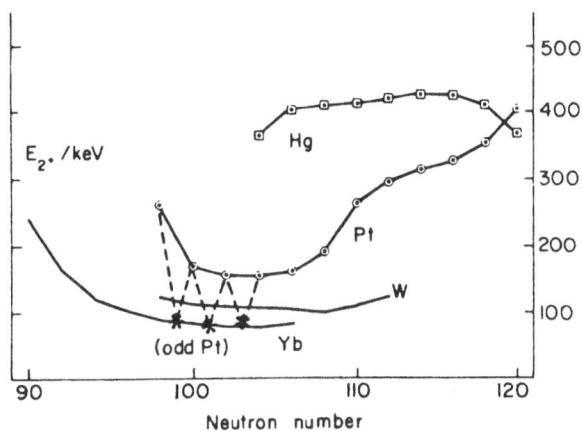

Figure 6. The 2_1^+-0_1^+ energies which are an inverse measure of the moment of inertia are given for the Yb, W, Pt and Hg nuclei.

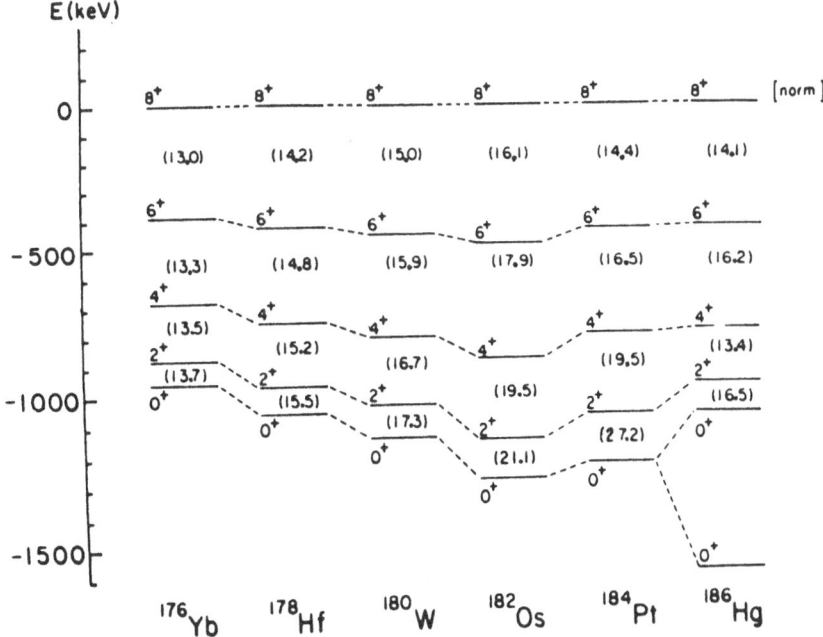

Figure 7. The systematics of the yrast states in the N = 106 isotones.

this parameter is essentially constant up the band for the well-deformed ^{176}Yb and ^{178}Hf as expected. In ^{186}Hg where the ground band is crossed by a well-deformed band at low spin^2, the members of the excited deformed band again have a nearly constant $\mathscr{J}$. In ^{184}Pt there is a factor of 2 change in $\mathscr{J}$ extracted from the 2-0 and 8-6 transitions with the 8-6 value having a rotational value similar to the well-deformed nuclei.

This small $\mathscr{J}$ from the 2-0 transition which makes the ground state not look well deformed is interpreted[35] as arising from an interaction of the 0_1^+ deformed ground state with the low lying 0_2^+ more near-spherical state with the 0_1^+ state being pushed down as found in ^{74,76}Kr. In the Pt isotopes, and in the excited deformed band in $^{184-188}$Hg Wood[35] suggests it is the $(\pi h_{9/2})^2$ configuration from above the Z = 82 closed shell which plays a major role in the deformation. This is again analogous to the proton and neutron $g_{9/2}$ configurations which are strongly down sloping with deformation to form the gap at N and Z = 38 in the Kr and Sr nuclei. Already the $h_{9/2}$ orbital in the odd gold isotopes was known to drop very low to near the ground state as N decreases to ^{185}Au (ref. 36). Lecture two[7] gives additional Au results.

A similar dropping of a high spin orbital to give rise to deformation and shape coexistence is reported in the proton closed shell, Z = 50 Sn nuclei. In 1974 we suggested[28] from our lifetime measurements of the low lying 0_2^+ state in ^{116}Sn (ref. 37) that this may be the head of a collective band with deformation quite different from the ground state. Later lifetime measurements[38] have shown that the 0_2^+ state lifetime is shorter, 45 ± 10 ps, than our value which had a relatively large error 150 ± 50 ps (ref. 37) to show even larger collective structure for the band head. (The suggestion[38] that we had measured the lifetime of the 0_3^+ state is not correct since that state is not populated in our work. However, our lifetime is a mixture of the 0_2^+ and 2^+ state built on it.) Subsequently, Bron et al.[39] discovered in ^{116}Sn a well-developed rotational band characteristic of a well-deformed shape built on the 0_2^+ state to establish shape coexistence in ^{116}Sn. Similar excited bands were found in ^{112,114} and ^{118}Sn. Again as in the Kr, Pt and Hg nuclei, the structure of these Sn deformed states is related to the filling of a high j orbital from above a shell or subshell closure. For Sn the structure of the deformed states is assigned $\{1/2[431]^2 \, 9/2^+[404]^2\}$ associated with promoting a pair of $g_{9/2}$ protons to the $g_{7/2}$ orbital from above the Z = 50 closed shell. This follows the similar interpretation for the classic double closed shell spherical ^{16}O nucleus where evidence for shape coexistence was first presented in terms of deformed excited states being 2-particle, 2-hole states that coexisted with the spherical ground state.[40]

We now have several theoretical approaches to shape coexistence in the nuclei discussed that include second minimum in the potential energy surfaces[3,8,11,41], a dynamic deformation method[42] and recently the Interacting Boson Model with configuration mixing[43]. Time does not permit discussion of other examples of shape coexistence that are being reported in different mass regions. Suffice it to say that shape coexistence is a much more general feature of the nuclear landscape than anyone supposed. Such coexistence offers an important bridge between models which describe near-spherical nuclei and ones that describe well-deformed nuclei. The different shapes seen in one nucleus can be understood in terms of different configurations and their mixings to give us a deeper insight into the interplay of single particle and collective degrees of freedom in the structure of nuclei.

Another exciting area of nuclear structure is what happens to a nucleus at higher and higher angular momentum. At very high angular momentum theoretical considerations suggest that almost all nuclei are deformed. The competition between single particle aligned configurations and collective rotations came into play as more and more pairs of particles form aligned configurations. In prolate nuclei, this can lead to a change to an oblate shape at high angular momentum; for example, above 40-50 ℏ in rare earth nuclei (see ref. 6 for further details of what can happen at very high angular momentum).

Finally, let us note the work of Bengtsson et al.[5] who also looked at nuclei as a function of angular momentum. In their work on the interplay of rotation, deformation and pairing they discuss another type shape coexistence which enters as a function of angular momentum in the traditionally considered near-spherical, vibrational rare earth nuclei with N ≤ 88. This is illustrated in Fig. 8. The well-known sharp transition from near-spherical to deformed shapes between N = 88 and 90 in the rare earth region is seen there for the ground state. What they point out is that as the angular momentum increases, the critical neutron number N at which the change from spherical type vibrational nuclei to deformed nuclei decreases and somewhere between

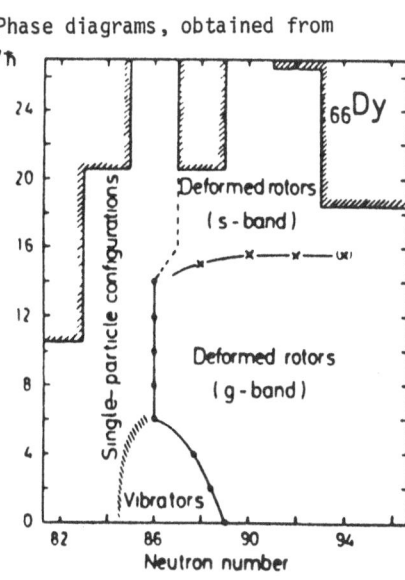
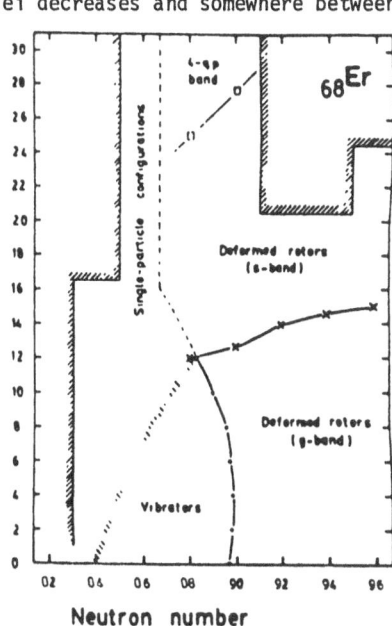

Figure 8. Phase diagrams, obtained from experimental quantities. Above the shadowed line, no experimental data are available.

spins of 6^+ and 12^+ in Gd, Dy and Er, these nuclei become deformed for N = 86. They are deformed at even lower spin at N = 88. So that in this classical spherical-to-deformed transition region, one finds another type of shape coexistence. Their results[6] simply underscore again that shape coexistence at low and intermediate spins indeed plays an important role in many regions of the periodic table and is giving us new insights into nuclear structure.

This work was carried out in part under contract with the U.S. Department of Energy, Contract No. DE-AS05-76ER05034.

REFERENCES

1. J.H. Hamilton et al., Phys. Rev. Letts. 35, 562 (1975); J.D. Cole et al., Phys. Letts. 37, 1185 (1976).
2. J.H. Hamilton, Proc. Int. Conf. on Selected Topics in Nuclear Structure, ed. V.G. Soloviev et al. (Dubna, USSR, 1976), Vol. II, p. 303; J.H. Hamilton, Nukleonika 24, 561 (1979).
3. J.H. Hamilton et al., Phys. Rev. Lett. 32, 239 (1974); J.H. Hamilton et al., Phys. Rev. Letts. 36, 340 (1976).
4. J.H. Hamilton et al., Nuclear Interactions, ed. B.A. Robson (Springer Verlag, Berlin, 1979); Phys. Energiae Fortis et Phys. Nuclearis 3, 355 (1979)(in Chinese).
5. R. Bengtsson et al., IV Int. Conf. on Nuclei Far From Stability, CERN 81-09 (1981), p. 509.
6. A. Bohr and B.R. Mottelson, Nuclear Structure (Benjamin, NY, 1975), Vol. II; and Nuclei At Very High Spin, ed. G. Leander et al., Physica Scripta 24 (1981).
7. J.H. Hamilton, elsewhere in these Proceedings.
8. M. Vergnes, Structure of Medium-Heavy Nuclei, ed. G.S. Anagnosta, Inst. Phys. Conf. Ser. 49, 25 (1980).
9. R. Lecomte et al., Phys. Rev. (in press).
10. K. Kumar, J. Phys. G. Nucl. Phys. 4, 849 (1978).
11. S. Frauendorf et al., Future Directions in Studies of Nuclei Far From Stability, ed. J.H. Hamilton et al. (North-Holland, Amsterdam, 1980), p. 133 (other references are given in this review); S. Frauendorf and V. V. Pashkevich, Phys. Lett. 55B, 365 (1975).
12. R.B. Piercey et al., Phys. Rev. Lett. 47, 1514 (1981); R.B. Piercey et al., Phys. Rev. 25, 1914 (1982).
13. P. Möller and J. R. Nix, Nucl. Chemistry Symp. on Nuclei Far From Stability (Las Vegas, March 1982), Am. Chem. Soc. 183 National Meeting, Abstracts Nucl. 16.
14. M. Barclay et al., private communication, Fall (1981).
15. M. Barclay et al., private communication Vanderbilt Univ., Spring (1982).
16. N. Sakamoto et al., Phys. Lett. 83B, 39 (1979).
17. R.L. Robinson et al., Phys. Rev. 21, 603 (1980).
18. Z. Z. Zhao et al., Phys. Energiae Fortis et Phys. Nuclearis 6, 255 (1982) (in Chinese).
19. E. Nolte et al., Z. Phys. Z268, 267 (1974).
20. R.B. Piercey et al., private communication, Vanderbilt University.
21. R. Bengtsson and S. Frauendorf, Nucl. Phys. A314, 27 (1979); ibid. A327, 139 (1979).
22. R.E. Azuma et al., Phys. Lett. 86B, 5 (1979).
23. H. Wollnik et al., Nucl. Phys. A291, 355 (1977).
24. F. Schussler et al., Nucl. Phys. A339, 415 (1980).
25. J.C. Hill et al., IV Int. Conf. on Nuclei Far From Stability, CERN 81-09 (1981), p. 443.
26. K. Sistemich et al., Z. Physik A289, 225 (1979).
27. R.K. Sheline et al., Phys. Lett. 41B, 115 (1972).
28. A.V. Ramayya and J.H. Hamilton, Gamma Ray Transition Probabilities, ed. S.C. Pancholi et al., (Univ. of Delhi Press, India, 1974), p. 125.

29. E. Creifetz et al., Phys. Rev. Lett. 25, 38 (1970).
30. E.R. Flynn et al., Nucl. Phys. A218, 285 (1974).
31. T. A. Kahn et al., Z. Physik A283, 105 (1977).
32. I. Ragnarsson, IV Int. Conf. on Nuclei Far From Stability, CERN 81-09 (1981), p. 434.
33. R. Bengtsson et al., IV Int. Conf. on Nuclei Far From Stability, CERN 81-09 (1981), p. 509.
34. E. Hagberg et al., Phys. Lett. 78B, 44 (1978); and Nucl. Phys. A318, 29 (1979).
35. J.L. Wood, IV Int. Conf. on Nuclei Far From Stability, CERN 81-09 (1981), p. 612.
36. E.F. Zganjar, Future Directions in Studies of Nuclei Far From Stability, ed. J.H. Hamilton et al., (North-Holland, Amsterdam, 1980), p. 49.
37. E. de Lima, M.S. Thesis (Vanderbilt University, Nashville, TN/USA, 1977).
38. R. Julin, Ph.D. Thesis, University of Jyväskyla (Jyväskyla, Finland, 1979).
39. J. Bron et al., Nucl. Phys. A318, 335 (1979).
40. H. Morinaga, Phys. Rev. 101, 254 (1956).
41. G. Gneuss and W. Greiner, Nucl. Phys. A171, 449 (1971); and P.O. Hess, J.A. Maruhn and W. Greiner, Future Directions in Studies of Nuclei Far From Stability, ed. J.H. Hamilton et al., (North-Holland, Amsterdam, 1980), p. 151.
42. K. Kumar, Future Directions in Studies of Nuclei Far From Stability, ed. J.H. Hamilton et al., (North-Holland, Amsterdam, 1980), p. 265.

Note in Proof: Lister et al., Phys. Rev. Lett. 49, 308 (1982), established the 278 keV 2^+ level of ^{78}Sr.

NEW DIRECTIONS IN STUDIES OF NUCLEI FAR FROM STABILITY WITH HEAVY IONS

J. H. Hamilton
Physics Department, Vanderbilt University
Nashville, TN 37235/USA

INTRODUCTION

In just over a decade since large scale efforts began in this field, studies of nuclei far from stability have become one of the major frontiers in nuclear research. Many new phenomena have been observed as unexplored regions have been probed along lines of constant N and constant Z over the last decade. Far off stability more exotic particle decay modes not seen near the valley of beta stability have been found with others being actively sought. Many fascinating new results have come as we are now able to trace systematically over large changes in neutron number, 20 to even 30 neutrons, the properties of both ground and excited collective and single particle states of a given element.

For example, with recent studies of ^{78,80}Sr the structure of the Sr nuclei now are traced from one new region of very strong ground state deformation with N = 38 (ref. 1) across the spherical N = 50 closed shell region to a second region of very strong deformation at N $\geq$ 60 (as also discussed in ref. 1). Neither of these two new regions of strong deformation were expected. Their suddenness and magnitude of the deformation are related to a reinforcing of deformation by both protons and neutrons! Other important discoveries have been made through such systematic studies; for examples, shape coexistence with bands of overlapping energy levels built on quite different deformations in the light mercury isotopes[2] and the sudden dropping of single particle states across the energy gap at the Z = 82 shell closure[3]. These broad systematics studies are offering challenging new tests for nuclear models.

Many techniques are being employed to study nuclei far from stability. Here we concentrate on some highlights in the use of heavy ions to probe nuclei far from stability. More details can be found in a recent, extensive review of this field.[4] In addition to probing nuclei far from stability along lines of count Z or N, nuclei at very high angular momenta are another important area of research far from the stable ground states.[5] In this paper we will look at some new directions being opened up by heavy ions, some of the new facilities and then some interesting new results. In the two following papers,[1,5] still other new results are considered. Only a limited number of the many different areas of research can be covered. One exciting area of proton decay will be covered in a seminar presented at this conference.[6]

HEAVY ION REACTIONS

The field of heavy ion reactions generally has been considered to exclude projectiles with A $\leq$ 4, however, recent ^{3}He and ^{4}He induced reactions that have included exotic heavier-ion transfers such as (^{4}He, ^{11}C) may be included. Most studies of nuclei far from stability have used fusion-evaporation reactions at 3-10 MeV/nucleon with HI deep-inelastic transfer collisions becoming more important. Future directions will include wider ranges of heavy ions and with much higher energies of 100 to 1,000 MeV/u at various new heavy ion laboratories.

Already recently a new process, projectile fragmentation has been shown to be a most promising new technique to make nuclei far off stability, see review of Symons.[7] In the fragmentation of a 212 MeV/u beam of ^{48}Ca by C and Be targets (900 mg/cm^2), 14 new isotopes were identified, ^{22}N, ^{26}F, ^{33,34}Mg, ^{35,36,37}Al, ^{38,39}Si, ^{41,42}P, ^{43,44}S and ^{44,45}Cl. This technique has some major advantages that include a resultant gain in efficiency can be as much as 10^6 over a typical low energy experiment. Experiments to check the predictions of the theoretical mass formulas close to the limit of stability become feasible with this technique. Here again one has only begun to explore the possibilities. Already existing data indicate that going to still heavier projectiles at similar E/A will give higher yields of isotopes still further from stability.

Recently exotic two step processes have been proposed.[8] Based on available beams and cross section data, several illustrative examples have been proposed.[8] An example of the production of a secondary radioactive target is

$$\alpha + {}^{206}Pb \rightarrow {}^{210-x}Po + xn \quad (x = 5 - 11)$$

and a second reaction with the radioactive target

$$\alpha + {}^{199-205}Po \rightarrow {}^{199}Rn + yn \quad (y = 4 - 10)$$

An example of the use of a secondary radioactive beam, the evaporation residues can be energetic enough to induce a second fusion reaction on a light target. For example, ^{141}Pr + ^{20}Ne $\rightarrow$ ^{153}Tm + 8n; ^{153}Tm + ^{24}Mg $\rightarrow$ ^{177}Tl*. The de-excitation products of ^{177}Tl* might be $^{173-174}$Tl, $^{173-174}$Hg and $^{171-173}$Au. Note ^{184}Tl is the lightest mass thallium decay which has been studied.[2] With transfer or fragmentation reactions at 10-200 MeV/u, one has other exotic possibilities to produce both proton rich and neutron rich secondary beams like ^{36}Ca or ^{16}O.

Massive transfer, also called partial fusion, reactions in which fast protons, deutrons and tritons are observed is also a promising new technique to look at high angular momentum states.[9,10] This will be discussed in more detail later.

We have not at all exploited the different possibilities to produce nuclei far to very far off stability with heavy ion fusion reactions. The primary reason for this is that the major efforts in this field have been at accelerators which were

limited in high intensity ion beams to A $\lesssim$ 20. The first universal heavy ion accelerator, UNILAC at GSI, has already shown the value of going to heavier projectiles
like [58]Ni (ref. 11). However, there too, only the surface has been scratched. Examples
of virtually unexplored reactions include 5-10 MeV/u beams like [28]Si, [32]S, [40]Ca on
targets like [40]Ca, [58]Ni, [144]Sm and [156]Dy. These and other beams are now available
at UNISOR from the new Holifield Heavy Ion Accelerator.

SOME NEW DIRECTIONS IN INSTRUMENTATION AND RESEARCH

Most of the efforts to study nuclei far from stability have involved magnetic isotope separators and in-beam gamma-ray spectroscopy. However, velocity filter devices like SHIP at GSI where the hold-up times are down to 10^{-6}s are opening up
many new studies. For example, Hofmann et al.[12] have bombarded targets of [107,109]Ag
nat,[108,110]Pd and [103]Rh with [58]Ni ions and with the aid of the velocity separator
SHIP identified eleven new isotopes including the first alpha emitting Ta isotopes
and the lightest known Re, W and Hf isotpes, [157-161]Ta, [161-164]Re, [160]W and [156]Hf.
A comparison of their alpha energies with the predictions of different mass formulas shows the agreement is best for the droplet model.[13] So while the formula
of Liran and Zeldes[14] gives the best overall fit, their formula does not correctly predict the alpha energies of the known nuclei farthese from stability in any
of the four cases. Such a velocity filter designed by the MIT group is nearly rea-
dy at Holifield. It is designed so it can be integrated into a total energy-mass
spectrograph at a later time. More sophisticated recoil-mass spectrometers are
planned at Daresburg and Michigan State (see review, ref. 1).

Another important area being opened up is measurements of magnetic dipole moments,
μ, of ground states and excited states. These moments can give unique signature of
the structure of a state. For example, in an odd-odd nucleus the structure may not
be apparent from the spin but the coupling of the odd proton and odd neutron magnetic moment may lead to a unique interpretation. In excited states, for example the
proton or neutron character of a rotation aligned state state may be established by
measuring the sign of μ.

With He refrigerators one can measure the angular distribution of γ rays from
oriented nuclei and extract information about the spins of the states. Unfortunately, nuclear orientation angular distribution (NO-AD) experiments are of no help
in determining both the moment μ and the spin I_p, although given I_p it is usually
possible to deduce μ. The orientation parameters B_2 which describes the state I_p
depends on the product μH (H is the magnetic field), but μH can only be deduced from
B_2 if I_p is known, since the dependence of B_2 on μH varies with I_p.

The solution to this dilemma is to perform nuclear magnetic resonance on the oriented
nuclei (NO-NMR). This procedure involves measuring the gamma-ray anisotropy as a

function of the NMR frequency. The measurement thus determines a frequency, rather than an anisotropy, and can therefore be done with great precision since all gamma-rays are identically affected. Moreover, NaI detectors may be used so that shorter counting intervals are possible. What is significant here is that the NO NMR experiment determines the product gH (g is the nuclear g-factor, μ/I_p), while the NO-AD experiment determines μH as a function of spin. Combining these two measurements gives the desired spin I_p. This is a rare and exciting technique because it gives a direct measurement of I_p. In most nuclear physics experiments spins are indirectly deduced rather than directly measured. Thus, this combination of measurements is a powerful technique, especially as one moves farther from stability where the structures are totally unknown.

Another advantage of the NO-NMR technique is that magnetic moments may be determined with great precision. Typically, NO-AD measurements may yield μ with an uncertainty of order 0.1 nuclear magneton, while NO-NMR can do at least an order of magnitude better. This increased precision can have important consequences for nuclear structure calculations; although a rough measurement of μ may be sufficient for a determination of the primary configuration of the state I_p, the important details of configuration mixing (eg., different K quantum numbers) can only be revealed through precise values of μ.

Such a He refrigerator is being placed on line at the Daresbury accelerator[15] and one has been proposed for the UNISOR facility. This is a powerful system for obtaining much important information including besides spins and moments, the multipole admixtures of transitions. Such admixtures, as seen in a later section, can be crucial in unraveling crucial structure questions.

As an example for excited states, from in-beam studies rotation aligned states have been seen in nuclei from A = 70 to 248. In ^{68}Ge triple forking was observed at 8^+ (ref. 16). In that work the two lowest 8^+ levels were interpreted as rotation aligned states built on two quasiproton and two quasineutron $(g_{9/2})^2$ configurations-the first case for both proton and neutron configuration built on the same orbit in one nucleus. Two quasiparticle-plus-rotor calculations supported that interpretation. More recently shell model calculations of Weeks et al.[17] suggest the two lowest 8^+ states arise from representation mixing and different alignments of a pair of $g_{9/2}$ neutrons only.

The magnetic moments of proton and neutron configurations have opposite signs. Thus, even a crude measurement of μ would select which explanation is correct. The transient field method can be used with heavy ion reactions to measure the magnetic moments of these two 8^+ states. The applicability of the method depends on the high recoil velocity of the residual nuclei imparted by the heavy ions and small angular scattering. We have carried out such a transient-field measurement on ^{68}Ge. A sandwich target 230 μg/cm^2 ^{40}Ca, 900 μg/cm^2 ^{nat}Fe and 50 μm Au backing was bombarded with 100 MeV ^{32}S ions. The reaction ^{32}S + ^{40}C $\rightarrow$ ^{68}Ge + 4p. The external magnetic field was 340 gauss and the transient field (extrapolated) about 17 M gauss. Two Ge(Li) detectors were used. This experiment was carried out at the Florida State Tandem in May and the data analysis is in progress. One clearly sees differences and it is only a matter of whether sufficient statistics were recorded. Some beam difficulties limited the energy and reduced the relative cross-section. Nevertheless, these data show these are feasible experiments and open up the possibility to determine the structure of many of these new bands seen in this region including the negative parity bands.

GAMMA-RAY-AND-ELECTRON SPECTROSCOPY -- ON-LINE AND IN-BEAM

Prior to the early 1970's detailed level structures were known for very few nuclei far from stability. Through in-beam, gamma-ray spectroscopy studies and measurements of radioactive decays with on-line isotope separators there has been an explosion in our knowledge of the level structures of nuclei far off stability. These two techniques can give very complementary results and the application of both techniques to a given nucleus has been important in many studies. It is not the quantity of data that have been generated in just the last few years alone, impressive as these are, but the exciting new vistas and understandings that they have exposed and brought about that have given these studies major significance. Indeed one of the challenges of the field is to extract from the literally hundreds of gamma rays which can occur in a single decay the nuggets which contain the significant new physics. Studies of nuclei as functions of N, Z and angular momenta I over wide ranges of those parameters are providing many fascinating new insights into the structure of nuclei. One now sees the importance and interplay of shell corrections, collective effects and single particle motions in individual nuclei throughout the periodic table. The shape of a nucleus no longer should be considered fixed but seen as an important measure of configuration mixing, with various types of shape coexistence now seen in many different nuclear mass regions.[1] Some theories are now challening experiments with detailed theoretical predictions of the level structures of unobserved nuclei far from stability. New heavy-ion facilities and new and improved techniques are opening up opportunities to scale numerous new mountain ranges far from the valley of beta stability in the next decade.

Two examples will illustrate the exciting results coming out of both on-line studies of radioactive decays and in-beam spectroscopy. Shape coexistence in the even-even Pt and Hg nuclei far off stability were described in the first paper.[1] Here, let us look at the odd A nuclei. Somewhat different but analogous types of shape coexistence are seen in the odd-A Au and Tl nuclei. The $h_{11/2}$ hole states built on weakly oblate Hg cores are relatively stable between ^{195}Au to ^{185}Au while the $h_{9/2}$ particle states built on the Pt cores reflect the well-known oblate-prolate shape change in the Pt ground states. Thus in ^{189}Au the $h_{11/2}$ triaxial-oblate and $h_{9/2}$ triaxial-prolate bands are close in energy to give another type of shape coexistence[18-20]. In the light Tl nuclei one finds $h_{9/2}$ proton particle states coupled to the even-even Hg cores which are slightly oblate and $h_{11/2}$ proton hole states coupled to nearly spherical Pb cores. The $h_{9/2}$ band shows strong coupling expected for the oblate Hg cores. While the $h_{11/2}$ states are quite low in the Au nuclei, they are considerably higher in Tl and were not known prior to recent UNISOR work.[21] The structure built on this $h_{11/2}$ excitation is quite different from that of the $h_{9/2}$ excitation and, as expected in the core-particle picture, indicates weak coupling to the nearly spherical Pb cores.

Recent studies of ^{187}Au by Zganjar et al.[22] provide a fascinating extension of our knowledge of shape coexistence in this region. As already noted, the $h_{9/2}$ particle states in the light Au isotopes built on the Pt cores are well established and similarly $h_{9/2}$ states in the odd Tl isotopes. The question is, can one find $h_{9/2}$ states in the odd Au and Tl isotopes coupled to excited 0_2^+ states in the Pt and Hg cores with quite different deformation from the ground states of the cores. The observation of such states would provide an exceptional opportunity to study particle-core coupling models since one would have the same particle coupled to different shaped cores in the same nucleus, eg., presumably slightly oblate ground states and strongly prolate excited 0_2^+ states in $^{184-188}$Hg and the reverse in $^{178-188}$Pt. In the even-even cores, very strong E0 electron decays are observed in the 0_2-0_1, 2_2-2_1, etc. transitions between the excited and ground bands because of their different shapes. Such E0 decays are considered signatures for transitions between two $h_{9/2}$ bands built on the 0^+ states with different shapes in the same core.

In the very complex spectra of these odd A nuclei (several hundred γ rays per decay), it can be difficult to extract the exciting physics from the mass of data. Zganjar et al.[22] developed very careful analysis techniques to unravel the complex spectra in the decays of ^{187}Hg and ^{187}Hg^m. From very high statistical, quality data (2 x 10^7 γ-γ-t events) they were able to advance gates one channel at a time over complex lines and unravel even very close lying cascade doublets like 298.8-271.7 and 299.6-271.2 keV. Studies of the decays of ^{187m,g}Hg at UNISOR resulted in a level scheme for ^{187}Au with 19 new levels below 1 MeV when compared to the most recent previous study.[23] Part of the results of their very careful analysis is shown in Fig. 1. A new, excited band is established with large E0 admixtures that indicate they are $\Delta I=0$ transitions to the earlier identified $9/2^-$ band that arises from the coupling of the $h_{9/2}$ particle and 0_1^+ ground state of ^{186}Pt. The new band, beginning at 443.5 keV, has all the expected properties and is assigned as arising

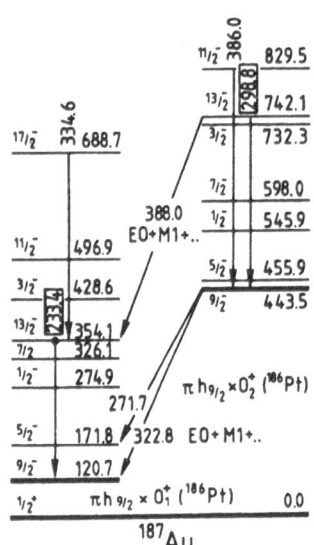

Figure 1. Level scheme for ^{187}Au.

from the coupling of the $h_{9/2}$ state to the excited 0_2^+ state in ^{186}Pt.

The $9/2^-$-$9/2^-$ spacing or the centroid shift between the bands of 320 keV is less than 0_2^--0_1^+ spacing of 472 keV in ^{186}Pt. This is interpreted[22,24] as a blocking effect of the odd $h_{9/2}$ particle when coupled to the well deformed states, which are assumed to be generated by putting a pair of protons in the $h_{9/2}$ orbital. The $(\pi h_{9/2})^2$ configuration then would contribute to the ground states of the Pt nuclei which are well deformed but to the excited 0_2^+ states which are well deformed in the Hg nuclei. It has been shown that indeed in the $h_{9/2}$ system of levels the deformation in ^{187}Au is considerably stronger that in $^{191-195}$Au nuclei.[20] Ekström et al.[20] suggest that both deformation and the neutron-proton interaction effect given by Goodman[25] play a role in the drop in energy of the $h_{9/2}$ orbital in the odd-A Au and Tl isotopes. Thus the $h_{9/2}$ odd proton in ^{187}Au blocks the 0_1^+ deformed state in Pt more than the 0_2^+ state to shift up the 0_1^+ energy and decrease the effective 0_1^+-0_2^+ energy gap in ^{187}Au, while in the Hg core it is the 0_2^+ state which is blocked more to increase the 0_1^+-0_2^+ energy gap in the Tl nuclei.[22,24] This increased gap in the Tl nuclei may explain why a band coupled to the 0_2^+ state of ^{188}Hg has not been found in ^{189}Tl. Thus the odd particle, in addition to testing particle-core coupling models, may also test the structure of the cores themselves.[24]

The very recent studies of the characteristics of massive transfer (also called partial fusion) reactions suggest that this is a very powerful new tool for studies of nuclei far off stability and to high spin[9,10]. These studies are considered in more detail in the last lecture, but let us summarize some results here. Yamada et al.[9] measured the γ-ray multiplicity in coincidence with fast forward, p, d, and t produced by 167-MeV ^{14}N on ^{154}Sm. These emissions are thought to arise in some type of pre-equilibrium process with the subsequent fusion with the target followed by neutron evaporation. Here we will not concern ourselves with the reaction mechanism but with its usefulness in nuclear structures studies. The average γ-multiplicity is measured to be $M_\gamma = 31$ for all yrast transitions up to spin 26 $\hbar$ in ^{158}Er (ref. 9). These data and the observed narrower mass distribution of the residual nuclei compared to that for compound nuclear reactions indicate that this partial fusion reaction occurs at a narrow, high ℓ-window. The deduced average angular momentum transferred was 63 ℏ which is comparable to the critical angular momentum, ℓ_{cr}, predicted for the fusion of ^{13}C and ^{154}Sm.

The same ^{14}N reaction on ^{150}Sm leads to ^{154}Er where discrete states have been observed up to spins of 35-36 ℏ (ref. 26). The γ rays from the known discrete states up to 36-36 ℏ were observed.[10] The relative intensities of the observed γ rays normalized to one for those from the low spin states are compared with similar data from a fusion reaction experiment (Fig. 2). Note that at the highest spins one has a relative enhancement of a factor of 10 in the massive transfer reaction. Thus, one should be able to extend the limits of our knowledge of discrete states to considerably higher spins than the present limits for fusion reactions.

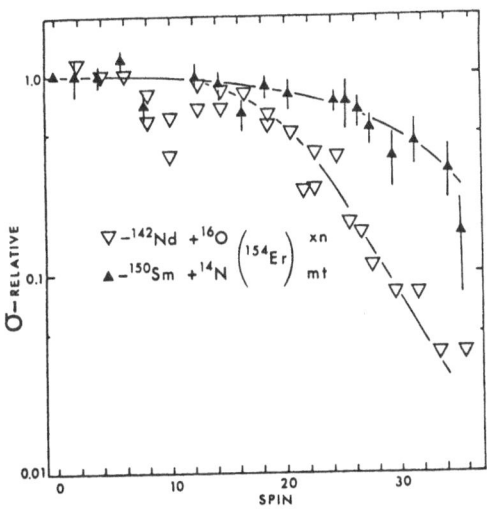

Figure 2. A comparison of γ ray intensities normalized at low spin for a fusion and partial fusion reactions.

Another new development for in-beam studies well off stability is the neutron multiplicity technique.[27] A Köln-Vanderbilt collaboration has shown that a full range of in-beam experiments can be done by this technique. As one goes farther from stability the heavy-ion reaction cross-sections drop rapidly. Charged particle evaporation that brings one closer to stability begin to dominate over neutron evaporation. Thus the neutron evaporation channels are too weak to be studied in any detail, if only γ-rays are recorded. A neutron multiplicity technique has been developed and used at Köln and at the Holifield Heavy Ion Research Facility. The system at Köln is shown in Fig. 3 and described briefly elsewhere.[27]

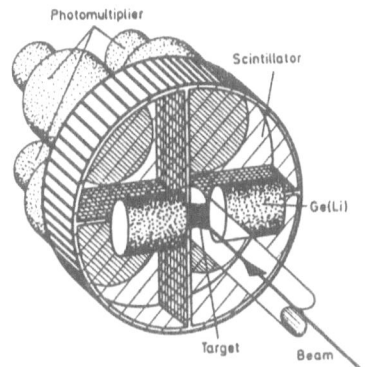

To get a high coincidence rate a 10 cm thick cylindrical neutron scintillation detector of 35 cm diameter filled with NE 213 was used. The detector is divided into four optically tight segments each driven by an independently working ultra

Figure 3. The neutron-multiplicity system at Köln.

fast photomultiplier. Neutron detection in the presence of γ-rays makes an n-γ discrimination inevitable as neutron detectors are usually sensitive to all kinds of radiation. The average neutron efficiency of the system for neutrons evaporated from a compound nucleus is estimated from several measurements to be greater than 0.6. A full range of experiments including n-n-γ, n-γ-γ, n-γ(θ) n-conversion electron and n-γ (plunger lifetime) experiments have been done.

The enhancements of the 2p-n and p-2n channels over the 3p channel in the reaction $^{58}Ni + ^{19}F \rightarrow ^{77}Rb^*$ are illustrated in Fig. 4. The 2-0 and 4-2 transitions in ^{74}Kr

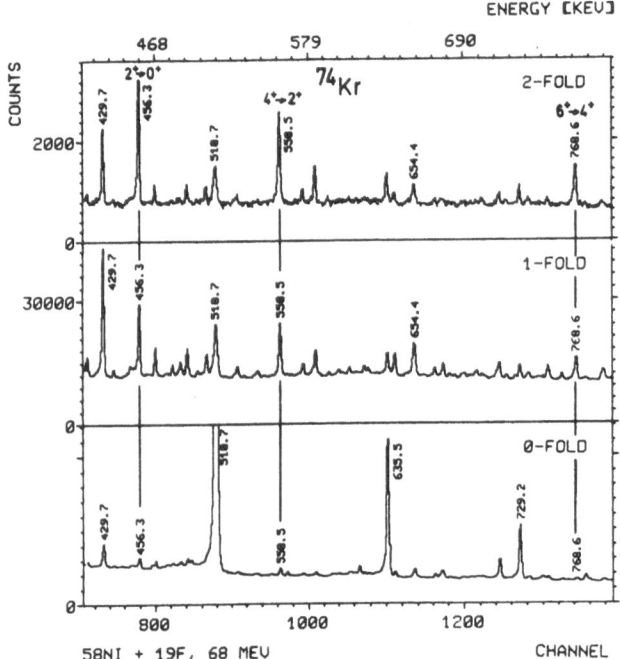

Figure 4. Neutron multiplicity gated singles spectrum of the reaction $^{58}Ni + ^{19}F \rightarrow ^{77}Rb^*$ (ref. 29).

are just barely seen in the singles spectrum. As discussed in the last lecture, in the n-n-γ and n-γ-γ spectrum we have identified transitions up to a tentative 20^+ state in ^{74}Kr. The lifetimes of the 2^+ and 4^+ states in ^{74}Kr were measured to be 28.7 ± 5.7 and 13.2 ± 0.7 ps (ref. 28). The B(E2) values for the decay of the 2^+ and 4^+ states are 78 ± 1.3 and 61.7 ± 3.2 single particle units. These data provide additional evidence for strong deformation in ^{74}Kr.

A final point of interest in this work is that as one looks at the intensity patterns of γ-rays associated with ^{74}Kr, there is a clear break in the pattern above 10^+. Below 10^+ there is a regular rise in intensity of each transition as its level spin decreases. This is the results expected for a fusion-evaporation reaction where there is side feeding into all members of the yrast cascade. However, above 10^+ the side feeding is essentially constant from 12^+ to a tentative 20^+. Our massive transfer work leads us to suggest that these levels are primarily populated by fast neutrons emitted in some pre-equilibrium process which leaves the nucleus in a high spin state with a narrow ℓ-window. We hope to test this by measuring the n-γ spectra as a function of neutron energy.

In another study at Holifield, Maguire from our group in cooperation with ORNL and Georgia State scientists have investigated the $^{16}O + ^{93}Nb$ reaction at 208 MeV (ref. 29). They looked at n,p and α in coincidence with fusion-evaporation residues. Maguire has analyzed the p and α data. The p and α experimental data are in agreement with predictions of statistical evaporation theory until one nears energies that correspond to the beam velocity. Sholders are seen on the high energy side at the beam velocity. This sholder can be explained by assuming quasi-elastic break-up of the projectile with the escape of a beam velocity α (or p) and fusion of the $\sim$156 MeV ^{12}C remnant. The analysis of the neutron data is in progress. The amount of complete fusion is the order of 10-20% here and is larger than that predicted by the Wilcynski model.

The understanding of these pre-equilibrium reactions is an important area. From the standpoint of this paper, there is a wide range of data accumulating that show the importance of these reactions with heavy ions and usefulness of them for studies of high spin states. This will be considered in more detail in the next paper.

This work was carried out in part under contract with the U.S. Department of Energy Contract No. DE-AS05-77ER05034.

References

1. J.H. Hamilton, elsewhere in these Proceedings.
2. J.H. Hamilton, Proc. Int. Conf. on Selected Topics in Nuclear Structure, ed. V. G. Soloviev et al. (Dubna,USSR, 1976), Vol. II, p. 303.
3. E.F. Zganjar, Future Directions in Studies of Nuclei Far From Stability, ed. J. H. Hamilton et al., (North-Holland, Amsterdam, 1980), p. 49.
4. J.H. Hamilton, Heavy Ion Collisions, ed. R. Bock, Vol. III (North Holland Publ. Co., in press.
5. J.H. Hamilton, (lecture 3) elsewhere in these Proceedings.
6. D. Schardt et al., elsewhere in this Vol.
7. J.M. Symons, Proc. IV Int. Conf. on Nuclei Far From Stability, CERN 81-09 (1981), p. 668.
8. Dufour et al., Phys. Rev. C23, 801 (1981).
9. H. Yamada et al., Phys. Rev. C24, 2565 (1981).
10. H. Yamada et al., private communication, Vanderbilt Univ. (1981); also H. Yamada and J.H. Hamilton, Proc. Int. Conf. on Dynamics of Nuclear Collective Motion (Mt. Fuji, Japan, July 1982).
11. D. Schardt et al., IV Int. Conf. on Nuclei Far From Stability, CERN 81-09 (1981), p. 168.
12. S. Hofmann et al., Z. Phys. A291, 53 (1979).
13. W.D. Myers and W.J. Swiatecki, Nucl. Phys. 81, 1 (1966).
14. S. Liran and N. Zeldes, At. and Nucl. Data Tables 17, 431 (1976).
15. N.J. Stone and W.D. Hamilton, Report of Daresbury Facility.
16. A.P. de Lima et al., Phys. Lett. 83B, 43 (1979); ibid, Phys. Rev. C23, 213 (1981).
17. K.J. Weeks et al., Nucl. Phys. A371, 19 (1981).
18. E.F. Zganjar, Future Directions in Studies of Nuclei Far From Stability, eds. J.H. Hamilton et al., (North-Holland, Amsterdam, 1980), p. 49.
19. V. Berg, Future Directions in Studies of Nuclei Far From Stability, eds. J.H. Hamilton et al. (North-Holland, Amsterdam, 1980), p. 353.

20. C. Ekström et al., Nucl. Phys. <u>A348</u>, 25 (1980).
21. L.L. Collins, et al. Phys. Rev., to be published.
22. E.F. Zganjar et al., <u>Proc. of Int. Conf. on Nuclei Far From Stability</u>, CERN 81-09 (1982), p. 630.
23. Bourgeois et al., Nucl. Phys. <u>A295</u>, 424 (1978).
24. J.L. Wood, IV Int. Conf. on Nuclei Far From Stability, CERN 81-09 (1981), p. 612.
25. A.L. Goodman, Nucl. Phys. <u>A287</u>, 1 (1977).
26. A.W. Sunyar, <u>Proc. Symp. on High Spin Phenomena in Nuclei</u>, ANL/PHY-79-4 (1979) p. 77.
27. J. Roth et al., in <u>Proc. of IV Int. Conf. on Nuclei Far From Stability</u>, CERN 81-09 (1981), p. 680.
28. J. Roth et al., to be published.
29. C.F. Maguire et al., private communication, Vanderbilt-ORNL, 1982.

MULTIPLE DISCONTINUITIES OF THE MOMENT OF INERTIA AT HIGH SPIN

J. H. Hamilton
Physics Department, Vanderbilt University
Nashville, TN 37235/USA

INTRODUCTION

What happens to a nucleus at high spins is one of the major fields of nuclear structure today as evidenced by the several recent international conferences in this field. In this lecture we will concentrate on two new approaches to study high spin states--partial fusion reactions and neutron multiplicity experiments. The latter has been used to study states to a record high spin for a light nucleus, tentatively 20^+ in ^{74}Kr (ref. 1-3). The former has been employed to observe third and higher discontinuities of $\mathscr{J}$ in ^{158}Er (refs. 4,5).

TWO DISCONTINUITIES OF $\mathscr{J}$ AT HIGH SPIN IN ^{74}Kr

Two discontinuities of the moments of inertia, $\mathscr{J}$, of the yrast cascades in ^{158}Er and ^{160}Yb (refs. 6,7), and very recently a third and apparently higher discontinuity of $\mathscr{J}$ in ^{158}Er (refs. 5, 8) have provided evidence for the persistence of collective rotations and the alignment of individual pairs of quasiparticles to much higher spins that previously thought for deformed rare earth nuclei. Recent evidence for strong ground state deformation was reported in ^{74,76}Kr based on data for states with $I^\pi \lesssim 10^+$ (ref. 9). With a newly developed neutron multiplicity technique,[1] we have observed the yrast cascade in ^{74}Kr to the highest spin reported for a medium-light nucleus, tentatively 20^+. Two discontinuities were observed in $\mathscr{J}$ of the yrast cascade in ^{74}Kr.

Backbending of $\mathscr{J}$ has been observed at 8^+ in ^{68}Ge (ref. 10) and ^{80}Kr (ref. 11) with bands built on three 8^+ states in ^{68}Ge and two 8^+ states in ^{80}Kr. These two discontinuities at 8^+ are interpreted as the crossing of rotation aligned bands built on both proton and neutron $(g_{9/2})^2$ configurations in ^{68}Ge. A similar interpretation is made[11] for ^{80}Kr.

The levels of ^{74}Kr were studied via the reaction ^{58}Ni(^{19}F, p2nγ)^{74}Kr at 66-68 MeV with >99% enriched, thick targets. A recently developed neutron-multiplicity-γ coincidence technique[1] was essential to separate the weak neutron evaporation channel to ^{74}Kr from the competing charged particle channels. Four large liquid scintillator neutron detectors especially designed to cover most of the forward 2π

solid angle were used to gate the γ-ray spectra. The overall neutron efficiency of about 30% made it possible to apply this technique not only to (n,γ) and (n,n,γ) coincidences, but also to (n,γ,γ) measurements, angular distributions, and recoil distance lifetime measurements.

The level scheme of ^{74}Kr was known to 8^+. In Fig. 1 is shown the levels established on the basis of our n-γ-γ and n-γ(θ) coincidence measurement with a beam energy of 66.5 MeV using a thick target and List-mode technique.[1,2] States up to a tentative 20^+ state were established by carefully and systematically varying the windows over the peaks and corresponding background cuts and by adding various coincidence gates on all the lower lying transitions.

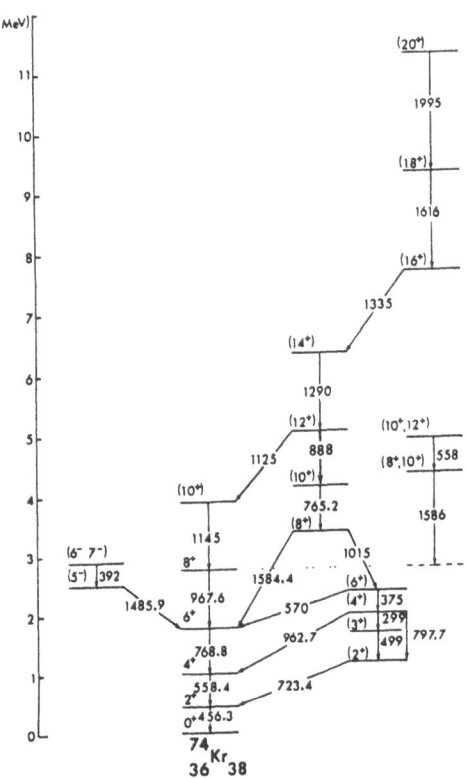

Figure 1. Level scheme of ^{74}Kr.

The multipolarities and mixing ratios of the γ-rays were deduced from an angular distribution measurement at the same energy. We measured spectra at the angles 0^0, 10^0, 75^0, 90^0, 115^0, 125^0, and 133^0 gated with one neutron. A γ-ray yield function was measured at 6 beam energies between 56 and 67 MeV. The yield and γ(θ) data and spectra obtained by gating with one and two neutron events and χ^2 fitting procedure led to the spin assignments displayed in Fig. 1.

In Fig. 2 is shown a plot of the angular momentum I(ω) as a function of ω where $\hbar\omega = E_\gamma((I + 1) \to (I - 1))/2$ for our data for ^{74}Kr and other recent results for ^{76,78}Kr (refs. 9, 12). There are three discontinuities seen in I(ω) for ^{74}Kr. The break at 2^+_1 has been interpreted as arising from an interaction of the strongly deformed 0^+ ground state and a low lying 0^+_2 near spherical excited state to push down the 0^+ energy (and enlarge the $2^+_1-0^+$ energy).[9] In both ^{74}Kr and ^{78}Kr there are very similar discontinuities in I(ω) and correspondingly in $\mathcal{I}$ above 10^+. There is only a slight up bend of I in ^{76}Kr. However, in ^{80}Kr two 8^+ and 10^+ states are seen and are interpreted as the crossing of a two quasiparticle rotational aligned band with $(g9/2)^2$ protons. In ^{74}Kr two 8^+ and 10^+ levels also seen. The branching

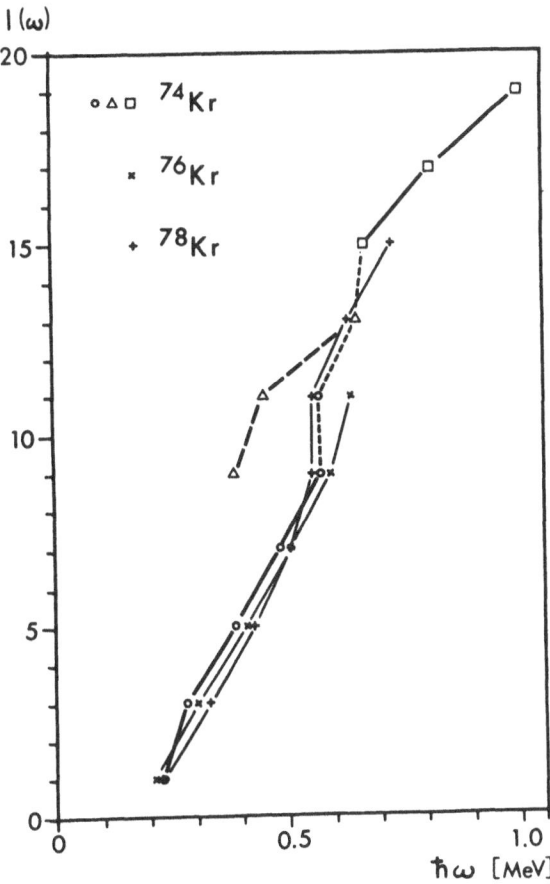

$I(\omega)$

^{74}Kr

^{76}Kr

^{78}Kr

0.5   1.0

$\hbar\omega$ [MeV]

Figure 2. Plot of the angular momentum $I(\omega)$ as a function of ω.

ratios and energies indicate the 8_2^+ and 10_2^+ levels form a band with the yrast 12^+ and (14^+) levels that cross the ground band above 10^+. This is in contrast to ^{80}Kr where the 8_1^+ and 10_1^+ levels are the two quasiparticle rotation aligned states. This first band to cross the ground band is interpreted as a two quasiparticle $(g_{9/2})^2$ configuration, but it is not clear whether it is a proton or neutron configuration.

The second discontinuity of $\mathscr{I}$ occurs above 14^+ with the (16^+), (18^+), and tentative (20^+) states showing similar alignment. Their extra alignment is about the same as the extra alignment in the first band, 2-3 units. No such break in I is seen in the tentatively assigned 16^+ level in ^{78}Kr (ref. 12) which has 4 more neutrons. These data suggest a block effect of the extra neutrons and indicate at least two of the four quasiparticles in the highest band in ^{74}Kr are neutrons. In summary, these data provide the first evidence for a second high spin discontinuity in $\mathscr{I}$ in a medium light nucleus and that collective rotations and the alignment of individual two quasiparticle configurations continue to very high energies in this region.

THIRD AND HIGHER DISCONTINUITIES OF $\mathscr{I}$ AT HIGH SPIN IN ^{158}Er

As discussed briefly in an earlier lecture,[3] the partial fusion reaction, PFR, in which a fast p,d, or t is emitted in the forward direction followed by the fusion of the remaining part of the projectile is a very promising new way to study nuclei at very high angular momentum. Yamada et al.[4,5] have studied this reaction by bombarding ^{150,154}Sm with 167 MeV ^{14}N from the Oak Ridge Cyclotron ORIC. The

evaporation p,d,t are separated in energy from those from PFR. By gating on the
fast p,d,t, the gamma ray multiplicity was measured in the ^{14}N reactions on ^{154}Sm
that lead to $^{157-159}$Er. The Ge(Li) spectrum showed that the ^{158}Er product was
dominant. The ^{158}Er reaction was more than 50% of the observed PFR and the gamma-
rays from the decay of states up to 36^+ were observed in ^{158}Er.

As shown in Fig. 3, the gamma ray multiplicity is constant as a function of the
spin of the yrast states to as high as statistics allow measurement, spin 26, for
all three reactions p,d, t that lead to ^{158}Er. A similar multiplicity was measured
for the reactions leading to ^{157,159}Er. These data indicate
no side feeding below I = 28.
The average M_γ is 31. This is
the highest M_γ presently observed.
The mass distribution of the residual nuclei also is appreciably
more narrow than that for compound
nuclear reactions. Both of these
data indicate that the PF reactions in which p,d, or t are
emitted are associated with a
narrow, high angular momentum
window.[4] The deduced average
angular momentum transferred is
$63\hbar$ which is comparable to the
critical angular momentum ℓ_{cr} predicted for the fusion of ^{13}C and
^{154}Sm.

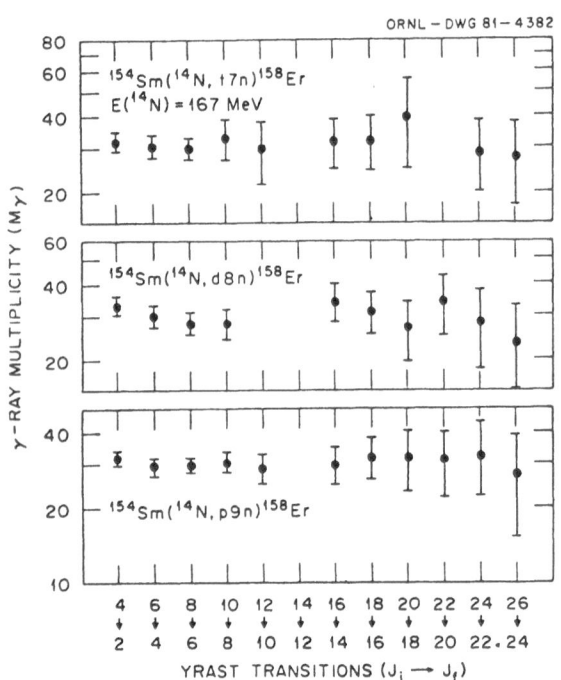

Figure 3. Measured γ-ray multiplicities.

With this high, narrow ℓ window,
the PF reaction should be a very
useful tool to study nuclei at high angular momentum. Further support for this is
seen in our ^{154}Er studies.[3] In our second experiment, the same PF reaction was
studied with three large, well colluminated NaI detectors to obtain E_γ-E_γ correlation spectrum.[5] The E_γ-E_γ correlation method is a powerful method. To show that it
is useful with NaI detectors we carried out very extensive Monte Carlo simulation
calculations. These calculations show that indeed with the order of one to five
million events if the background is reasonably low that one can easily study the
valley to quite high spins, the order of 50 $\hbar$.

The measured and simulated spectra reproduced the known first three backbends of
$\mathscr{I}$ in ^{158}Er very nicely. The third backbend at 1.1 MeV was in fact observed in

our data (see Fig. 4) prior to its observation from discrete gamma-ray energies.[8]

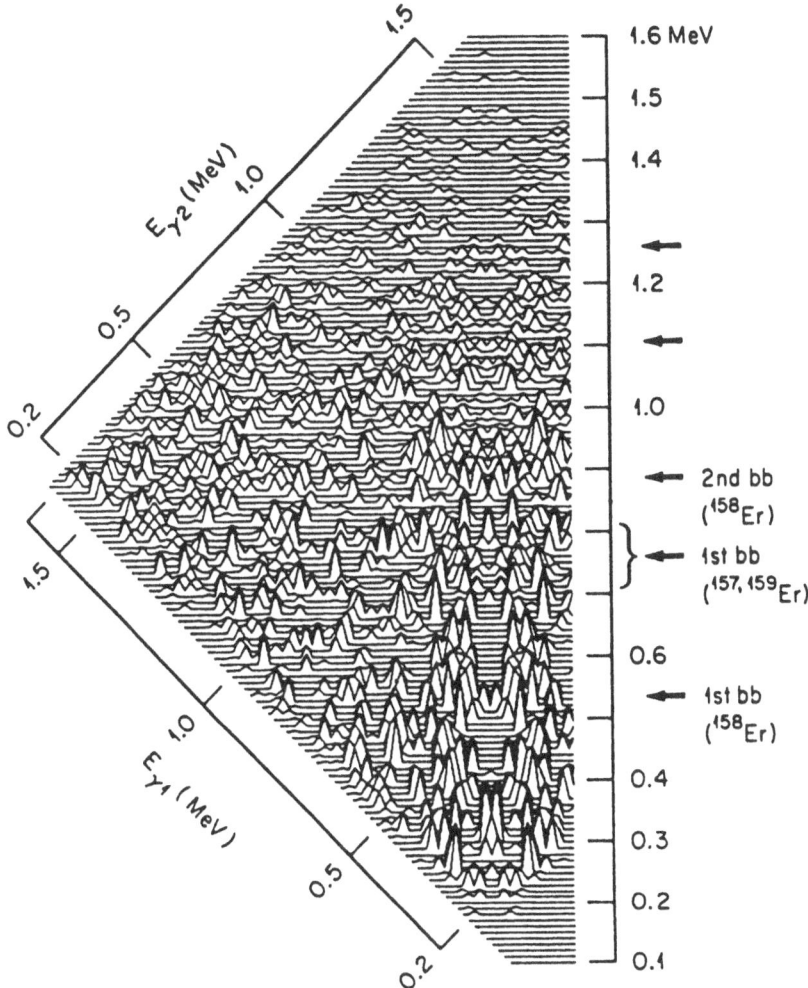

Figure 4. E_γ-E_γ correlations in [157-159]Er.

These data with the Copenhagen itteration method applied to subtract background are from five million events as compared with the 50 million required for Ge(Li) work. The difference in numbers of events required to obtain significant spectra is the essentially five times higher photo peak efficiency for NaI compared to Ge(Li) de-tectors and the fact that the PF reactions lead to a high ℓ-window so that many of the side feeding γ rays at lower spins are not present, eg., the background is much reduced.

Note in addition to the third backbend of $\mathscr{J}$, one sees evidence for other bridges at 1.24 and perhaps 1.46 MeV (ref. 5). These bridges are in reasonable agreement with Cranked Shell Model calculations of the next crossings produced from the alignment of the next pairs of quasiparticles.[13] The most important information from these data is the observation that the valley continues to an energy of 1.46 MeV. This valley indicates the continuation of collective excitations to this energy--the highest energy observed to date. These data support the continuation of collective motion up to a spin about 48ħ. The breakoff of the valley about 1.46 MeV can be the signal for the predicted change of phase from super fluid to normal states associated with a quenching of the pairing field.[14] This is predicted to lead to a change to an oblate shape where the angular momentum is carried by pairs of aligned particles. As discussed in more detail in another paper (Yamada and Hamilton),[5] we believe these data provide evidence for this phase change and that the change is rather sudden.

In summary we conclude[5] that these data show that indeed partial fusion reactions are a powerful new tool for the studies of the structure of nuclei at very high spins. Third and higher discontinuities of $\mathscr{J}$ are seen in ^{158}Er. The valley characteristic of collective rotations extends to about 1.46 MeV where it apparently terminates. This termination may signal the sudden onset of a phase transition to an oblate shape.

REFERENCES

1. J. Roth et al., in Proc. of IV Int. Conf. on Nuclei Far From Stability, CERN 81-09 (1981), p. 680.
2. J.H. Hamilton et al., Proc. IV Int. Conf. on Nuclei Far From Stability, CERN 81-09 (1981'), p. 391.
3. J.H. Hamilton, elsewhere in this Vol.
4. H. Yamada et al., Phys. Rev. C24, 2565 (1981).
5. H. Yamada et al., to be published; and H. Yamada and J.H. Hamilton, Int. Symp. on Dynamics of Nuclear Collective Motion (Mt. Fuji, Japan) (July, 1982).
6. I.Y. Lee et al., Phys. Rev. Lett. 38, 1454 (1977).
7. L.L. Riedinger et al., Phys. Rev. Lett. 44, 568 (1980).
8. J. Burdge et al., Phys. Rev. Lett. 48, 539 (1982).
9. R.B. Piercey, et al., Phys. Rev. Lett. 47, 1514 (1981).
10. A.P. de Lima et al., Phys. Lett. 83B, 43 (1979); and Phys. Rev. C23, 213 (1981).
11. D.L. Sastry et al., Phys. Rev. C23, 2086 (1981).
12. H.P. Hellmeister et al., Phys. Lett. 85B, 34 (1979); Nucl. Phys. A332, 241 (1979).
13. S. Frauendorf, Phys. Scri. 24, 349 (1981).
14. A. Bohr and B.R. Mottelson, Nuclear Structure, Vol. 2 (Benjamin, NY, 1975).

RECENT RESULTS ON NUCLEI FAR FROM STABILITY
IN THE MASS REGION A ≃ 70

A. V. Ramayya[+]

Physics Department, Vanderbilt University, Nashville, Tennessee 37235
Institut fur Kernphysik, Johann Wolfgang Goethe-Universitat
zu Frankfurt, Germany

and

J. Eberth

Institut fur Kernphysik, Universitat zu Koln, Germany

I. INTRODUCTION

Recent advances in experimental techniques, of in-beam γ spectroscopy for the detection of γ rays from evaporation residues induced by heavier ions, have made possible to investigate the level structures of nuclei far from stability. The nuclei in the mass region A ≃ 70 are very interesting because they exhibit wide variety of band structure and are also transitional nuclei. Earlier studies have revealed highly collective band structures[1] of the following types: 1) Ground state band, 2) One or two even spin positive parity bands with 8^+ as band heads, 3) Odd and even spin negative parity bands, 4) $\Delta J = 1$ even-parity bands with 2^+ as band heads, 5) Coexistence of spherical and deformed bands. Not all these types of bands have been observed in all the nuclei in this mass region. Various nuclear models, such as rotation-vibration model, two-quasiparticle-plus-rotor model, interacting-boson model have been used to describe these bands. However, the applicability of these models depend on the availability of accurate experimental energy levels, spins and parities, reduced transition probabilities, quadrupole moments and g-factors for the levels of these nuclei and, if possible, up to very high angular momentum. In the present paper we discuss some of the advances made in in-beam gamma-ray spectroscopy techniques and the results obtained in this mass region. Also, we discuss briefly our present understanding of these results.

II. EVAPORATION CODES

For the identification of nuclei, hitherto unknown, one often depends upon the excitation functions and cross-bombardment techniques. One is often faced with the question, what nuclei can we produce with heavier ions? Can we produce neutron deficient nuclei with N ≲ Z in this mass region, with sufficient cross-section? Recently Pulhofer[2] and Blann[3] have written computer codes (CASCADE, MBII) to calculate the cross-sections for evaporation residues obtained from heavy-ion induced fusion reactions. In these calculations, the projectile and target form a compound nucleus in statistical equilibrium and the Hauser-Feshbach formula together with the statistical nuclear model are applied in order to calculate the intensities of various decay chains.

In Fig. 1 we have shown the calculations performed with the CASCADE code for the reaction ^{32}S on ^{46}Ti. The calculations performed with MBII code are not appreciably different. These calculations indicate that the resulting compound nucleus will predominantly evaporate charged particles leading to the residual nuclei near the line of stability, while the cross-section for evaporation of neutrons leading to neutron deficient nuclei are small. In fact the calculations predict the cross-section of the order of 1 mb or less for N ≲ Z nuclei in this mass region. For the reaction ^{32}S on ^{46}Ti the calculations seem to predict the energy dependence of various reaction channels quite well. Also, the relative cross-sections of various channels seem to be approximately correct. However, these codes have several parameters which can be adjusted for fine tuning. From these calculations and

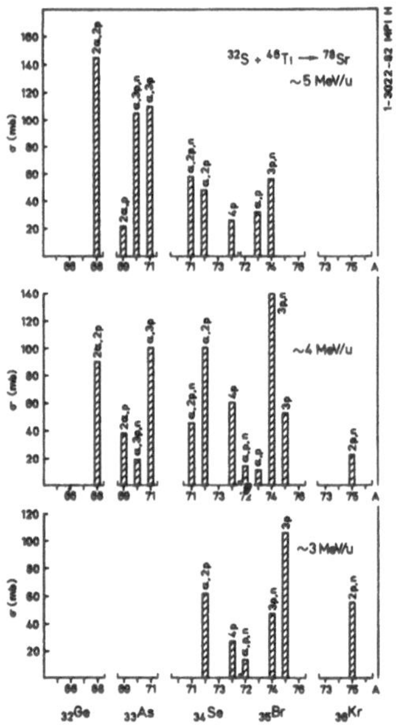

Figure 1. Cascade calculations for ^{32}S ^{46}Ti
   at different bombarding energies.

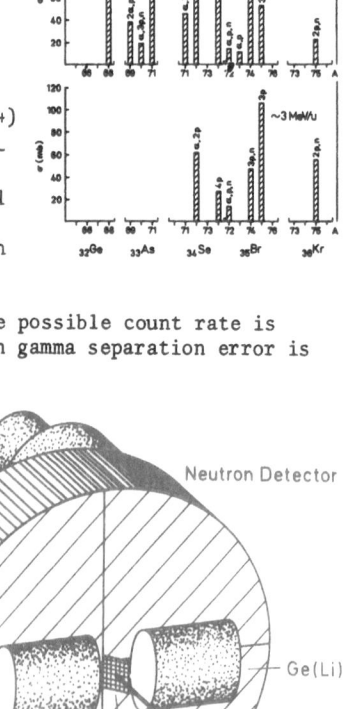

experimental data, one can realize immediately,
that as we increase the bombarding energy, we
jump into lower Z region by evaporating more
and more α particles and/or protons instead of
evaporating neutrons. By bringing in more
energy with heavier ions, we do not seem to
reach nuclei with N ≤ Z with appreciable cross-
sections. However, the calculations have to be
further tested for very low cross-section
channels.

III. NEW EXPERIMENTAL TECHNIQUES AND RESULTS

   Recently an efficient n-γ coincidence technique[4]
has been developed to separate the weak neutron channels from competing charged particle channels. The
neutron detector (Fig. 2) is a 35 x 10 cm cylindrical
liquid scintillator (NE 213) which is divided into
four independent optically light tight segments, each
coupled to an independently working fast photomultiplier. The pulse shape discrimination is used to
separate neutron signals from gamma ray signals. The possible count rate is
5 x 10^5 counts per second per segment and the neutron gamma separation error is
2000:1. However, one can achieve
the same objective of suppressing
the particle channels by using independent detectors also. Recently,
for the investigation of level scheme
of ^{75}Kr, we have used up to 8 neutron
detectors. With multiple neutron detectors, one can also determine the
multiplicity of the evaporated neutrons, which in turn helps to identify
the residual nuclei. Figure 3 shows
an example of gamma spectrum from our
earlier studies in ^{74}Kr. More
recently we have investigated the
residual nuclei produced in the reaction ^{32}S on ^{46}Ti using the above
technique. With this technique we
have been able to identify for the
first time, the γ-rays of ^{75}Kr. For
an easy identification, one can construct a new spectrum by dividing the
neutron gamma coincidence spectrum by
gamma spectrum and multiplying the
resulting spectrum by a factor of
1000. An example of such a spectrum
is shown in Fig. 4. The γ-rays
corresponding to the neutron channels

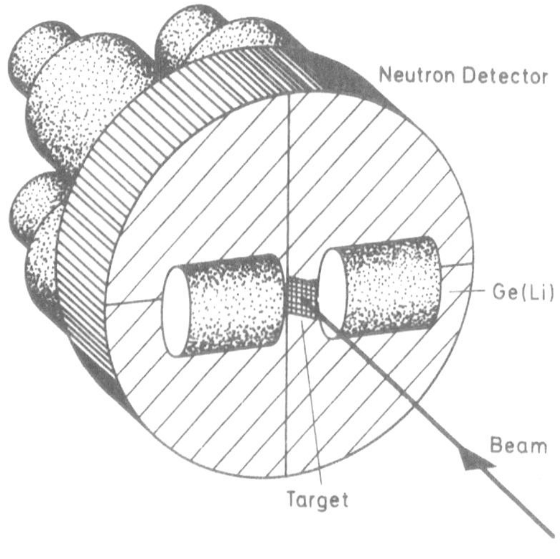

Figure 2. Efficient neutron detector system.

will appear as real peaks and the gamma rays corresponding to the charged particle
channels will appear as negative peaks. Because of the great efficiency of neutron
detector system, the system can be combined with other systems such as plunger,
mini-orange for other spectroscopy measurements.
   Recently we have used eight neutron detectors in conjunction with two Ge(Li)
detectors to construct the energy level diagram[5] of ^{75}Kr. These results are shown
in Fig. 5. The neutron detector is coupled to a plunger to measure the lifetimes

Figure 3. Singles γ-ray spectrum and γ-ray spectrum in coincidence with 1-fold and 2-fold neutrons.

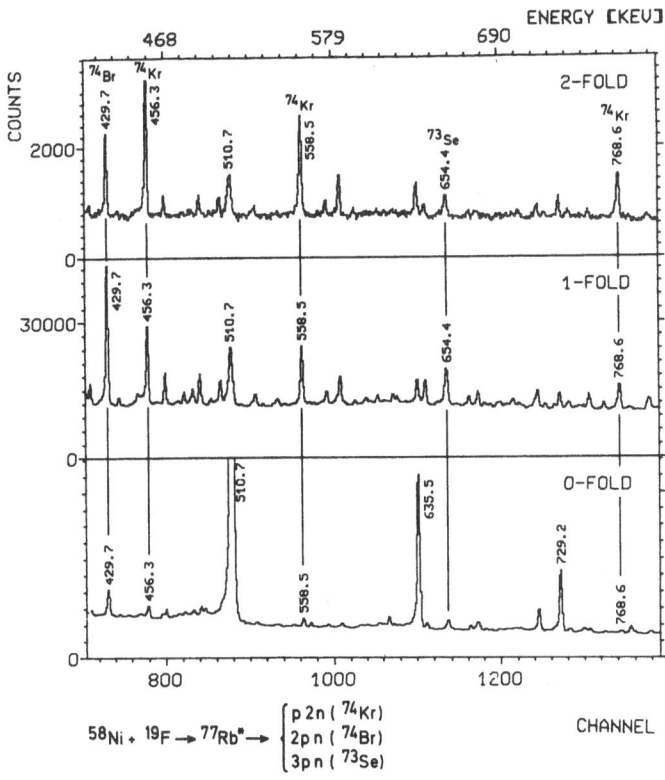

$$^{58}Ni + {}^{19}F \rightarrow {}^{77}Rb^* \rightarrow \begin{cases} p\,2n\ (^{74}Kr) \\ 2p\,n\ (^{74}Br) \\ 3p\,n\ (^{73}Se) \end{cases}$$

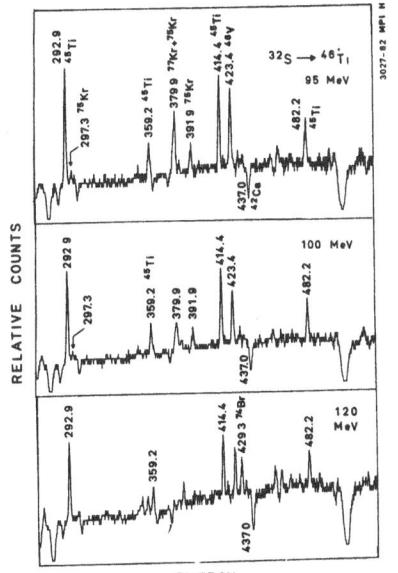

Figure 4. Gamma-ray spectrum in coincidence with neutrons is divided by singles γ-ray spectrum and multiplied by 1000.

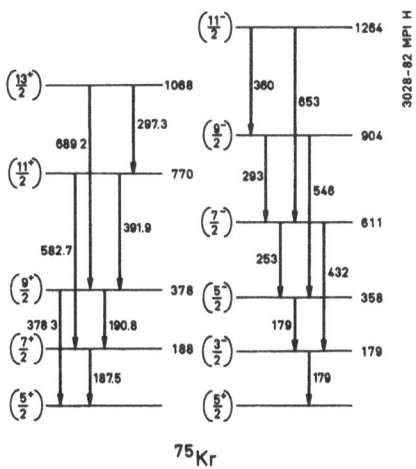

Figure 5. Level scheme of ^{75}Kr.

of nuclear levels. A schematic diagram of such an arrangement is shown in Fig. 6.
Specifically developed plunger[6] is used to mea-
sure the lifetimes of levels in ^{74}Kr. The distance between target and stopper is varied by a
combination of two techniques; for distances larger
than 50 μm, a micrometer screw with a resolution of
1 μm is used while for smaller distances an additional piezo crystal is used which provides a
linear motion up to 15 μm. The voltage for the
piezo crystal (0-1000 V) is supplied by a fast HV
operational amplifier with a rise time of 1V/μs.
This allows us to use the piezo crystal as part of
a fast feedback loop for distance stabilization
during the measurement. The capacity between
target and stopper determines the frequency of a
quartz oscillator which is stabilized to a reference oscillator by a phase locked loop circuit.
By this way the system compensates for all thermal
effects of the plunger mechanism and all mechanical vibrations due to vacuum pumps up to several

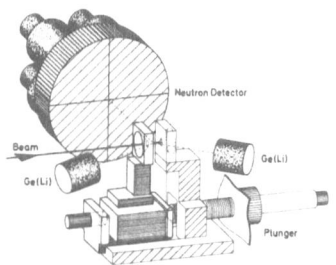

Figure 6. Schematic arrangement of plunger and neutron
detector.

hundred Hz and keeps the target-stopper distance constant within less than 0.1 μm.
The lifetimes of levels in ^{74}Kr are measured[7,8] using the above technique and the
results are shown in Table I. Since the data is not corrected for side feeding
transitions, the results of higher spin states should be considered as tentative.
The γ-ray spectrum in coincidence with neutron detector as a function of target
stopper distance is shown in Fig. 7.

TABLE I.

E_x (keV)	Transition	τ (ps)	B(E2) e^2 fm^4
456.3	$2^+ \to 0^+$	28.8 ± 5.7	1432^{-237}_{+353}
1014.7	$4^+ \to 2^+$	13.2 ± 0.7	1139^{-57}_{+64}
1783.5	$6^+ \to 4^+$	4 ± 1	760^{-152}_{+253}
2751.1	$8^+ \to 6^+$	1 ± 1	~962

The The neutron detector is also coupled to a mini-orange spectrometer to measure
conversion electrons in coincidence with neutrons. Recently these techniques are
extensively used to investigate the levels in even-A Kr isotopes and the odd mass
^{73}Se and ^{75}Kr. The results of these investigations and our present understanding
of these nuclei will be discussed below.

IV. DISCUSSION OF RECENT RESULTS ON NEUTRON DEFICIENT ISOTOPES

The experimental evidence, such as B(E2) values on neutron deficient Kr isotopes, indicate that these nuclei are deformed in the ground states[9,10]. The
$2^+ \to 0^+$ energy spacing is, however, relatively large. To explain the relatively
large $2^+ \to 0^+$ energies in ^{74,76}Kr and large B(E2)'s Piercey et al.[10], proposed coexistence model, similar to the one proposed earlier for Se isotopes. In case of
^{74,76}Kr isotopes an interaction between 0^+ deformed ground state and the excited
spherical 0^+ state pushes down the deformed ground state, whereas in ^{72,74}Se isotopes an interaction between the excited spherical 2^+ state and deformed 2^+ state is
assumed. The origin of shape coexistence for N = 38 nuclei can be attributed to the
gaps in single particle spectrum and the number of protons which delicately balance

whether a deformed shape or near spherical shape is lowest in nuclei in this mass region. As Z moves away from 28 or 50, when N = 40, the level density for a spherical shape becomes very high and the minimum of the proton deformation energy moves to deformed shapes. The deformed structure is generated from the spherical ones by transferring two pairs of neutrons from the $f_{5/2}$ to the $g_{9/2}$ shell and (vice-versa) may be an important source for the coupling of different structures. Further, it was also pointed out that by scaling the extracted unperturbed $2^+ \rightarrow 0^+$ energies in ^{74,76}Kr by $A^{5/3}$, the deformation of the ground state is slightly larger than super deformation recently reported[11] for ^{100}Sr. It is not yet clear whether these deformed states are prolate or triaxial.

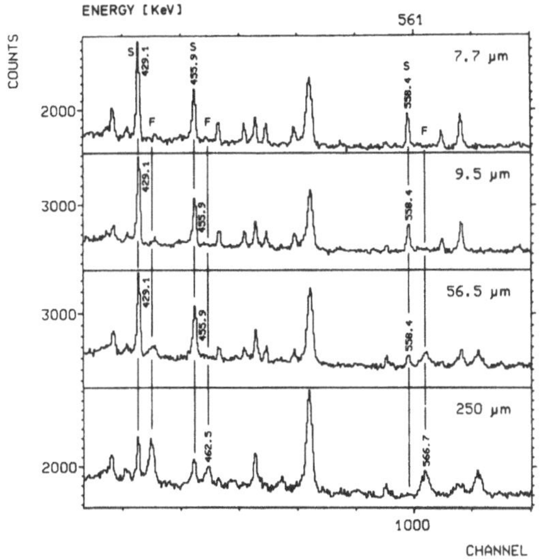

To get better understanding and to provide a pictorial insight Siewert

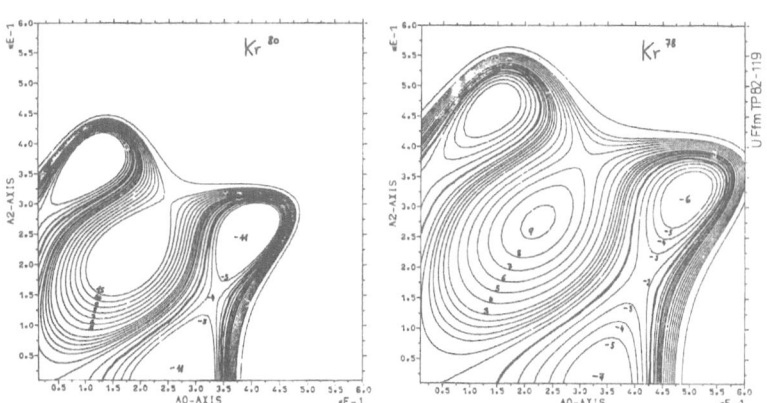

Figure 7. Gamma-ray spectrum in coincidence with neutrons as a function of target-stopper distance.

et al.[12], have calculated the collective potential energy surface (PES) directly from the experimental low energy spectra of even-A Kr isotopes as a function of collective coordinates. Only quadrupole deformations are considered in detail. The method is discussed by Gneuss and Greiner[13]. In this method collective Hamiltonian is diagnolized in the basis of the 5-dimensional quadrupole oscillator. From the collective PES one can obtain quite easily more information. For example, the depths of the minimum in the PES gives the stability of the corresponding nuclear shape; the valleys in the surface indicate which changes of shape are favored. The PES is plotted as a function of collective coordinates a_0 and a_2.

The PES's for ^{80,78}Kr are shown in Fig. 8 and the cuts of the surface along $\gamma = 0^0$ and 40^0 are shown in Figs. 9 and 10. The comparison between the experimental and the calculated energy spectra and the B(E2) values are shown in Figs. 11 and 12. By looking at the PES's one can conclude that these nuclei are nearly γ unstable with $\gamma = 40^0$. The zero-point

Figure 8. Potential energy surfaces for ^{78}Kr and ^{80}Kr.

energy is shown as a thick line in Fig. 8. The deformation parameter β of ^{78}Kr is larger than that of ^{80}Kr. The calculated B(E2) value of $6^+ \rightarrow 4^+$ transition is much larger than the experimental value in both the cases which suggests that one should take into account the interaction between single particle degrees of freedom and the collective degrees of freedom even at low spins such as 6^+.

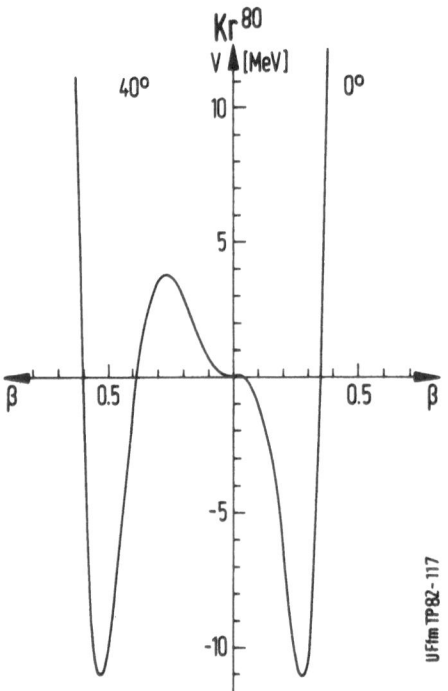

Figure 9. Cuts of the PES surface along γ = 0° and γ = 40° for ⁸⁰Kr.

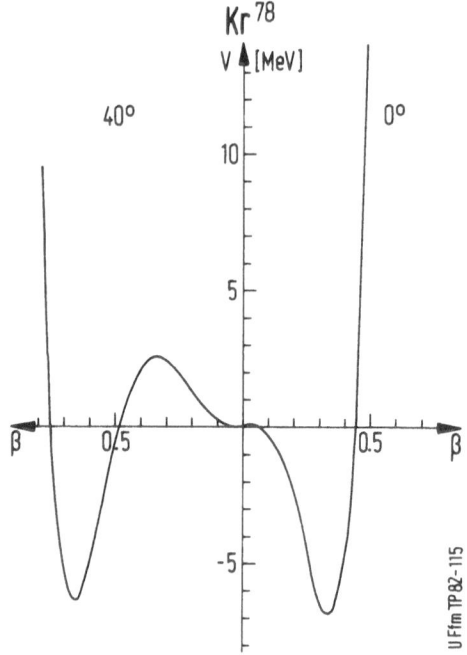

Figure 10. Cuts of the PES surface along γ = 0° and γ = 40° for ⁷⁸Kr.

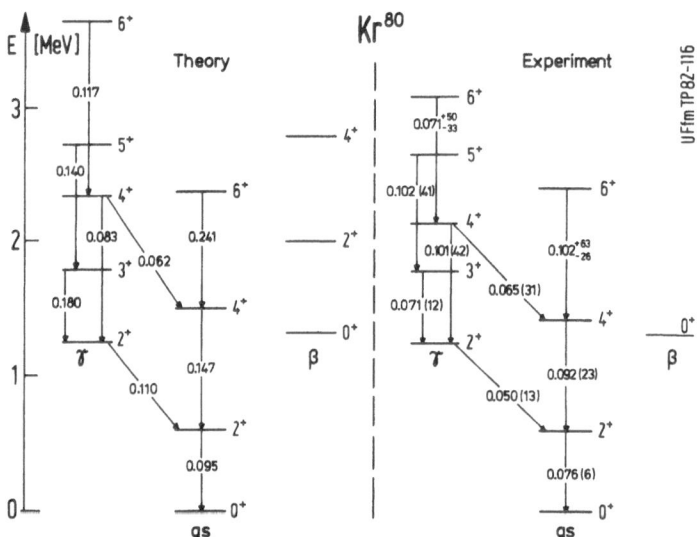

Figure 11. Comparison of experimental and theoretical results for ⁸⁰Kr.

Figure 12. Comparison of experimental and theoretical results for ^{78}Kr.

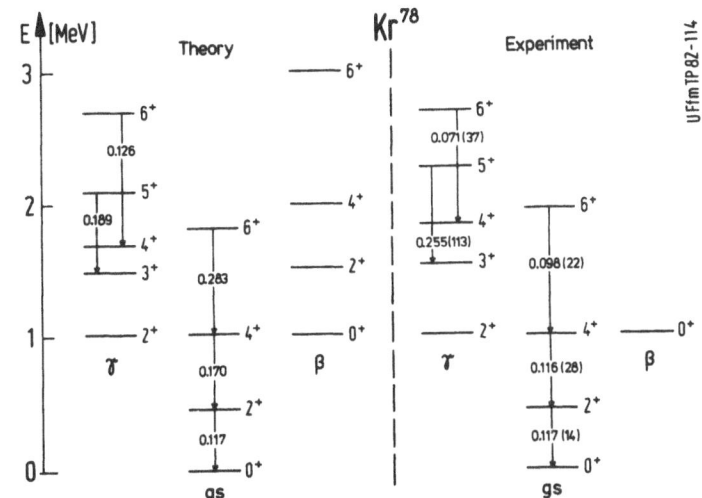

The preliminary calculations are also performed for ^{76}Kr. The results indicate that the PES has two minima along the pro-late axis, one at $\beta = 0$ and the other at $\beta = 0.4$. The ground state wave function is distributed over the two minima and the main contribution is from the minimum at $\beta = 0.4$. The 0^+ member of the β band is located in the first minimum. Because of the large deformation the 0^+ member of the β-vibrational band is lowered considerably and interacts with the ground state and pushes down the 0^+ energy. Because of this interaction one does not observe a very low $2^+ \to 0^+$ energy for strongly deformed nuclei in Kr isotopes. By far the Kr isotopes are the strongly deformed nuclei.

In order to explain the negative parity bands, with band head at 5^- and the positive parity bands with 8^+ as band head, the two quasiparticle-plus-rotor calculations have been performed by Soundranayagam et al.[14] for Kr isotopes. For nuclei with Z and N far from the magic numbers, the interaction between particle and collective degrees of freedom becomes important. As a result, the pairing effects are reduced and one is likely to observe Coriolis decoupled rotational bands at higher excitations. In these calculations an axially symmetric rotor is employed. The calculations in general have the following parameters: λ (Fermi energy), Δ (gap energy), δ (deformation parameter), $\mathscr{J}$ and c of the VMI model, k and μ of the shell model and 20 basis states. Even though the calculations have seven parameters, these parameters have inherent physical meaning and are not quite arbitrary. The Fermi energy is determined by the δ calculated from the B(E2) value of the 2^+ level and Nilsson diagram, $\mathscr{J}$ and c are determined by fitting the ground state band to VMI model, and the gap energy is approximately equal to $12/A^{1/3}$. Of course, one has to perform 4 different set of calculations to obtain positive and negative parity proton and neutron two-quasiparticle excitations. In these four different sets of calculations, the parameters are varied to obtain the best fit.

The calculations for Kr isotopes indicate that the two-quasi-particle levels for protons are lower than that of neutron two-quasi-particle levels by approximately 0.5-1.0 MeV. The experimental and calculated energies are compared in Fig. 13. Best fit is always obtained for two quasi-proton levels. Examination of the wave functions obtained from the best fit reveals that all but one (the second 6^- state of ^{80}Kr) of the negative parity two quasi-proton states contain up to 99.8% of the $(g_{9/2}, f_{7/2})$ configurations. In the wave function of the second 6^- state of ^{80}Kr the $(g_{9/2}, f_{5/2})$ and $(g_{9/2}, p_{3/2})$ configurations constitute 92% and 8% respectively. Thus, large contributions to the wave functions of the negative parity states in these nuclei come from the Nilsson configurations $7/2^-$ (303) of negative parity, $3/2^+$(431) and $1/2^+$ (440) of positive parity.

The calculations also suggest that the positive parity bands with 8^+ as band heads is the result of coupling two decoupled $g_{9/2}$ protons to an axially symmetric rotor. In the case of ^{80}Kr the calculations indicate that the K = 0 component (corresponding to complete alignment) is 49% and is about 40% for K = 1. These results are an indication that the classical picture of complete alignment is not fully applicable to nuclei which have to satisfy the quantum mechanical condition.

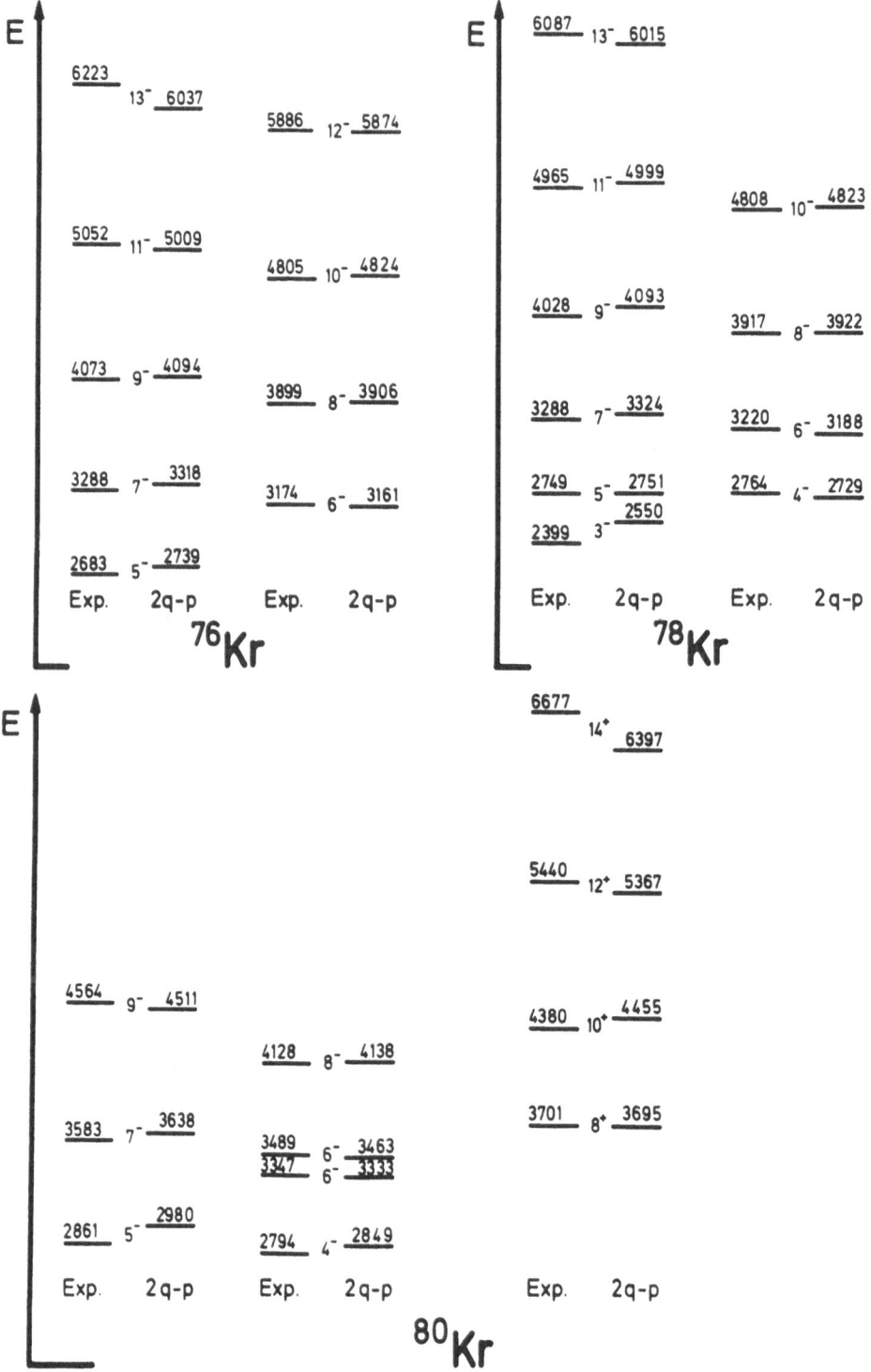

Figure 13. Two quasi-particle-plus-rotor calculations for Kr isotopes.

These types of calculations can be tested further only when the experimental data such as g- factors of these states becomes available.

In order to get better understanding of the nuclei in this region, the neighboring odd mass ^{73}Se (ref. 15) and ^{75}Kr (ref. 5) nuclei are also investigated. The properties of the low spin states in ^{72}Se and ^{74}Se are interpreted earlier in terms of a second minima[16] in the PES along the prolate axis. In this model the excited 0^+ states and the higher spin states are members of a rotational band in the second minimum at larger deformation and coexist with the vibrational states, with mixing of the vibrational and rotational states at 2^+. The neighboring ^{73}Se should also show similar behavior to that of ^{72}Se and ^{74}Se. The level scheme[15] of ^{73}Se is shown in Fig. 14 and the transition probabilities are given in Table II.

Figure 14. Level scheme of ^{73}Se.

From the reduced transition probabilities of positive and negative parity levels one can extract a deformation of 0.31 for the negative parity levels and much smaller deformation of 0.22 for the positive parity band. The positive parity band is further interpreted as decoupled $g_{9/2}$ band. The observation of two different deformations supports the interpretation[10,16] in terms of a coexistence model for ^{72}Se and ^{74}Se. In this model deformed states are developed from the spherical states by scattering a pair of neutrons from the fp-shell to the $g_{9/2}$ orbital. The negative parity band (energies and B(E2)'s) can be reproduced rather well by a simple rigid rotor model with $\beta = 0.31$.

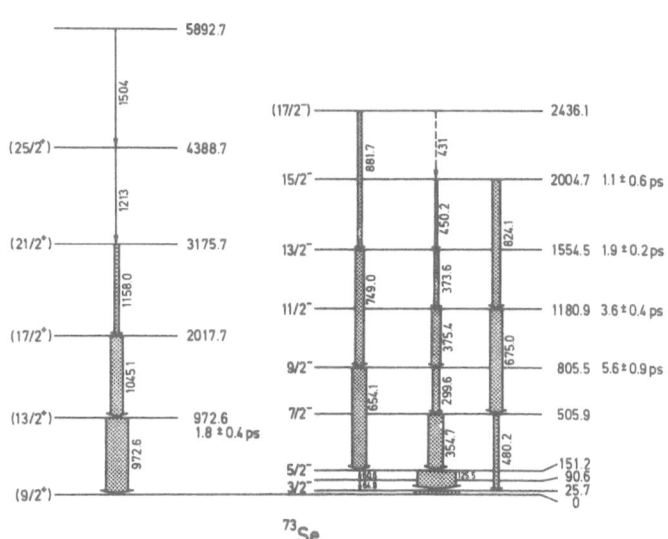

^{73}Se

TABLE II.

E_x	Transition	τ(ps)	B(E2) $\left[e^2 \text{ fm}^4\right]$	
			Exp.	Calculation
2004.7	$15/2^- \to 11/2^-$	1.1 ± 0.6	1335 ± 790	1222
1554.7	$13/2^- \to 9/2^-$	1.9 ± 0.2	1220 ± 180	1152
1180.9	$11/2^- \to 7/2^-$	3.6 ± 0.4	850 ± 110	1048
805.5	$9/2^- \to 5/2^-$	5.6 ± 0.9	760 ± 120	883

Our preliminary level scheme[5] of ^{75}Kr also indicates the existence of positive and a negative parity band (Fig. 5). However, the experiments are not yet completed. The level structure of ^{75}Kr looks quite similar to that of ^{77}Kr. The negative parity bands of both ^{75}Kr and ^{77}Kr can be fitted to a rigid-rotor motor model rather well

In conclusion, the new techniques developed recently offers many possibilities to carry out detailed investigations on nuclei far from stability. The recent results on light mass Se and Kr isotopes can be understood in terms of a second minimum along the prolate axis in the PES. In the case of Se isotopes the ground state 0^+ is located in the spherical minimum and the excited 0^+ state is located in the deformed minimum. In the case of ^{76}Kr, the ground state is mainly located in the deformed minimum and the excited 0^+ state is located in the spherical minimum. Furthermore, the light mass Kr isotopes are gamma soft with $\gamma = 40^\circ$. The negative parity bands and the other positive parity bands can be reproduced well with proton two-quasiparticle-plus-rotor type calculations. The investigations of the neighboring odd isotopes further support the conclusions reached earlier.

Research at Vanderbilt University is supported by a Department of Energy Research Grant, Contract No. DE-AS05-76ER05034.

+Invited seminar speaker.

REFERENCES

1. J. H. Hamilton, A. V. Ramayya and R. L. Robinson, Proc. Int. Conf. on Nuclear Interactions, Canberra, Australia, 1978, ed. B. A. Robinson in Lecture Notes in Physics (Springer, New York, 1979), Vol. 92, p. 253.

2. F. Puhlhofer, Nucl. Phys. A280 (1977) 267.

3. M. Blann, University of Rochester, Private Communication.

4. J. Roth, L. Cleemann, J. Eberth, T. Heck, W. Neumann, M. Nolte, R. B. Piercey, A. V. Ramayya, J. H. Hamilton, Conf. on Nuclei Far From Stability (Helsingør, Denmark, 1981).

5. M. A. Herath-Banda, A. V. Ramayya, L. Cleemann, J. Eberth, J. Roth, T. Heck, W. Koenig, B. Martin and K. Bethge, to be published.

6. L. Cleemann, J. Eberth, W. Neumann, N. Wiehl, V. Zobel, Nucl. Inst. and Meth. 156 (1978) 477.

7. J. Roth, Ph.D. thesis, University of Köln.

8. J. Roth, L. Cleemann, J. Eberth, A. V. Ramayya, R. B. Piercey and J. H. Hamilton, to be published.

9. N. Sakamoto, S. Matsuki, K. Ogino, Y. Kadota, T. Tanabe and Y. Okuma, Phys. Lett. 83B (1979).

10. R. B. Piercey, A. V. Ramayya, J. H. Hamilton, X. J. Sun, Z. Z. Zhao, R. L. Robinson, H. J. Kim, John C. Wells, Phys. Rev. C25 (1982) 1941; ibid. Phys. Rev. Lett. 47 (1982) 1524.

11. R. E. Azuma, G. L. Borchert, L. C. Carraz, P. G. Hansen, B. Jonson, S. Mattson, O. B. Nielsen, G. Nyman, I. Ragnarsson and H. L. Ravn, Phys. Lett. 86B (1979) 5.

12. M. Siewert, J. Maruhn, W. Greiner, A. V. Ramayya, to be published.

13. G. Gneuss and W. Greiner, Nucl. Phys. A171 (1971) 449.

14. R. Soundranayagam, S. Ramavataram, A. V. Ramayya, J. H. Hamilton and R. L. Robinson, Phys. Rev. C25 (1982).

15. G. S. Li, W. Neumann, L. Cleemann, J. Eberth, T. Heck and J. Roth, to be published.

16. R. B. Piercey, A. V. Ramayya, R. M. Ronningen, J. H. Hamilton, V. Maruhn-Rezwani, R. L. Robinson and H. J. Kim, Phys. Rev. C19 (1979) 1344.

FUSION AND COMPOUND NUCLEI DECAY FOR LIGHT AND INTERMEDIATE-MASS SYSTEMS:

^{24}Mg, ^{28}Si + ^{12}C ; ^{24}Mg + ^{24,26}Mg ; ^{28}Si + ^{24}Mg, ^{28,29,30}Si

S. GARY[†] and C. VOLANT

DPh-N/BE, CEN Saclay, 91191 Gif-sur-Yvette Cedex, France

Abstract

Complete fusion or residue distributions have been measured with several experimental techniques for the following systems: ^{24}Mg, ^{28}Si + ^{12}C ; ^{24}Mg + ^{24,26}Mg ; ^{28}Si + ^{24}Mg, ^{28,29,30}Si. The range of energies spreads from 1 to 3 times the Coulomb barrier. None of these systems presents structures in its fusion excitation function as pronounced as the ones observed for lighter entrance channels. The intensities of the residue distribution are generally in agreement with the predicitons of the multiple deexcitation code CASCADE. The shape and absolute values of complete excitation functions are compared to macroscopic models and to the predicitons of a coupled-channel approach of the reaction, which gives a more satisfactory account of experimental data.

A full paper on this subject can be found in Phys. Rev. C25 (1982) 1877.

———————————

[†]Present address: *Commisariat à l'Energie Atomique, 94190 Villeneuve-Saint-Georges, France.*

NUCLEAR 'MOLECULAR' STATES[*]

U. Mosel[+], O. Tanimura and R. Wolf

Institut für Theoretische Physik, Universität Giessen
6300 Giessen, West Germany

Abstract

In the first part of this paper some models for the description of inelastic scattering of light heavy ions are discussed based on a reaction-theoretical approach. In the second part the structure aspects
of a nucleus-nucleus collision are treated based on an analysis of
two center correlation diagrams.

[*]Work supported by BMFT and GSI Darmstadt

[+]Invited Lecturer

I. Introduction

The problem of nuclear molecular states is now over 20 years old[1])
but has still not been conclusively solved. It is until now still
unclear whether nuclear molecules really do exist or whether the
phenomena, like e.g. regular resonance structures, that in the past
have been taken as proof for their existence are, for example, due
to simple angular momentum window effects only.

In these lectures some of the more recent developments in this field
will be covered, both theoretical and experimental. After a short
discussion of some of the experimental phenomena that will be treated
we will try to review very briefly the models previously used. We
will then come to a discussion of our present understanding of well
matched channels and the physical mechanism that is behind the resonances in these.

Next will follow a discussion of mismatched channels. We will try to
argue there that the mismatched channels are a much more sensitive
tool to study the question of molecules as the structure aspects are
not completely overshadowed by trivial angular momentum window effects. In these studies it emerges simultaneously that the concept of
an equivalent local potential is a powerful tool to understand the
results of coupled channel calculations in physical terms.

The discussion of the reaction-theoretical models of nuclear molecules will be finished with a summary of the weakness of these models.
In particular, the alignment is a sensitive additional quantity that
has been measured recently. It will be shown that up to now no calculation is able to reproduce this alignment correctly.

In a next step some structural properties of nuclear molecules will
be discussed, based on relatively old two center shell model calculations done by Chandra and Mosel[2]). They allow one to interpret
the results of recent experiments on α-emission in heavy ion collisions in a natural and very physical way.

The physical effects that we will be concerned with in these lectures are the appearances of gross structures in the excitation
functions of single and mutual inelastic scattering between rela-

tively 'light' heavy ions, like e.g. $^{12}C+^{12}C$ or $^{16}O+^{16}O$. Such data have become increasingly available over the last few years; they all show regular gross structures at higher energies with a width of roughly 1 MeV (see refs. 3, 4 for example). Superimposed on these broad "resonances" are fine structures that are not understood yet and will not be treated in this article.

II. Molecular Models

II.1 Coupled Channels Descriptions

The standard - and quite successful - descriptions of the gross structures observed in heavy ion scattering were all based on the coupled channels formalism. This formalism underlies both the pioneering work of Imanishi[5], the double-resonance mechanism of Scheid and Greiner[6] and the Band Crossing Model (BCM) developed by Abe et al.[7]. In particular the latter model allows one to determine easily the channels that are expected to couple most strongly with the elastic one. These are just the ones that in a E vs J(J+1) plot cross the elastic band and that are thus well matched in the original, unperturbed potentials (see fig. 1). Thus this model clearly relies on a weak coupling

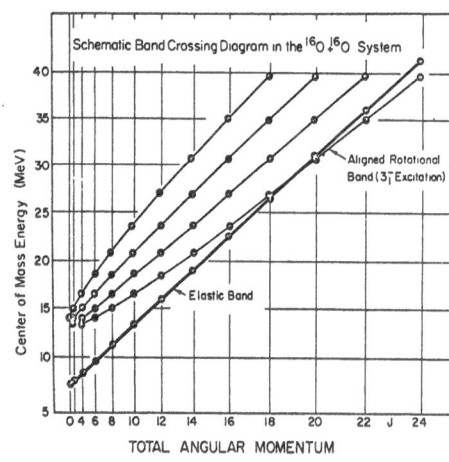

Fig. 1: Relevant bands for $^{16}O+^{16}O \rightarrow ^{16}O+^{16}O(3^-)$ inelastic scattering (from ref. 8). Only the aligned band crosses the elastic band. The band belonging to the 0_2^+ excitation is not shown; it would lie about 6 MeV above the elastic band and parallel to it.

assumption in the sense that matching or no matching is determined on the basis of the diagonal potentials alone. It is also immediately clear that a badly mismatched channel, like e.g. $^{16}O+^{16}O\rightarrow^{16}O+^{16}O(0_2^+)$, is not expected to resonate because the 0_2^+ band never crosses the elastic band.

For the well matched channels, however, the BCM is quite successful; it does give a series of bumps in the angle integrated cross sections that are dominated by successive partial waves (see fig. 2 from ref. 8).

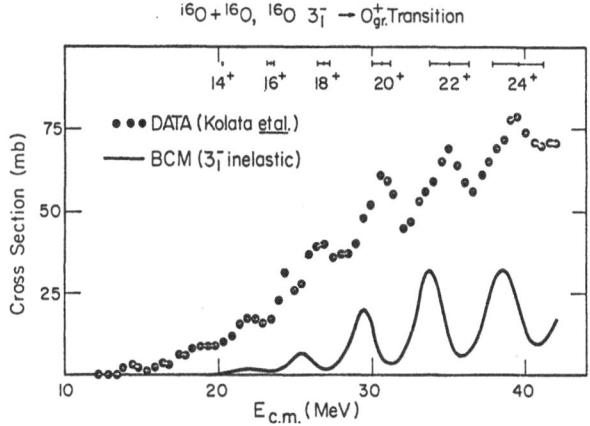

Fig. 2: Cross section for $^{16}O+^{16}O\rightarrow^{16}O+^{16}O(3^-)$ as calculated in the band crossing model (from ref. 8). The data are from Kolata et al. (ref. 34), they contain all the processes that deexcite through the 3^- state.

II.2 Strong Absorption Model

The observation that in the BCM the prediction of structure relies on both a weak coupling assumption and on a matching argument leads one to use the completely nonresonant Austern-Blair-Hahne (AB) prescription for a description of these data[9]. This has indeed been done by Phillips et al.[10] who obtain a good agreement with the measured single excitation cross sections for the reaction $^{16}O+^{16}O\rightarrow^{16}O+^{16}O(3^-)$ as far as the positions of the gross structure are concerned. Again the gross structures obtained are due to single partial waves; their presence obviously depends crucially on sufficiently narrow angular momentum windows since otherwise the structures would be smeared out.

In the AB method the inelastic scattering amplitude is related to derivatives of the S-matrix for the elastic channel. The S-matrix used by Phillips et al., however, does not give a good fit to the elastic scattering at forward angles. If, on the other hand, the exact elastic channel S-matrix as obtained from a best fit potential is used then the agreement of the AB with the data is considerably worse: the AB method no longer gives the prominent gross structures in the excitation functions[11,12]. The reason for this disagreement is simply that the AB model relies on strong absorption but that the potentials describing the scattering of relatively light heavy ions are all surface transparent and thus weakly absorbing.

II.3 DWBA-Descriptions

The next better model (in terms of increasing sophistication) is the DWBA. Here Tanimura has performed some very nice studies by fitting the one step process in the DWBA and then predicting the two step process in the same approximation[12]. In this way an unsatisfactory feature of many DWBA calculations, i.e. the large number of free parameters, is bypassed.

In this model it is assumed that the unperturbed potential is the empirical optical potential as in the conventional single or two-step DWBA: The unperturbed potential includes most of the effects of the couplings with inelastic and/or reaction channels and higher order effects are also taken into account by an adjustment of the empirical coupling strength. The T-matrix is then written up to the second order term in the interaction V as

$$T_{ji}^{J} = \tilde{T}_{i}^{J}\delta_{ji} + <u_{j}^{(-)}V_{ji}u_{i}^{(+)}> + \sum_{m(i\neq j)} <u_{j}^{(-)}V_{jm}G_{m}V_{mi}u_{i}^{(+)}>. \qquad (1)$$

Here the lower index specifies the state $((I_1,I_2)I,L,J)$ belonging to the channel (I_1,I_2), where $I_1(I_2)$, I, L and J represent the spin of nucleus 1(2), the channel spin, the orbital and total angular momenta, respectively. The T_{ji}^{J} in Eq. (1) is related to the S-matrix as

$$S_{ji}^{J} = \delta_{ji} - 2\pi i T_{ji}^{J} . \qquad (2)$$

In Eq. (1) $\tilde{T}_{i}^{J}$ denotes the T-matrix of the unperturbed Hamiltonian, $u_{i}^{(\pm)}$ the distorted wave function with incoming (-) or outgoing (+)

boundary condition and G_m the Green function of the empirical optical model Hamiltonian. The interaction V for the inelastic collective excitation is assumed to be of the conventional form:

$$V(r) = \sum_i R_i (dV_1^{nuc}(r)/dR_i) \sum_\mu \alpha_{\lambda\mu}^{(i)} Y_{\lambda\mu}(\hat{r}_i) ,$$ (3)

where $V_1^{nuc}(r)$ denotes the real part of the nuclear distorting potential in the elastic channel, R_i the potential radius of nucleus i (i = 1 and 2) and $Y_{\lambda\mu}(\hat{r}_i)$ the spherical harmonic. The symbol $\alpha_{\lambda\mu}$ denotes the usual phonon operator, $\alpha_{\lambda\mu} = (S_\lambda/\lambda)(a_{\lambda\mu}+(-)^\mu a_{\lambda-\mu}^\dagger)$, for the vibrational model and is transformed into the deformation parameter β_λ for the rotational model. It should be noted that there is in principle another second order term proportional to d^2V/dR^2. Since the second derivative of the potential, however, changes sign at the nuclear radius the contribution of this term to the cross section is much smaller and is, therefore, neglected.

The first, second and third terms in Eq. (1) represent the T-matrices of the elastic, single inelastic and mutual inelastic scatterings, respectively. In the last term the indices i, j and m are all different, which forbids the second order process that one nucleus is excited and then de-excites back to the same state, as in the conventional two-step process method. This process is assumed to be contained in the empirical optical potentials that are used in these calculations with slight modifications only for the inelastic channels. In this sense also the method used here is not the lowest order approximation to the exact coupled channels method but contains inherently a summation over many higher order contributions.

The total angle - integrated cross section for the inelastic excitation to the states with angular momenta I_1' and I_2' in target and projectile, respectively, from the ground states I_1 and I_2, can be written as

$$\sigma(I_1',I_2') = \sum_J \sigma_J(I_1',I_2')$$

$$= \frac{\pi}{k_1^2}(\delta_{I_1 I_2} \delta_{A_1 A_2}+1) \sum_{J,j\neq 1} (2J+1)\left|2\pi i T_{j1}^J\right|^2 ,$$ (4)

where $\sigma_J(I_1', I_2')$ denotes the contribution of the scattering states with total angular momentum J to the angle-integrated cross section $\sigma(I_1', I_2')$. The quantity k_1 denotes the wave number in the initial channel and A_1 and A_2 the mass numbers of target and projectile, respectively.

In the following some of the results that have been obtained on inelastic scattering will be described; the results for transfer reactions can be found in Tanimura's paper[12].

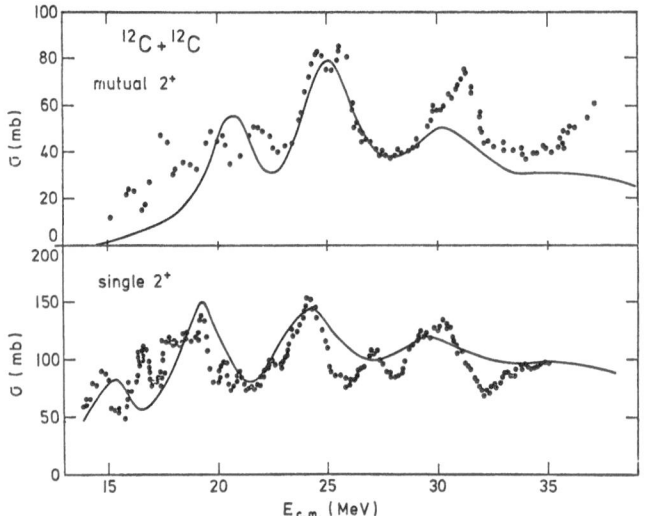

Fig. 3a: Angle-integrated cross sections for single and mutual excitation of the 2^+ state in $^{12}C + ^{12}C$ (from ref. 12)

Fig. 3a and 3b show the cross sections for the 2^+ and 3^- excitations in $^{12}C + ^{12}C$ and $^{16}O + ^{16}O$, respectively. The predictions for $^{12}C + ^{12}C$ and $^{16}O + ^{16}O$ reproduce well the gross structures in both the single and the mutual excitation channels and produce the correlation between them. Fig. 3b also shows the energy behavior of the σ_J defined in Eq. (4) in $^{16}O + ^{16}O$ scattering together with the spin values. Obviously the predicted peaks are ascribed mainly to the single grazing J contributions. The same holds good also in the other scatterings: In $^{12}C + ^{12}C$ the peaks of the gross structures at $E_{c.m.} \sim$ 15, 19, 24, and 30 MeV are due to J = 10, 12, 14, and 16, respectively. These spin assignments agree with the traditional ones. However, a note of caution is appropriate here since these assignments

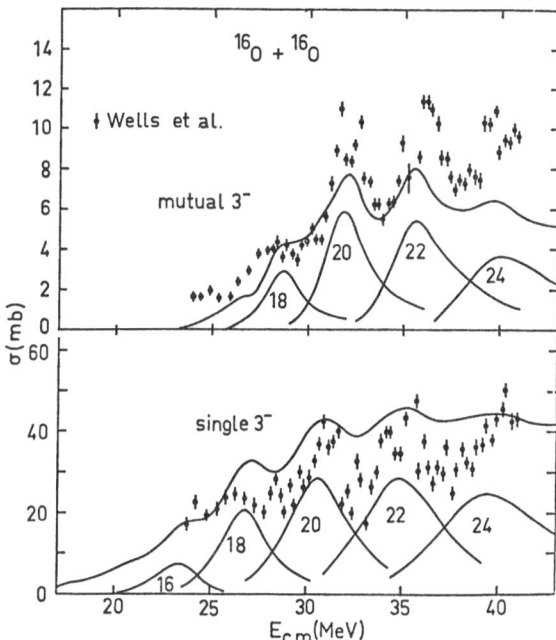

Fig. 3b: Angle-integrated cross sections for single and mutual excitation of the 3$^-$ state in $^{16}O+^{16}O$ (from ref. 12)

are based on fits of excitation functions only and not on careful analyses of angular distributions[13].

One should remark that there is no free parameter in the prediction of the mutual excitation cross section. Good agreements with the magnitudes of the mutual excitation cross sections are the signature for the consistency of the model used.

The agreement reached is obviously quite good. On the contrary, the same calculations give a rather bad description of the 0_2^+ channel that is mismatched by 3$\hbar$. Here the gross structure is missed completely[12]. This is an indication that the structure aspect of these excitations is not correctly described. On the other hand, the structure in the well matched channel is primarily determined by angular momentum windows due to simple kinematical matching. Any particular structure effects do not show up against this dominant, but trivial, background.

In the following few paragraphs the physical reason for the gross

structure in the DWBA cross sections is discussed. Its basis is a
kinematical matching condition for the real part of the total diagonal
potential of the unperturbed Hamiltonian:

$$V_i^{total}(r) = V_i^{nuc}(r) + V_i^{Coul}(r) + \frac{\hbar^2}{2\mu_i} \frac{L(L+1)}{r^2} - Q_i, \tag{5}$$

where $V_i^{nuc}(r)$ and Q_i represent the real part of the nuclear distorting
potential and the Q-value, respectively, in the channel i. The left
part of Fig. 4 shows the r-dependences of $V_1^{total}(r)$, $V_2^{total}(r)$ and
$V_5^{total}(r)$ for the $^{12}C+^{12}C$ system, where the indices 1, 2 and 5 denote
the elastic channel, the aligned coupling state of the single 2^+ channel and that of the mutual 2^+ channel, respectively. These three
states are of key importance because only these are well matched with
one another. The three channels support a zero-node resonance for
J≤16 as denoted by the horizontal lines. For J≤14, two aligned coupling
states resonate at nearly the same energy, resulting in the correlations between the single and mutual cross sections. For J≥16 the resonance energies do not coincide any more and the correlation would
be expected to disappear. This is, of course, a consequence of the
potentials used and thus experiments at higher energies would be very
important for a check of these ion-ion potentials: persistence of
the structure to higher energies would indicate a need for deeper potentials.

The right part of Fig. 4 shows the energy dependence of $|u_i^{(+)}(R_0)|^2$
for the same states as in the left part of the figure, where R_0 is
the potential radius. The functions exhibit clearly a resonant behavior. As seen from Eq. (3) the coupling interactions have a peak at
$r = R_0 = 5.9$ fm with a narrow width of $a_0 = 0.35$ fm and behave similar
to a δ-function. Since $u_i^{(+)}(r)$ for the grazing angular momenta J_g varies
smoothly as a function of r near $r = R_0$, Eq. (1) and Eq. (4) show that
the energy variations of $\sigma_{J_g}(2^+,0^+)$ and $\sigma_{J_g}(2^+,2^+)$ are due mainly to
those of $|u_2^{(-)}(R_0)u_1^{(+)}(R_0)|^2$ and $|u_5^{(-)}(R_0)G_2(R_0,R_0)u_1^{(+)}(R_0)|^2$, respec-
tively. It was found that the Green's function $|G_2(R_0,R_0)|^2$ varies
much more smoothly than $|u_2^{(+)}(R_0)|^2$ as a function of energy. Therefore,
we conclude that the prominent gross structures in the inelastic cross
sections have their origin in those of $|u_f^{(+)}(R_0)|^2|u_1^{(+)}(R_0)|^2$ where
the index f specifies the aligned coupling state in the exit channel.
Exactly this same mechanism is responsible also for the appearance of
gross structures in the transfer channels[12].

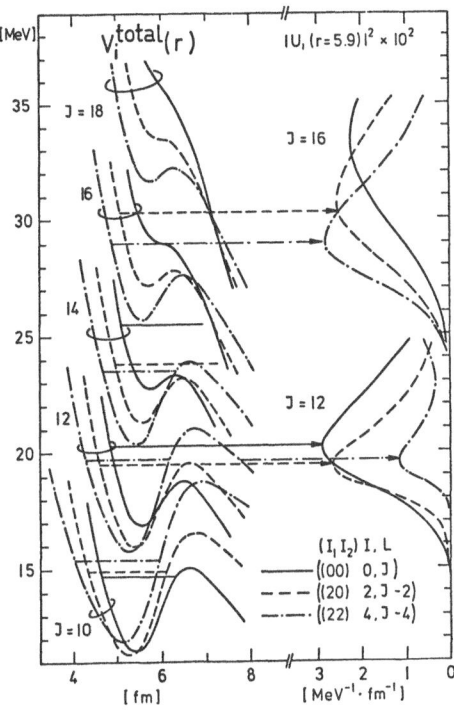

Fig. 4: Potentials for different channels for $^{12}C + ^{12}C$ are shown on the left. On the right side the squares of the radial wave functions at a fixed R = 5.9 fm are shown as a function of energy (from ref. 12).

On the other hand, for the mismatched scatterings, $|u_f^{(+)}(R_0)|^2$ shows a resonance-like behavior at specific J values smaller than J_g, while $|u_i^{(+)}(R_0)|^2$ has already decreased and has small values without any structure at the same energies and angular momenta. Therefore, the product $|u_f^{(+)}(R_0)|^2|u_i^{(+)}(R_0)|^2$ becomes small but still exhibits gross structures as a function of energy. This results in gross structures of the cross sections with very small absolute values and at angular momenta significantly below the grazing ones.

II.4 Strong Coupling Model: Mismatched Channels

As mentioned earlier the normal coupled channel calculations like, for example, those by Abe et al. are basically also weak coupling models and can, therefore, also not explain why the 0_2^+ channel in $^{16}O + ^{16}O$, for example, for which excitation functions are available[4], resonates in phase with the elastic channel. Obviously such an effect could be explained in the band crossing model only if the coupling were so strong that not the diagonal potentials but instead some

"diagonalized" potentials that contain the effects of the channel cu-
plings would have to be considered.

It is, therefore, extremely gratifying that a calculation based on a
strong coupling model[14) is indeed able to give a good description of
the single 3⁻ and the 0₂⁺ cross section[15) (see fig. 5). The coupling

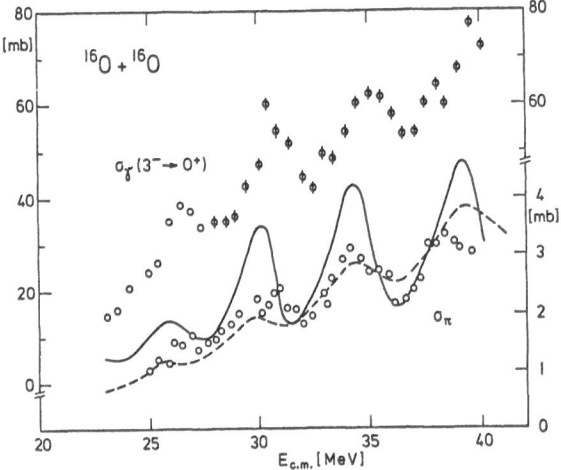

Fig. 5: Angle-integrated
cross sections for
the 0⁺ and the 3⁻
excitation in
¹⁶O+¹⁶O (from ref.
15). Data for the
3⁻ channel are from
ref. 34, they con-
tain contributions
also from other
channels that de-
excite through the
3⁻ state.

matrix elements used in this calculation have been obtained from a
folding model; they are considerably larger than the standard gradient
type couplings and also shifted somewhat towards smaller distances.

A physically transparent explanation for this success is obtained by
calculating the local equivalent potentials, i.e. the potentials that -
when used in a normal optical model calculation - give exactly the
same wave function as the coupled channel calculations. The coupled
channel equations have the form

$$\left[-\frac{\hbar^2}{2\mu} \frac{d^2}{dr^2} + U_{ii} - E \right] u_i(r) = - \sum_{j \neq i} V_{ij} u_j(r) \tag{6}$$

Here the $u_i(r)$ are the radial wave functions for the channel i. The
potentials U_{ii} contain in addition to the diagonal coupling potentials
V_{ii} also the centrifugal parts, the Coulomb and absorptive potentials

and the excitation energy of channel i.

The polarization potential ΔV for channel i that contains all the channel coupling effects is then defined by the equation:

$$\left[- \frac{\hbar^2}{2\mu} \frac{d^2}{dr^2} + U_{ii} - E \right] u_i(r) = - \Delta V_i(r) u_i(r) \tag{7}$$

and the trivially equivalent local potential (TELP) is given by:

$$V_i^{TELP} = U_{ii} + \Delta V_i \tag{8}$$

The quantities ΔV can easily be computed once the cc-equations have been solved since the coupled channel wave functions $u_i(r)$ and the whole lhs of eq. (7) are known then.

These potentials V^{TELP} are shown in fig. 6 together with the diagonal

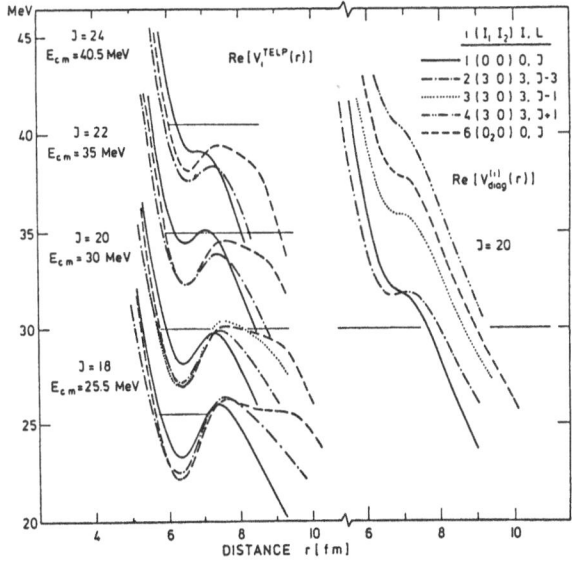

Fig. 6: Diagonal potentials U_{ii} (eq. 6) for J=20 are shown on the right for several channels of $^{16}O+^{16}O$. On the left the trivially equivalent local potentials V^{TELP} (eq. 8) are given for the energies and total angular momenta indicated.

potentials for J=20 in the right hand part of the figure. The remarkable result is that they differ very significantly from the original diagonal potentials: The diagonal potentials, for example, for the angular momentum J=20, are shifted by more than 30 %. In addition the equivalent potentials all show pockets that can support resonances and they all become very similar to each other so that the resonances practically all lie on top of each other. In this case the nondiagonal couplings thus have had a decisive influence[15]. At the same time this example serves as a nice illustration for the insight that may be gained by an inspection of the equivalent local potentials[16].

II.5   General Discussion of Polarization Potentials

The polarization potentials discussed in section II.4 contain all the effects of the couplings between the different channels. In a macroscopic language they describe how the nuclei will deform when they approach each other. The information on nuclear stiffnesses etc. is contained in the transition strengths and thus enters into ΔV through the coupling potentials.

These same potentials can also be obtained in a quite different, though, of course, completely equivalent way. Following Feshbach's theory[17] one can introduce a projection operator P that projects on a particular channel. The optical potential in this channel then is given by:

$$U = V_{PP} + V_{PQ} \frac{1}{E-H_{QQ}+i\varepsilon} V_{QP} \tag{9}$$

where V_{PQ}, e.g., stands for PVQ (Q=1-P). The second term on the right hand side of this equation is the polarization potential ΔV. Since the propagator $G = [E-H_{QQ}+i\varepsilon]^{-1}$ is nonlocal, ΔV is nonlocal itself. It can also be seen that in addition ΔV is energy-dependent (through G) and, since all matrix elements are angular momentum coupled, also dependent on the total angular momentum J.

For a realistic calculation of the polarization potentials one would like to start with a system in which all the strong inelastic excitations can be treated explicitly and in which furthermore no break-up channels are open. Then the total flux goes either into these inelastic states or into compound nuclear excitations whose influence can be well described by an optical model potential. Such a situation is

indeed given in the case of $^{12}C+^{12}C$, for which we have performed calculations[18].

The $(0^+,0^+)$, $(0^+,2^+)$, $(0^+,0_2^+)$, $(2^+,2^+)$, $(0^+,3^-)$, and $(0^+,4^-)$ states
were included in the calculation. These states exhaust at $E_{c.m.} \approx 30$ MeV
almost all of the reaction cross section aside from fusion[19]. The
absorption into the many dense-lying compound nuclear states is
described by a usual imaginary potential that contains a J-dependence.
This J-dependence decreases the imaginary potential with increasing J
and is necessary for the grazing angular momentum to have enough flux
left in the surface region. At the grazing angular momentum we expect
strong transitions to the strongly coupled collective states and
therefore, a transition from volume absorption (due to fusion) to
surface absorption (due to excitations).

The polarization potential is nonlocal and can be calculated with the
Green's function[20]:

$$G_{ij}(r,r') \sim - \sum_k \begin{cases} f_{ik}(r)h_{jk}(r') & r<r' \\ h_{ik}(r)f_{jk}(r') & r>r' \end{cases} \qquad (11)$$

where f_{ik} and h_{ik} are the regular and irregular solutions of the cc-
equations, in which the elastic state is eliminated, with the asymptotic behavior:

$$f_{jk}(r) \xrightarrow[r\to\infty]{} \frac{1}{\sqrt{k_j}} \frac{1}{2} (H_j^{(-)}\delta_{jk} - H_j^{(+)}S_{jk})$$

$$h_{jk}(r) \xrightarrow[r\to\infty]{} \frac{1}{\sqrt{k_j}} H_j^{(+)}\delta_{jk} \quad . \qquad (12)$$

Then the nonlocal potential is given by:

$$v^J(r,r') = \sum_{i,j} v_{0i}^J(r)G_{ij}^J(r,r')v_{j0}^J(r') \qquad (13)$$

This nonlocal potential is strongly angular momentum dependent with a
large nonlocality range at low angular momenta while the potential
becomes more local near the grazing angular momentum.

The local polarization potential however is much more interesting. The equivalent local polarization potential can be calculated in two different ways:

$$\Delta V^J_{TELP}(r) = \int V^J(r,r')\psi^J_0(r')dr'/\psi^J_0(r) \tag{14}$$

or

$$\Delta V^J_{TELP}(r) = \sum_{i \neq 0} V^J_{0i}(r)\psi^J_i(r)/\psi^J_0(r) \tag{15}$$

(the latter is the method of eq. 7). If done correctly, both expressions do, of course, agree. The local polarization potentials calculated in this way are still J dependent and complex.

Fig. 7 shows an example of the nonlocal polarization potential (ab-

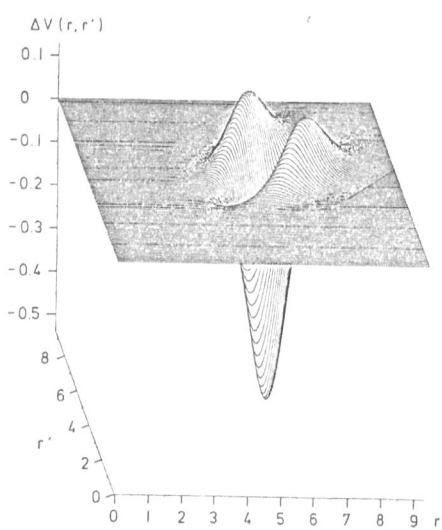

Fig. 7: Nonlocal imaginary part of ΔV for $^{12}C + ^{12}C$ for $J=6$.

sorptive part) whereas fig. 8 exhibits the J-dependence of these potentials ΔV at a fixed energy. It can be seen that the polarization potentials are negative and have a Gaussian form peaked in the nuclear surface region. At low angular momenta the imaginary part of

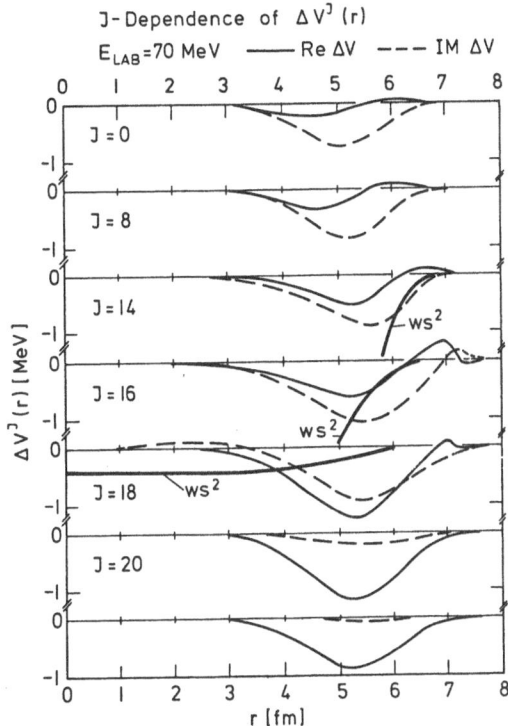

J-Dependence of $\Delta V^J (r)$

$E_{LAB}=70$ MeV —— Re ΔV - - - IM ΔV

Fig. 8: Real and imaginary parts of ΔV for several angular momenta for $^{12}C + ^{12}C$ as a function of r.

ΔV is dominant. This imaginary potential reaches its maximum depth near the grazing angular momentum and disappears rather quickly beyond J_{gr}. The real part of ΔV, which has the maximum depth again around J_{gr}, is also present beyond J_{gr}. The physical picture is that at impact parameters where a real excitation can happen, especially around the grazing collision impact parameter, the mutual influence of the fields of the two nuclei is strongest with a corresponding strong absorption into these states. Beyond the grazing impact parameter, however, this mutual distortion is too weak to produce real transitions; the virtual transitions, however, generate the real part of the polarization potential.

These effects are large: For a typical shallow potential of a central depth of 17 MeV the polarization potential is about 1 MeV in the important surface region where the Woods-Saxon itself is rather small.

The J dependent background imaginary potential is nearly zero around J_{gr}. Therefore, almost the whole imaginary potential for partial waves close to J_{gr} comes from this polarization effect. It should be mentioned that the J dependence of ΔV is rather insensitive to the J dependence of the background potential.

These studies are to our knowledge the first comprehensive investigations of the effects of nuclear polarizability on heavy ion potentials. The effects , as has been shown, can be quite large. The use of free propagators $G^{(0)}$ instead of the correct Green function is not justified in this situation.

II.6   Open Problems : Alignment

All the models discussed above ascribe the gross structures in the cross sections to relatively narrow angular momentum windows around the grazing partial waves. They also all rely on a matching argument, either in the original potentials or in the equivalent local potentials and they all agree in their prediction that the fully aligned substates of any reaction channel should be the most prominent ones.

Because of this predicted dominance of the fully aligned substates in which the intrinsic angular momentum is oriented parallel to the orbital angular momentum one expects that the alignment, which can be measured, shows prominent peaks at the peaks of the cross sections.

The alignment has been measured by now by two different groups[21,22]. Although the data show some discrepancies they agree in two aspects: First, they do show gross structures in the alignments that are correlated with those in the cross sections and second, the measured alignments are smaller than expected and are nearly compatible with an equal distribution of all possible substates. Calculations of the alignment in several different models[23] all give large alignments, as expected, and thus do not describe the data as can be seen from the comparison in fig. 9. This latter point is a complete puzzle up to now.

Somewhat more encouraging is a new measurement by the Munich group for the mutual excitation channel[24]. In this measurement the alignment of one of the two ^{12}C nuclei averaged over all possible orienta-

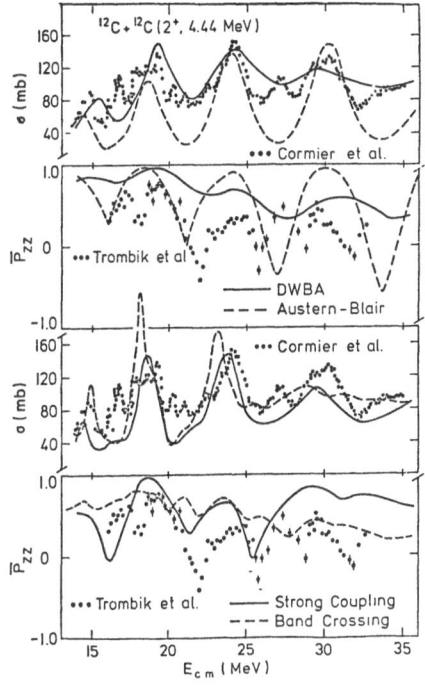

Fig. 9: Measured (ref. 22) and calculated (ref. 23) angle
averaged alignments $\bar{P}_{zz}$
together with inelastic cross
sections (from ref. 23).

tions of the other is measured. The data do show two prominent peaks
at $E_{c.m.} \approx 24$ MeV and 30 MeV that by far exceed the equal occupation
limit and are in line with a sticking limit picture (see section III).
Since, however, the alignment of the other ^{12}C is not measured, but
averaged over, one has to be somewhat careful with any final conclusion
even though the experiments certainly do look very promising.

III.   Shape Isomers as Molecular States?

In the preceding chapters we have tried to describe some of the reaction-theoretical models for a description of nucleus-nucleus interactions. The structure of the particular nuclei involved entered only
through the empirical transition strengths that fix the nondiagonal
coupling matrix elements.

In this chapter a different approach is chosen that relies on structure considerations for a dinuclear system. In atomic physics the

value of so called correlation diagrams that describe the energy levels of the electrons as a function of the nucleus-nucleus distance is well established[25]. These diagrams exhibit crossings and avoided (or quasi) crossings at which electron excitations during a collision can take place.

In exactly the same way the nuclear two-center level diagrams may be a useful tool to understand what is actually happening during a nucleus-nucleus collision. Fig. 10 shows the diagram appropriate for a

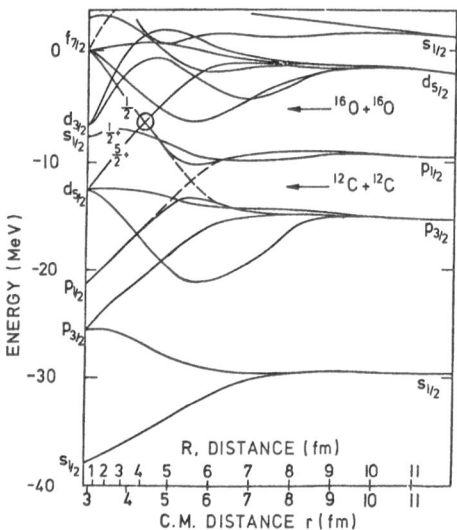

Fig. 10: Correlation diagram of symmetric heavy ion systems (from ref. 2).

symmetrical $^{12}C+^{12}C$ or $^{16}O+^{16}O$ collision[2]. One can easily see that ^{12}C nuclei that approach each other get distorted but not destroyed in their structure down to a distance of about 6 fm. There a quasi-crossing of the occupied $p_{3/2}$ shell with a state originating in the excited $p_{1/2}$ shell (i.e. another N=1 state) takes place. In a fast process there is a large probability for making the jump up into the higher state and through another crossing into the $f_{7/2}$ shell of the ^{24}Mg nucleus[2]. The state that is formed in ^{24}Mg is thus a simple excited quartet state with an α-cluster (because of spin-isospin degeneracy) in the $f_{7/2}$ shell and a quartet hole state in the $p_{1/2}$ shell. This is exactly the configuration of a highly deformed shape isomeric state in ^{24}Mg at deformation $\varepsilon \approx 1.26$, $\gamma = 42^\circ$ obtained by

Leander and Larsson[27] in a Nilsson model calculation which is indeed reached by the dynamical path as outlined above[2].

The speculation is thus that this state is populated in a $^{12}C+^{12}C$ collision. Indeed it can be shown that the particular dynamical path discussed above is not a consequence of the specific two center model used but can easily be constructed as the diabatic path in a level scheme that is constructed by using only the Pauli principle[26]. That this state is furthermore connected to the question of nuclear molecules is indicated by the decay pattern expected for such a state and its observation in recent experiments[28].

The ^{24}Mg shape isomer is highly excited and is obviously expected to decay by α-emission to an excited ^{20}Ne state with a quartet hole in the p-shell and a quartet in the sd-shell. This state is the intrinsic state underlying the excited rotational band in ^{20}Ne with a band head at about 6-7 MeV as was shown by Harvey[29] and by Arima[30]. These considerations would also predict a continuation of the α-decay chain into the excited 0_2^+ state in ^{16}O.

It is extremely encouraging to see the results by Cosman et al.[28,31] on the $^{12}C+^{12}C \rightarrow ^{20}Ne^* + \alpha$ channel that indeed do show that the α-decay goes primarily to just these excited states in ^{20}Ne. This observation rules out the possibility that the α-decay originates from a compound nucleus since such a process would favor the higher Q-values connected with a ground state transition. The experiments of Cosman et al. also show that the α-decay channel shows prominent gross structures that are correlated with those in the inelastic scattering channels thus implying a connection between these two phenomena. The same features have also been observed by Katori et al. in their studies of $^{12}C+^{16}O$ (ref. 32).

If indeed the shape isomeric state in ^{24}Mg is populated in a $^{12}C+^{12}C$ collision then, of course, an angular momentum transfer from relative motion into intrinsic motion has to take place since for the formation of a single ^{24}Mg nucleus sticking of the two ^{12}C nuclei must occur. The sticking limit for the angular momentum transfer $\Delta \ell$ is well known; for a symmetrical system it is:

$$\Delta \ell = \frac{2}{7} \ell_i$$

where ℓ_i is the angular momentum in the entrance channel[33]. In the energy range in question where the grazing angular momenta are between $\ell \sim 12$ and $\ell \sim 18$ this sticking limit implies that a mutual 2^+ excitation ($\Delta\ell=4$) in the $^{12}C+^{12}C$ system takes place and this particular channel should thus be a signature and most sensitive to the formation of a shape isomeric molecules as pointed out in some more detail in ref. 26.

IV. Summary

The main lesson that one may learn from the studies discussed in these lectures is that the gross structures observed in inelastic heavy ion scattering are due to an overlap of enhancements of the wave functions in the incoming and outgoing channels. This conclusion holds for all models that are presently being used for a description of these phenomena. If these "enhancements" are due to true resonances of the nucleus-nucleus system or to simple angular momentum window effects is very hard to decide in the presence of absorption.

The mismatched channels, however, lie clearly outside the trivial windows and are thus more sensitive to the underlying structure. Here indeed an encouraging simultaneous description of the 3^- and the 0_2^+ channels in $^{16}O+^{16}O$ scattering has been reached in a strong coupling calculation. It will be a challenge for the future to see if such a model can indeed give a consistent description of other mismatched channels as well.

Ultimately, the question of nuclear molecules cannot be answered by reaction theory alone. The structure information that is contained in nuclear correlation diagrams must clearly be used as well. In the last chapter of these lectures we have discussed some features of α-decay channels that may be understood in such a framework. It will be exciting to see if more details of these correlation diagrams, like e.g. the rise of a s.p. level into the continuum, can be explored in the future.

References

There are two recent, comprehensive proceedings of conferences on heavy ion molecular states available that can be consulted for more details both on experiments and on theory:

I)   Nuclear Molecular Phenomena, N. Cindro ed.,
     (North Holland, Amsterdam) 1978

II) Resonances in Heavy Ion Reactions, K.A. Eberhard ed.,
     Lecture Notes in Physics, Vol. 156 (Springer, Berlin) 1982

1)   D.A. Bromley et al., Phys. Rev. Lett. $\underline{4}$ (1969) 365

2)   H. Chandra and U. Mosel, Nucl. Phys. $\underline{A298}$ (1978) 151

3)   T.M. Cormier et al., Phys. Rev. Lett. $\underline{40}$ (1978) 924
     T.M. Cormier, in II), p. 95

4)   W.S. Freeman et al., Phys. Rev. Lett. $\underline{45}$ (1980) 1479
     P. Paul, in II), p. 161

5)   B. Imanishi, Nucl. Phys. $\underline{A125}$ (1969) 33

6)   W. Scheid et al., Phys. Rev. Lett. $\underline{25}$ (1970) 176

7)   T. Matsuse et al., Progr. Theor. Phys. $\underline{59}$ (1978) 1009

8)   Y. Kondo et al., Phys. Rev. $\underline{C22}$ (1980) 1068

9)   N. Austern and J.S. Blair, Ann. Phys. (NY) $\underline{33}$ (1965) 15
     F.J.W. Hahne, Nucl. Phys. A104 (1967) 545

10) R.L. Philipps et al., Phys. Rev. Lett. $\underline{42}$ (1979) 566

11) O. Tanimura, in II), p. 372

12) O. Tanimura, University of Giessen preprint 1982, to be published

13) O. Tanimura et al., Univ. of Giessen preprint 1982, to be published

14) B. Imanishi, in I), p. 379
     O. Tanimura and B. Imanishi, Phys. Lett. $\underline{80B}$ (1979) 340
     O. Tanimura and T. Tazawa, Phys. Rep. $\underline{61}$ (1980) 253

15) O. Tanimura and U. Mosel, Phys. Rev. $\underline{C24}$ (1981) 321

16) M.A. Franey and P.J. Ellis, Phys. Rev. $\underline{C23}$ (1981) 787

17) H. Feshbach, Ann. Phys. (NY) 19 (1962) 287

18) R. Wolf et al., Univ. Giessen, to be published

19) B.R. Fulton et al., Phys. Rev. C21 (1980) 198

20) R. Wolf et al., Z. Phys. A305 (1982) 179

21) K.A. Erb, Proc. Int. Conf. Resonant Behavior of Heavy Ion Systems, Aegean Sea, Greece, 1980

22) W. Trombik, in II), p. 297

23) O. Tanimura and U. Mosel, Phys. Lett. B (1982), in press

24) W. Trombik, seminar at this summerschool and to be published

25) U. Fano and W. Lichten, Phys. Rev. Lett. 14 (1965) 627

26) U. Mosel, in II), p. 358

27) G. Leander and S.E. Larsson, Nucl. Phys. A239 (1975) 93

28) E.R. Cosman, in II), p. 112

29) M. Harvey, Nucl. Phys. A202 (1973) 191

30) A. Arima et al., Phys. Rev. Lett. 25 (1970) 1043

31) E.R. Cosman et al., MIT preprint 1981 and Phys. Rev. C21 (1980) 2111

32) K. Katori et al., Phys. Rev. C21 (1980) 1387

33) R. Bass, Nucl. Phys. A231 (1974) 45

34) J.J. Kolata et al., Phys. Rev. C19 (1979) 2237

THE EXCITATION AND DECAY OF ISOSCALAR GIANT RESONANCES

A. VAN DER WOUDE

Kernfysisch Versneller Instituut
Groningen, The Netherlands

THE EXCITATION AND DECAY OF ISOSCALAR GIANT RESONANCES.

In the past few years several review papers have been published on the subject of giant resonances, see for instance references 1-3. In these references the gross features of these collective modes are already described in detail. The present contribution will focus on some new developments and especially on those topics which have not yet been discussed in general review papers on giant resonances before and/or which may be of special interest to the field of heavy ion research.

The topics to be covered are:

(i)   Survey of our present knowledge on isoscalar giant resonances. Attention will be paid to the problems involved in the analysis of inelastic hadron scattering data. One of the main open questions, namely the behaviour of the monopole resonances in light nuclei will be discussed in some detail.

(ii)   The excitation of giant resonances by inelastic heavy ion scattering. It will be shown that at present no new information on giant resonances has come out of heavy ion scattering. Future prospectives, when higher energy heavy ion beams with better quality become available will be discussed.

(iii) Damping and decay of giant resonances. The trend observed for the width of the giant resonances will be summarised. An extensive discussion of the fission width of giant resonances in ^{238}U will be presented.

1. SURVEY OF OUR PRESENT KNOWLEDGE ON ISOSCALAR GIANT RESONANCES.

1.1.   Predictions.

Qualitative the main features of giant resonances can be understood from a schematic microscopic model as illustrated in figure 1 for a closed-shell nucleus. Giant resonances are collective 1p-1h excitations which on account of the parity $(-)^L$ can be arranged in the following scheme:

monopole	$L=0$			$2\hbar\omega$		
dipole	$L=1$		$1\hbar\omega$			
quadrupole	$L=2$	$0\hbar\omega$		$2\hbar\omega$		
octupole	$L=3$		$1\hbar\omega$		$3\hbar\omega$	
hexadecapole	$L=4$	$0\hbar\omega$	$2\hbar\omega$			$4\hbar\omega$

The transitions with $0\hbar\omega$ correspond to the low-lying inner-shell transitions if we do not have a closed-shell nucleus.

Due to the residual p-h interactions, isoscalar giant resonances will be located at a lower and isovector resonances at a higher excitation energy than the

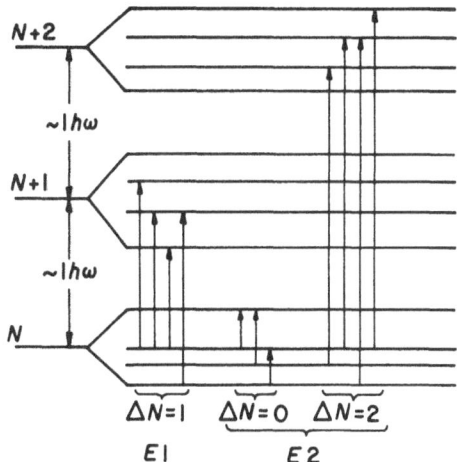

$N+2$

$\sim 1\hbar\omega$

$N+1$

$\sim 1\hbar\omega$

N

$\underbrace{\Delta N=1}$ $\underbrace{\Delta N=0}$ $\underbrace{\Delta N=2}$

$E1$ $E2$

Fig.1.1. Schematic representation of E1 and E2 single-particle transitions between shell model states.

corresponding p-h excitation. Thus the $2\hbar\omega$, $\Delta T=0$, $L \geqslant 2$ resonances are expected at about $65A^{-1/3}$ MeV instead of $82A^{-1/3}$ and the $1\hbar\omega$ $\Delta T=1$ GDR resonance at $75A^{-1/3}$ MeV instead of $41A^{-1/3}$ MeV. The $2\hbar\omega$, $L=0$ resonance, the giant monopole or the so-called breathing mode, is of special interest because its excitation energy E_x is directly connected to the nuclear compressibility K_A:

$$E_x^{0+} = \frac{\hbar^2}{m}\left(\frac{K_A}{\langle r^2\rangle_{g.s.}}\right)1/2 \tag{1}$$

By measuring the A-dependence of K_A, in principle one should be able to extract a value for the nuclear matter compressibility K_∞ $\lim_{A\to\infty} K_A$. The breathing mode energy is actually the only direct method to determine this quantity K_∞ which is of fundamental importance for the understanding of nuclear matter. However, as will be discussed, deriving K_∞ from the present data is not as straight forward as one might have hoped.

The schematic arrangement shown above already indicates one serious experimental problem: different multipole strength functions like the $2\hbar\omega$, $\Delta T=0$ with $L=2,4,\ldots$ ones will be located at approximately the same excitation energy. In fact, as we will see, it is still a major experimental problem to decompose the $\Delta T=0$, $2\hbar\omega$ resonance into its different multipole components. This feature is illustrated in figure 2 which shows an example of a theoretical (α,α')-spectrum for 120 MeV α's on ^{208}Pb. Since the α-particle itself is an isoscalar particle, only isoscalar resonances are excited. The $2\hbar\omega$ bump consists of a mixture of $2^+, 4^+$ and 0^+ strength. The situation becomes even more complex if also isovector resonances will be excited as is the case in (e,e') scattering. Then the $2\hbar\omega$ $\Delta T=0$ bump and $1\hbar\omega$ $\Delta T=1$ bump (GDR) mix and the resulting spectrum is even more complicated.

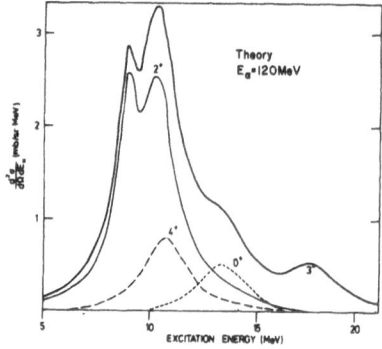

Fig.1.2. Theoretical (α,α) — spectrum for ^{208}Pb at E = 120 MeV and 14° scattering angle[4].

1.2. Experimental techniques for studying the isoscalar giant resonances.

1.2.1. The tools.

It is clear that in order to unravel the expected complicated spectrum one should use a probe which is as selective as possible. By far the most selective one is the photon. Absorption measurements with real photons are mainly sensitive to E-1 strength because of the low momentum transferred. In order to study the L≠1 multipoles one has to use other probes and for the study of the isoscalar ones, inelastic hadron scattering like (α,α') or $(^3He,^3He')$ has been used extensively. DWBA predictions for (α,α') excitations are shown in figure 3[5]. In agreement with the Blair phase rule the L=0,2 and 4 angular distributions are approximately in phase and out of phase with the L=1,3,...distributions. The main difference between L=0 and the L=2,4 distributions is in the small angle behaviour 0 < θ < 5°. The unequivocal identification of L=0 strength in inelastic scattering requires scattering experiments at very small angle. Figures 2 and 3 also shows that distinguishing between L=2 and 4 strength in (α,α') experiments will be difficult. They nearly completely overlap in excitation energy while their angular distributions only differ substantially around θ ~ 8° where the L=2 distribution shows a rather shallow minimum which does not occur in the L=4 one.

It is then clear that for detecting L=0 strength one has to perform measurements at very small scattering angles, preferably 0° included. Experimentally this is only possible by using magnetic devices like a magnetic spectrograph so that beam and scattered particles can be separated spatially. A very careful set-up of the beam-guiding system is required in order to avoid a halo of lower energy particles around the beam. The technique has been pioneered by the Texas A&M group [1,5,6] and also has been extensively been used at Grenoble[7,8]. Recently also the

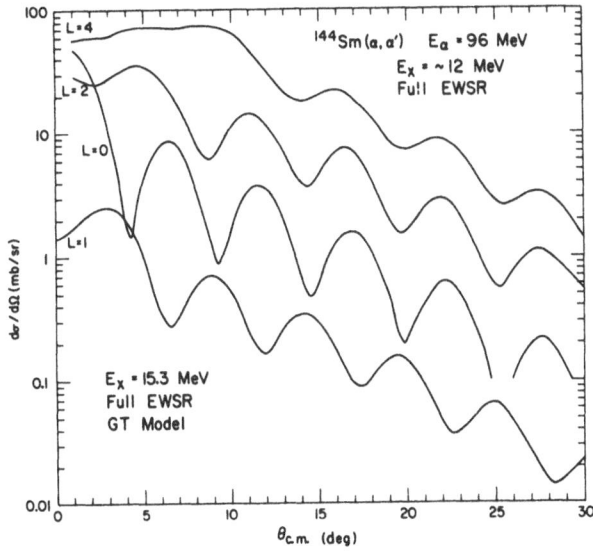

Fig.1.3. Angular distributions for inelastic scattered α-particles calculated in DWBA[5].

Groningen group has succeeded in measuring (α, α') at $0°$ and used it to measure the decay modes of the L=0 strength in ^{40}Ca[9].

1.2.2. The data and their analysis.

Figure 4 shows a typical set of (α, α') spectra and a possible decomposition in continuum and multipole strength[10]. Besides the large bump around $E_x \sim 14$ MeV due to GQR and GMR excitation and the one around $E_x \sim 7$ MeV due to excitation of the $1\hbar\omega$, L=3 resonance (Low Energy Octupole Resonance) there is some

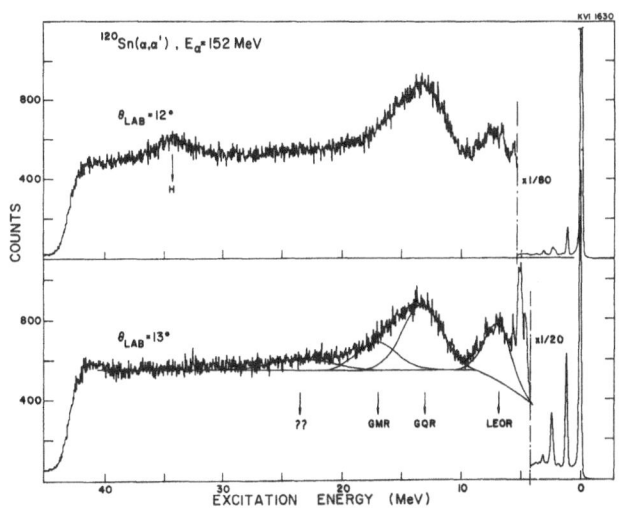

Fig.1.4. $^{120}Sn(\alpha, \alpha')$ -spectrum at E_α = 152 MeV. A possible decomposition into a continuum and in different multipole contributions is indicated[10].

evidence for a broad bump around $E_x \sim 22$ MeV which very likely is due to L=3, 3ℏω excitation (HEOR). From this spectrum one also clearly recognizes the main problems for data analysis: how to draw the continuum and what is the shape of the resonances.

Usually one assumes for sferical nuclei Gaussian or Breit-Wigner like strength distribution for each resonance. Such an assumption is probably justified because: (i) It holds for the ΔT=1 GDR[11] and since the damping mechanisms for giant resonances are expected to be similar, it should also hold for the GQR and GMR (ii) It works well for decomposing the spectra taken at 0° where the monopole is at a maximum and at an angle of about 4° where the monopole contribution is at a minimum. It certainly does not hold for lighter nuclei where the bump characteristic for excitation of the 2ℏω, ΔT=0 strength is dissolved into many different fragments[12].

Subtracting the continuum underlying the resonances is largely a question of taste since no reliable theoretical guidance is yet available. Processes which can contribute to the continuum in addition to the rather obvious instrumental background are: (i)Quasi-elastic scattering processes like (α,α'p) or (α,α'n) (ii)Multistep excitation of the dense spectrum of complicated states of all multipolarities and (iii)Processes like[13]

$$\alpha + A \rightarrow (A-1) + He^{5^*}_{\rightarrow \alpha + n} \text{ or } Li^{5^*}_{\rightarrow \alpha + p}$$

Although some progress has been made in understanding the shape and magnitude of the spectra at high excitation energy, in the region of main interest, just a few MeV above the neutron separation energy, there is still a lot of ambiguity. This is especially serious when one wants to unravel L=2 and 4 contributions: their angular distributions differ mainly in the minima and the cross section at those angles is most sensitive to the uncertainties in continuum subtraction. At present the situation is such that in most (α,α') and also (^{3}He,^{3}He') scattering experiments one can not distinguish between L=2 and L=4 contributions to the 2ℏω bump. Recently it has been shown that in 200 MeV (p,p') experiments there is a sensitivity to L=4 admixtures[14,15]. For ^{120}Sn[14] and ^{208}Pb[15] the presence of a small amount of L=4 strength in the 2ℏω bump was observed indeed.

A typical series of spectra illustrating the presence of L=0 strength in ^{90}Zr is shown in figure 5, which is taken from reference 7. Again the reader can easily judge for himself that although the overall features – the decomposition in two Gaussian like contributions with the L=0 one going through a minimum at about θ ∼ 5° – certainly holds, the choice of a different continuum could be equally well justified and will largely effect the cross sections extracted.

Besides the experimental uncertainties illustrated above, there are also uncertainties involved in the analysis of the cross sections and in their conversion to multipole strength.

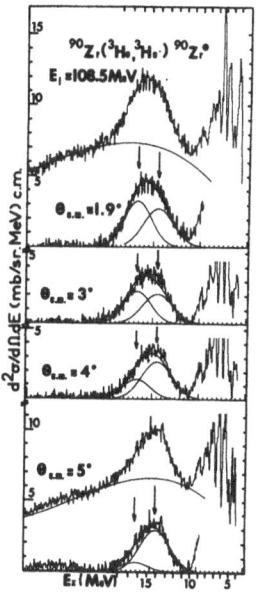

Fig.1.5. Small-angle (^{3}He,^{3}He') inelastic scattering data on ^{90}Zr at E(^{3}He) = 108 MeV. The $E_x \sim$ 16 MeV bump has a minimum around θ = 5°, which identifies it as a ΔL=0 transition[7].

In general from the cross section angular distribution one derives a mass deformation parameter β_L^M using a DWBA analysis with collective form factors[1]. Here the usual and major assumption is that the vibrational model form factor corresponds to the real one, which is not necessarily the case.

The resonance strength is expressed as a fraction of the appropriate sumrule limit according to:

$$S_L(IS) = (\beta_L^m)^2 \, E_x / \tilde{S}_L$$

where for L > 0 the sumrule limit $\tilde{S}_L$ for a uniform mass distribution and isoscalar excitations is given by:

$$\tilde{S}_L = (\hbar^2/2m \, R^2) \, (4\pi/3A) \, L(2L + 1)$$

$\tilde{S}_L$ has been derived for the transition operator $r^2 \, Y_L^M$, but the real transition operator may have a substantial different R-dependence (see for instance ref.1). This seems to be an especially serious problem for high energy (p,p') scattering where using the standard analysis only about half the L=2 strength is found compared to (α,α') experiments[14,15].

For the monopole strength nearly all inelastic hadron scattering data are analysed by assuming a transition density obtained from radial scaling of the density distribution[14]:

$$\delta \, \rho(r) = \alpha \left(3 \, \rho_0(r) + r \, \frac{\delta \rho_0(r)}{\delta r} \right) \qquad (2)$$

The deformation parameter α can be expressed as a fraction of the sumrule limit:

$$\alpha^2 \, E_x = (\frac{\hbar^2}{2m}) \, \frac{1}{A \langle r^2 \rangle_{g.s.}}$$

Strictly speaking this procedure is only justified if the sumrule is exhausted by a single state at E_x. The associated transition potential $\delta \, U(r)$ is then in nearly all cases taken as:

$$\delta U(r) = 3 \, U_o(r) + r \, \frac{\delta \, U_o(r)}{\delta r} \tag{3}$$

where $U_0(r)$ is the optical potential.

In principle a better way would be to use a folding potential in which for instance the transition potential $\delta U(r)$ is obtained from:

$$\delta U_F(r) = \int \delta \, \rho(r') \, V_{N\alpha} (|r-r'|) \, dr' \tag{4}$$

where $V_{N\alpha}$ is a nucleon–alpha interaction. Using $\delta U(r)$ or $\delta U_F(r)$ in the DWBA calculations makes a very large difference in the monopole strength extracted from the data[10], while for the GQR the effect is well with in the experimental uncertainties. This indicates that at present the analysis of the monopole strength is still ambiguous.

1.3.   Summary of information on giant resonances.

The present status of the information on giant resonances is shown in table 1[2]. For more detailed information the reader is referred to some recent review papers. Here one should remark that: (i) The GQR parameters in nearly all cases are derived under the assumption that the whole $2\hbar\omega$ bump centered around $E_x \sim 65A^{-1/3}$ is solely due to L=2 excitation: L=4,6, contributions are not taken into account. (ii) For the monopole strength one generally has used the non–folded transition potential given by formula 3. (iii) There is rather convincing evidence[14,17-20] for the existence of $3\hbar\omega$, L=3 strength but a precise determination of the actual strength is hampered by the uncertainties associated with continuum subtraction. (iv) The effect of the splitting of the multipole strength in deformed nuclei has been discussed elsewhere[1-3].

Table 1

			Heavy nuclei (A ≳ 60)			Deformed		Light nuclei (A ≲ 40)		
ΔL	ΔT	Shell	Status	E_x(MeV)	$S^{a)}$	theor.	exp.	Status	$S^{a)}$	Remarks
E0	0	2ℏω	***	~80 $A^{-1/3}$	(50–100) %	yes	*	???	–	1977
	1	2ℏω	–	–	–	–	–	–	–	–
E1	0	1ℏω	–	–	–	–	–	–	–	spurious
	1	1ℏω	***	32 $A^{-1/3}$+20.6 $A^{-1/6}$	100 %	yes	***	fragmented	–	1947
E2	0	2ℏω	***	~65 $A^{-1/3}$	(50–100) %	yes	**	fragmented	(30–50) %	1971–72
	1	2ℏω	*	~110 $A^{-1/3}$			**			
E3	0	1ℏω	**	~32 $A^{-1/3}$ or fragmented			*	fragmented	~15 %	
		3ℏω	*							
	1		?			?	?	?		
E4	0	2ℏ	*	~65 $A^{-1/3}$	?	?	?	?	–	
		4ℏ	?					?	–	

*** firmly established

** evidence

* data can be analyzed in a way which is not inconsistent with expectations

a) sum rule strength in the giant resonance region

1.4.   The problem of the monopole strength in light nuclei and the nuclear incompressibility.

Probably the most puzzling question left is the behaviour of the monopole resonance for nuclei $A \lesssim 100$. To illustrate this the monopole strength and excitation as summarized in reference 8 have been plotted in figures 6 and 7. Here the strength has been derived using a transition potential given by formula 3. The trend as displayed in these figures is that for lighter nuclei the strength observed becomes only a fraction of the total sumrule strength and that the excitation energy of this small fraction starts to deviate strongly from a $A^{-1/3}$ dependence.

The behaviour of E_x versus $A^{-1/3}$ is not unexpected. It can be shown (see reference 20 and references therein) that K_A, as defined in formula (1) is related to K_∞ , the compression modulus for nuclear matter $(A \rightarrow \infty)$ by:

$$K_A = K_\infty + K_\Sigma A^{-1/3} + K_\tau (\frac{N-Z}{A})^2 + K_c Z^2 A^{-4/3} \qquad (5)$$

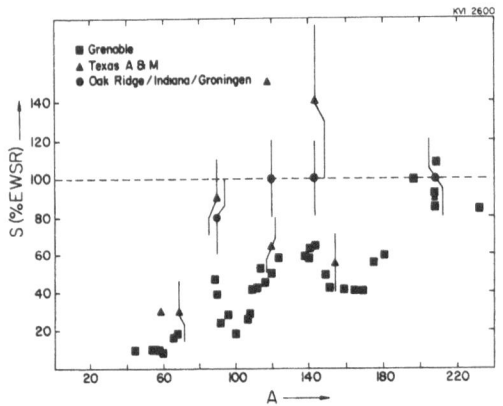

Fig.1.6. Systematics of the monopole strength expressed as a fraction of the $\Delta T=0$, E0 energy-weighted sumrule strength[8].

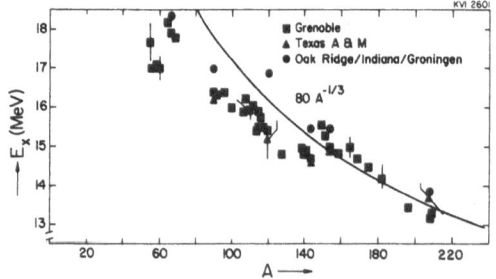

Fig.1.7. Systematics of the excitation energy E_x for the monopool resonance[8].

where K_Σ , K_τ and K_c are related to the second derivatives of the corresponding coefficients in the nuclear mass formula. For light nuclei the $A^{-1/3}$ term becomes large and deviations from the simple E_x versus $A^{-1/3}$ law result. In order to obtain K_∞ from K_A, one has to do a multiparameter fit to the available data, but such a procedure would only make sense if the strength at $E_x = (\hbar^2/2m)(K_A/\langle r^2 \rangle)^{1/2}$ exhausts an appreciable fraction of the sumrule. And as figure 1.6 shows, this is certainly not true for light nuclei with A<100, that is exactly for those nuclei of which the K_A-values would weight most in a multi-parameter fit. Thus the values deduced for K_∞ obtained through such a multi-parameter-fit are at present quite ambiguous.

The next best approach is to try to reduce the number of free parameters in formula (5) by calculating some of the coefficients or by establishing some kind of relation between them using a nuclear model. In this way Treiner et al[21] show that the data for heavy nuclei are consistent with K_∞ = 220±20 MeV. Clearly it would be of great importance to solve the puzzle of the missing monopole strength in light nuclei either by showing that the standard DWBA approach is not valid or by determining experimentally that a large amount of E0 strength is present indeeed but has escaped detection uptill now.

References section 1.

1. Proceedings Giant Multipole Resonance Topical Conference, Oak Ridge 1980, editor F.E. Bertrand (New York: Harwood Academic Publishers)
2. J. Speth and A. van der Woude, Rep.Prog.Phys. 44(1981)719
3. F.E. Bertrand, Nucl.Phys. A354(1981)129c
4. J. Wambach, V.A. Madsen, G.A. Rinker and J. Speth, Phys.Rev.Lett. 39(1977)1443
5. D.H. Youngblood, P. Boguchi, J.D. Bronson, W. Gary, Y-W Lui, and C.M. Rozsa, Phys.Rev. C23(1981)1997
6. D.H. Youngblood, C.M. Rosza, J.M. Moss, D.R. Brown and J.D. Bronson, Phys.Rev.Lett. 39(1977)1188; C.M. Rozsa, D.H. Youngblood, J.D. Bronson, Y-W Lui and U. Gary, Phys.Rev. C21(1980)1252
7. M. Buenerd, C. Bonhomme, D. Lebrun, P. Martin, J. Chauvin, G. Duhamel, G. Perrin and P. de Saintignon, Phys.lett. 84B(1979)305
8. See also M. Buenerd, The Giant Monopole Resonance in Nuclei ISN 81.38. Lectures given at the International Nucleus Physics Workshop, Trïeste, October 5-30, 1981
9. S. Brandenburg et al. private communication
10. F.E. Bertrand, G.R. Satchler, D.J. Horen, J.R. Wu, A.D. Bacher, G.T. Emery, W.P. Jones, D.W. Miller and A. van der Woude, Phys.Rev. C22(1980)1832
11. B.L. Berman and S.C. Fultz, Rev.Mod.Phys. 47(1975)713
12. K. van der Borg, M.N. Harakeh and A. van der Woude, Nucl.Phys. A365(1981)243
13. D.R. Brown, I. Halpern, J.R. Calarco, P.A. Russo, D.L. Hendrie and H. Homeyer, Phys.Rev. C14(1976)896
    D.R. Brown, J.M. Moss, C.M. Rozsa, D.H. Youngblood and J.D. Bronson, Nucl.Phys. A313(1979)157
14. F.E. Bertrand, E.E. Gross, D.J. Horen, J.R. Wu, J. Tinsley, D.K. McDaniels, L.W. Swenson, and R. Liljestrand, Phys.Lett. 103B(1981)326
15. C. Djalali, N. Marty, M. Morlet and A. Willis, Nucl.Phys. A380(1982)42
16. G.R. Satchler, in Elementary Modes of Excitation in Nuclei, edited by A. Bohr and R.A. Broglia (North Holland, Amsterdam, 1977)
17. H.P. Morsch, M. Rogge, P. Turek, and C. Mayer-Böricke, Phys.Rev.Lett. 45(1980)337
18. T.A. Carey, W.D. Cornelius, N.J. DiGiacomo, J.M. Moss, G.S. Adams, J.B. McClelland, G. Pauletta, C. Whitten, M. Gazzaly, N. Hintz, and

C. Glashausser, Phys.Rev.Lett. <u>45</u>(1980)239

19. T. Yamagata, S. Kishimoto, K. Yuasa, K. Iwamoto, B. Saeki, M. Tanaka,
     T. Fukuda, I. Miura, M. Inoue, and H. Ogata, Phys.Rev. <u>C23</u>(1981)937
20. B. Bonin et al., Preprint Saclay 1982
21. J. Treiner, H. Krivine, O. Bohigas and J. Martirell, Nucl.Phys. <u>A371</u>(1981)253

2. Heavy Ions and Giant Resonances

2.1. Introduction

Ever since it was suggested that giant resonance excitation may play an important role in the fast energy dissipation observed in deep inelastic heavy ion collisions, there has been a great interest in the topic "giant resonance excitation by heavy ions". In GR excitation n $\hbar\omega$ energy can be dissipated in a single step process, which could easily amount to 20-30 MeV. Although this simple single step process can not account for the whole energy dissipation of $\gtrsim$ 100 MeV observed in deep-inelastic heavy ion reactions, it can be viewed as first step in a complicated series of interactions[1]. Thus experiments designed to look for GR excitation by heavy ions are of direct interest to the field of heavy ion nuclear physics.

But vice versa, one might also expect some definite advantages in giant resonance research by using heavy ion inelastic scattering. In the first place from simple kinematics one can expect that in inelastic, peripheral heavy ion reactions a relatively large amount of angular momentum will be transferred to the nucleus and thus that the excitation of high spin giant resonances will be relatively favoured. This could not only be used to find the high (L > 2) L $\hbar\omega$ components but also might be of great help in for instance disentangling the L-distribution in the 2 $\hbar\omega$ bump.

Secondly the continuum underlying the spectrum obtained in light ion inelastic scattering, which, as discussed before, is the main limitation in an accurate quantitative determination of giant resonance parameters, is expected to be considerably reduced in inelastic heavy ion scattering. Especially the contribution from direct knock-out should be greatly reduced.

For previous reviews on the subject see references 2 and 3.

The data

Since nearly all experiments have been performed for ^{208}Pb this discussion will be limited to the data taken on that nucleus - see table 1.

Table 1.

Survey of heavy ion inelastic scattering experiments for giant resonance
excitations on ^{208}Pb

Projectile	E/A (MeV)	Reference
^{6}Li	26	4
^{9}Be	27	5
^{12}C	16.6	6
^{14}N	19	7
^{16}O	19.7	8

Typical spectra are shown in figures 1 to 5. The experimental situation can
be summarized as follows: (i) In each of the spectra shown one recognizes the bump
around 65 A$^{-1/3}$ MeV (~ 11 MeV) due to 2 ℏω g.r.excitation. (ii) The ^{16}O-spectrum

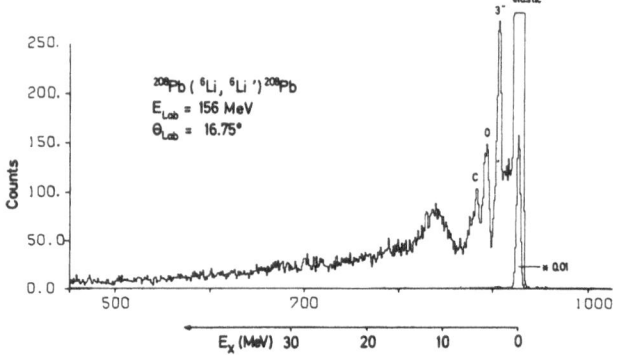

Fig.2.1. Inelastic
spectrum measured in
the ^{208}Pb(^{6}Li,^{6}Li')
reaction[4].

Fig.2.2. Inelastic spectrum measured in the ^{208}Pb
(^{9}Be,^{9}Be') reaction. The dashed line below the
GQR shows the assumed background. The dotted peak
at 19 MeV excitation energy is a DWBA estimate for
the strength observed in ref.8. The double arrow
shows the region were (^{9}Be,^{10}B→^{9}Be+p) and
(^{9}Be,^{10}Be→^{9}Be+n) sequential pickup decay processes
(left and right arrow respectively) are expected.
From reference 5.

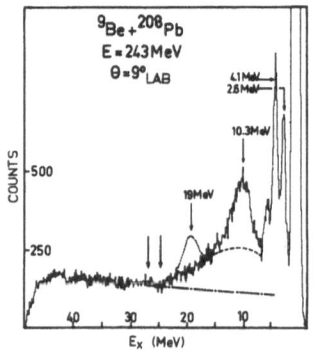

Fig.2.3. Spectra of the $^{208}Pb(^{12}C,^{12}C')$
reaction at $E_{^{12}C}$=200 MeV. For comparison
a spectrum of $^{12}C^{208}Pb(\alpha,\alpha')^{208}Pb$ at E_α is
presented. The right-hand side shows the
low excitation retion ot the $^{208}Pb(^{12}C,^{12}C')$
^{208}Pb reaction at θ_{lab}=23.5°. From ref.6.

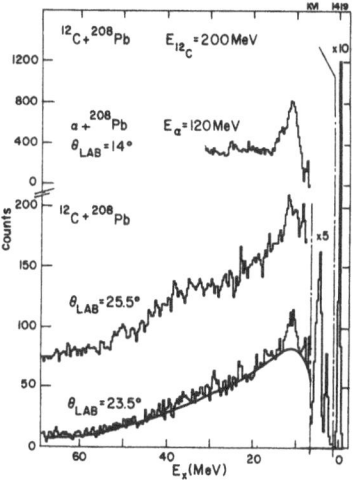

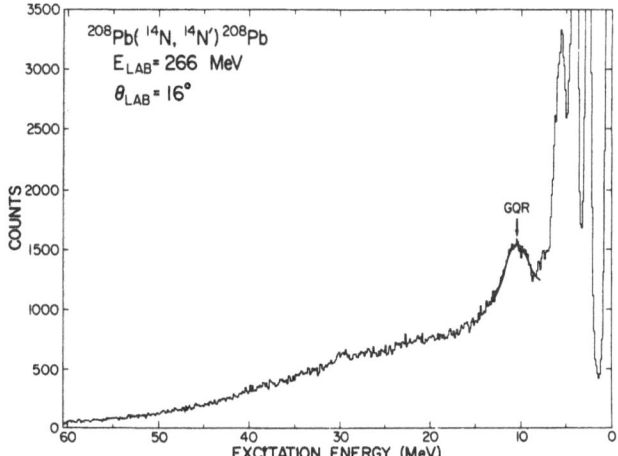

Fig.2.4. Inelastic ^{14}N
scattering spectrum for
^{208}Pb. The GQR position
in indicated by the
arrow; the solid line
shows a gaussian
calculated with GQR
parameters from α-
scattering. From
reference 7.

shows additional structures at higher excitation energy, which are not seen in any
of the other spectra. This rules out the possibility that they are due to L=3 and/or
L=5 giant resonance excitation as was suggested[8]. Most likely they are due to
processes like

$$^{16}O + ^{208}Pb \rightarrow ^{17}O^* + ^{207}Pb^* \text{ or } ^{17}F^* + ^{207}Tl^*$$
$$\downarrow \qquad\qquad \downarrow$$
$$^{16}O + n \qquad\qquad ^{16}O + p$$

These processes are known to be able to create "giant-resonance" like bumps and have
already extensively been discussed for instance in connection with inelastic α-
scattering. (iii) Angular distributions as shown for instance in figure 6 are
compatible with a L=2, 100% EWSR excitation. (iv) Some evidence for selectivity in

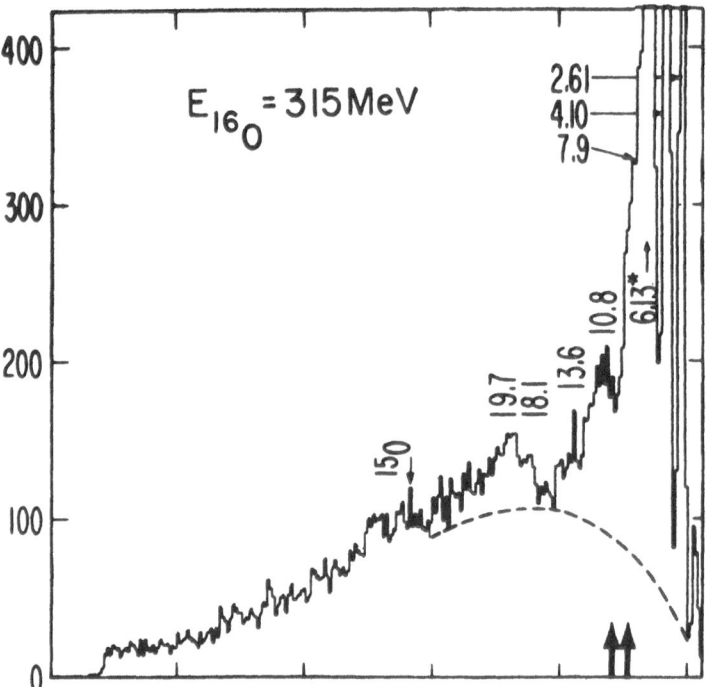

Fig.2.5. Inelastic ^{16}O scattering spectrum for ^{208}Pb. The dashed line shows the assumed background. From reference 8.

angular momentum is shown in figure 3 where the (α,α') and $(^{12}C,^{12}C')$ spectra are displayed simultaneously. The L=0 resonance located at E_x=13.9 MeV does not seem to be excited in the $(^{12}C,^{12}C')$ reaction. This presumably is due to the large angular momentum mismatch for L=0 excitation in $(^{12}C,^{12}C')$. (v) The continuum underlying the $2\hbar\omega$ bump can be thought off as to exist out of two components. One component is rather low and flat, while the other component is much larger at the GQR $2\hbar\omega$-excitation energy but decreases with increasing excitation energy and is nearly absent at higher excitation energies like $E_x \sim$ 25 MeV for ^{9}Be and $\sim$ 60 MeV for ^{12}C, ^{14}N and ^{16}O.

It is tempting to identify the low and flat continuum at high excitation energies with the expected low continuum due to knock-out. But than the question arises to what processes the other, unexpected, bell shaped part could be due to. It is clear that from the experimental point of view this is the crucial question to answer, since with the large continuum around one will never be able to obtain data which convincingly show the existence of additional higher multipole giant resonances.

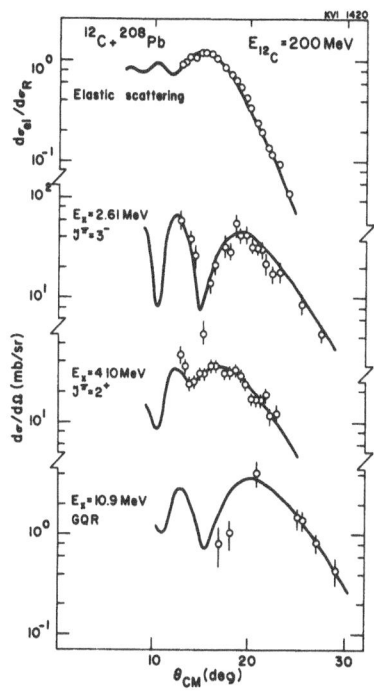

Fig.2.6. Angular distributions for the elastic scattering and target excitations in the $^{208}Pb(^{12}C,^{12}C')^{208}Pb$ reactions at E_{12C}=200 MeV. The solid lines are the results of DWBA calculations.

2.2. Discussion

From the data presented one is tempted to conclude that heavy ion scattering is not a very promising tool for studying giant resonances. The continuum below the giant resonances is not small while the cross section for GR excitation is smaller than in for instance (α,α') scattering. Moreover the angular distributions are not characteristic for the multipolarity excited.

The last point is illustrated in figure 7. It shows that L=2,3 and 4 excitations have for all practical purposes identical angular distributions.

The processes responsible for the excitation of the "new" bell-shaped continuum is not quite clear. In case of the $(^{12}C,^{12}C')$ reaction it was observed[6] that the continuum underlying the GQR, integrated from 10 to 55 MeV excitation increases from $d\sigma/d\Omega$ = 20 mb/sr at θ_{lab}= 27° to $d\sigma/d\Omega$ = 270 mb/sr at θ_{lab} = 14°. In order to find out whether the continuum could be explained by inelastic excitation of higher multipole giant resonances, calculations were performed assuming a 100% exhaustion for the giant octupole (3^-) and hexadecapole (4^+) resonances. For θ_{lab} = 20° the sum of the cross sections of the 3^- and 4^+ giant resonances, located at E_x = 15 MeV, would amount only to about 15 mb/sr while the continuum cross section between 10 and 55 MeV excitation is 140 mb/sr at the same angle. So clearly the bulk

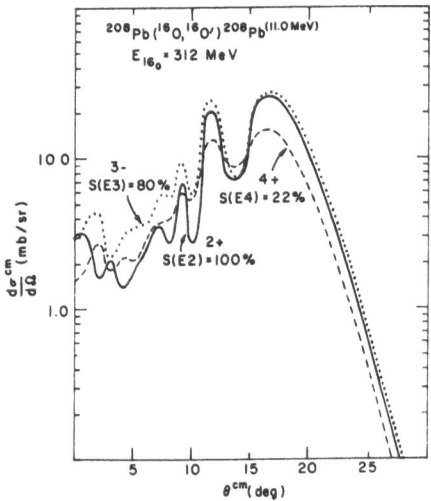

Fig.2.7. DWBA calculations for inelastic transitions of different multipolarities for the scattering of ^{16}O on ^{208}Pb at 312 MeV (from reference 2).

of the continuum must be due to other processes. Strong mutual excitation in the ^{13}C + ^{207}Pb channel was observed ($d\sigma/d\Omega$ = 150 mb/sr at θ_{lab}= 20°) for particle bound states in ^{13}C. Since one might expect that particle unbound states are excited with comparable strengths one can easily account for the observed continuum cross section. Thus it seems that mutual excitation processes (transfer and subsequent breakup) are more important than the excitation of multipole resonances. As remarked before, these processes are probably also responsible for the resonance like bumps observed in the ^{16}O-spectrum of figure 5.

In general one may expect that multistep processes play an important role in these reactions. In fact multiphonon excitations are supposed to be important in the damping mechanism in deep inelastic reactions as already mentioned before.

Thus at present there is no advantage in using heavy ions for GR studies. However, there is some hope that in the future, when good quality (light) heavy ion beams become available, this situation may change.

This hope is based on the expectation that by going from 20-30 MeV/A to say 40-100 MeV/A the cross sections with which giant resonances will be excited, is increased by about an order of magnitude, that the angular distributions will show more of a diffraction like pattern and also that the bell shaped continuum moves up to higher excitation energies and becomes broader.

The fact that the cross-section increases with increasing bombarding energies can be easily checked by performing the appropriate DWBA calculations. Figure 8 shows the result of such a calculation[2]. As this figure shows, the increase in cross section with increasing bombarding energy is not specific for heavy ions but also true for light ions like α-particles. It can be understood by considering the classical analogue: the response of a system of damped harmonic oscillators (the surface oscillations) to a short pulse (the projectile-target interaction). The shorter the time duration of the pulse the higher the frequency with which the system responds. The classical relation $\Delta \nu \Delta t \sim 1$ for the frequency spectrum of a pulse translates into $\Delta E/h \sim v/d$ where v is the projectile velocity and d the interaction distance. So projectile velocity and excitation energy have to be matched and as the DWBA calculations show, good matching for the 2 $\hbar\omega$ excitations is obtained if $E/A \sim 50-100$ MeV.

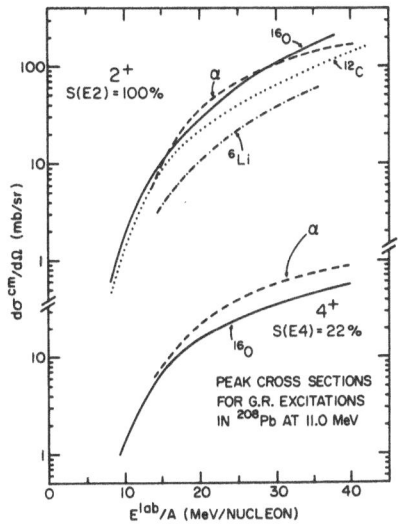

Fig.2.8. DWBA predictions of the energy dependence of the peak cross sections for giant resonance excitations in ^{208}Pb at 11 MeV. From reference 2.

The interaction between projectile and target has two components, the long-range Coulomb and short-range nuclear interaction. Our simple picture tells us that for low projectile velocities only the nuclear interaction has to be taken into account. But with increasing velocity (energy) the Coulomb excitation also becomes more important and may become even larger than the nuclear interaction. As calculated in reference 3 this actually occurs in inelastic ^{12}C scattering at 1 GeV on ^{208}Pb. It implies that at such energies also isovector resonances will be excited with an appreciable cross section and that the isospin selectivity which is so useful in experiments like (α,α') scattering gets partially lost. On the other hand DWBA calculations predict that some sensitivity to angular momentum will be retrieved.

Finally if indeed the bell shaped continuum is largely due to pick-up followed by sequential decay process the resonance like bumps due to this process will move to higher excitation energies. This fact is already well known from (α,α') scattering[9]. It can be understood from simple kinematic considerations which predict that for sufficiently high projectile energies, so that binding energies can be neglected, the apparent excitation energy ΔE of the bump will be centered roughly around $\Delta E_{exc} \sim E_{proj}/(A + 1)$. Also at higher bombarding energies the bump will be smeared out. Thus the result of these two effects together will be that it will be easier to locate the real resonances.

The assumption that the bell-shaped continuum is due to the process (transfer + sequential decay) explains why the bell-shaped continuum is very similar for the three reactions $(^{12}C, ^{12}C')$, $(^{14}N, ^{14}N')$ and $(^{16}O, ^{16}O')$. In the $(^{9}Be, ^{9}Be')$ experiment though it seems that the bell-shaped continuum is sharper and located at lower excitation energies, contrary to what one would expect from the (pickup + decay) picture. Also ^{9}Be is only weakly bound and its low lying excited states are already particle unstable so that mutual excitation processes will not show up in the ^{9}Be-spectrum. Thus in this case the bell-shaped continuum must be due to other processes. This suggests that the bell-shaped structure observed in this reaction might be due indeed to excitation of the $L < 5$ multipole strength supposed to be present between 10 and 20 MeV of excitation energy. However, as will be clear from figure 2, it will be very hard to disentangle such a continuum in the appropriate multipole contributions and to get quantitative information of their strength distribution. Probably all we will be able to say from such experiments is that the observed phenomena are compatible with GR excitation.

2.3. Beyond $2\hbar\omega$ with heavy ions.

The real challenge for GR excitation by heavy ions is to find high multipole excitations at high excitation energies, well beyond $2\hbar\omega$. Some evidence that high multipoles can be excited indeed comes from a $(^{40}Ca + ^{40}Ca)$ experiment performed at Orsay[10-12] with beam energies of 160, 284 and 400 MeV. In this reaction the existence is observed of broad structures at high total excitation energies in the energy spectra of some projectile like fragments emitted in the collision.

Representative data at 400 MeV are shown in figure 9. For different exit channels rather pronounced structrures are observed at about 25, 50, 80 and 120 MeV excitation energy: the structure at 50 MeV is also observed for a bombarding energy of 284 MeV. Furthermore it was noticed that these structures only show up in the energy spectra of nuclei resulting from the transfer of at most a few nucleons while their angular distribution is decreasing exponentially with angle. These observations imply that these bumps are formed during a short interaction time. They show all the characteristics of giant resonance excitation followed or preceeded by a transfer of a few nucleus or nucleon clusters:

(i) Their apparent excitation energy is independent of bombarding energy and angle

(ii) The higher excited bumps are stronger excited at higher bombarding energies

Fig.2.9. c.m. kinetic energy spectra for Ca
isotopes produced at θ_{lab} = 10° in the reaction
^{40}Ca + ^{40}Ca at 400 MeV, a) inelastic scattering,
b)^{39,41,43}Ca energy spectra. The arrows
indicates the corresponding total excitation
energy calculated for a two-body reaction. From
reference 11.

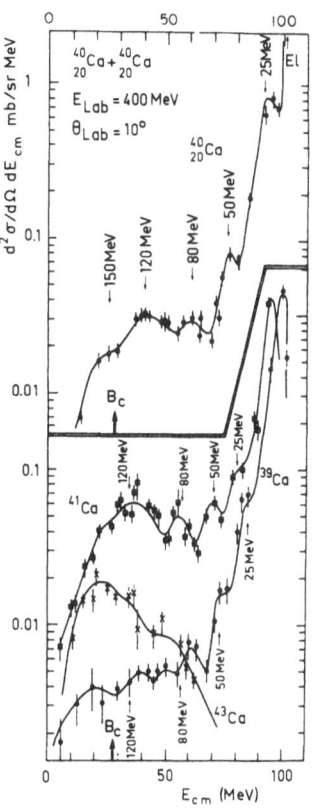

(iii) They are the result of a fast process. But as the discussion in the previous
section has shown bumps can also be introduced by reactions like (particle pickup +
sequential decay). The first process occurs during the short interaction time
($\sim 10^{-22}$sec) while a statistical evaporation of the resulting excited fragments
takes a longer time: $\sim 10^{-21}$sec. Making certain assumptions on (i) the yield
distribution of the fragments (ii) the partition of excitation energy between the
fragments and (iii) the details of light particle emission, Hilcher et al.[13] have
shown that in the ^{40}Ca + ^{40}Ca system at 400 MeV one expects bumps to be created by
such processes. The result of such a calculation is shown in figure 10. From these
calculations it is clear that from inclusive spectra only one will not be able to
determine in a quantitative way to what extent the processes, (giant resonance
excitation + particle transfer) or (particle transfer + sequential decay) contribute
to the observed spectra.

In addition one should remark that: (i) An extensive search for structure at
E_x > 20 MeV has been conducted for ^{40}Ca in the past mainly by using inelastic α-

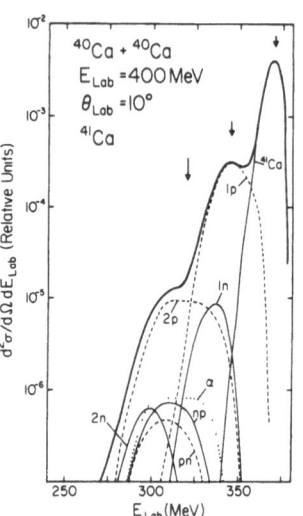

Fig.2.10. The heavy line is the calculated laboratory energy spectrum of ^{41}Ca. The remaining curves
represent the individual contributions to the
production of ^{41}Ca from various primary fragments.
The curve labeled ^{41}Ca corresponds to the yield of
primary fragment ^{41}Ca that has survived particle
decay. The other curves are labeled by the particles
emitted from the primary fragments to produce the
final fragment ^{41}Ca. The vertical arrows indicate the
positions of the bumps or shoulders in the
experimental energy spectrum. From reference 13.

particle scattering at bombarding energies E/A > 30 MeV. None was found. However,
these experiments were mostly designed to look for the L=0 monopole-strength and
thus the conditions to observe high L multipole strength were certainly not optimal.
(ii) As was argued before the bombarding energy of 5-10 MeV/A in the cm system is
amazingly low for exciting in a direct step high lying structures at $E_x \gtrsim 25$ MeV.
This is even more so for the recent performed ^{40}Ca + ^{40}Ca at an bombarding energy of
160 MeV[12] lab in which a bump at 17 MeV was excited corresponding to the known GQR
structures.

It is then clear that although it is tempting to associate the observed bumps
with collective mode excitation, there still remain many open questions like their
unusual strong excitation for the relative velocities involved. On the other hand if
true it would open up a whole new field in giant resonance research.

References section 2

1. R.A. Broglia et al., Phys.Rev.Lett. 41(1978)25
2. C.K. Gelbke, Topical conf. on giant resonances, Oak Ridge 1979, ed. by F.E.
   Bertrand (Harwood New York 1980)
3. D. Lebrun et al., Comptes Rendus, Deuxième Cilloque Franco-Japonnais de
   Physique Nucleaire avec des Ions Lourds.
   M. Buenerd, XXth Int. National Meeting on Nuclear Physics, January 1982, Bormio
4. H.J. Gils et al., Phys.Lett. 68B(1977)427
5. D. Lebrun et al., Phys.Rev. C23(1981)552
6. R. Kamermans et al., Phys.Lett. 82B(1979)221
7. U. Gary et al., Phys.Lett. 93B(1980)39
8. P. Doll et al., Phys.Rev.Lett. 42(1979)366
9. A. Kiss et al., Phys.Rev.Lett. 37(1976)1188
10. N. Frascari et al., Phys.Rev.Lett. 39(1977)918
11. N. Frascari et al., Z. Phys. A294(1980)167
12. Y. Blumenfeld et al., Phys.Rev. C25(1982)2116
13. D. Hilchert et al., Phys.Rev. C20(1979)556

3.   The Width and Decay-Properties of Giant Resonances

3.1.   Introduction

Giant resonances are located at excitation energies well above the particle emission threshold so that they mainly decay by particle emission. Macroscopically the decay properties of giant resonances are reflected in the width Γ, as determined from an analysis of single scattering data. More information can be obtained though from studying the decay properties of these resonances, that is from studying the energy spectrum and angular distribution of the emitted particles. Such experiments require measuring the emitted particle and the inelastically scattered particle in coincidence.

The total width Γ of a giant resonance can be written as the sum of three different contributions:

$$\Gamma_{tot} = \Gamma\!\uparrow + \Gamma\!\downarrow + \Delta\Gamma \qquad\qquad (3.1)$$

Here $\Delta\Gamma$ is due to the fact that not all the collective (1p-1h) strength, which acts as a doorway for the giant resonance, is necessarily concentrated in one single state as would follow from a simple schematic model. More realistic RPA calculations indicate that even for spherical nuclei, the instrinsic 1p-1h strength can be already fragmented. This is especially true for the higher L-multipole strength[1]. In deformed nuclei there is an additional effect due to deformation (see for instance reference 2 and references therein) which for the GDR can even give rise to a splitting of the resonance[3].

The $\Gamma\!\uparrow$ component is due to the coupling of the (1p-1h)-state in the nucleus A to the continuum, resulting in a semi-direct decay to the hole states in the (A-1) nucleus. The $\Gamma\!\downarrow$ component is due to the fact that the simple (1p-1h) doorway state mixes with the more complicate (2p-2h) state of the nucleus. The (2p-2h) states again couple to the continuum or to more complicated (3p-3h) states. This process continues in time till finally a totally equilibrated system is reached in which the excitation energy is equally divided over all possible degrees of freedom and which decays like a compound nucleus:

$$\Gamma\!\uparrow \qquad \Gamma\!\downarrow \underbrace{\hspace{5cm}}_{\Gamma\!\downarrow\uparrow} \qquad\qquad \Gamma\!\uparrow\!\downarrow$$

(1p-1h) + $\rightarrow$ (2p-2h) + $\rightarrow$ (3p-3h) + ...$\rightarrow$ compound nucleus
$\rightarrow$ time

Here $\Gamma\!\downarrow\uparrow$ represents pre-equilibrium decay so that $\Gamma\!\downarrow = \Gamma\!\downarrow\uparrow + \Gamma\!\downarrow\downarrow$. Calculations indicate that in heavy nuclei $\Gamma\!\uparrow < \Gamma\!\downarrow$, at least for L > 1 giant resonances, so that the largest contribution to Γ is from the spreading width $\Gamma\!\downarrow$. Macroscopically the damping of a collective nuclear vibration is determined by the nuclear viscosity.

Assuming a 2-body viscosity Nix and Sierk[4] calculate that for

L > 2 resonances $\Gamma\downarrow_L = c(L)A^{-2/3}$ where $c(L)$ depends on the multipolarity L of the resonance. In principle the viscosity obtained for the small amplitude collective oscillations like the giant resonances should be related to those of large amplitude motions as encountered in fission and deep-inelastic collisions. A few microscopic calculations have been performed for $\Gamma\downarrow$. If one assumes that the coupling matrix element $\langle(2p-2h)|V|(1p-1h)\rangle$ and the level distance D of the (2p-2h) states are constant, $\Gamma\downarrow = 2\pi\ V^2/D$. In a recent calculation[5] one finds for ^{40}Ca that $\bar{V}^2 \sim 0.1$ MeV, $D\sim (1/5)$ MeV so that $\Gamma\downarrow \sim 3$ MeV while for ^{208}Pb $\bar{V}^2 \sim 0.02$ MeV, $D \sim (1/20)$ MeV so that $\Gamma\downarrow \sim 2.5$ MeV. Special effects can be expected[5,6], if the main mechanism of spreading of the (1p-1h) state is due to p(h) -fonon coupling. Neutron decay of such a state would lead to final states in the (A-1) nucleus which correspond to neutron holes coupled to a fonon.

In principle the different contributions can be determined by measuring the neutron decay spectrum of the giant resonance. In the extreme case of (semi-) direct decay only well defined hole states in the (A-1) nucleus will be populated while in the other extreme case of complete equilibrium the neutron spectrum will be characteristic for a statistical decay as can be calculated from a Hauser-Feshbach kind of calculation. Unfortunately in reality the situation is much more complicated. Hole-states will also be populated in statistical decay especially if the energy available for particle (neutron) decay is only a few MeV as is the case for instance for the GQR in ^{208}Pb. Moreover practically nothing is known about the neutron spectrum resulting from pre-equilibrium decay. Typically all what one can do is to determine experimentally to what extent a measured neutron spectrum deviates from what one calculates from a Hauser-Feshbach calculations. This deviation would then correspond to a non-statistical component, which is not necessarily equivalent to the semi-direct ($\Gamma\uparrow$) component since also pre-equilibrium decay may give rise to such deviations. On the other hand, the Hauser-Feshbach like component does not necessarily result from a statistical decay only, since pre-equilibrium decay may give rise to a neutron spectrum resembling statistical decay.

Deviations from an evaporation spectrum have been found for the GDR in ^{208}Pb[6] and the GQR in ^{119}Sn[7]. In both cases it is only a minor fraction about 20%. For the GQR of ^{119}Sn it was moreover shown[7] that the shape of the non-statistical component is consistent with the assumptions that only the pure hole-states are populated in ^{118}Sn which suggests a semi-direct decay. These experiments are thus consistent with the theoretical expectations that for the GDR and GQR $\Gamma\uparrow \ll \Gamma\downarrow$ but they do not yet show to what extent pre-equilibrium or statistical decay occur.

Such information can in principle be obtained for the actinide nuclei where statistical decay does not only occur through neutron evaporation but also through fission. In fact if we make the reasonable assumption that the time scale involved in the fission of a system with an excitation energy of 10-20 MeV and with low angular momentum is large compared to the spreading time, fission can only occur from a fully equilibrated system. Thus by measuring the fission width one has a direct measure of the $\Gamma\!\downarrow\!\downarrow$ component. In fact at present it is the only way this component can be measured – see section 3.2.

In light nuclei (A $\lesssim$ 40) the situation is quite different in the sense that here also charged particle decay and especially α-decay can occur and in some cases is even favoured over neutron-decay. Moreover it turns out that there is a large semi-direct component. For more details see references 6, 8 and 9.

The total width Γ can be measured from single scattering experiments provided that the different overlapping multipoles can be adequately separated. It has been argued above that experimentally $\Gamma\!\uparrow \ll \Gamma$ for the GQR and GDR. Also the contribution of $\Delta\Gamma$, in formula (3.1) is small especially for L < 2 and for non-deformed nuclei, so that for the resonances $\Gamma \sim \Gamma\!\downarrow$.

The systematics of Γ versus E_x, the excitation energy, is shown in

Fig.3.1. Systematics of the width Γ of quadrupole and monopole resonance as a function of the excitation energy E_x of the resonance. The full line represents the systematics for the giant dipole resonance[11].

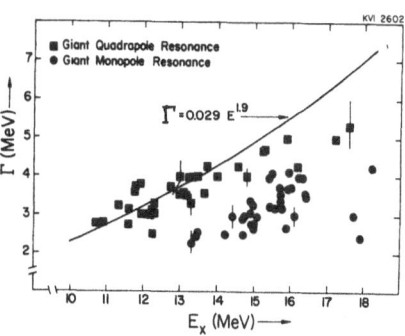

figure 1. The data are taken from a tabulation of reference 10. Several features can be seen (i) The width of the GQR is, if plotted versus E_x, very similar to the one found for the GDR, which is well represented by[16] $\Gamma(GDR) = 0.029\ E_x^{1.9}$. This suggests that the spreading mechanism for the GDR and the GQR are very similar (ii) Since for the GQR, $E_x \sim A^{-1/3}$, the width $\Gamma(GQR) \sim A^{-2/3}$, as expected from a macroscopic two-body calculation[4]. (iii) The width of the GMR is appreciably smaller, by about 1.0 to 1.5 MeV than the corresponding widths of the GDR and the GQR. This suggests that for the GMR there is some effect which prevents the speading of the (1p-1h) doorway state into the (2p-2h) states. Qualitatively such a trend agrees with what has been calculated for ^{208}Pb[12]: $\Gamma\!\downarrow$ (GQR) $\sim$ 2.7 MeV and $\Gamma\!\downarrow$ (GMR) $\sim$ 0.8 MeV, although quantitatively there is still a rather large disagreement with the experimental values of $\Gamma(GQR) \sim$ 3.0 MeV and $\Gamma(GMR) \sim$ 2.5 MeV.

3.2.   The fission decay of giant resonances in the actinide nuclei

3.2.1.   Introduction

As mentioned already before, for actinide nuclei a fully equilibrated highly excited system can not only decay by neutrons but also by fission. Thus by measuring the fission probability $P_f(E_x,A)$ of a giant resonance (or of any highly excited state) one has a direct measurement of $\Gamma_{\downarrow\downarrow}$ [13]:

$$\Gamma_{\downarrow\downarrow}/\Gamma_{tot} = P_f/P_f^C \qquad (1)$$

where $P_f^C(E,A)$ is the fission probability of the compound nucleus A at an excitation energy E_x. P_f^C is approximately known, for instance from neutron capture experiments. For ^{238}U one finds that $P_f^C \sim 0.2$ at $E_x \sim 10$ MeV[14]. One can also calculate $P_f^C(E_x,A)$ assuming that the probability of the two competing processes, neutron decay and fission, is determined by the appropriate level densities: for neutron decay the level density of the (A-1)nucleus, for fission decay the level density at the saddle point of the A-nucleus[13]. These calculations are model dependent in the sense that the fission probability will depend on the number of different saddle points, each with its own symmetry properties, that are available for the fission process. An extra complication arises if $E_x>B_{nf}$, the threshold for second-chance fission. In that case after neutron emission the (A-1)nucleus may still have enough energy to fission. This will result in an increase in P_f. A similar effect occurs if $E_x>B_{2nf}$. Figure 2 shows the fission probability calculated along these lines for ^{232}Pa[13]:

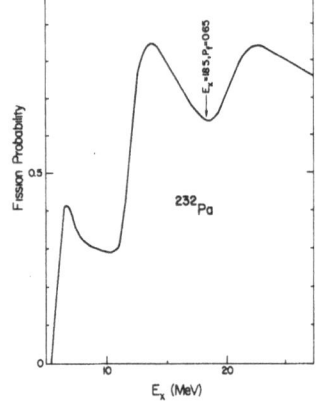

Fig.3.2. Fission probability as a function of excitation energy for the compound nucleus ^{232}Pa as obtained from a statistical model calculation.

the level densities and fission barriers of the A,A-1 and A-2 nuclei are tuned to give for each nucleus the correct, experimentally known, fission probability for $E_x<B_{nf}$. Such calculations are believed to be accurate to ±20%. Note that if the excitation energy is increased past the $E_x=B_{nf}$ or $E_x=B_{2nf}$ threshold $P_f(E_x)$ shows a kind of resonance-like increase. These "resonances" in P_f occur in a region where one might also expect the $3,4,\ldots\ldots$ $\hbar\omega$ resonances to occur, which makes it

complicated if not impossible to extract in a reliable way P_f for these resonances. Since P_f can be determined only with an accuracy of about 20%, also $(\Gamma\downarrow\downarrow/\Gamma)$ can be only determined by this method to ±20%. The important point to stress though is that although limited in its accuracy it is the only direct way to determine $(\Gamma\downarrow\downarrow/\Gamma)$ and as such even a number with limited accuracy is already very important.

3.2.2.   The fission probability of the GDR

The fission probability of the GDR can be determined from a simulataneous measurement of the (γ,n), $(\gamma,2n)$, (γ,nf)..cross sections[3,15,16]. Typical data are shown in figure 3 for ^{238}U [5]. Because of the selectivity of the absorption process, (see section 1), these data are thus characteristic for $\Delta L=1$, $\Delta T=1$ excitation. The values of $P_f(E_x)$ for a number of nuclei is shown in figure 4. The fission probability $P_f(E_x,A)$ rises sharply at the threshold $B_f(\Delta T=1,\Delta L=1)$ for fission, and either drops down or stays constant above B_n because of the competition between fission and neutron evaporation, depending on whether $\Gamma_n/\Gamma_f \gg 1$ or < 1 . From say 8 MeV on it remains approximately constant to $E_x=B_{nf}$ after which it nearly doubles in values. By using a special detection technique Caldwell et al.[16] were able to distinguish for $E_x>B_{nf}$ between first chance fission P_f and second-chance fission P_{nf} and they found that $P_f \sim P_{nf}$ for $E_x>B_{nf}$. Thus as an example the fission probability $P_f(E_x)$ for the GDR is (0.22 ± 0.02) for $7<E_x<16$ MeV for the nucleus ^{238}U. Using relation (1) and a value of $P_f^C \sim 0.2$ one has to conclude that within the uncertainties involved $(\Gamma\downarrow\downarrow/\Gamma_{tot}) \sim 1$ for the GDR. Thus for ^{238}U, the GDR decays mainly statistically, that is like a compound nucleus. This agrees with the fact that the measurement of the neutron decay spectrum shows only a very small direct component, as discussed in section 3.1.

3.2.3.   The fission probability of the $2\hbar\omega$ isoscalar resonances

The topic of the fission probability of the $2\hbar\omega$ isoscalar resonances like the GQR has been and to some extent still is, a controversial one. In this section we will review the results of the different kind of experiments on this subject and draw some conclusions from the data that have been obtained up till now.

Three different kinds of experiments have been reported, nearly all of them on ^{238}U. These are inclusive: (e,f) experiments from which probabilities $P_f(GQR)$ like 40-70%[17] or 0-25%[18,19] have been reported, $(\alpha,\alpha'f)$ experiments from which P_f values ranging from about 5% to 25% have been deduced[20/23] and finally $(e,e'f)$ experiments[25,26] which seem to indicate a non-resonant E2 strength distribution.

3.2.3.1. The electron fission (e,f) experiments on ^{238}U

In these experiments[17-19] an ^{238}U target is bombarded with an electron (positron) beam and the fission yield is measured as a function of the electron (positron) energy. The field produced at the nuclear site by the passing electron is decomposed into radiation multipoles, the effect of which on the nucleus can be compared with that produced by real, electromagnetic radiation. The virtual photon

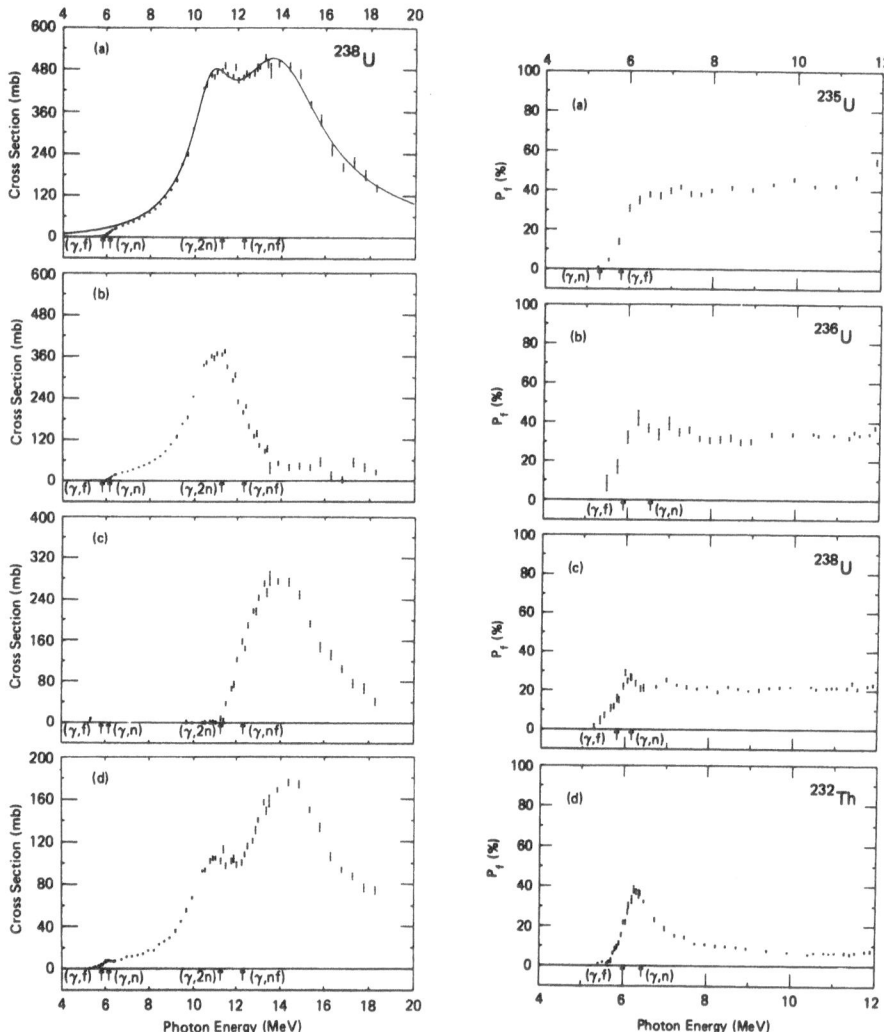

Fig.3.3. Photonuclear cross sections for ^{238}U: (a) $\sigma(\gamma,tot)$, with a Lorentz-curve fit; (b) $\sigma(\gamma,n)$; (c) $\sigma(\gamma,2n)$; (d) $\sigma(\gamma,F)$

Fig.3.4. Fission probability P_f for the GDR

spectrum can be calculated using DWBA methods. Assuming that only E1 and E2 absorption have to be taken into account and since E1-photoabsorption cross section is known from real photon absorption measurements it is in principle possible to determine the E2 one from a measurement of the total cross section. The advantage of this method for studying E2 contributions is that in the virtual photon spectrum the E2 is more intense than the E1 radiation, which is not true for a real photon spectrum. Thus in principle this method should be more sensitive to E2 contributions.

An additional experimental degree of freedom in this method is to compare e^- and e^+ results: because of the Coulomb distortion the ratio of E1 to E2 intensity in the virtual photon spectrum is different for e^- and e^+ and thus the ratio σ^-/σ^+ is sensitive to the assumed E-1 and E-2 contributions.

A summary of the results obtained by the Sao-Paulo group[17] and the Giessen group[18,19] are shown in figure 5. Clearly there is an experimental problem: the absolute cross sections from the Sao Paulo group (open circles) are about twice as large as the Giessen-results, the full dots and hatched bars in figure 5. The calculated cross-section for E-1 excitation only, is shown by the hatched area. This figure shows that from the Sao Paulo data one would have to conclude that there is

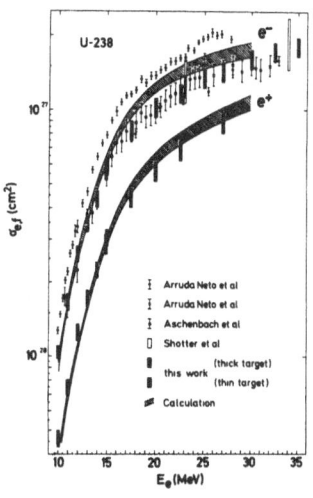

Fig.3.5. Comparison of absolute electron- and positron-induced fission cross sections for ^{238}U. The error bars include both systematic and statistical errors. Hatched areas: DWBA calculation for a pure E1 excitation including the systematic uncertainties in the $\sigma_{\gamma f}$ values. From reference 19.

an additional, presumably E2 contribution. On the other hand the Giessen data do not show any evidence for additional strength. A similar conclusion one can draw from the ratio σ^-/σ^+ [19]. In fact, in both sets of data the E1 contribution alone is already overestimating the measured effects and this casts some doubt on the whole method.

3.2.3.2. Fission decay of isoscalar resonances excited by inelastic hadron
scattering

The fission decay of the $2\hbar\omega$ isoscalar giant resonances in ^{238}U, centered
around $E_x = 65\ A^{-1/3}MeV$ (~ 10.5 MeV for ^{238}U) has been studied in a number of
inelastic alpha scattering coincidence experiments, which are summarized in table 1.

Table 1. Summary of $^{238}U(\alpha,\alpha'f)$ experiments

Particle	Energy (MeV)	Scattering angle	Fission detector angle θ_f	P_f	Ref.
α	120	$(18\pm3)\,^\circ$	many angles in- and out-of plane	$\leqslant 11\%$	20 21
α	152	11.5°, on resonance 17.5°, off resonance	recoil angle recoil angle	25%	23
α	172	13°	several angles	$\sim <5\%$	22

Also here there is a large discrepancy between the experimental numbers, as quoted
by the authors. In this case however this is not so much due to conflicting
experimental results but rather to a different interpretation of data which are in
essence in agreement with each other.

In the 120 MeV $(\alpha,\alpha'f)$ experiment[20,21] the inelastically scattered alpha
particles were detected in a magnetic spectrograph at $(18\pm3)\,^\circ$. A complete angular
distribution of the fission fragments in- and out- of scattering plane was measured,
so that the total fission probability can be calculated. The Jülich-Bonn group,
measured $(\alpha,\alpha'f)$ coincidence spectra for six, in-plane, fission fragment angles. In
the 152 MeV α-experiment of the Oak Ridge-Indiana group a different set-up was
chosen: the fission fragments were only measured at one angle corresponding to the
recoil direction, which is an axis of rotational symmetry for the fission fragment
angular distribution[21]. However in this experiment the inelastically scattered α-
particles where measured at 11.5° where the bump due to isoscalar $2\hbar\omega$ resonance
excitation is strongly and at 17.5°, where it is not at all excited respectively
(see figure 10). So from this experiment one can deduce $dP_f(\theta_f,E_x)/d\theta_f)_{\theta_f=0^\circ}$ for
the bump from the difference between the fission coincidence spectra measured on
(11.5°) and off (17.5°) resonance.

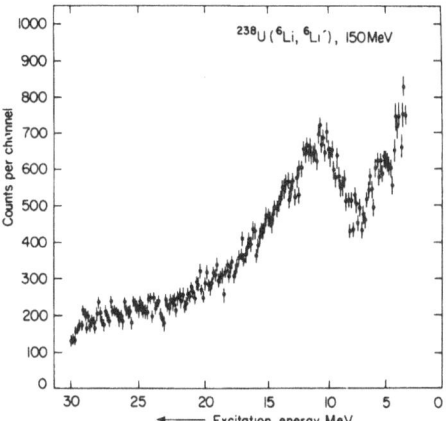

Fig.3.6. Giant resonance structure observed for the $^{238}U(^6Li,^6Li)$ reaction at 150 MeV. The spectrum has been corrected for slight ^{12}C and ^{16}O contaminants. From reference 24.

As is shown in figures 6, 9 and 10, a giant resonance-like structure is clearly excited in inelastic hadron scattering on ^{238}U: it peaks at around 11 MeV and has a width of about 4 MeV. As we have shown in section 1 it is presumably due to excitation of $2\hbar\omega$ $\Delta L=0,2,4.....$ strength concentrated around $E_x\sim 11$ MeV.

Figures 7 shows some results of the $^{238}U(\alpha,\alpha,'f)$ experiment at 120 MeV[20]. In contrast with the singles spectrum the fission coincidence spectrum summed for all angles around the recoil axis is featureless for $8 \lesssim E_x \lesssim 12.5$ MeV, the region where one might have expected the bump as observed in the singles sepctrum to show up. Remarkably enough, the spectrum summed over the range $\theta_f = 50$-$90°$ with respect to the recoil axis does show some kind of bump peaking around $E_x\sim 9.5$ MeV. Such a bump, also shows up in the $(^6Li,^6Li'f)$ spectrum which is measured[24] at an average fission-fragment angle of 65° with respect to the recoil axis, to which also an out-of-plane spectrum (measured at 90°) is added. This spectrum is also shown schematically in fig.7. The excitation energy of this bump at about 9.5 MeV is definitely below the excitation energy where the bump observed in inelastic scattering peaks.

Figure 8 shows the total fission probability $P_f(E_x)$ as obtained from these 120 MeV (α,α') data. It is clear that in the region $9.5 \lesssim E_x \lesssim 12.5$ MeV $P_f(E_x)$ has a lower than normal (≈ 0.22) fission probability. One possible interpretation would be to assume that the underlying continuum has a normal fission probability and that the dip observed in $P_f(E_x)$ would be caused by the fact that the GR-bump has a low fission probability. Another interpretation would be to assume that the continuum underlying the resonance has a fission probability decreasing with increasing excitation energy. This would be the case for instance if a substantial part of this continuum is due to a quasi-elastic neutron knock-out process.

Figure 9 shows a sample of the data obtained from the 172 MeV 238U$(\alpha,\alpha'f)$

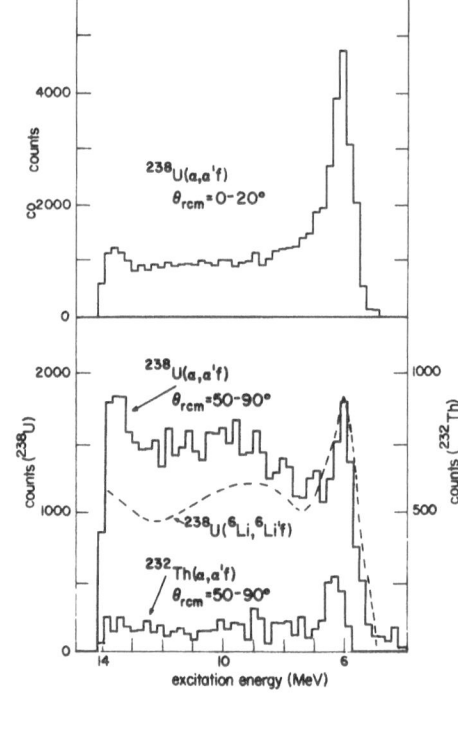

Fig.3.7. Fission-coincident
α-particle spectra squeezed
and summed around the recoil
axis (0-20°) and backwards with
respect to the recoil axis (50-90°).
Also shown as a dashed curve are the
results from the (^{6}Li,^{6}Li'f)
experiment[24], normalized to the
(α,α'f) data at 6 MeV. The dotted
curve in the lower part of the
figure indicates a lower limit for
the contribution of the continuum.

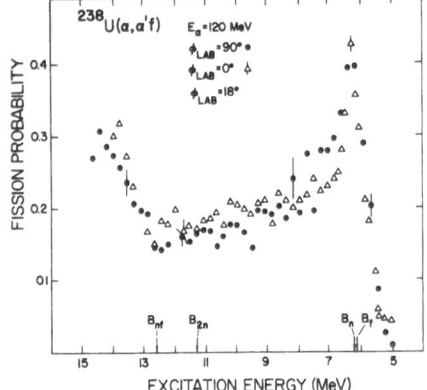

Fig.3.8. Fission probability as a function
of excitation energy determined from out-of-
plane and in-plane data. From reference 21.

experiment[22]. The 2ℏω-bump, strongly excited in the singles (α,α')-spectrum, is
only weakly reflected as a small bump at E_x ~ 10 MeV in the fission coincident
spectrum. This little bump has a fission-fragment angular distribution which is
peaked sideways with respect to the recoil axis[22]. The total fission probability of
the 2ℏω bump was estimated to be < 5%.

Such a small bump at E_x ~ 10 MeV is also observed[23] in the 152 MeV on-
resonance (α,α'f) data - see figure 10. It is interesting to compare the values for
$dP_f(\theta_f E_x)/d\theta_f$ for the on- (11.5°) and off-resonance (17.5°) data - the main
difference is a dip in the fission probability at about 10 < E_x < 12 MeV, very

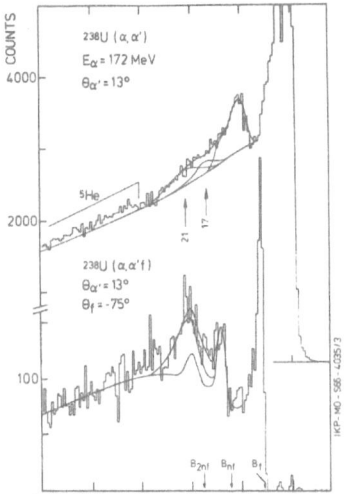

Fig.3.9. Comparison of 13°(α,α')
spectrum with the coincidence
(α,α'f) spectrum at θ_f=-75°. From
reference 21.

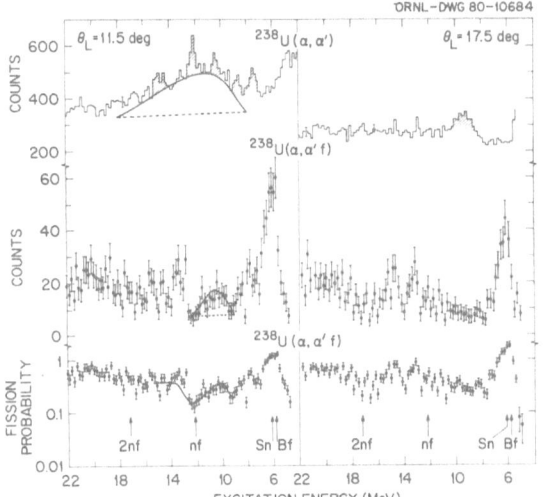

Fig.3.10. Top: Inelastic alpha
spectra from the reaction
$^{238}U(\alpha,\alpha')$ for the indicated
angles. Contaminants are shown
cross hatched. At 11.5°, the
deduced shape of the giant
resonance peak and the underlying
continuum are shown as solid
curves. Middle: Spectra of alpha
particles is measured in
coincidence with fission
fragments. In the 11.5° spectrum
an enhancement at $E_x \sim 10$ MeV is
indicated by the smooth curve.
Bottom: Fission probability for
the (α,α'f) reaction. B_f is the
fission threshold; S_n is the
neutron separation energy for
^{238}U; nf and 2nf indicate the
energies for the onset of second
and third chance fission. From
reference 23.

similar to what was observed for the 120 MeV (α,α'f) data - see ref. 21. This
indicates that this dip is associated with the resonance - bump in the singles
spectrum and not with the underlying continuum.

Thus all the (α,α'f) experiments on ^{238}U seem to be consistent in the
sense that they indicate that only the low-energy part of the isoscalar 2ℏω-bump
fissions, and that the part around $E_x \sim 11.5$ MeV, corresponding to the peak cross-
section in the singles spectrum, has a small (< 0.22) fission probability.

3.2.3.3. <u>Fission decay of giant resonances: the ^{238}U (e,e'f) experiment</u>

Recently two ^{238}U (e,e'f) experiments have been performed using the new, high current, high dutycycle electron accelerator facilities at Illinois and Stanford. Such experiments clearly have a big advantage over the inclusive (e,f) experiments discussed before. With respect to the $(\alpha,\alpha'f)$ experiments the main differences are that: the (e,e'f) experiments are much more selective in the excitation of low multipole strength, and that both, electric isoscalar and isovector strength can be excited.

Table 2

Comparison between (e,e') and (α,α') experiments under the conditions valid for the fission decay experiments

	E_e	θ_e	q^{-1} (fm^{-1})	ΔT	ΔL	θ_f	references
Illinois	46.5	60	0.36			$(90\pm20)°$	
	56.9	60		0,1	<2	$(180\pm20)°$	25
(e,e'f)	67.1	60					
	67.1	80	0.59				
Stanford	80	40	0.28	0,1	<2	several angles	26
(e,e'f)	118		0.41				
$(\alpha,\alpha'f)$			0		<8		

Also in the (e,e') experiments it is impossible to seperate E-0 and E-2 strength, since they have the same q-dependence. Moreover it is hard to deduce absolute fission probabilities since the singles (e,e') cross-sections are dominated by the radiation tail from elastic scattering events.

Figure 11 shows the result of the Illinois experiment[25]. These results were obtained by assuming that only E1 and E0/E2 strength is present. For each 100 keV the data were analyzed separately with respect to their q-dependence, using calculated values for the form factors and the known (γ,f) (=E1) contribution. Similarly a preliminary result of the Stanford data is shown in figure 12.

Several features emerge from these data: (i) The E2/E0 distribution as shown in figure 11 is flat and featureless from 7 MeV < E_x < 11.5 MeV and does not show any pronounced resonant behaviour. In the Stanford data, figure 12, there is in addition evidence for a bump around 9 MeV which is not at all or only weakly seen in the Illinois data. (ii) In both experiments the E2/E0 strength distribution from

Fig.3.11. The E2/E0 strength in the
fission channel as determined from
the Illinois experiment[25], and the
measurured angular asymmetry.

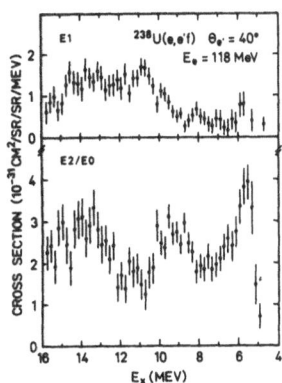

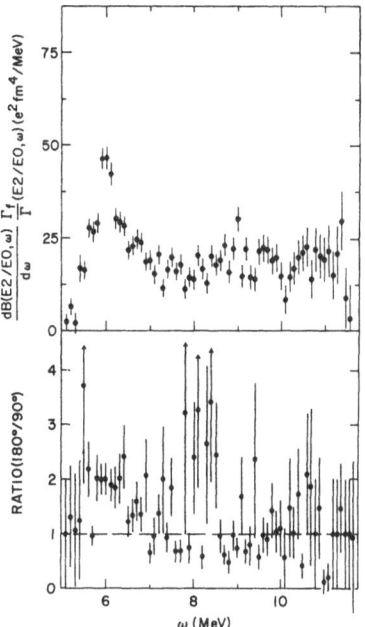

Fig.3.12. The tentative decomposition of the
$^{238}U(e,e'f)$ coincidence cross section at 118
MeV into the E1 and E0/E2 contributions.
Reference 26.

$E_x \sim$ 7 MeV to about $E_x \sim$12 MeV represents about 10% of the E2 EWSR. If one assumes
that $(\Gamma_f/\Gamma_{tot}) \sim 0.22$ like has been found for the GDR, this would represent ~50% of
the E2 EWSR. (iii) If the bump observed at $E_x \sim$ 9 MeV in the Stanford data survives
a final analysis its absence or much weaker excitation in the Illinois data is
probably due to the fact that the contribution of fission fragments detected at
large angles to the recoil axis is relatively much larger for the Stanford than for
the Illinois experiment. This might indicate that this bump has an angular
distribution peaking at large angles with the respect to the recoil axis, similar to
what was observed in the hadron-induced experiments (section 3.2.3.2.).

3.2.3.4. Discussion

The previous discussions on the experimental results can be summarized as follows:
(I) The large GQR fission branch ($P_f \gtrsim 40\%$) for ^{238}U suggested by Arrudo Neto et
al.[16] is not confirmed by any of the other (e,f), ($\alpha,\alpha'f$) or (e,e'f)

experiments.

(II) None of the coincidence experiments shows evidence for a resonance behaviour corresponding to what is observed in the single inelastic hadron scattering experiment around $E_x = 65A^{-1/3}$ MeV.

(III) The (e,e'f) experiments show evidence for a continuum of E2/E0 strength for $7 \lesssim E_x \lesssim 12.6$ MeV . If this would be pure E-2 strength, it would exhaust about 10% of the E2 EWSR, which with a fission probability $P_f \sim 0.22$ would amount to about 50% of the E2 EWSR. One should note here that this corresponds to an upper limit: the analysis has been performed assuming that no E3 strength is present, and that all the detected strength is due to E-2 excitation and not partly due to E0 excitation. Also part of the strength could be isovector.

(IV) There is evidence from the $(\alpha,\alpha'f)$, the (e,e'f) and also the $(^6Li,^6Li'f)$[24] experiments that there is a resonance in the fission channel around $E_x \sim 9.5$ MeV which has an fission fragment angular distribution peaking sideways with respect to the symmetry (=recoil) axis. This resonance is located at a lower excitation energy than the maximum of the bump observed in inelastic hadron scattering, but it could correspond to fission of the low energy branch of this bump.

The interpretation of these data is still a puzzle. From inelastic hadron scattering experiments we know that isoscalar L=0,2,4.. strength is concentrated, although we do not exactly know how much L=0,2 or 4 strength is actually present in this bump. We do not find such a concentration back in the fission coincident spectrum. Fragmented E2/E0 strength is observed though in the fission channel. Does this imply that concentrated strength that is due to a collective excitation, does have a low fission probability while fragmented non-collective strength has a normal one? And why would the fission probability be different for isoscalar and isovector resonances? Recently Zawischa and Speth[24] have performed an RPA calculation for ^{238}U including all states $3\hbar\omega$ below and above the Fermi-edge, in which a rather large concentration of 2^+ strength is found around $E_x \sim 10$ MeV. It is interesting to note that the K=0 L=0 and L=2 strength are mixed around $E_x \sim 9$ MeV and that this component is much weaker than the K=1 and K=2 L=2 strength around $E_x \sim 10$ MeV. If we assume that the K quantum number is conserved during the process: excitation of resonance → deformation to the saddle point → scission than this would predict that the coincidence fission spectrum would show a bump around $E_x \sim 10$ MeV which is relatively stronger for angles off the recoil axis. This is what is observed indeed. On basis of the calculated strength it would also imply that the bump corresponding to what is seen in the (e,e') experiment of reference 26, figure 12 corresponds to a fission probability $P_f \sim 5\%$ for the K=1,2 components.

References section 3

1. J. Wambach, F. Osterfeld, J. Speth and V. Madsen, Nucl.Phys. A324(1979)77
2. J. Speth and A. van der Woude, Rep.Progr.Phys. 44(1981)719
3. B.L. Berman and S.C. Fultz, Rev.Mod.Phys. 47(1975)713
4. J.R. Nix and A.J. Sierk, Phys.Rev. C21(1980)396,
   see also Phys.Rev. C25(1982)1068
5. F. Bortingnon and R.A. Broglia, Nucl.Phys. A371(1981)405
6. L.S. Cardmann, Nucl.Phys. A354(1981)173c
7. K. Okada et al, Phys.Rev.Lett. 48(1982)1382
8. G.J. Wagner in Giant Multipole Resonances, ed. F.E. Bertrand,
   Harwood Academic Field (1980)251
9. K.T. Knöpfle, Lecture Notes in PHysics 108(1979)311
10. M. Buenerd, The giant monopole resonance in nuclei ISN 81,38, Grenoble 1981.
    Lectures given at the International Nuclear Physics Workshop, Triёste 1981
11. R. Bergère, Lecture Notes in Physics 61(1977)1
12. G.F. Bertsch, P.F. Bortignon, R.A. Broglia and C.H. Dasso,
    Phys.Lett. 80B(1979)161
13. M.N. Harekeh, Lecture Notes on Physics 158, page 236
14. R. Vandenbosch and J.R. Huizenga, Nuclear Fission Academic Press, New York
    1973; S. Bjørnholm and J.E. Lynn, Rev.Mod.Phys. 52(1980)725
15. A. Veyssière, A. Beil, R. Bergère, T. Carlos, A. Lepretre and K. Kernbath,
    Nucl.Phys. A199(1973)45
16. J.T. Caldwell, E.J. Dowdy, B.L. Berman, R.A. Alvarez and P. Meyer,
    Phys.Rev. C21(1980)1215
17. J.D.T. Arruda Neto and B.L. Berman, Nucl.Phys. A349(1980)483
    J.D.T. Arruda Neto, S.B. Herdade, B.L. Berman and I.C. Nascimento,
    Phys.Rev. C22(1980)594 and references therein
18. J. Aschenbach, R. Haag and H. Krieger, Z.Phys. A292(1979)285
19. H. Ströher, R.D. Fischer, J. Drexler, K. Huber, W. Kneissl, R. Ratzek,
    H. Ries, W. Wilke and H.J. Maier, Phys.Rev.Lett. 47(1981)318
20. J. van der Plicht, M.N. Harakeh, A. van der Woude, P. David and J. Debrus,
    Phys.Rev.Lett. 42(1979)1121
    J. van der Plicht, M.N. Harakeh, A. van der Woude, P. David, J. Debrus,
    H. Janssen and J. Schulze, Nucl.Phys. A346(1980)349
21. R. De Leo, M.N. Harakeh, S. Micheletti, J. van der Plicht, A. van der Woude,
    P. David and H. Janssen, Nucl.Phys. A373(1982)509
22. H.P. Morsch, Lecture Notes in Physics 158(1982)254
23. F.E. Bertrand, J.R. Beene, C.E. Bernis Jr., E.E. Gross, D.J. Horen, J.R.Wu and
    W.P. Jones, Phys.Lett. 99B(1981)213
24. A.C. Shotter, C.K. Gelbke, T.C. Awes, b.B. Bark, J. Mahoney, T.J.M. Symons and
    D.K. Scott, Phys.Rev.Lett 43(1979)569
25. D.H. Dowell, L.S. Cardmam, P. Axel, G. Bolem and S.E. Williamson,
    Phys.Rev.Lett. 49(1982)113
26. K. van Bibber et al., Lecture Notes in Physics 158(1982)278
27. D. Zawischa and J. Speth, Lecture Notes in Physics 158(1982)231

SPECTROSCOPY OF SUPERHEAVY QUASIMOLECULES AND QUASIATOMS

U. Müller, N. Aboul-El-Naga, J. Reinhardt, T. de Reus,
P. Schlüter, M. Seiwert, G. Soff[+], K.H. Wietschorke,
B. Müller, and W. Greiner[*]
Institut für Theoretische Physik
der Johann Wolfgang Goethe-Universität,
Robert-Mayer-Straße 8-10, Postfach 111 932,

D-6000 Frankfurt am Main, West Germany

+) Gesellschaft für Schwerionenforschung mbH,
Planckstraße 1, Postfach 11 05 41,

D-6100 Darmstadt 11, West-Germany

ABSTRACT

Collisions of very heavy ions at energies close to the Coulomb barrier are discussed as a unique tool to study the behaviour of the electron-positron field in the presence of strong external electromagnetic fields. Theoretical predictions for K-hole production, δ-electron and positron emission rates are compared with experimental results. Emphasize is laid on effects due to nuclear reactions close to the Coulomb barrier. Internal electron-positron pair production from electric monopole transitions is investigated for atoms with charge numbers up to $Z \approx 170$. The effect of field theoretical corrections to the binding energy of the deepest bound state, here the self-energy, is also discussed.

1. INTRODUCTION

Collisions of very heavy ions with bombarding energies close to the Coulomb barrier offer us the possibility to perform a spectroscopy of electronic states in transient superheavy systems within a charge range of $100 \lesssim Z \lesssim 190$ ($_{92}$U + $_{98}$Cf). The pivotal question to answer is whether the binding energy of the strongest bound electronic state can reach or exceed twice the electron rest mass.

An important task in this connection is the investigation of electron excitation processes in superheavy systems. Two possible excitation mechanisms are visualized in Fig. 1. As an example we consider the schematic level structure for a head-on collision of Pb + Cm with a total charge $Z = 178$ at 5.9 MeV/u bombarding energy. Binding energies of the adiabatic 1sσ- and 2sσ- states[1] are displayed versus the internuclear separation R. The turning point between incoming and outgoing trajectory is given by $R_{min} = 17$ fm.

A typical one-step excitation process is the direct ionization from the 1sσ-state to the positive continuum. In addition the multi-step processes play an important

---

[*]Invited speaker at the Conference on "Fundamental Aspects in Heavy Ion Physics", La Rabida, Spain, June 1982

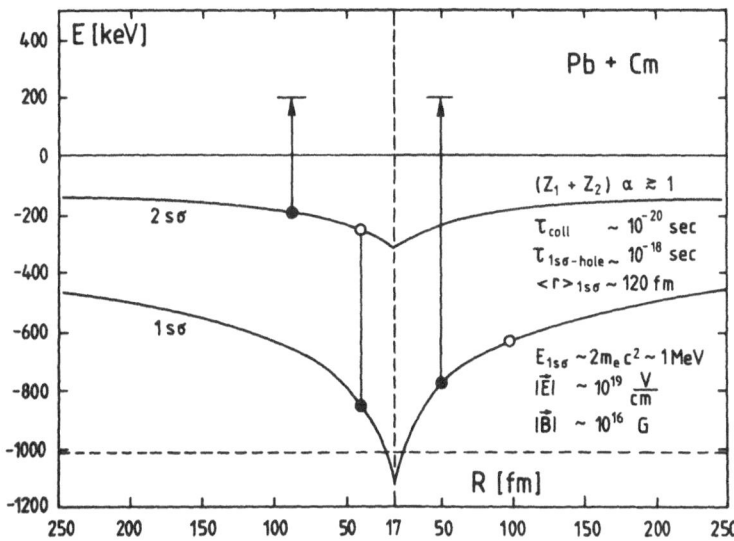

Fig. 1: Binding energies of the 1sσ- and 2sσ-state versus internuclear separation for the Pb+Cm system. One-step and two-
step electron excitation processes are indicated qualitatively.
On the right hand side characteristic numbers describing the
collision are given.

role. Actually they form the dominant part of inner shell vacancy production probabilities. An example for a two-step process is depicted in Fig. 1, where a 2sσ-
electron is ionized to the continuum. The remaining hole is filled by the 1sσ- electron which leaves a vacancy in the K-shell. Since initial and final state of both
processes are indistinguishable, they have to be treated coherently. This is achieved
by solving coupled channel equations for the electron occupation amplitudes of the
quasimolecular adiabatic states.

Strong relativistic effects are best visible in the wavefunctions. Fig. 2
shows the radial density $r^2|\psi(r)|^2$ of the 1s - electron in the field of an extended
nucleus. The Schrödinger hydrogenic 1s-wavefunction is given by $\psi_{1s}(r) = 2(Z/a_B)^{3/2}$.
e^{-rZ/a_B}, where $a_B = \lambda_e/\alpha = 52918$ fm is the Bohr radius. The axes in Fig. 2 have
been scaled appropriately in order to make the shape of the nonrelativistic density
independent of the charge Z. Deviations therefore are solely due to relativistic
effects. While the shape of $r^2|\psi(r)|^2$ does not alter much in the region of the ordi-
nary periodic table, the wavefunction becomes severely contracted as Zα approaches
or even exceeds the value 1. Although a complete collapse of the wavefunction' is
averted, the scaled Bohr radius a_B/Z then loses its importance and the only relevant
length scale is determined by the nuclear radius R. The inset in Fig. 2 gives the
electron density at the origin, $|\psi(0)|^2$, as a function of Z. Apart from the Z^3-
dependence known from the nonrelativistic solution the density increases by additional
three orders of magnitude going to Z = 180.

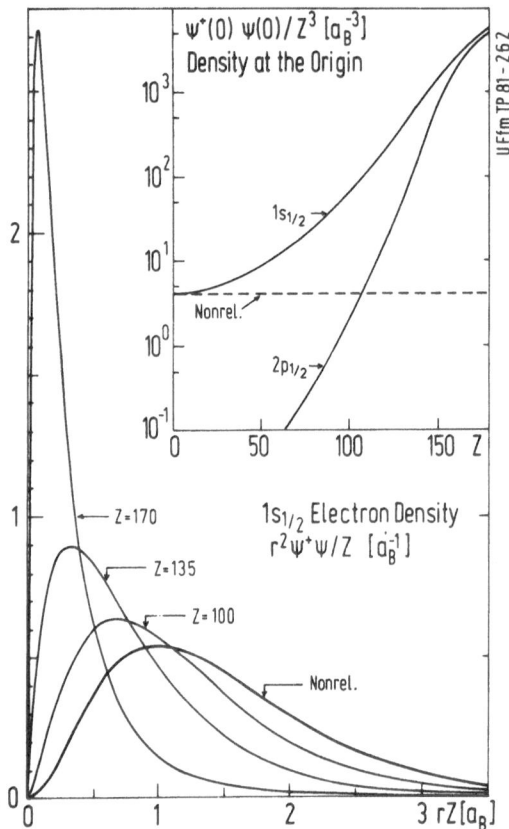

Fig. 2: The radial electron density of the $1s_{1/2}$-state in superheavy atoms with nuclear charges Z = 100, 135, and 170. For comparison the nonrelativistic Schrödinger wavefunction is drawn. The axes are scaled with powers of Z in order to make the nonrelativistic density independent of the charge. The inset shows the drastic increase of the electron density at the origin over its **nonrelativistic** value. The wavefunctions of the $1s_{1/2}$ and $2p_{1/2}$-states become very similar in the region $Z\alpha > 1$.

The figure also displays the electron density at the origin for the $2p_{1/2}$-wavefunction. For a nonrelativistic p-wave, characterized by the sharp angular momentum $\ell = 1$, this value is exactly zero. The $2p_{1/2}$-Dirac bispinor, however, carries a mixture of $\ell = 0$ (upper component) and $\ell = 1$ (lower component) orbital angular momentum. In atoms with $Z\alpha > 1$ the 'small' and 'large' components of the wavefunction become comparable in magnitude. As a consequence the distinction between $s_{1/2}$- and $p_{1/2}$-states is diminished. This explains the steep rise of the $2p_{1/2}$-density as well as the strong increase of its binding energy as a function of Z.

The shrinkage of the 1s-wavefunction at short internuclear distances is a typical feature in collisions of very heavy ions. Sticking to our example Pb + Cm at E_{lab} = 5.9 MeV/u we find a radial expectation value for the $1s\sigma$-orbital of about 120 fm at the distance of closest approach. Due to this extreme localization (within the Comptonwavelength of an electron, $\lambdabar_e$ = 386 fm) transfer of high momenta components to the strongly bound electrons resulting from the nuclear motion becomes possible. Therefore δ-electrons up to kinetic energies of 2 MeV can be observed experimentally[2], whereas the classical allowed maximum energy transfer to an electron at rest is about 10 keV.

After a short theoretical survey concerning electron excitation processes we discuss the bound state solution of the two-centre Dirac equation in chapter 3 and deal with the self-energy of electrons in critical fields in chapter 4. In the following we present numerical results concerning K-hole and δ-electron production rates and compare them with some experimental data. After this we present a few new results concerning positron creation in deep inelastic nuclear collisions with special regard to nuclear models. Finally we discuss the possibility of internal electron-positron pair creation in supercritical compound systems.

2. THEORETICAL DESCRIPTION OF ELECTRON EXCITATIONS

In this chapter we summarize the essential equations upon which our calculations are based. We emphasize that in superheavy quasimolecules a full relativistic treatment of the electron motion is absolutely necessary since the combined charge $Z = Z_P + Z_T \gtrsim 1/\alpha$ and the electron binding energies become comparable with the electron rest mass or even exceed it. If we want to describe the dynamical evolution of an electron involving two colliding nuclei we thus have to solve the time-dependent two-centre Dirac equation

$$i\partial/\partial t \; \Phi_i \; (\vec{R}(t)) = H_{TCD} \; (\vec{R}(t)) \; \Phi_i \; (\vec{R}(t)) \; . \tag{2.1}$$

H_{TCD} is the relativistic two-centre Hamiltonian which depends sensitively on the internuclear separation $\vec{R}(t)$. For this reason it is useful to expand the total wave function Φ_i into Born-Oppenheimer states ϕ_j which are represented by the stationary molecular states (c.f. chapter 3). The expansion reads

$$\Phi_i \; (\vec{R}(t)) = \sum_j \; a_{ij}(t) \; \phi_j(\vec{R}(t)) \; \exp \; \{-i\chi_j(t)\}, \tag{2.2}$$

where the sum includes an integration over continuum states with positive and negative energies. The phase factors χ_j are chosen as

$$\chi_j(t) = \int^t dt' \; <\phi_j \; (\vec{R}(t')) \, | \, H_{TCD} \; (\vec{R}(t')) \, | \, \phi_j(\vec{R}(t'))> \; . \tag{2.3}$$

Inserting the expansion of eq. (2.2) into eq. (2.1) and projecting with stationary eigenfunctions we obtain the following set of first-order coupled differential equations for the occupation amplitudes $a_{ij}(t)$

$$\dot{a}_{ij}(t) = -\sum_{k \neq j} \; a_{ik}(t) \; <\phi_j | \partial/\partial t | \phi_k> \; \exp \; \{+i(\chi_j - \chi_k)\}, \tag{2.4}$$

with the initial condition $a_{ij}(-\infty) = \delta_{ij}$.

After splitting the time derivative operator in terms of a radial and a rotational coupling and neglecting the latter one, we solve the coupled equations (2.4) by simultaneous numerical integration. In the independent-particle approximation excitations of the many-electron system are described by incoherent summation over one-electron transition probabilities[3]. The number of particles created in a state above the Fermi level, p>F, up to which the quasimolecular levels are initially filled, is

$$N_p = \sum_{k<F} |a_{kp}(\infty)|^2 \qquad (2.5)$$

and the number of holes, e.g. positrons, in a state below the Fermi level, q<F, is

$$N_q = \sum_{k>F} |a_{kq}(\infty)|^2 . \qquad (2.6)$$

For the number of correlated particle-hole pairs $N_{p,q}$ one has to calculate

$$N_{p,q} = N_p \cdot N_q + |\sum_{k<F} a_{kp}^* a_{kq}|^2 . \qquad (2.7)$$

This expression is relevant for instance for the calculation of a δ-electron distribution which should be measured in coincidence with $1s\sigma$-vacancy formation. The first term $N_p \cdot N_q$ just corresponds to a multiplication of the independent measured number of created $1s\sigma$-vacancies and the number of created δ-electrons. The last term is the exchange term due to the Pauli exclusion principle. Eq. (2.7) also holds for particle-particle or hole-hole correlations if the sign of the second term is changed.

However, in a measurement there is usually no distinction between a δ-electron with spin up and a $1s\sigma$-vacancy with spin down etc. For a comparison with measured data we therefore have to perform a summation. This leads[4] for the total number of particle-hole pairs to

$$N_{p,q}^{tot} = N_{p\uparrow q\downarrow} + N_{p\downarrow q\uparrow} + N_{p\downarrow q\downarrow} + N_{p\uparrow q\uparrow} = 2(N_{p,q} + N_p N_q) . \qquad (2.8)$$

A further complication arises if it is impossible to distinguish between $1s\sigma$- and $2p_{1/2}\sigma$-vacancy formation as is the case in symmetric colliding systems. Here the vacancy sharing produces asymptotically K-vacancies resulting from both $1s\sigma$ and $2p_{1/2}\sigma$-excitation. Also in the case of δ-electrons of energy E a separation of $Es\sigma$ and $Ep_{1/2}\sigma$-states cannot be performed easily. As final result of this averaging we obtain for the number of δ-electrons with energy E (p=E) measured in coincidence with a K-vacancy (q=$1s_{1/2}$)

$$N_{p,q}^{tot} = N_{E,1s\sigma} + N_{E,2p1/2\sigma} , \qquad (2.9)$$

with

$$N_{E,1s\sigma} = 2(N_{Es\sigma,1s\sigma} + N_{Es\sigma} \cdot N_{1s\sigma}) + 4N_{Ep1/2\sigma}N_{1s\sigma} ,$$

$$N_{E,2p1/2\sigma} = 2(N_{Ep1/2\sigma,2p1/2\sigma} + N_{Ep1/2\sigma}N_{2p1/2\sigma}) + 4N_{Es\sigma}N_{2p1/2\sigma} .$$

3. BOUND STATE SOLUTIONS OF THE TWO-CENTRE DIRAC EQUATION

In this chapter we briefly discuss solutions of the stationary Dirac equation for a two-centre Coulomb potential. We start with the representation of the Dirac equation in spherical coordinates

$$[i\gamma_5\sigma_r\left(\frac{\partial}{\partial r} + \frac{1}{r} - \frac{\beta}{r}\hat{K}\right) - \frac{Z_1e^2}{\left|\vec{r} - \vec{R}/2\right|}$$

$$-\frac{Z_2e^2}{\left|\vec{r} + \vec{R}/2\right|} + \beta m_e]\ \phi_\mu(\vec{r}) = E\phi_\mu(\vec{r})\ . \tag{3.1}$$

The spin-orbit operator is defined by the relation

$$\hat{K} \equiv \beta(\vec{\sigma}\vec{\ell} + 1)\ . \tag{3.2}$$

It is known that the two-centre Dirac equation is more difficult to handle than its nonrelativistic counterpart, since it is not separable in any orthogonal coordinate system. The most accurate and flexible approach at present is based on a multipole expansion of the wave function[1,5,6]

$$\phi_\mu(\vec{r}) = \sum_K \phi_{\mu K}(\vec{r}) = \sum_K \begin{pmatrix} g_K(r)\ \chi_K^\mu \\ if_K(r)\ \chi_{-K}^\mu \end{pmatrix}\ . \tag{3.3}$$

$f_K(r)$ and $g_K(r)$ are the radial wave functions; the spinor spherical harmonics are given by

$$\chi_K^\mu = \sum_{m=\pm 1/2} (\ell, \tfrac{1}{2}, j, \mu-m,m)\ Y_\ell^{\mu-m}(\theta,\phi)\chi^m\ . \tag{3.4}$$

$-K$ is the eigenvalue of $\hat{K}$ and connected to the angular momentum

$$K = \begin{cases} \ell & \text{for } j=\ell-\frac{1}{2} \\ -\ell-1 & \text{for } j=\ell+\frac{1}{2} \end{cases}, \tag{3.5}$$

$j = |K|-\frac{1}{2}$ being the total angular momentum. The magnetic quantum number μ is the projection of the total angular momentum on the axis connecting the two nuclei (z-axis). After expanding the two centre potential into multipoles

$$V(\vec{r},\vec{R}) = \sum_{\ell=0}^\infty V_\ell(r,R)P_\ell(\cos\theta)\ , \tag{3.6}$$

a set of coupled differential equations for the radial wave functions is derived, which can be solved by numerical methods.

In the following we discuss the relativistic correlation diagram for the asymmetric system Pb+Cm. In Fig.3a we plot on a double logarithmic scale the 21 lowest σ-(solid lines) and π-(dashed lines) states for R = 16 fm to R = 3000 fm. The dashed-dotted lines denote the energies for extended nuclei. The energy levels in the boxes I and II between 300 and 1000 fm are presented separately on a linear scale in Figs. 3b and 3c.

Relativistic effects lead to various striking consequences: So the relativistic splitting between the states $2p_{3/2}\sigma$ and $2p_{1/2}\sigma$ amounts to more than the electron rest mass. Furthermore at the critical distance $R_{cr} \approx 26$ fm (for pointlike nuclei) the binding energy of the $1s\sigma$-state exceeds twice the electron rest mass, i.e., this state 'dives' into the negative energy continuum (c.f. chapter 6). The sudden energy

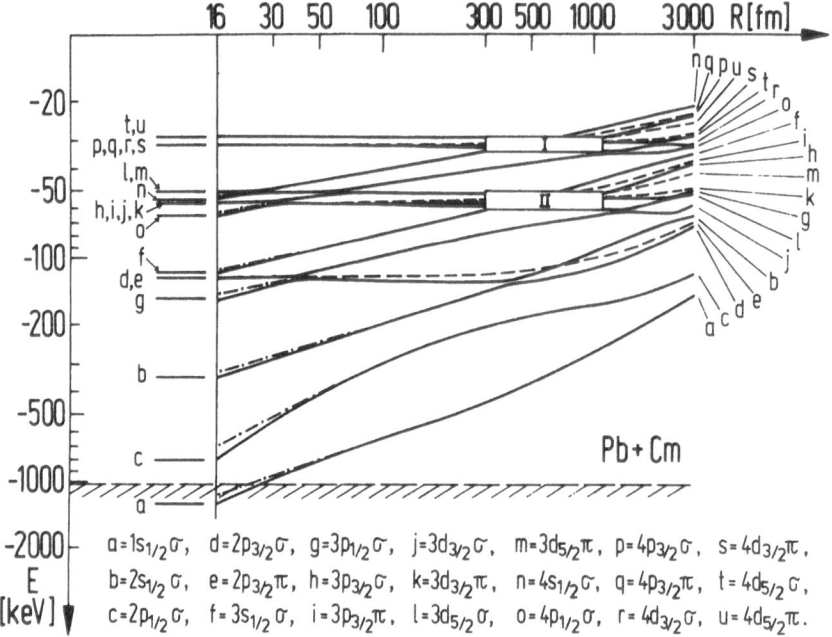

a = $1s_{1/2}\sigma$,   d = $2p_{3/2}\sigma$,   g = $3p_{1/2}\sigma$,   j = $3d_{3/2}\sigma$,   m = $3d_{5/2}\pi$,   p = $4p_{3/2}\sigma$,   s = $4d_{3/2}\pi$,

b = $2s_{1/2}\sigma$,   e = $2p_{3/2}\pi$,   h = $3p_{3/2}\sigma$,   k = $3d_{3/2}\pi$,   n = $4s_{1/2}\sigma$,   q = $4p_{3/2}\pi$,   t = $4d_{5/2}\sigma$,

c = $2p_{1/2}\sigma$,   f = $3s_{1/2}\sigma$,   i = $3p_{3/2}\pi$,   l = $3d_{5/2}\sigma$,   o = $4p_{1/2}\sigma$,   r = $4d_{3/2}\sigma$,   u = $4d_{5/2}\pi$.

Fig. 3a

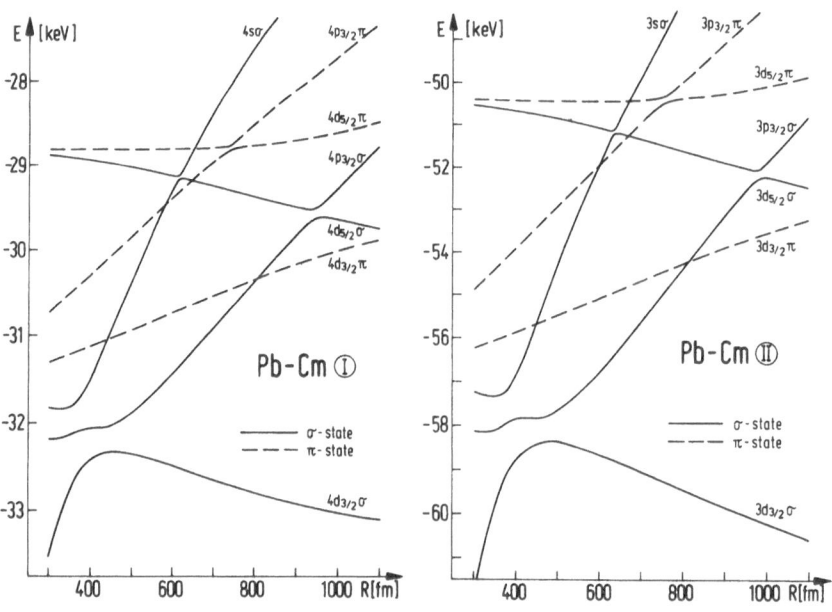

Fig. 3b

Fig. 3c

change of the strongest bound states near the distance of closest approach is indeed very impressive. The strongest energy change is found for the $2p_{1/2}\sigma$-state. Its binding energy increases by 56.4% from R = 100 fm to R = 16 fm. The variation of the $1s\sigma$-binding energy amounts to about the electron rest mass in a two-centre distance interval of ΔR = 84 fm. The nuclear extension lowers the binding energy of the $1s\sigma$-state by 10% and that of the $2p_{1/2}\sigma$-state by 15%.

4. THE SELF-ENERGY OF ELECTRONS IN CRITICAL FIELDS

The K-electron binding energy E_{1s} increases strongly as a function of the nuclear charge Z. For Z = 150, E_{1s} amounts to about the electron rest mass and hence one enters the truly relativistic domain. For $Z \geq 170$ the binding energy exceeds twice the electron rest mass and the K-shell electron becomes a resonance imbedded in the negative energy continuum, which opens the possibility of spontaneous positron production[7-9].

The major motivation of these investigations was the question whether field theoretical corrections, such as vacuum-polarization and self-energy, may prevent such an extraordinary strong binding. These processes are visualized by the Feynman diagrams in Fig. 4.

The dominant vacuum-polarization contribution is provided by the attractive Uehling potential. Its influence on electronic binding energies for superheavy systems has been calculated by various authors[7,10,11]. For the critical nuclear charge Z_{cr} the Uehling potential leads to an energy shift $\Delta E_{VP}^{(n=1)}$=-11.8keV[11], thus Z_{cr} will decrease by 1/3 of a unit. The remaining vacuum-polarization effects in lowest order of the fine structure constant α but in all orders of $(Z\alpha)^n$ were evaluated by M. Gyulassy[12] and by Rinker and Wilets[13]. These authors made use of the angular momentum decomposition of the electron propagator in spherically symmetric potentials that was developed by Wichmann and Kroll[14]. The obtained energy shift of $\Delta E_{VP}^{(n>1)}$= +1.15keV[12] is very small compared with the total K-shell binding energy of 1 MeV.

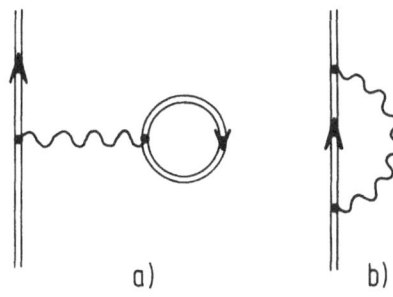

Fig. 4: Feynman diagrams for the lowest order vacuum-polarization (a) and self-energy (b). The double lines indicate the exact propagators and wave functions in the Coulomb field of a nucleus.

a)   b)

Electronic self-energy corrections for high-Z systems have been first studied in the pioneering work of Brown et al. [15-17]. In these theoretical investigations the traditional expansion[18] of the Feynman diagrams in powers of the coupling constant $(Z\alpha)$ of the external field was avoided. This method was further refined and success-fully applied in computations of electron energy shifts in high-Z elements by Desiderio and Johnson[19], who allowed for a realistic nuclear charge distribution as well as the electron-electron interaction in the Hartree-Fock approximation. The pre-cise analysis of self-energy corrections by P. Mohr[20] is based on the Coulomb poten-tial for point-like nuclei. Due to the singular nature of the potential these calcu-lations are restricted to nuclear charges below $Z = \alpha^{-1} \sim 137$. Cheng and Johnson[21] continued the calculations of Ref. 19 up to $Z = 160$, where a repulsive energy shift for K-shell electrons of $\Delta E_{SE} = +7.3$ keV was found.

In our calculations we employed the methods developed by Desiderio and Johnson[19], which may be slightly simplified by restriction to K-shell electrons. We performed our calculations for hydrogen-like systems. The external potential energy $V(x)$ is determined by the nuclear charge distribution, for which a homogeneously charged sphere with a radius $R = 1.2\ A^{1/3}$ fm has been assumed.

To check our computer code, we computed the self-energy contribution to the K-shell binding energy for various nuclear charges from Z=80 (mercury) up to Z=160 and

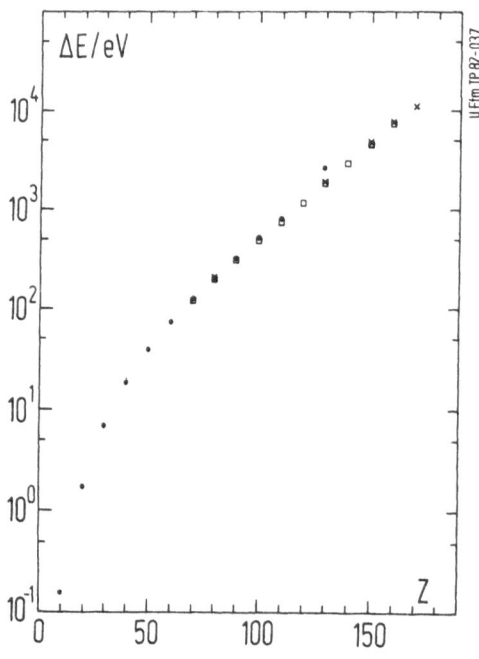

Fig. 5: The self-energy shift ΔE of K-shell electrons as a func-tion of the nuclear charge Z. The calculations are performed to first order in α but to all orders in the coupling constant $(Z\alpha)$ of the external field. The dots denote the numerical results of P. Mohr[20] for 1s-electrons in the Coulomb field of point-like nuclei. The squares represent the values obtained by Cheng and Johnson[21] for a Hartree-Fock potential and extended nuclei. The results of the present calculations[23] for extended nuclei are in-dicated by crosses.

found good agreement with the published values of Refs. 20 and 21 (Fig. 5). Our calculation for increased $Z=169 \lesssim Z_{cr}$ yielded ΔE_{SE} = 10.839 keV. For the critical nuclear charge Z=170 we adjusted the nuclear mass number and hence the nuclear radius such that the K-electron energy eigenvalue differed only by 10^{-3} eV $\sim 10^{-9}$ E_{1s}^{b} from the border line of the negative energy continuum. As the most important result we found an energy shift of ΔE_{SE} = 10.989 keV $\pm$ .3 keV, which still represents only a 1% correction to the total K-electron binding energy. Therefore it may safely be neglected in investigations of ionization probabilities[22] in superheavy quasimolecular systems. If one adds to this the vacuum-polarization calculations of Refs.11 and 12 for critical external potentials, the total energy shift due to radiative corrections of order α amounts only to 300 eV. This tiny effect is at present far outside of any measurable consequences. The various calculations for the self-energy correction of K-electrons in high-Z atoms are summarized [23] in Fig. 5. On a logarithmic scale the energy shift is displayed versus the nuclear charge Z. For Z $\geq$ 70 it is well described by an exponential increase.

We conclude that radiative corrections as vacuum-polarization or self-energy may not prevent the K-shell binding energy from exceeding $2mc^2$ in superheavy systems with Z > Z_{cr} $\sim$ 170.

5. RESULTS FOR THE VACANCY FORMATION AND δ-ELECTRON PRODUCTION IN SUPERHEAVY QUASI-MOLECULES

Binding energies as well as the radial expectation value $<r_{1s\sigma}>$ play an important, but contrary role when calculating K-hole formation and δ-electron production. Whereas a strong increase in the binding energy hinders, e.g., the ionization of an electron from the K-shell - thus producing a K-hole and, eventually, a δ-electron - the strong decrease of $<r_{1s\sigma}>$ leads to high Fourier frequencies in the pulse spectrum - thus stimulating K-shell ionization. Therefore one can predict a maximum for K-hole production rates, $P_{1s\sigma}$, around Z $\sim$ 170.

Subsequently we investigate this effect more systematically for K-hole formation as well as for δ-electron production rates. For this purpose we evaluated K-vacancy production rates for various combined nuclear charges Z=134 up to Z=184. We present coupled channel calculations stressing the Z-dependence of $P_{1s\sigma}$(b) in Fig. 6. Part a shows K-hole production rates for charges Z$\leq$164. We observe an increasing slope with rising binding energy. Vacancy formation increases especially for small impact parameters as a consequence of the sharply localized 1sσ-state. The latter effect allows the transfer of high Fourier frequencies to the K-shell originating from the nuclear motion, thus leading to increased K-hole production. In Fig. 6b combined nuclear charges beyond Z=164 are considered. In this region the ionization probabilities decrease due to the extremely strong binding energy which amounts to nearly twice the electron rest mass.

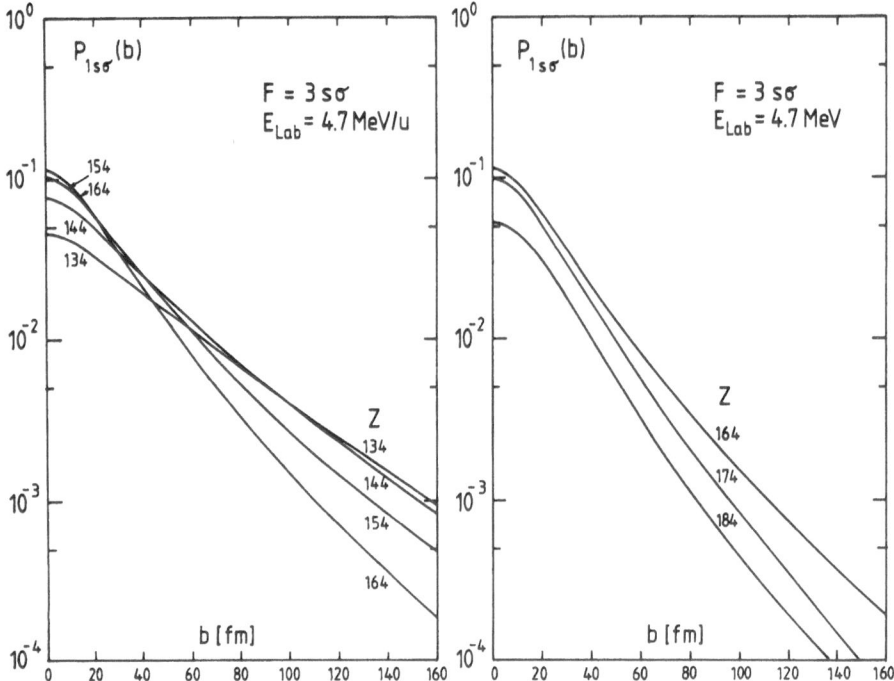

Fig. 6: Number of created 1sσ-vacancies versus impact parameter b. The different curves belong to different total charges Z. Asymmetric systems are assumed. Part (a) for Z≤164, part (b) for Z≥164.

Another observable quantity where the influence of 1sσ-binding energies and the radial expectation value $<r_{1s\sigma}>$ can be demonstrated is the differential emission probability of δ-electrons as a function of the kinetic electron energy. In Fig. 7 cross sections for δ-electrons, measured in coincidence with 1sσ-vacancies, are displayed for systems with united charges Z=134 up to Z=184. In the region of united charge Z<164 the δ-electron cross section rises with increasing Z. The high-kinetic part with E_{e^-}>500 keV grows faster than the low-kinetic one. This is due to the stronger localization of the 1sσ-orbit which allows for high momenta transfer as described above. But if the united charge exceeds Z~170 the influence of the rapidly increasing 1sσ-binding energy becomes dominant. Thus the emission cross section even diminishes which is shown for Z=184.

Therefore it would be highly desirable to measure more systematically K-hole formation and δ-electron cross sections in dependence on the united charge Z in order to verify the interplay of strong binding and sharp localization in ionization processes.

In the following we present a few comparisons between theoretical predictions and experimental results for inner-shell ionization and δ-electron production probabilities. A large variety of theoretical results are already published in Refs.

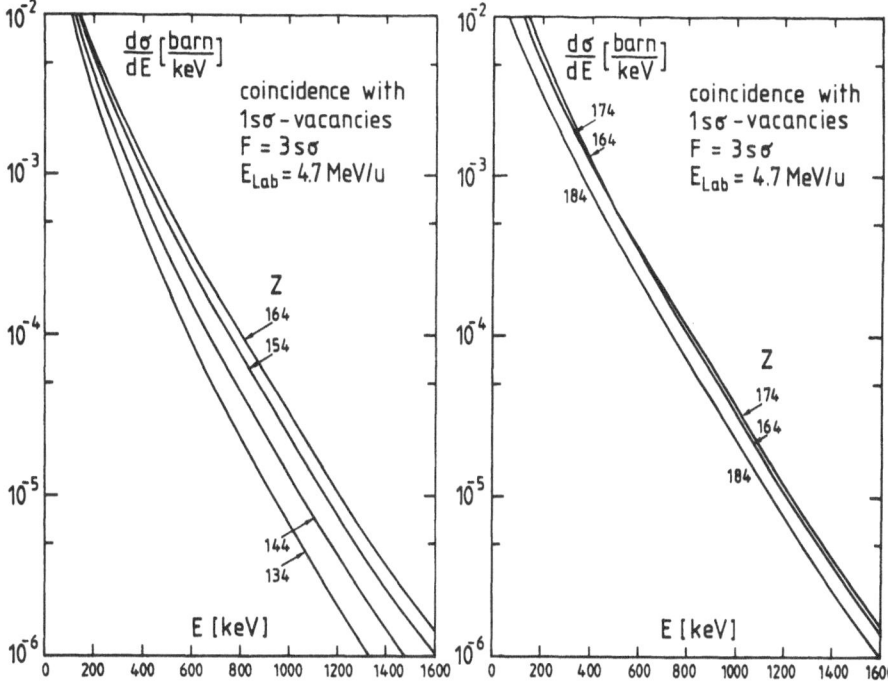

Fig. 7: δ-electron distribution in coincidence with 1sσ-vacancies versus the kinetic electron energy for different combined charges Z.

3, 4, 24 and 25. For corresponding experimental ones we refer to Refs. 2, 22, 26-30.

The number of created 1sσ-vacancies per collision is calculated according to eq. (2.6) for the systems Xe + Au and Xe + Pb. In our calculations we chose F=3 corresponding to the last bound state occupied initially. The results for Xe+Au are plotted in comparison with experimental data of D. Liesen et al.[27,28] in Fig. 8a. We find fair agreement up to impact parameters b≲90 fm. Beyond this value the measured data are underestimated. Similar agreement is achieved for the dependence on the bombarding energy E at fixed impact parameters b which is shown in Fig. 8b.

A system of comparable combined charge is Xe+Pb. Experimental data of Anholt et al.[29]   are in remarkable agreement with our coupled channel results for $P_{1s\sigma}(b)$ at E_{lab}=4.6 MeV/u (solid line in Fig. 9). At E_{lab}=7.2 MeV/u we only find fair agreement for small impact parameters. Furthermore K-vacancy production within first order time-dependent perturbation theory is presented (dashed line). In comparison to our coupled channel results these values are off by a factor 3 - 5, stressing that it is indispensable to take into account multistep excitations. Scaling laws which are based on simplifying approximations within the framework of first-order time-dependent perturbation theory are useful to determine the dependence of 1sσ-vacancy production on kinematical parameters $\{b, E_{ion}\}$. They fail, however, in predicting the measured data quantitatively.

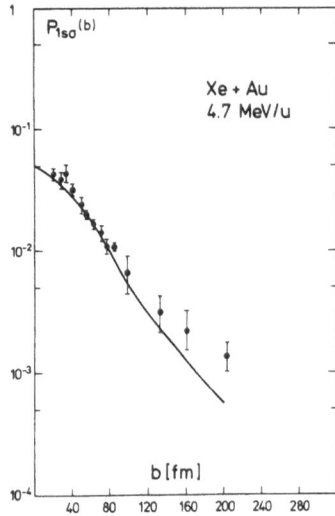

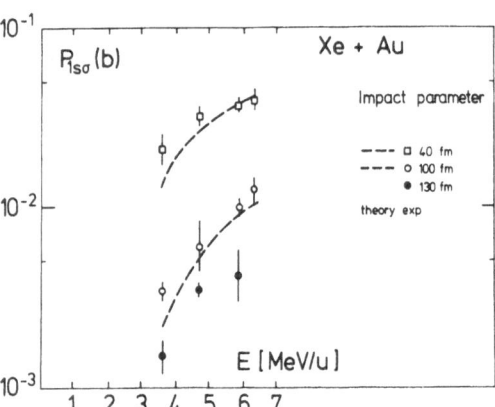

Fig. 8: (a) Number of created 1sσ- vacancies per Xe+Au collision at 4.7 MeV/u laboratory energy as a function of impact parameter b. Also shown are experimental data of D. Liesen et al.[27,28]. (b) Bombarding energy dependence of the number of created 1sσ- vacancies in the Xe+Au system[28] for some specific impact parameters.

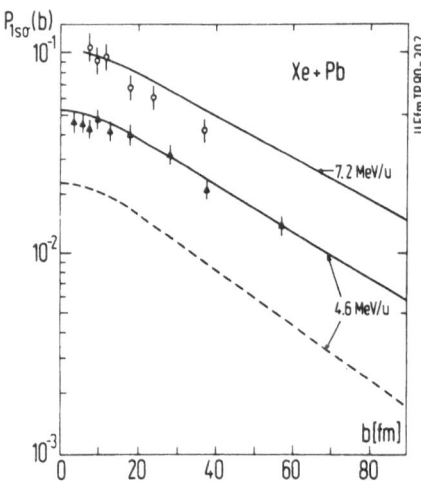

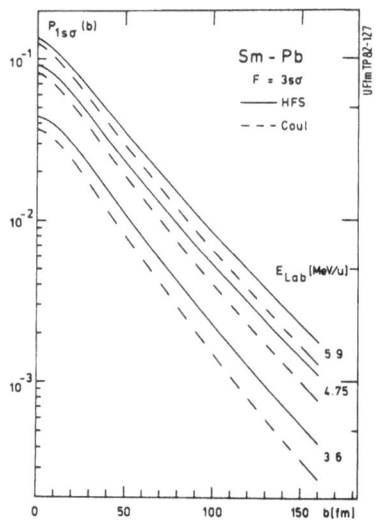

Fig. 9: Number of created 1sσ- vacancies per collision versus classical impact parameter b for the Xe+Pb system (solid lines). Various bombarding energies are considered. The figure also displays a result obtained within first-order time-dependent perturbation theory (dashed line). The experimental data there are taken from R. Anholt et al.[29].

Fig. 10: The system Sm+Pb is investigated concerning K-hole production at three different bombarding energies (E_{lab} = 3.6 MeV/u, 4.75 MeV/u, and 5.9 MeV/u). The solid lines represent HFS-results.

We have to stress again that contributions due to rotational coupling and the influence of electron screening[31] have been neglected. To estimate the effect of screening we have performed coupled channel calculations using basis states obtained in the Hartree-Fock-Slater model. Fig. 10 shows a comparison of K-hole production rates in Sm+Pb collisions at different bombarding energies using non-screened and HFS-screened electron states. Due to the reduction of binding energies[32] we obtain an increase in vacancy production. The effect is best seen at smaller bombarding energies: Since multistep processes become more important at lower lab-energies, the changes in the matrix elements thus lead to stronger effects. The increase of $P_{1s\sigma}$ is between 20% (for b ~ 0) and more than 50% (for b ~ 150 fm) comparing Coulomb- and HFS- values in 3.6 MeV/u collisions.

Finally we turn to a short comparison between theoretical predictions and experimental results for δ-electron emission in heavy-ion collisions. In Fig. 11 the double differential electron cross section $d^2\sigma/d\Omega dE$ is displayed where the triangles denote the total spectrum of emitted δ-electrons. The experimental data representing coincidence measurements with K-vacancies[2,26] are in good agreement with our theoretical calculations.

6. POSITRON CREATION IN SUPERCRITICAL QUASIMOLECULES

For a nuclear charge $Z \simeq 173$ the 1s-state enters the negative energy continuum, $E_{1s} < -mc^2$; the binding energy has reached the threshold where spontaneous pair production becomes possible. At this point the spectrum of eigenstates of the Dirac equation

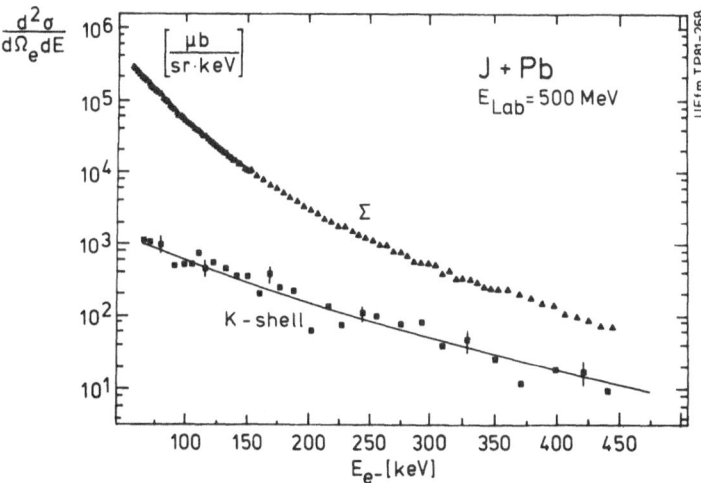

Fig. 11: Total δ-electron distribution (triangles) and δ-electrons in coincidence with K-vacancies for the system J+Pb at E_{lab}=500 MeV versus kinetic electron energies are results of Refs. 2,26. The latter data are compared with absolute values of coupled channel calculations.

is subject to a characteristic change: The 1s-state becomes a resonance in the lower energy continuum. In a Gedankenexperiment it is possible to increase the charge of a bare nucleus from $Z < Z_{cr} \simeq 173$ to $Z > Z_{cr}$, the unoccupied 1s-state will then be filled under emission of two (due to spin-degeneracy) positrons. The new stable ground state of the system consists of the nucleus plus two electrons in the K-shell; it is called the charged vacuum[8,33,34]. The experimental exploration of this new phenomenon would constitute an important test of the theory of Quantum Electrodynamics (QED) in the region of strong fields.

At present the unique way of extending the periodic table of elements, where these atomic effects can be pursued, lies in heavy ion collisions. In such scattering experiments, however, the dynamics of the collision becomes extremely important. The time-scale must be sufficiently long to allow the electron (positron) to adjust to the variation of the combined Coulomb field of the two nuclei. Since the typical velocities required to bring the nuclei close together are about $v/c \simeq .1$, the adiabatic picture is meaningful only for electrons in relativistic motion.

In this chapter we will concentrate on positron creation. In particular, it is our aim to give an adequate description for positron production in supercritical collision systems, where $Z_T + Z_P$ exceeds 173. In such collisions the resonance property of the 1sσ-state must be handled with care. In Ref.9 a formalism is developed, which avoids those difficulties and moreover has heuristic value for the interpretation of the positron creation process. The method is based on the observation that the continuum wavefunction of the supercritical system at resonance energy $E_p = E_{res}$ is quite similar to the discrete 1sσ-state in the subcritical case except for an oscillating tail – small in amplitude – reaching out to infinity. This structure reflects the occurrence of a tunneling process through the gap separating the particle- and antiparticle solutions of the Dirac equation. Apart from the asymptotic behaviour the 1sσ-wavefunction retains much of its identity, e.g. the strongly localized charge distribution having the extension of the atomic K-shell. Many properties, e.g. the radial matrixelements for ionization, may be continued smoothly to the supercritical region just by neglecting the tail of the wavefunction.

This idea can be used to develop a general method to treat resonance scattering. In this context Wang and Shakin[35] introduced a projection formalism for resonances in the nuclear continuum shell model: After having defined a normalizable quasi-bound wavefunction ϕ_R, a new continuum $\tilde{\varphi}_{Ep}$ is constructed which spans a subspace orthogonal to ϕ_R and replaces the old continuum ϕ_{Ep}. The modified continuum states satisfy the original Dirac equation supplemented by an inhomogeneous term containing an integral over the solution $\tilde{\varphi}_{Ep}$

$$(\Pi - E_p) \, |\tilde{\varphi}_{Ep}> - <\phi_R|H|\tilde{\varphi}_{Ep}>|\phi_R> . \tag{6.1}$$

If the states ϕ_R and $\tilde{\varphi}_{Ep}$ are used as part of the basis in eq. (2.2) the 1sσ-state couples to the positron continuum by two separate coupling operators

$$\dot{R} <\tilde{\varphi}_{Ep}|\partial/\partial R|\phi_R> + i/\hbar <\tilde{\varphi}_{Ep}|H|\phi_R> . \tag{6.2}$$

The second matrixelement arises since ϕ_R and $\tilde{\varphi}_{Ep}$ are not exact eigenstates of the two-centre Hamiltonian H. It does not depend on the nuclear motion and leads, in the static limit $R(t) = \text{const} < R_{cr}$[32], to an exponential decay of a hole prepared in ϕ_R with the width

$$\Gamma = 2\pi \; |<\tilde{\varphi}_{Eres}|H|\phi_R>|^2 \; . \tag{6.3}$$

The developed formalism thus has led to the emergence of 'induced' and 'spontaneous' positron coupling, the latter resulting from the presence of an unstable state ϕ_R in the expansion basis. In practice, however, this does not result in a threshold behaviour at the border of the supercritical region. Firstly both coupling matrixelements enter via their Fourier transforms depending on the time development of the heavy ion collision. Their contributions have to be added coherently so that in a given collision there is no physical way to distinguish between them. Secondly in Coulomb collisions the rapid variation of the quasimolecular potential causes significant contributions from the dynamical coupling, whereas the period of time for which the internuclear distance $R(t)$ is less than R_{cr} is usually very short ($\sim 10^{-21}$ sec) compared to the decay time of the $1s\sigma$-resonance ($\sim 10^{-19}$ sec).

Therefore, the predicted production rates and energy spectra of positrons continue smoothly from the subcritical to the supercritical region. Qualitative deviations of the positron production rate in supercritical collision systems are expected only under favourable conditions: Since the 'spontaneous' and 'dynamical' couplings exhibit a different functional dependence on the nuclear motion, an increase in collision time can be expected to provide a clear signature for supercritical collisions. Therefore Rafelski et al.[36] suggested the study of positron emission in heavy ion reactions at bombarding energies above the Coulomb barrier, where the formation of a di-nuclear system or of a compound nucleus would eventually lead to a time delay within the bounds of the critical distance R_{cr}. During this delay time T the spontaneous decay of the $1s\sigma$-resonance, by filling dynamically created K-shell holes under emission of positrons, might be strongly enhanced.

Before we will reflect on experimental results for positron creation, achieved at the Gesellschaft für Schwerionenforschung (GSI) in Darmstadt, and their comparison with theory, we first want to discuss the main theoretical predictions. While calculations performed in the framework of perturbation theory describe much of the physics involved in the excitation processes, the growing importance of multi-step processes again calls for the solution of the coupled channel problem[9]. We have integrated the system of differential equations (2.4) using the monopole approximation including up to 8 bound states and ~15 states in the upper continuum, separately for the angular momentum channels $\kappa = +1$ and $\kappa = -1$.

Positron emission rates increase very fast with total nuclear charge, flattening somewhat for the highest Z-values. If parametrized by a power law $(Z_T + Z_p)^n$ the power takes values of 20 down to 13, if a Fermi-level at $F = 3s\sigma$, $4p_{1/2}\sigma$ is assumed, or $n \simeq 29$ for bare nuclei (F=0). Mainly responsible for this effect is the contribution of the 1s-state which in normal collisions (F>O) is suppressed due to the small K-vacancy probability. If the K-shell is empty it becomes the dominant final state for pair production. This clearly reflects the strong coupling between the 1s-state and the antiparticle continuum which it approaches and even enters in the supercritical region. In all cases investigated the channels κ= -1 and κ= +1 contribute about equally to the total result.

Now we turn to the discussion of experimental results and their comparison with the described predictions. But first we want to point out a major problem in analysing the experimental data. Already for bombarding energies well below the Coulomb barrier ($E \sim .8E_c$) the nuclei can be excited by Coulomb excitation, the emitted photons with energy above 1022 keV can undergo pair conversion. Thus one has to measure simultaneously the γ-spectrum and fold it with the conversion coefficients. Here one has to know - or to assume - the γ-ray multipolarity. Monopole conversion cannot be handled by this method. Up to now, all conclusions on positron production in heavy-ion collisions had to rely on the described procedure for background subtraction.

The first generation of experiments established the dependence of positron excitation rates on the kinematic conditions as well as on the combined charge Z_u. Fig. 12 shows results of Kozhuharov et al.[37] for three collision systems, Pb+Pb, U+Pb, and U+U, at 5.8 MeV/u bombarding energy, measured with an orange-type β-spectrometer. The positron production probability in an energy window $E(e^+)$ = 490$\pm$50 keV is shown as a function of projectile cm-scattering angle Θ_{cm}. The nuclear background is subtracted. Because target and projectile cannot be distinguished in the experiment, the theoretical values have been symmetrized with respect to forward and recoil scattering. The Z-dependent increase is well described, which spans an order of a magnitude while $\Delta Z/Z$ is only 12%. Also the shape of the theoretical curves is in quite good agreement with the experimental data. On the other hand the absolute magnitude of the theoretical values is generally too high.

Using another type of experimental set-up, a solenoidal spectrometer, Backe et al.[38] obtained integrated and differential positron probabilities for various impact parameters for the heavy-ion collision systems Pb+Pb, Pb+U, U+U, and U+Cm. As in the experiment described above general agreement with theory is found. In the Pb+Pb case and, for smaller distances of closest approach, even in Pb+U and U+U collisions the data seem to agree also in absolute values, in contrast to the experiment discussed above. But in the heaviest accessible system U+Cm (Z_u=188) and for larger distances R_{min} theory again has a tendency to overestimate the measured data. The experimental slopes are somewhat steeper than predicted, while the Z-dependence is overestimated. For given kinematic parameters {E,b} the Z-dependence of the total positron probabilities is given by $Z^{\sim 17}$, for lighter systems as well as for heavier ones[39].

From these data no qualitative signature for the 'diving' of the 1sσ-state in U+U, U+Cm collisions can be extracted, in agreement with theoretical predictions.

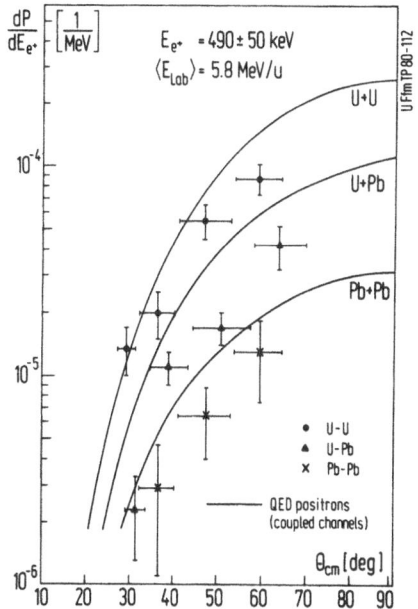 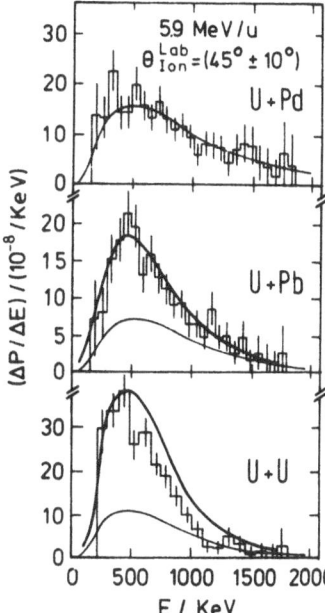

Fig. 12: For 5.8 MeV/u Pb+Pb, Pb+U, and U+U collisions positron production
probability in an energy window E + = 490±50 keV is displayed as
a function of projectile scattering angle Θ_cm. The experimental
data are taken from Kozhuharov et al.[37]. Here, the nuclear background is subtracted.

Fig. 13: Spectra of emitted positrons in 5.9 MeV/u collisions measured by
Backe et al.[39,40] in coincidence with ions scattered in the angular
window Θ_lab=45°±10°. The spectrum in the lightest system, U+Pd, is
explained by nuclear pair conversion alone (light line). In the
U+Pb and U+U systems the sum of nuclear and calculated atomic positron production rates (full lines) is displayed.

More sensitive information can be obtained by the measurement of energy spectra of

positrons detected in coincidence with the scattered ions. Their knowledge is most

useful if one wants to find deviations hinting to the positron creation mechanism.

Fig. 13 shows the first published positron spectra of Backe et al.[39,40] for

three collision systems, U+Pd, U+Pb, and U+U, at 5.9 MeV/u bombarding energy; the

ions are detected in an angular window $\theta_{lab}=45^{0+}_{-10}{}^{0}$. For U+Pd (Z=138) no atomic

positrons are expected, the data can be fully accounted for by nuclear conversion

(light curve). Extrapolating this procedure to the U+Pb system (light curve) the

sum of background and calculated QED positron rates (full curve) is in excellent

agreement with the observed emission rates. In the spectrum of the supercritical

U+U system ($E_{1s} \simeq 1200$ keV for $R_{min} \simeq 21$ fm) some deviations are seen.

A possible source which could cause deviations in the shape of the positron

spectra from the results presented so far will be discussed in the following. To

obtain theoretical predictions for positron production it is essential to include

the 'spontaneous' coupling for collisions where the 1sσ-state joins the lower continuum. If it is left out of the calculation the resulting positron spectra would be
strongly altered: The 'induced' radial coupling is changed at the same time as the
spontaneous coupling becomes important. Both contributions add up coherently and
cannot be observed separately[9]. A promising strategy to get a clear qualitative signature for the diving process will be to modify the time structure and to select
heavy-ion collisions with prolonged nuclear contact time[36]. Such nuclear reactions
are expected to occur at energies close to or above the Coulomb barrier. The nuclear
delay time T should provide a handle to distinguish supercritical systems.

Using a schematic model[41] for the trajectory we have performed coupled channel
calculations for the four heavy-ion collision systems Pb+Pb, Pb+U, U+U, and U+Cm,
corresponding to Z_{united} = 164, 174, 184, and 188, respectively. Independently of
assumptions on the incoming and outgoing path, of dissipation of nuclear kinetic
energy or angular momentum, and of the position of the Fermi level, all positron
spectra exhibit the following features:

In subcritical collision systems ($Z_T+Z_P\lesssim173$) a delay time T causes modulations
in the positron spectrum with a width $\Delta E=2\pi\hbar/T$. In Fig. 14a positron spectra are

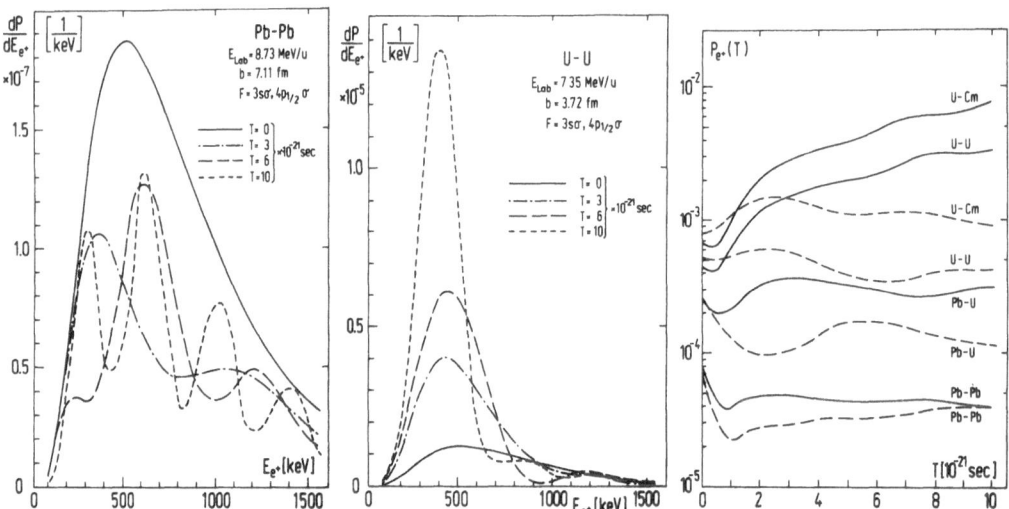

Fig. 14: Spectra of positrons created in subcritical (part a) and supercritical
(part b) heavy-ion collisions assuming grazing Coulomb trajectory (full
lines) and nuclear reactions leading to delay times T=3·, 6·, and 10·10^-21
sec, resp., using a schematic model for the trajectory[41]. Whereas for the
lighter collision systems modulations in the positron spectra are present,
a distinct peak at the 'resonance' energy $E_{1s\sigma}(R_{min})$ builds up for systems
with Z_u>173.- Part c: Probability for the emission of positrons, integrated over kinetic positron energy, as a function of reaction time T.
Full lines: Contribution of s-partial waves, dashed lines: $p_{1/2}$-partial
waves' contribution. For the supercritical systems U+U and U+Cm $P_{e^+,s}$
increases strongly with T.

displayed for a Pb+Pb collision (E_{lab}=8.73 MeV/u, b=7.11 fm, F=3sσ, $4p_{1/2}$σ) with delay times T=0 (pure Rutherford scattering), 3·, 6·, and $10 \cdot 10^{-21}$ sec. The modulations are due to interference effects in much the same way as predicted for the δ-electron spectra in deep inelastic heavy-ion collisions[42].

In addition to the oscillatory interference patterns an enhancement of positron production in time-delayed supercritical collisions is observed, where the binding energy of the lowest bound states exceeds the value $2mc^2$. For long delay times a distinct peak in the positron spectrum is found at the location of the supercritical bound state resonance (binding energy minus $2mc^2$) due to the spontaneous pair-creation mechanism. A detailed analysis of the spectra reveals that this peak emerges gradually as $Z_u = Z_T + Z_P$ exceeds Z_{cr}. Positron spectra for the supercritical system U+U (E_{lab}=7.35 MeV/u, b=3.72 fm, F=3) are shown in Fig. 14b. With increasing delay time the position of the maximum drifts slowly from the kinematic maximum to the 'resonance energy', which depends on the combined charge, the separation of the two nuclei and on the nuclear charge distribution[41].

The integrated positron emission probability P_{e^+} as a function of T for collision systems Pb+Pb, Pb+U, U+U, and U+Cm is displayed in Fig. 14c where the solid lines denote s-state-, the dashed lines $p_{1/2}$-state contributions. The absolute values, $P_{e^+}(T=0)$, should not be compared since the impact energies are chosen differently. However, the variation of P_{e^+} with delay time T is of particular interest. In the investigated subcritical systems, $P_{e^+}(T)$ remains roughly unchanged or is even reduced by up to a factor two due to destructive interference between incoming and outgoing path. In the supercritical systems the probability rises approximately in linear relation with T, which reflects the increasing contribution of the spontaneous decay. In the schematic model its slope depends on the separation of the two nuclei, R_{min}, in general it depends also on the time-dependent charge distribution during nuclear contact. At small values of T interference effects seem to prevail even in supercritical collisions.

However, for any chosen set of experimental parameters, the nuclear reaction time T may (and will) not be sharp but distributed over a certain range with a time distribution function f(T). As an assumption we took a Gaussian centered at $\bar{T}$

$$f(T) = \frac{1}{\sqrt{2\pi\tau}} \exp\left(-\frac{(T-\bar{T})^2}{2\tau^2}\right) . \tag{6.4}$$

The resulting positron spectra for parameters $\bar{T} = 16 \cdot 10^{-21}$ sec and τ = 0 or τ = $2 \cdot 10^{-21}$ sec are displayed in Fig. 15 for a head-on U+U collision at E_{lab} = 5.9 MeV/u. If we consider the 'subcritical' p-states only (part a) we observe that the oscillations disappear already for τ = $2 \cdot 10^{-21}$ sec. Thus we conclude that the appearance of several oscillations in subcritical lepton spectra can be expected only for sufficiently sharp nuclear reaction times. Also in the total spectrum (part b) the oscillations are damped out for increasing τ. But most striking is the occurrence of the dominant nonvanishing first peak, which originates from the spontaneous part of the positron production mechanisms.

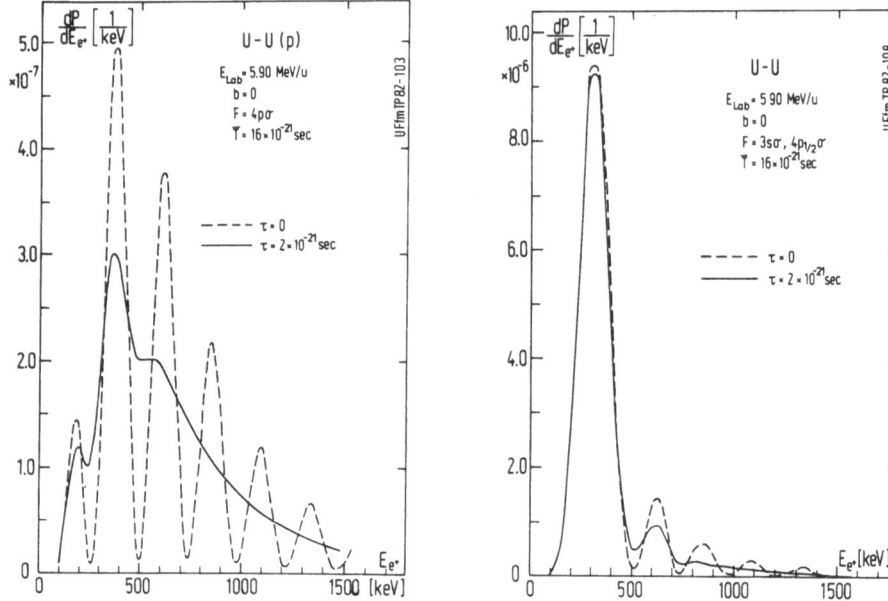

Fig. 15: Differential positron production probability versus kinetic positron energy E_e+ in a central U+U collision at E_{lab} = 5.9 MeV/u, assuming a Gaussian nuclear reaction time distribution centered at $\bar{T}$ with a width τ. (a) p-states only representing a subcritical system, (b) including also the contribution of the s-states. Most striking is the appearance of the first pronounced peak which originates from spontaneous positron production.

The results described so far were obtained within the schematic model for the nuclear motion. It facilitates a systematic study of the time delay effect and allows for an investigation of the conceptually interesting limit of large sticking times. To analyse a given experiment, however, the employed nuclear trajectories should be consistent with the elastic and inelastic heavy-ion scattering data. Many reaction models with different degrees of refinement have been discussed in the literature. We have calculated trajectories with the macroscopic model of Schmidt et al.[43], which accounts for neck formation. Strong deviations from Coulomb trajectories are found, energy loss up to ~30% (for b~0) can be obtained. The change of the positron spectrum for collisions with varying degree of nuclear contact is demonstrated in Fig. 16. Part a shows the modified U+U-trajectories R(t) for several orbital angular momenta from ℓ=0 head-on to ℓ=500$\hbar$ near grazing collisions. The corresponding positron spectra, given in Fig.16b, show a gradual enhancement at E_e+ = 500 keV. As expected a longer delay time ΔT, compared to pure Rutherford scattering, causes an increase of positron production in the s-channel. On the other hand the change in kinematics causes a drift to lower kinetic energies in the $p_{1/2}$-partial wave spectra due to destructive interference. Both effects taken together lead to an enhancement of the maximum and a drift towards lower energies also in the total spectrum.

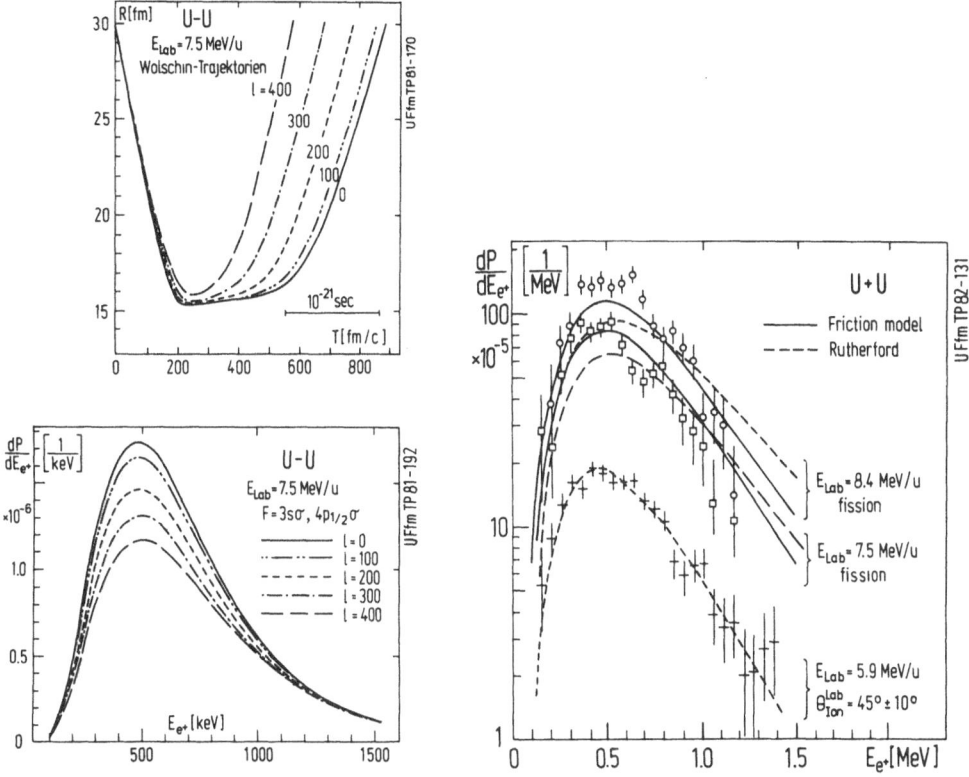

Fig. 16: (a) Nuclear trajectories calculated in the friction model[43] for 7.5 MeV/u
U+U collisions at various values of the orbital angular momentum ℓ between
0 and 400ħ. (b) Energy spectra of positrons emitted in the collisions
shown in (a). The results for the angular momentum channels s and $p_{1/2}$
have been added. (c) Comparison of theoretical predictions for U+U at
E_{lab}=5.9 MeV/u, 7.5 MeV/u, and 8.4 MeV/u, with experimental data of Backe
et al.[44], as described in the text.

Measurements by Backe et al.[44] seem to indicate such tendencies: In U+U and
U+Cm collisions at energies above the Coulomb barrier positron spectra have been
measured in coincidence with fission fragments in order to get a signature for close
nuclear contact. The analysis shows an enhancement of dP/dE$_{e^+}$ at lower kinetic energies in qualitative agreement with Fig. 16b. For a quantitative comparison one
has to integrate the theoretical impact parameter-dependent positron spectra over all
values of b which lead to a nuclear reaction, weighted by the corresponding probability w(b) to induce nuclear fission. Performing the integration with a primitive
weight factor w(b) = 1 for b<b$_{grazing}$ and w(b)=0 elsewhere, there remains, however,
an energy shift of ~50 keV in the experimental data in comparison with the theoretical curves, which might be due to electron screening effects[31]. Furthermore, as
mentioned above, the theoretical values have to be reduced by an overall factor ~ 2/3.

Fig. 16c shows the experimental data for U+U collisions at E_{lab} = 5.9 MeV/u, 7.5 MeV/u, and 8.4 MeV/u, in comparison with theoretical results excluding electron screening, but reduced by the factor mentioned above. Dashed lines indicate pure Rutherford scattering trajectories, whereas the solid lines display spectra calculated with the modified trajectories of Fig. 16a. One can conclude that for an even better agreement longer delay times ΔT ($\sim 2 \cdot 10^{-21}$ sec) are needed. Further investigations along these lines seem to be very promising, both to establish the mechanism of positron production and to deduce the nuclear reaction time scale.

As another interesting theoretical problem one might speculate about the presence of nuclear collisions with very long reaction times. How would positron spectra look like if, at a given scattering angle, a superposition of Rutherford scattering and long-lasting nuclear reactions is assumed? If, for the sake of simplicity, nuclear scattering with definite delay time T is assumed, a relative fraction q can be used as a measure of the relative cross section of reactions, leading to long contact times, compared to pure Rutherford cross section. As positron production is very strongly enhanced for reaction times larger than 10^{-20} sec, a peak superimposed on the smooth spectrum of positrons, emitted in the much more frequent 'distant' Coulomb collisions, could be observable. A rough estimate shows that for long delay times a peak may be prominent even if the differential reaction cross section is less than 1% of the Rutherford cross section.

To obtain the full shape of the positron spectrum, an additional assumption about the angular distribution $d\sigma^N/d\theta$ of the nuclear reaction component is needed. We shall discuss two extreme simplified cases: (i) isotropic break-up of the compound system, (ii) strong focussing of the reaction fragments into a narrow angular window in the c.m. system. If the reaction products were emitted isotropically, a line should be observable in the positron spectrum at all ion scattering angles. However, it would be most pronounced at $\theta_{c.m.} = 90^{\circ}$, being suppressed at other angles relative to the elastic scattering cross section. As an example Fig. 17 shows spectra

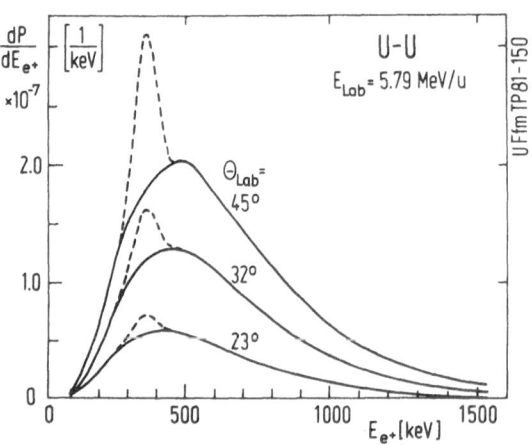

Fig. 17: Spectra of positrons emitted in 5.79 MeV/u U+U collisions in coincidence with a scattered nucleus for three selected lab ion angles. The fully drawn curves are calculated assuming Rutherford scattering only. The dashed lines show the effect of an additional nuclear reaction with a lifetime $T = 4 \cdot 10^{-20}$ sec. A relative fraction of $q = 2.4 \cdot 10^{-3}$ reactions per elastically scattered ion (at 45°) has been assumed.

of positrons, reduced by a factor ~ 2/3, for 5.79 MeV/u U+U collisions for several
fixed scattering angles. A long lived nuclear reaction component (dashed lines) with
$T = 4 \cdot 10^{-20}$ sec and admixture $2.4 \cdot 10^{-3}$ (at $\theta_{c.m.} = 90^0$) has been added under the assumption of an isotropic distribution in the reaction plane. The integration for the nuc-
lear reaction contribution was performed using a simple ad hoc classical trajectory.
The internuclear distance $R(t)$ was prescribed to shrink to a minimum value (smaller
than 2a!) required to produce a 1s-resonance at 370 keV kinetic energy. This distance
was kept fixed long enough to obtain a contact time of $4 \cdot 10^{-20}$ sec.

If the emitted reaction fragments are focussed under a certain scattering angle,
a detailed quantitative analysis depends strongly on the strength of the focussing
effect. Such an effect could be produced, if the life-time T of the nuclear molecular or compound system depended strongly on the angular momentum ℓ carried into the
reaction. This is not an unreasonable assumption since it is evident that T must
vanish for very large values of ℓ.

Thus spontaneous positron production could also prove to be important for the
study of nuclear reactions. Although several other methods[45,46,42] have been proposed
to investigate the reaction time in deep inelastic nuclear collisions or the life
time of compound nuclei by its influence on atomic processes, positron spectroscopy
may be the most sensitive method (in very heavy systems) because the delay could result in an increase of the yield at resonance by one or two orders of magnitude.

Two experimental groups[47,48] recently have performed experiments with U+Th, U+U,
and U+Cm at energies close to the Coulomb barrier. Contrary to the results of Backe
et al.[44] their positron spectra seem to show remarkable structures. C. Kozhuharov,
P. Kienle et al.[47] measured at 5.8 MeV/u beam energy positron spectra in coincidence
with ions scattered into various narrow angular windows, using an orange-type β-spec-
trometer. A preliminary analysis shows sharp maxima in the spectra that are most
pronounced under laboratory scattering angles around $\theta_{lab} \sim 45^0$. In the U+U measurement the position of the peak was found to be located at ~370 keV with a width of about
~90 keV (FWHM), for U+Th the effect is less well established. After subtraction of a
smooth background the number of positrons per detected ion emitted in the peak is
roughly 10^{-5} at $\theta \sim 45^0$. If the observed structure is of quasimolecular origin, it must
be produced in very long-lasting nuclear reactions. If one compares the experimental
width with spectra from coupled channel calculations based on the schematic sticking
model a minimum value $T \sim 4 \cdot 10^{-20}$ sec is required. Should the observed line width have
instrumental origins or be due to additional broadening effects the reaction time T
would have to be even longer. Taking the coupled channel results the probability for
positron production in a delayed ($T = 4 \cdot 10^{-20}$ sec) collision should be $\sim 4.7 \cdot 10^{-3}$, which
have to be compared with the observed probability $\sim 10^{-5}$. Thus a fraction of q=
$P_{e^+}(exp)/P_{e^+}(theor) \simeq 2 \cdot 10^{-3}$ delayed collisions per elastically scattered ion is sufficient to produce the observed effect (at 45^0). This number should serve only for
a general orientation, since it depends on the details of the model assumed.

Similar features were detected in the experiment of H. Bokemeyer, J.S. Greenberg et al.[48] in U+U and U+Cm collisions. When displaying their data for U+U for different narrow scattering angle windows Θ_{lab} the differential positron spectra exhibit prominent peak structures at different kinetic energies E_{e^+}.

One might suppose that those peaks are caused by nuclear background processes. However, nuclear transitions of, e.g., multipolarity E1 or E2 should also be observable in the emitted photon spectra, provided that proper Doppler shift corrections are performed. Whether the peaks can be caused by E0-processes will be discussed later on. On the other hand, if a sharply focussed nuclear reaction takes place it is not surprising that an experiment not triggering for the optimal kinematic conditions might smear out any evidence for structure. Further investigations are needed to settle this question. Should the observed phenomena indeed be caused by reactions with a very long time scale, this would have far reaching consequences for the physics of nuclear systems in the superheavy region. More about those aspects in the last section of this chapter.

The superheavy compound system may not be a static object. Thus we investigated the influence of the internal dynamics of the dinuclear system on the energy spectra of the emitted positrons. The following model calculation can be understood as a classical picture for the conversion process of internal pair creation in the supercritical quasimolecule. The 'internuclear' separation R(t) is supposed to oscillate around the distance of closest approach R_0 of the Coulomb trajectory for a fixed duration T

$$R = R_0 \ (1 - \alpha_0 \ \sin(2\pi\nu t)). \tag{6.5}$$

In Fig. 18 positron spectra are plotted for a central U+U collision at $E_{lab} =$ 6.2 MeV/u. In order to demonstrate the qualitative effect we have restricted the calculations to couplings between s-states only. The parameters in eq. (6.5) are fixed by $\alpha_0 = .25$ and $\nu = .125 \cdot 10^{21} \ sec^{-1}$. Various sticking periods between T=0 and T=108$\cdot 10^{-21}$s are considered in Fig. 18a . Fig. 18b shows on a linear scale the computed positron spectrum for T=36$\cdot 10^{-21}$s. As in Fig. 14b we find the dominant 'spontaneous peak'. But in addition a second pronounced peak appears at about $E_{e^+} =$ 800 keV. This reflects the fact that part of the vibrational energy of the dinuclear system is transferred to the emitted lepton.

We now turn to the influence of a nuclear time delay in deep-inelastic heavy ion reactions on δ-electron spectra. Within the schematic trajectory model[41] we calculated δ-electron spectra in the head-on heavy-ion collision U+U, where the nuclei stick together up to 10^{-20} sec. Fig. 19 shows the spectra of emitted δ-electrons choosing a delay time T=0 (part a) and T=10^{-20}s (part b). A coincidence with created 1sσ-vacancies is assumed. The dashed-dotted and the short-dashed lines represent the accidental coincidence term and the exchange term of eq. (2.7), respectively. Their sum is denoted by full lines. The oscillatory pattern is part b is caused by the real correlation between δ-electron and the 1sσ-vacancy. The incoherent term

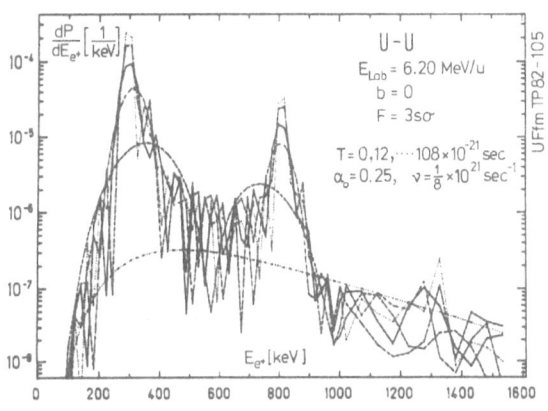
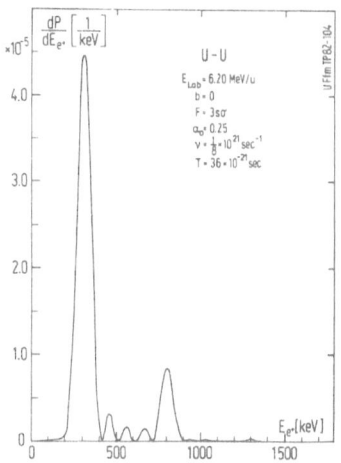

Fig. 18: Positron spectra in a U+U head-on collision calculated under the assumption
that the distance between the nuclear centres oscillates around the distance
of closest approach R_o of the Coulomb trajectory (c.f. eq. (6.5)). Nuclear
reaction times T = 0, 12, 36, 60, 84, and $108 \cdot 10^{-21}$ sec are considered. The
longest duration T corresponds to the most pronounced 'spontaneous peak',
etc. Note the logarithmic scale in part a).

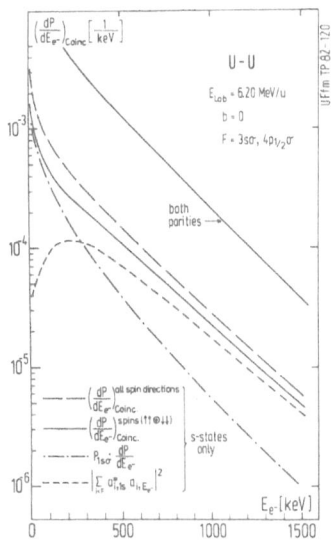
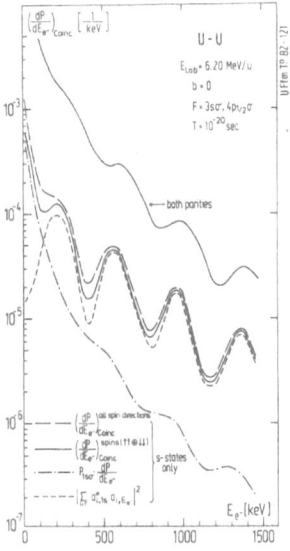

Fig. 19: Differential emission probability of δ-electrons in dependence on the kinetic electron energy for the system U+U at E_{lab} = 6.2 MeV/u and b=0. A
coincidence with 1sσ-vacancies is assumed, according to eqs. (2.7)-(2.9).
(a) No nuclear sticking time, (b) T = 10^{-20} sec.

$N_p \cdot N_q$ produces only a smooth distribution, reflecting the fact that the measurement of δ-electron spectra alone in very high-Z systems will not be sufficient to measure reaction times.

However, because there is usually no experimental discrimination of spin direction (c.f. eq. (2.8)), one has to account for accidental coincidences (long-dashed lines). Up to now only excitations of s-state electrons are considered. But since in the symmetric system U+U also p-states contribute the measurable spectra have to be calculated according to eq. (2.9), represented by the upper full lines in Fig. 19 ('both parities'). Thus one would expect, if at all, observable oscillatory structures in δ-electron spectra only for asymmetric collision systems.

As discussed above the structures in the positron experiments of Kozhuharov et al.[47] and Bokemeyer et al.[48] might be interpreted in terms of the spontaneous decay of the vacuum due to the formation of a superheavy quasimolecule or quasiatom with Z = 184. The lifetime of such a compound system must then be in the order of magnitude of 10^{-20}sec or even longer. Are such reaction times consistent with nuclear physics? We first want to discuss very briefly the formation of such a superheavy quasimolecule while afterwards we speculate about the possibility that a quasiatom with nuclear charge Z=184 might be stable.

We have to study the interference of Coulomb and nuclear interaction between both nuclei for the incoming channel: With decreasing two-centre distance R the increased repulsion due to the Coulomb potential is lowered by the nuclear force just at a distance of a few fm before the classical touching point of the two surfaces. If the attractive interaction is strong enough, a 'pocket' occurs in the (total) nuclear potential, which is responsible for a more or less stable configuration of a nuclear molecule with small overlap. We call this pocket 'adhesion minimum', because the molecule is formed only due to macroscopic forces. Shell effects do not play an important part at this stage and are neglected up to now. Of course, the nuclear potential depends strongly on the deformation and orientation of both nuclei.

To calculate the potential we use the proximity potential[49] modified in such a way that we are able to consider nuclei with quadrupole and hexadecupole deformations in various orientations. The result for two ^{238}U-nuclei is displayed in Fig. 20, where we use a hexadecupole deformation β_4 = .106[50]. A minimum occurs for such orientations where two flat parts of the hexadecupole deformed surfaces face each other. Obviously such minima are strongly dependent on the value of β_4 and will vanish with decreasing hexadecupole deformation (Fig. 21).

Preliminary results of a calculation due to a double folding model[54] using the M3Y - force of Ref. 55 show, in contrast to the results presented above, a much deeper minimum, but only for the nuclear orientation 'top on top' $(\theta_1=\theta_2=0)$. However, this difference is not surprising: The original proximity potential is tested for experimental fusion cross sections, whereas the M3Y - interaction fits especially elastic scattering data, i.e., the effective ranges of the two models are completely different

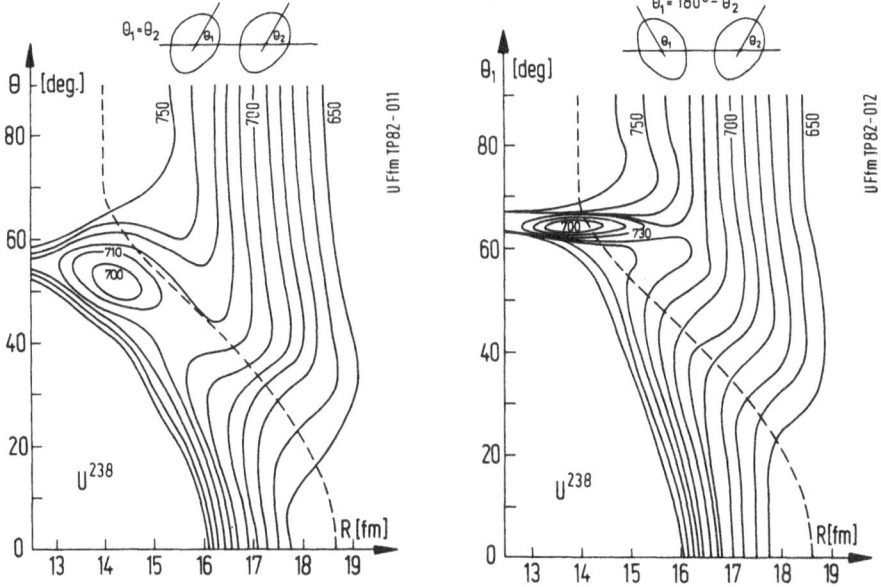

Fig. 20: Nuclear potential of two ^{238}U nuclei ($\beta_2=.264$, $\beta_4=.106$) as function of the two centre distance R and the orientation angles Θ_1 and Θ_2 in the proximity approximation. The dashed line indicates the touching point of the two surfaces.

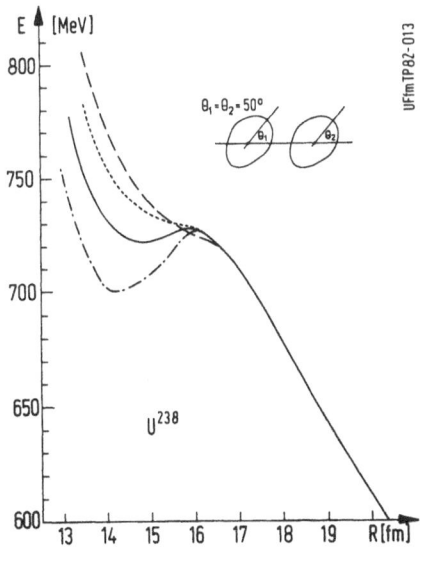

Fig. 21: The dependence of the nuclear potential on the hexadecupole deformation as function of R for the orientation angles $\Theta_1=\Theta_2=50°$. The deformation parameters are $\beta_2=.277$, $\beta_4=.013$ (dashed line) from myonic X-rays[51], $\beta_2=.226$, $\beta_4=.052$ (dotted line) from proton scattering[52], $\beta_2=.261$, $\beta_4=.087$ (solid line) from electron scattering[53], and $\beta_2=.261$, $\beta_4=.106$ (dashed-dotted line) from Coulomb excitation[50].

and equivalent to a Yukawa range of .626 fm and 1.1 fm, respectively. However, both calculations indicate that there is a reasonable chance for the formation of a super-heavy nuclear quasimolecule due to the mechanism described above called adhesion minimum.

If there is little energy and angular momentum in the initial superheavy molecular system and the potential pocket is deep enough, maybe a superheavy nucleus is produced nearly cold and with little angular momentum. We want to discuss now the possibility that such a giant nucleus with Z=184 will be stable. A microscopic-macroscopic method is used to determine its potential surface. For the macroscopic part we perform an extrapolation of the atomic mass formula of the droplet model[56,57] towards the high-Z region. Due to the deficiency of the giant nucleus in neutrons and to the large Coulomb energy pushing the protons to the surface region of the nucleus we have to assume a counterpoising asymmetry force. This modification does not cause any detectable effects in the familiar part of the periodic table, however, for actinides it results in a ~1% deviation for the binding energy.

We extrapolate the shell model as well using the single particle Hamiltonian

$$H = -\hbar^2/2m \; \Delta + 1/2m\omega_o^2 r^2 - 2\hbar\omega_o \; \kappa \; (\vec{\ell} \cdot \vec{s} - 1/2\mu \vec{\ell}^2) - m\omega_o^2 r^2 \; (\sqrt{5} \; \sum_\mu \alpha_{2\mu} \times Y_{2\mu}). \qquad (6.6)$$

To calculate the parameters κ and μ of the $\vec{\ell} \cdot \vec{s}$- and $\vec{\ell}^2$ term, respectively, two methods are taken into account. The first one describes κ, μ as function[58] of the mass number $A^{1/3}$

$$\kappa_p = .0606 + .003A^{1/3}, \quad \mu_p = .638 + .003A^{1/3},$$

while in the second method κ is a function of the main quantum number N_{osc}[59], i.e., there are different values of κ for different shells

$$\kappa_p = .18 \left(1/2(N_{osc}+1) \cdot (N_{osc} + 2) \right)^{-1/3},$$

and μ_p equals .62, i.e. constant for all shells.

Fig. 22a shows the single particle levels calculated by the first method, yielding Z=182 as a magic number, whereas the second one displays also Z=164 as magic number (Fig. 22b). In Ref. 60 Z=164 was seen as the only magic number in this region due to a cut-off in Hilbert space, i.e., N_{osc} was restricted to $N_{osc} < 8$, whereas in the present calculation the main quantum number N_{osc} is increased up to 11.

The Strutinsky method is used to calculate the shell corrections being important for spherical and small deformed shapes of nuclei. The summation of shell corrections and macroscopic energy yields the potential surface (Fig.23), showing the possibility to find a relatively stable nucleus with Z=184 due to a barrier height in the order of 15 MeV. The calculation of half-lives against fission, α-decay, and electron capture is still in progress[61].

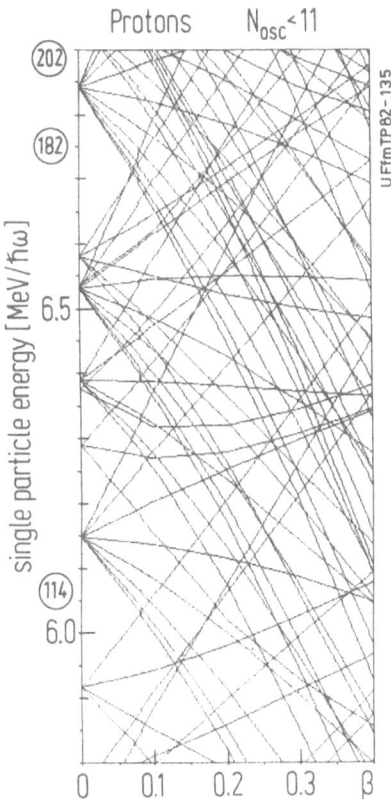

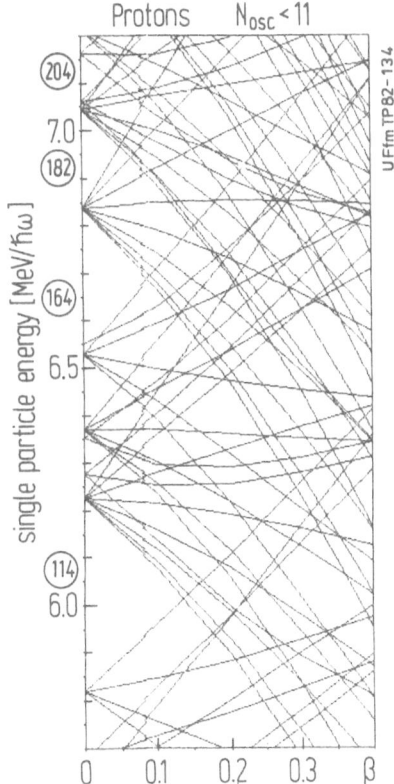

Fig. 22: Proton single particle levels in the superheavy region with parameters κ, μ as function[58] of $A^{1/3}$ (part a), and with κ depending on N_{osc} and μ being constant for all shells[59] (part b). The encircled numbers represent the predicted magic numbers in the superheavy region.

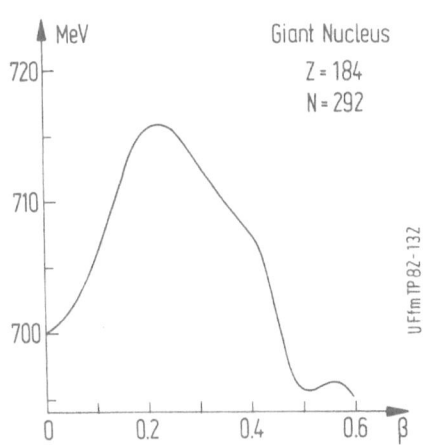

Fig. 23: Potential surface of a giant nucleus with Z=184 and N=292, i.e. the compound uranium + uranium - nucleus. The maximum shell correction, calculated due to the shell structure in Fig. 22a, results for β=0, leading to a barrier height of ~ 15 MeV.

7. CONVERSION PROCESSES IN SINGLE ATOMS AND IN SUPER-CRITICAL COMPOUND SYSTEMS

In collisions of very heavy ions such as U+U with E_{lab} > 3 MeV/u both nuclei are strongly Coulomb excited. For bombarding energies at about the Coulomb barrier trans- fer reactions or even deep inelastic nuclear reactions can take place which lead to additional excitations of the nuclei. This internal excitation energy may be irra- diated by a photon or may be transferred to a bound electron or to an electron of the negative energy continuum, which leads to ionization and electron-positron pair crea- tion, respectively. The latter process requires nuclear transition energies ω larger than twice the electron rest mass.   Nuclear EO-transitions are characterized by the absence of single photon emissions, because a photon must carry at least one unit of angular momentum. (We use $\hbar = c = m_e = 1$.)

The basic processes under investigation are depicted schematically in Fig. 24. The nucleus which undergoes EO-transition is labelled by its initial and final state angular momenta J_i, $J_f = J_i$ and eigenenergies E_i, $E_f = E_i - \omega$. Process a) describes the electron-positron pair creation. An electron of the negative energy continuum ($\varepsilon = -E < -mc^2$) with Dirac quantum number κ is lifted to the positive energy continuum. The final state energy obviously amounts to $E' = \varepsilon + \omega$ whereas the angular momentum quantum number remains unchanged. Since neither the initial electron state energy nor the final state energy is fixed one expects a continuous energy distribution for the emitted positrons. Process b) indicates the conversion of a K-shell electron ($n=1$, $l=0$, $j=\frac{1}{2}$, $\kappa=-1$) with energy eigenvalue $E_{1s_{1/2}}$. Thus bound states with definite energies are involved. Energy conservation then simply causes monoenergetic lepton emission for a fixed nuclear transition energy ω. Process c) symbolizes monoenerge- tic positron production. Here an electron of the negative energy continuum is ex- cited to a bound state, e.g. to the $1s_{1/2}$-state. This represents a rather rare pro- cess[62], which can be neglected completely.

Thus we focus our attention on the calculation of

i)   the pair conversion coefficient β, defined as the ratio of the pair production probability (process a)) compared with that of photon emission for a specific nuclear transition with energy ω. Since the energy of the electron and the positron takes continuous values we may express β also as integral of the differential pair conver- sion coefficient $d\beta/dE$. The lower bound of the integral is determined by the rest mass of the electron, which corresponds to vanishing kinetic energy, while the upper bound is given by the nuclear transition energy ω minus m_e.

ii)   the conversion coefficient α, defined as the ratio of the probabilities of inner-shell vacancy formation (process b)) and photon emission. In particular this mechanism is important for low energy nuclear transitions.

iii)   the ratio η of the two conversion probabilities for electron-positron pair creation and for the ionization of bound state electrons. As is shown in Ref. 62 this ratio is completely determined by the density of the electron wavefunctions at the nuclear origin, thus being independent of the nuclear wave function.

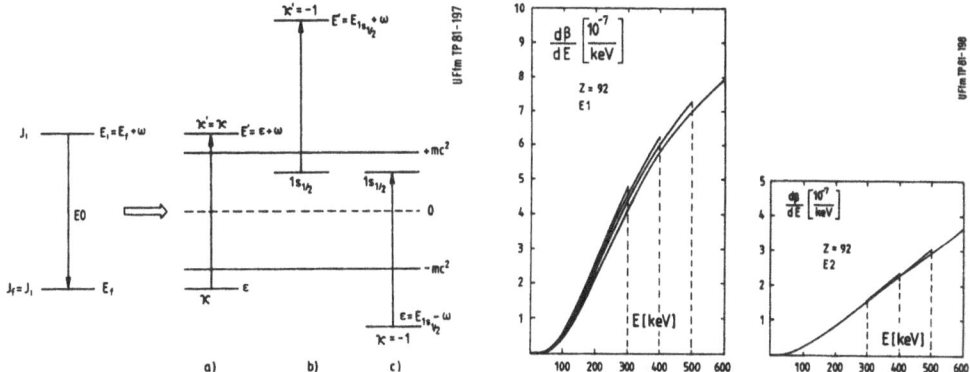

Fig. 24: Schematic representation of electron conversion processes accompanying nuclear E0-transition from a state $\{E_i, J_i\}$ to a state $\{E_f = E_i - \omega, J_f = J_i\}$. a) Electron-positron pair production leading to a continuous energy distribution of positrons and electrons. b) Conversion of K-shell electrons - a monoenergetic electron-production mechanism. c) Monoenergetic positron-production - a negligible process.

Fig. 25: Differential conversion coefficient $d\beta/dE$ with respect to the kinetic positron energy E for nuclear E1- and E2- transitions in $_{92}$U. Nuclear transition energies ω = 1323 keV, 1423 keV, 1523 keV, and 1623 keV are considered, corresponding to maximum kinetic positron energies of E_{max} = 300 keV, 400 keV, 500 keV, and 600 keV, respectively.

A major motivation for the investigation of internal pair creation results from the test of quantum electrodynamics in the presence of strong external fields. A large contribution to the total positron production rate in supercritical collisions stems from internal pair conversion following nuclear Coulomb excitation. For the experimental separation of both processes it is therefore necessary to know precisely the pair conversion coefficient. Contrary to E0-processes nuclear transitions of multipolarity E1 or E2 should also be observable in the emitted photon spectra provided that proper Doppler shift corrections are performed.

We computed the differential conversion coefficient $d\beta/dE$ for nuclear E1 and E2 transitions. For the nucleus $_{92}$U the energy distributions of emitted positrons is depicted in Fig. 25. Nuclear transition energies of 1323 keV, 1423 keV, 1523 keV, and 1623 keV are considered, which correspond to maximum positron energies of E_{max}^{kin} = 300 keV, 400 keV, 500 keV, and 600 keV, respectively.

For comparison we show in Fig. 26 the equivalent differential conversion coefficient $d\eta/dE$ for nuclear E0 transitions. As bound state only the atomic K-shell has been taken into account. The conversion probability of higher bound states is at least one order of magnitude smaller.

A numerical integration of $d\eta/dE$ yields the dimensionless ratio η. For Z = 92 η is presented in Fig. 27 as function of the nuclear transition energy. For ω = 1,423 keV we find that the probability for K-shell conversion is 556 times larger

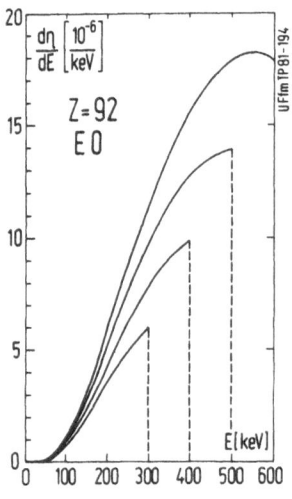
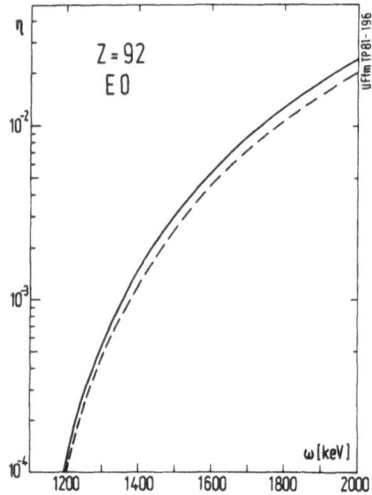

Fig. 26: Differential conversion probability ratio dη/dE with respect to the kinetic positron energy E for nuclear EO-transitions in U. The same transition energies as in Fig. 25.

Fig. 27: The ratio η versus the nuclear transition energy ω for Z = 92. The dashed
line is obtained by R. Thomas[63].

than the e^+, e^- pair creation probability. In the considered range of transition

energies η varies from 10^{-4} to $2 \cdot 10^{-2}$. The dashed line represents results of R.

Thomas[63]. Both calculations are in fair agreement.

For completeness we show in Fig. 28 the K-shell conversion coefficient α_K as

function of the nuclear transition energy ω. Nuclear transitions of multipolarities

E1, E2, and M1 in $_{92}$U are considered. The data are taken from table 64. Thus for

ω>600 keV the ionization of the K-shell is suppressed typically by a factor of 100

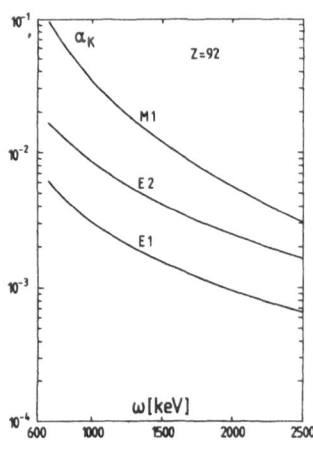

Fig. 28: The conversion coefficient α_K as function
of the nuclear transition energy ω.
Nuclear transitions of multipolarity E1,
E2, and M1 in $_{92}$U are considered. The
data are taken from table 64.

compared to the emission of a single photon.

Now we can discuss the possibility whether the observed structures in positron spectra[47,48] originate from nuclear E0-transitions or not. One convincing argument against this interpretation is related to the shape of the e^+-energy distribution. According to Fig. 26 the halfwidth of the spectra should be at least $\Delta E > 150$ keV. However, the observed structure is much narrower. The second argument is connected with the energy distribution of the emitted δ-electrons. In Ref. 62 it was shown that if the observed structures are caused by nuclear E0-transitions one should also observe a distinct peak in the δ-electron distribution. Furthermore sharp peaks in the energy distribution of emitted positrons with a halfwidth of less than 100 keV may not originate from nuclear E0-conversion processes.

We now turn to the discussion of electron-positron pair conversion in supercritical compound systems. A supercritical nucleus Z=184 which undergoes a transition with $\omega > 2$ during the nuclear reaction period T may transfer this excitation energy to one of the electrons in the negative energy continuum. The remaining hole is defined as positron. But also the K-shell electron can be lifted to the upper continuum. If T is longer than the spontaneous decay width of the K-shell resonance the K-vacancy will be filled again leading to spontaneous positron emission.

For the nuclear charge distribution a homogeneously charged sphere with a radius $R_n = 10.88$ fm has been assumed. In Fig. 29a we show the differential pair conversion coefficient $d\beta/dE$ as function of the positron energy E. Nuclear transitions with $\omega = 4$ and of multipolarity E1 are considered. Striking is the appearance of the pronounced peak at $E = E_{res}$. The dashed curves represent the various contributions of electron states with $\kappa = -1$ and $\kappa \neq -1$. As expected the resonance shows up only in the ($\kappa = -1$)-channel.

Similar ratios $d\beta/dE$ are obtained for multipolarities E2 and M1 (Fig. 29b). In addition maximum values for multipolarities E3, M2, and M3 are given. Obviously lepton emission predominantly occurs for nuclear transitions with magnetic multipolarities.

In the following table we present for the multipolarities E1, E2, M1 the total conversion coefficient β following from a numerical integration of $d\beta/dE$. The second line always gives percentage contribution of the $\kappa = -1$ channel.

ω	E1	E2	M1
4	1.74E-2	3.26E-1	4.03
	95.4%	97.1%	96.2%
6	2.17E-2	3.21E-1	1.11
	92.1%	96.2%	94.2%
8	2.04E-2	2.28E-1	4.43E-1
	88.7%	95.3%	92.0%

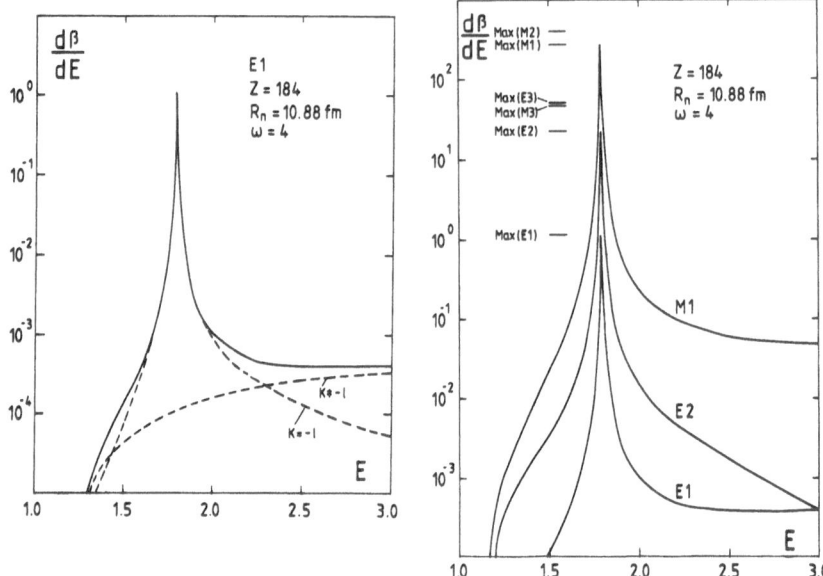

Fig. 29: a) Differential pair conversion coefficient dβ/dE as function of the positron energy E. Nuclear transitions with ω=4 and of multipolarity E1 in a
supercritical nucleus Z=184 with a radius R_n = 10.88 fm are considered.
The dashed curves denote the various contributions of the electron angular
momentum states. -b) As in part a), but including nuclear transition of multipolarities E2 and M1. In addition maximum values for various multipolarities are given.

Thus especially for magnetic transitions the total probability of lepton production
may exceed the photon emission probability even for ω > 2. However, one has to emphasize that the contribution of the discussed conversion processes to the positron
yield in a realistic heavy ion collision depends strongly on the specific nuclear
structure of the compound system.

8. CONCLUSIONS

We have studied the mechanisms of ionization and pair production in collisions
of very heavy ions within the framework of a dynamical theory of excitation based on
the quasimolecular picture. Massive relativistic effects in the wavefunctions of
bound state electrons, caused by the coherent action of the Coulomb field generated
by the two nuclei, are reflected in the excitation rates.

We discussed systematically the charge dependence of vacancy formation and δ-
electron production rates open to experimental verification. The steep increase at
small impact parameters was explained by the strong localization of the 1sσ-state.
Furthermore we predict a maximum for vacancy and δ-electron production rates around
$Z_T + Z_P$ = 170 due to the interplay of the extremely strong 1sσ-binding energy and
the radial expectation value $<r_{1sσ}>$. Comparing our coupled channel results with

experimental data we pointed out that the inclusion of multi-step processes is indispensable to obtain reasonable absolute values, since results obtained within first order time-dependent perturbation theory underestimate the measured data by factors 3-5. The influence of electron screening was investigated with respect to vacancy formation. We found an increase of vacancy production up to 50% for K-holes investigating Sm + Pb.

Considering vacancy formation we find in general fair agreement with experimental data except for small impact parameters and distant collisions, where we realize slight deviations from the experimental results. The present discrepancies still need clarification. Also the calculated δ-electron spectra are partly in excellent agreement with measured values of Kozhuharov and König et al.[26].

Concerning the increase of the $1s\sigma$-binding energy beyond twice the electron rest mass we have shown that the effect of field theoretical corrections to the binding energy is only in the order of 1%. Thus for $Z > 173$ it may not prevent the K-shell electron becoming a resonance imbedded in the negative energy continuum.

We have developed a theory which properly takes into account the resonance character of the 'dived' 1s-state. The results of our coupled channel calculations indicate that no sharp threshold effects are to be expected at the border of the supercritical region, in accordance with the notion of dynamical collision broadening. Collisions above the Coulomb barrier characterized by a sufficiently long reaction time may help to identify the process of spontaneous positron production.

The experiments performed so far have convincingly established the predicted strong increase of positron production in collisions with very high total nuclear charge and its concentration on close collisions. There remain some discrepancies with theory in absolute magnitude for the highest-Z systems and in the slope of $P^{e+}(b)$. On the other hand experimental results of Backe et al.[44], Kozhuharov et al.[47], and Bokemeyer et al.[48] seem to be very promising. Comparing theoretical predictions, calculated along realistic heavy-ion trajectories, with positron spectra, measured in coincidence with fission fragments[44], one may deduce nuclear delay times ΔT in the order of $\sim 2 \cdot 10^{-21}$ sec. Moreover, if the experimental data of Refs. 47, 48 are confirmed and if no other explanation of the positron line in the region 300-400 keV can be found, then one is led to the conclusion that this already constitutes the quantitative proof for the existence of a spontaneous change of the ground state (the 'vacuum decay'). The possibility that the structures in the positron spectra might be caused by E0-conversion due to nuclear deexcitation has been discussed. The results of chapter 7 offer a way to decide whether the observed structures can be explained by background or not.

The experimental findings open up a multitude of interesting questions and speculations to nuclear physics, the central questions being: how can the nuclear reaction mechanism be described, and: are there meta-stable superheavy nuclei in the range Z~160-190? More experimental and theoretical investigations are required before definite conclusions can be drawn.

ACKNOWLEDGEMENTS

We acknowledge very fruitful discussions with P. Armbruster, H. Backe, K. Bethge, H. Bokemeyer, F. Bosch, J.S. Greenberg, A. Gruppe, P. Kienle, W. Koenig, Ch. Kozhuharov, M. Krämer, D. Liesen, P. Mokler, D. Schwalm, J. Schweppe, and P. Senger concerning their experiments. We are grateful to W. Betz for making his calculations on the two-centre Dirac equation available to us.

This work was supported by the Bundesministerium für Forschung und Technologie (BMFT) and the Deutsche Forschungsgemeinschaft (DFG). One of us (G.S.) acknowledges the support of the DFG-Heisenberg Programm.

REFERENCES

1. G. Soff, W. Greiner, W. Betz, and B. Müller, Phys. Rev. A20 (1979) 169
2. C. Kozhuharov, in: 'Physics of Electronic and Atomic Collisions', S. Datz, ed., North-Holland, Amsterdam, 1982, p. 179
3. J. Reinhardt, B. Müller, W. Greiner, and G. Soff, Phys. Rev. Lett. 43 (1979) 1307
4. G. Soff, J. Reinhardt, B. Müller, and W. Greiner, Z. Physik A294 (1980) 137
5. J. Rafelski and B. Müller, Phys. Rev. Lett. 36 (1976) 517
6. W. Betz, Thesis, Universität Frankfurt, 1980
7. W. Pieper and W. Greiner, Z. Physik 218 (1969) 327
8. Ya. B. Zeldovich and V.S. Popov, Sov. Phys.-Usp. 14 (1972) 673
9. J. Reinhardt, B. Müller, and W. Greiner, Phys. Rev. A24 (1981) 103
10. F.G. Werner and J.A. Wheeler, Phys. Rev. 109 (1958) 126
11. G. Soff, B. Müller, and J. Rafelski, Z. Naturforsch. 29a (1974) 1267
12. M. Gyulassy, Phys. Rev. Lett. 33 (1974) 921
13. G.A. Rinker and L. Wilets, Phys. Rev. A12 (1975) 748
14. E.H. Wichmann and N.M. Kroll, Phys. Rev. 101 (1956) 843
15. G.E. Brown and G.W. Schaefer, Proc. Roy. Soc. (London) A233 (1956) 527
16. G.E. Brown, J.S. Langer, and G.W. Schaefer, Proc. Roy. Soc. (London) A251 (1959) 92
17. G.E. Brown and D.F. Mayers, Proc. Roy. Soc. (London) A251 (1959) 105
18. G.W. Erickson and D.R. Jennie, Ann. Phys. 35 (1965) 271 and 447
19. A.M. Desiderio and W.R. Johnson, Phys. Rev. A3 (1971) 1267
20. P.J. Mohr, Ann. Phys. 88 (1974) 26 and 52
21. K.T. Cheng and W.R. Johnson, Phys. Rev. A14 (1976) 1943
22. D. Liesen, P. Armbruster, F. Bosch, S. Hagmann, P.H. Mokler, H.J. Wollersheim, H. Schmidt-Böcking, R. Schuch, and J.B. Wilhelmy, Phys. Rev. Lett. 44 (1980) 983
23. G. Soff, P. Schlüter, B. Müller, and W. Greiner, Phys. Rev. Lett. 48 (1982) 1465
24. G. Soff, B. Müller, and W. Greiner, Z. Physik A299 (1981) 189
25. T. de Reus, U. Müller, J. Reinhardt, P. Schlüter, K.H. Wietschorke, B. Müller, W. Greiner, and G. Soff, preprint, GSI-81-38
26. F. Güttner, W. Koenig, B. Martin, B. Povh, H. Skapa, J. Soltani, Th. Walcher, F. Bosch, and C. Kozhuharov, Z. Physik A304 (1982) 207
27. D. Liesen, P. Armbruster, H.-H.Behncke, and S. Hagmann, Z. Physik A288 (1978) 417
28. D. Liesen, P. Armbruster, H.-H. Behncke, F. Bosch, S. Hagmann, P.H. Mokler, H. Schmidt-Böcking, and R. Schuch, in: 'Electronic and Atomic Collisions', N. Oda and K. Takayanagi, eds., North-Holland, Amsterdam, 1980, p. 337
29. R. Anholt, W.E. Meyerhof, and C. Stoller, Z. Physik A291 (1979) 287

    R. Anholt, W.E. Meyerhof, C. Stoller, and J.F. Chemin, in: 'Proc. of the 7th Internat. Conf. on Atomic Physics', Abstracts, D. Kleppner and F.M. Pipkin, eds., Cambridge, Mass., 1980
30. J.S. Greenberg, H. Bokemeyer, H. Emling, E. Grosse, D. Schwalm, and F. Bosch, Phys. Rev. Lett. 39 (1977) 1404
31. T. de Reus et al., to be published; c.f. also Ref. 25
32. K.-H. Wietschorke, B. Müller, W. Greiner, and G. Soff, J. Phys. B12 (1979) L31
33. J. Rafelski, B. Müller, and W. Greiner, Nucl. Phys. B68 (1974) 585
34. L.P. Fulcher and A. Klein, Phys. Rev. D8 (1973) 2455
35. W.L. Wang and C.M. Shakin, Phys. Lett. 32B (1970) 421
36. J. Rafelski, B. Müller, and W. Greiner, Z. Physik A285 (1978) 49

37. C. Kozhuharov, P. Kienle, E. Berdermann, H. Bokemeyer, J.S. Greenberg, Y. Nakayama, P. Vincent, H. Backe, L. Handschug, and E. Kankeleit, Phys. Rev. Lett. 42 (1979) 376

38. H. Backe, L. Handschug, F. Hessberger, E. Kankeleit, L. Richter, F. Weik, R. Willwater, H. Bokemeyer, P. Vincent, Y. Nakayama, and J.S. Greenberg, Phys. Rev. Lett. 40 (1978) 1443

39. H. Backe, W. Bonin, W. Engelhardt, E. Kankeleit, M. Mutterer, P. Senger, F. Weik, R. Willwater, V. Metag, and J.B. Wilhelmy, GSI Scientific Report 1979, GSI 80-3 (1980) 101

40. H. Backe, W. Bonin, W. Engelhardt, E. Kankeleit, M. Mutterer, P. Senger, F. Weik, R. Willwater, V. Metag, and J.B. Wilhelmy, 'Positron Production in Heavy Ion Collisions', preprint, 1979

41. J. Reinhardt, U. Müller, B. Müller, and W. Greiner, Z. Physik A303 (1981) 173

42. G. Soff, J. Reinhardt, B. Müller, and W. Greiner, Phys. Rev. Lett. 43 (1979) 1981

43. R. Schmidt, V.D. Toneev, and G. Wolschin, Nucl. Phys. A311 (1978) 247

44. H. Backe, W. Bonin, E. Kankeleit, M. Krämer, R. Krieg, V. Metag, P. Senger, and J.B. Wilhelmy, GSI Scientific Report 1981, in press

45. W.E. Meyerhof, R. Anholt, J.F. Chemin, and C. Stoller, in: 'Proc. of the XVIII. Internat. Winter Meeting on Nucl. Phys.', Bormio, Italy, 1980, p. 688, and references herein

46. U. Müller, J. Reinhardt, G. Soff, B. Müller, and W. Greiner, Z. Physik A297 (1980) 357

47. E. Berdermann, F. Bosch, M. Clemente, F. Güttner, P. Kienle, W. Koenig, C. Kozhuharov, B. Martin, B. Povh, H. Tsertos, W. Wagner, and Th. Walcher, GSI-Scientific Report 1980, GSI 81-2 (1981) 128

48. H. Bokemeyer, H. Folger, H. Grein, S. Ito, D. Schwalm, P. Vincent, K. Bethge, A. Gruppe, R. Schulé, M. Waldschmidt, J.S. Greenberg, J. Schweppe, and N. Trautmann, GSI Scientific Report 1980, GSI 81-2 (1981) 127

49. J. Błocki, J. Randrup, W.J. Swiatecki, and C.F. Tsang, Ann. Phys. 105 (1977) 427

50. F.K. McGowan, C.E. Bemis, J.L.C. Ford, W.T. Milner, R.L. Robinson, and P.H. Stelson, Phys. Rev. Lett. 27 (1971) 1741

51. J.P. Davidson, D.A. Close, and J.J. Malanify, Phys. Rev. Lett. 32 (1974) 337

52. R.M. Ronningen, R.C. Melin, J.A. Nolen, G.M. Crawley, and C.E. Bemis, Phys. Rev. Lett. 47 (1981) 635

53. T. Cooper, W. Bertozzi, J. Heisenberg, S. Kowalski, W. Turchinetz, C. Williamson, L. Cardman, S. Fivozinsky, J. Lightbody, and S. Penner, Phys. Rev. C13 (1976) 1083

54. V. Oberacker, private communications

55. G.R. Satchler and W.G. Love, Phys. Rep. 55 (1979) 183

56. W.D. Myers and W.J. Swiatecki, Ann. Phys. 55 (1969) 395 and 84 (1974) 186

57. W.D. Myers, At. Data Nucl. Data Tables 17 (1976) 411

58. C. Gustafson, Proc. Internat. Conf. on the Properties of the Nuclei far from the Region of β-Stability, Leysin, 1970, CERN 70-30, p. 654

59. P.A. Seeger and R.C. Perisho, LA 3751, Los Alamos, 1967

60. J. Grumann, U. Mosel, B. Fink, and W. Greiner, Z. Physik 228 (1969) 371

61. N. Aboul-El-Naga et al., to be published

62. G. Soff, P. Schlüter, and W. Greiner, Z. Physik A303 (1981) 189

63. R. Thomas, Phys. Rev. 58 (1940) 714

64. F. Rösel, H.M. Fries, K. Alder, and H.C. Pauli, At. Data Nucl. Data Tables 21 (1978) 291

LIST OF SHORT CONTRIBUTIONS BY THE PARTICIPANTS

M. ALBIŃSKA   Investigation of Fusion in the ^{20}Ne + ^{24}Mg System.

J. ALBIŃSKI   Barrier and Internal Contributions to Light- and Heavy-Ion Elastic
               Scattering.

M. BARRANCO   The Nucleus-Nucleus Interaction Potential at Finite Temperature Calcu-
               lated in the First Approximation of the Energy Density.

C. BECK   Resonant-like Structure in the ^{30}Si Compound Nucleus.

E. N. GAZIS   Residual Nucleus Formation by Different Exit Channels.

V. HNIZDO   Quadrupole Effects and the Folding Model.

D. KONNERTH   Scattering and Reactions of ^{12}C + ^{14}C.

J. KUNZ   Current Distributions for Collective Motion in Heavy Nuclei.

K E G LOEBNER Sub-Barrier Fusion Measurements for ^{32,36}S on Mo, Ru and Pd-Isotopes.

M A NAGARAJAN Elastic Scattering of Weakly Bound Heavy Ions.

H. NISHIOKA   Spin-Orbit Interaction Induced by Heavy Ion Projectile Excitation.

A. V. RAMAYYA Investigation of δ-Electron Spectra with Br Ion on Pb.

F. RAMI   Fission-like Process in the ^{32}S + ^{76}Ge System.

E. TOMASI   Influence of Collective Degrees of Freedom on Fusion.

F. TONDEUR   A Few Predictions about the Shell Structure and Deformation of Nuclei
               far from Stability.

W. TROMBIK   Spin Alignment in ^{12}C + ^{12}C Inelastic Scattering.

LIST OF PARTICIPANTS

Malgorzata M ALBINSKA	Institute of Physics, Technological University Ul. Podchorazych 1, 30-084 KRAKOW, Poland
Janusz ALBINSKI	Institute of Nuclear Physics Ul. Radzikowskiego 152, 31-342 KRAKOW, Poland
Clara ALONSO	Dept. Física Atómica y Nuclear, Facultad de Física Universidad de Sevilla, SEVILLA, Spain
María Victoria ANDRES	Dept. Física Atómica y Nuclear, Facultad de Física Universidad de Sevilla, SEVILLA, Spain
José Miguel ARIAS	Dept. Física Atómica y Nuclear, Facultad de Física Universidad de Sevilla, SEVILLA, Spain
Antonio S BAEZA	Dept. Física Fundamental, Facultad de C. Físicas Universidad de Valencia, BURJASOT (Valencia), Spain
Richard A BALDOCK	Nuclear Physics Laboratory Keble Road, OXFORD OX1 3RH, U.K.
David W BANES	Nuclear Structure Group, Department of Physics Univ. Birmingham, P.O.B. 363, BIRMINGHAM, U. K.
Manuel BARRANCO	Dept. Física Atómica y Nuclear, Facultad de Física Universidad de Sevilla, SEVILLA, Spain
Manuel BARRANCO	Facultad de Física, Universidad de Barcelona Diagonal 645, BARCELONA 28, Spain
Christian J R BECK	Centre de Recherches Nucléaires, Groupe P.N.I.N. 23 rue du Loess, 67037 STRASBOURG, France
Brigitte BILWES	Centre de Recherches Nucléaires, Basses Energies B.P. 20, 67037 STRASBOURG, France
Francisco BRIEVA	Departamento de Física, Fac. C. Físicas y Matemáticas Univ. Chile, Casilla 5487, SANTIAGO, Chile
Ricardo A BROGLIA	Niels Bohr Institute Blegdamsvej 17, 2100 COPENHAGEN, Denmark
Enrique BUENDIA	Dept. Física Atómica y Nuclear, Facultad de C. Físicas Universidad de Granada, GRANADA, Spain
José DIAZ MEDINA	Dept. Física Fundamental, Facultad de C. Físicas Universidad de Valencia, BURJASOT (Valencia), Spain
José Ignacio ESCUDERO	Dept. Física Atómica y Nuclear, Facultad de Física Universidad de Sevilla, SEVILLA, Spain
Amand FAESSLER	Univ. Tübingen, Institute of Theoretical Physics Auf der Morgenstelle 14, 7400 TÜBINGEN, W. Germany
José L FERRERO	Dept. Física Fundamental, Facultad de C. Físicas Universidad de Valencia, BURJASOT (Valencia), Spain
María Isabel GALLARDO	Dept. de Física Atómica y Nuclear, Facultad de Física Universidad de Sevilla, SEVILLA, Spain
Francisco J GALVEZ	Dept. Física Atómica y Nuclear, Facultad de C. Físicas Universidad de Granada, GRANADA, Spain
Evangelos N GAZIS	National Technical Univ., Physics Laboratory B Zografou Campus, ATHENS-624, Greece
Salvador GIL	Nuclear Physics Laboratory GL-10 University of WA, SEATTLE WA-98195, U.S.A.

Adriano GOBBI	Gesellschaft für Schwerionenforschung mbH Postfach 110541, 6100 DARMSTADT 11, W. Germany
Joaquín GOMEZ CAMACHO	Dept. Física Atómica y Nuclear, Facultad de Física Universidad de Sevilla, SEVILLA, Spain
Walter GREINER	Institut für Theoretische Physik der Universität Frankfurt-M Postfach 111932, D-6000 FRANKFURT-MAIN 11, W. Germany
Joseph H HAMILTON	Physics Department, Vanderbilt University NASHVILLE Tennessee-37235, U.S.A.
Kurt M HARTMANN (new address)	Sektion Physik der LMU München Am Coulombwall 1, D-8046 GARCHING, W. Germany
Vladimir HNIZDO	Department of Physics University of the Witwatersrand, JOHANNESBURG, South Africa
Peter E HODGSON	Nuclear Physics Laboratory Keble Road, OXFORD OX1 3RH, U.K.
Giuseppina IMME'	Istituto di Fisica dell'Università Corso Italia 57, 95129 CATANIA, Italy
Andrzej KOBOS	Nuclear Physics Laboratory Keble Road, OXFORD OX1 3RH, U.K.
Dieter KONNERTH	Sektion Physik der LMU München Am Coulombwall 1, D-8046 GARCHING, W. Germany
Jutta KUNZ	Institut für Theoretische Physik I Heinrich-Buff-Ring 16, 6300 GIESSEN, W. Germany
Giuseppe E LANZA	I.N.F.N. Sezione di Catania Corso Italia 57, 95129 CATANIA, Italy
John S LILLEY	Daresbury Laboratory, Science Research Council DARESBURY Warrington WA4 4RD, U.K.
Adriano de LIMA	Departamento de Física da Universidade 3000 COIMBRA, Portugal
Luigi LO MONACO	I.N.F.N. Sezione di Catania Corso Italia 57, 95129 CATANIA, Italy
K E Gunther LOEBNER	Sektion Physik der LMU München Am Coulombwall 1, D-8046 GARCHING, W. Germany
Manuel LOZANO	Dept. Física Atómica y Nuclear, Facultad de Física Universidad de Sevilla, SEVILLA, Spain
Gonzalo MADURGA	Dept. Física Atómica y Nuclear, Facultad de Física Universidad de Sevilla, SEVILLA, Spain
Arturo MANDLY	Dept. Física Atómica y Nuclear, Facultad de Física Universidad de Sevilla, SEVILLA, Spain
Joan MARTORELL	Departamento de Física Nuclear, Facultad de Ciencias Universidad de P. de M., PALMA DE MALLORCA, Spain
Ulrich MOSEL	Justus Liebig Universität, Fachbereich Physik Institut für Theoretische Physik, D-6300 GIESSEN, W.G.
Mangalam A NAGARAJAN	Daresbury Laboratory, Science Research Council DARESBURY Warrington WA4 4RD, U. K.
Christian NGÔ	Department de Physique Nucléaire, C.E.N. Saclay-Bt 34 91191 GIF-SUR-YVETTE, France
Hidetoshi NISHIOKA	Department of Physics, University of Surrey GUILDFORD Surrey, U. K.

N Peter A OLANDERS
Dept. of Math Physics, Lund Institute of Technology
Box 725, S-220 07 LUND, Sweden

Ruggero PENGO
Sektion Physik der LMU München
Am Coulombwall 1, D-8046 GARCHING, W. Germany

Adolfo PERUJO
Dept. Física Atómica y Nuclear, Facultad de Física
Universidad de Sevilla, SEVILLA, Spain

Alfredo POVES
Dept. Física Teórica C-XI, Universidad Autónoma de Madrid
Canto Blanco, MADRID 34, Spain

José M QUESADA
Dept. Física Atómica y Nuclear, Facultad de Física
Universidad de Sevilla, SEVILLA, Spain

Giovanni RACITI
Laboratorio Nazionale del Sud
Corso Italia 57, 95129 CATANIA, Italy

Akumuri V RAMAYYA
Physics Department, Vanderbilt University
NASHVILLE Tennessee-37235, U.S.A.

Fouad RAMI
Centre de Recherches Nucléaires, Groupe P.N.I.N.
67037 STRASBOURG Cedex, France

Miguel A RESPALDIZA
Dept. Física Atómica y Nuclear, Facultad de Física
Universidad de Sevilla, SEVILLA, Spain

Herminia ROMAN
Dept. Física Atómica y Nuclear, Facultad de Física
Universidad de Sevilla, SEVILLA, Spain

José ROS
Dept. Física Teórica, Facultad de Ciencias
Universidad de Granada, GRANADA, Spain

Carlos ROSSI
Istituto di Fisica
via Marzolo 8, PADOVA 35100, Italy

Filipe D SANTOS
Centro de Física Nuclear da Universidade de Lisboa
Av. Gama Pinto 2, 1699 LISBOA, Portugal

G Raymond SATCHLER
Oak Ridge National Laboratory
P.O.B. X, OAK RIDGE Tennessee 37830, U.S.A.

Dieter SCHARDT
G.S.I. Darmstadt
Postfach 110541, D-6100 DARMSTADT 11, W. Germany

Martin SEIWERT
Institut für Theoretische Physik der Universität Frankfurt-M
Postfach 111932, D-6000 FRANKFURT-MAIN 11, W. Germany

Peter J SMITH
Oliver Lodge Laboratory, Dept. of Physics
University of Liverpool, P.O.B. 147, LIVERPOOL L69 3BX, U.K.

Egle TOMASI
D Ph N MF - C.E.N. Saclay
91191 GIF-SUR-IVETTE Cedex, France

Francois TONDEUR
Université Libre de Bruxelles, Phys Théor et Math
Campus de la Plaine, Blv. du Triomphe, B-1050 BRUXELLES,
Belgium

Werner TROMBIK
Sektion Physik der LMU München
Am Coulombwall 1, D-8046 GARCHING, W. Germany

Adriann VAN DER WOUDE
Kernfysisch Versneller Instituut
Rijksuniversiteit Groningen, GRONINGEN, The Netherlands

Javier VIÑAS
Departamento de Optica, Facultad de C. Físicas
Univ. Barcelona, Diagonal 645, BARCELONA 28, Spain

Claude VOLANT
D Ph N/BE - C.E.N. Saclay
B.P.N. 2, 91191 GIF-SUR-IVETTE Cedex, France

P. Ring, P. Schuck

The Nuclear Many-Body-Problem

1980. 171 figures. XVII, 716 pages
(Texts and Monographs in Physics)
ISBN 3-540-09820-8

This book, while covering a fair amount of physical observations, stresses the methodology and technical aspects of the different theories presently used in the description of the nucleus. The authors present the more modern theories such as Boson expansions, generator coordinates, time-dependent Hartree-Fock method, and semiclassical models which so far have found only limited mention in textbooks. The book also covers subjects like the liquid drop and the shell model, both presented in a updated version in, for example, rotations and random phase approximation. The full presentation of mathematical details, illustrated by observational data, will help the student fully understand the present view on the nuclear many-body problem.

Contents: The Liquid Drop Model. – The Shell Model. – Rotation and Single-Particle Motion. – Nuclear Forces. – The Hartree-Fock Method. – Pairing Correlations and Superfluid Nuclei. – The Generalized Single-Particle Model (HFB Theory). – Harmonic Vibrations. – Boson Expansion Methods. – The Generator Coordinate Method. – Restoration of Broken Symmetries. – The Time Dependent Hartree-Fock Method (TDHF). – Semiclassical Methods in Nuclear Physics. – Appendices A–F. – Bibliography. – Author Index. – Subject Index.

Zeitschrift für Physik A

Atoms and Nuclei

 EPS Europhysics Journal

Editor in Chief: H. A. Weidenmüller, Heidelberg

Editorial Board: P. Armbruster, E. Bodenstedt, H.-J. Gerber, A. Goldhaber, I. V. Hertel, M. Lefort, B. Povh, G. zu Putlitz, J. Specht

Zeitschrift für Physik appears in three parts:
  A: Atoms and Nuclei
  B: Condensed Matter
  C: Particles and Fields
Each part my be ordered separately.

Coordinating Editor for Zeitschrift für Physik, Sections A, B and C is O. Haxel, Heidelberg

Atomic Physics:
Properties of atoms and molecules · Spectra of inner and outer shells · µ-mesic, pionic and other exotic atoms · Hyperfine interactions · Atomic and molecular collisions · Atomic studies with heavy-ion collisions · Theory

Nuclear Physics:
Properties of nuclei · Nuclear structure and reactions · Heavy-ion reactions · Fission · Hadron-nucleus interactions · Theory

Special Features: Rapid publication (3–4 months for original articles, 6 weeks for short notes); no page charge; back volumes available, also in microform.

Language: More than 95% English.

Articles: Original reports and short notes.

Subscription information or sample copy upon request.

Springer-Verlag Berlin Heidelberg New York

Lecture Notes in Physics

Selected Issues from

Lecture Notes in Mathematics